Lecture Notes in Computer Science 859

Edited by G. Goos, J. Hartmanis and J. van Leeuwen

Advisory Board: W. Brauer   D. Gries   J. Stoer

Thomas F. Melham   Juanito Camilleri (Eds.)

Higher Order Logic Theorem Proving and Its Applications

7th International Workshop
Valletta, Malta, September 19-22, 1994
Proceedings

Springer-Verlag

Berlin Heidelberg New York
London Paris Tokyo
Hong Kong Barcelona
Budapest

Series Editors

Gerhard Goos
Universität Karlsruhe
Postfach 69 80, Vincenz-Priessnitz-Straße 1, D-76131 Karlsruhe, Germany

Juris Hartmanis
Department of Computer Science, Cornell University
4130 Upson Hall, Ithaka, NY 14853, USA

Jan van Leeuwen
Department of Computer Science, Utrecht University
Padualaan 14, 3584 CH Utrecht, The Netherlands

Volume Editors

Thomas F. Melham
Department of Computing Science, University of Glasgow
17 Lilybank Gardens, Glasgow, Scotland, G12 8QQ

Juanito Camilleri
Department of Computer Studies, University of Malta
University Heights, Msida, Malta

CR Subject Classification (1991): F.4.1, I.2.2-3

ISBN 3-540-58450-1 Springer-Verlag Berlin Heidelberg New York

CIP data applied for

© Springer-Verlag Berlin Heidelberg 1994
Printed in Germany

Typesetting: Camera-ready by author
SPIN: 10479015   45/3140-543210 - Printed on acid-free paper

Preface

This volume is the proceedings of the international workshop on *Higher Order Logic Theorem Proving and its Applications*, held at the Foundation for International Studies in the University of Malta during 19–22 September 1994. The workshop is the seventh in a series of annual meetings that began in 1988 with a small user group meeting for the HOL theorem prover and has since grown into a full-scale international workshop. Although the main subject of these meetings is still the HOL system, the focus has widened over the years to include the application, design, and theory of any mechanized proof system for higher order logic.

The forty-two papers submitted were mostly of a high standard. All submissions were fully refereed, each paper being read by at least three reviewers appointed by the programme committee. Twenty-seven papers were selected for presentation as full research contributions at the workshop; all these papers appear in this proceedings. The workshop has a strong tradition of providing an open venue for the discussion and sharing of preliminary results, particularly those of students; a further twelve submissions were therefore accepted for informal presentation in a poster session.

The workshop organizers were especially pleased that Professor N. G. de Bruijn, Dr. Thomas Forster, and Dr. Keith Hanna accepted invitations to be guest speakers at the workshop. All three invited speakers also very kindly produced written contributions for inclusion in this proceedings.

We wish to thank the Bank of Valletta, Malta, who generously provided financial support for the workshop. We also gratefully acknowledge the support of the University of Malta, our host for the workshop, and of the Computing Science Department at Glasgow University. Jackie Attard Montalto and Yasmeen Staff gave invaluable assistance in matters of local organization.

Tom Melham
Juanito Camilleri

Conference Organization

Programme Chair

Dr. Thomas Melham
Dept. of Computing Science
University of Glasgow
17 Lilybank Gardens
Glasgow, Scotland, G12 8QQ
e-mail: tfm@dcs.glasgow.ac.uk

Workshop Chair

Dr. Juanito Camilleri
Dept. of Computer Science & AI
University of Malta
University Heights
Tal-Qroqq, Malta
e-mail: juan@unimt.mt

Programme Committee

By providing detailed and thoughtful reviews of the papers, the programme committee made a significant contribution to the quality of both the conference programme and this proceedings.

Flemming Andersen (TDR)
Richard Boulton (Cambridge)
Shui-Kai Chin (Syracuse)
Elsa Gunter (AT&T)
John Herbert (SRI)
Jeffrey Joyce (UBC)
Gilles Kahn (INRIA)
Ramayya Kumar (FZI)

Tim Leonard (DEC)
Karl Levitt (UC Davis)
Paul Loewenstein (SUN)
Tom Melham (Glasgow)
Tom Schubert (Portland State)
David Shepherd (INMOS)
Phil Windley (BYU)
Joakim von Wright (Åbo Akademi)

Additional Reviewers

The programme committee gratefully acknowledges the constructive assistance of the following additional reviewers:

Jim Alves-Foss, Penny Anderson, Myla Archer, Gregory Benson, Michael Butler, Rachel Cardell-Oliver, Anand Chavan, Paul Curzon, Nancy Day, Michael Donat, Gilles Dowek, Bill Findlay, Jens Godskesen, Mike Gordon, Cordelia Hall, Mark Heckman, Shahid Ikram, Jang-Dae Kim, Jens Kristensen, Finn Kristoffersen, Juin-Yeu Lu, Chet Murthy, Monica Nesi, Karsten Nyblad, Jørgen Nørgaard, Kim Dam Petersen, Jimmi Pettersson, Rob Shaw, Helene Wong, Cui Zhang, Zheng Zhu.

Contents

LCF Examples in HOL

Sten Agerholm*

BRICS***, Computer Science Dept., Aarhus University, DK-8000 Aarhus C, Denmark

Abstract. The LCF system provides a logic of fixed point theory and is useful to reason about nontermination, recursive definitions and infinite-valued types such as lazy lists. Because of continual presence of bottom elements, it is clumsy for reasoning about finite-valued types and strict functions. The HOL system provides set theory and supports reasoning about finite-valued types and total functions well. In this paper a number of examples are used to demonstrate that an extension of HOL with domain theory combines the benefits of both systems. The examples illustrate reasoning about infinite values and nonterminating functions and show how domain and set theoretic reasoning can be mixed to advantage. An example presents a proof of correctness of a recursive unification algorithm using well-founded induction.

1 Introduction

The LCF system [GMW79, Pa87] is a theorem prover based on a version of Scott's Logic of Computable Functions (a first order logic of domain theory). It provides the concepts and techniques of fixed point theory to reason about nontermination and arbitrary recursive (computable) functions. For instance, it has been successfully applied to reason about infinite data structures and lazy evaluation [Pa84b]. On the other hand, the HOL system [GM93] supports set theoretic reasoning. It has no inbuilt notion of nontermination, all functions are total, and only primitive recursive definitions are supported. It has mainly been used for reasoning about finite data structures and terminating primitive recursive functions.

In a way, extending HOL with domain theory as described in [Ag93, Ag94] corresponds to embedding the logic of the LCF system within HOL—this extension is called HOL-CPO in this paper. Thus, any proof conducted in LCF can be conducted in HOL-CPO as well, and axioms of LCF theories can be introduced as definitions, or derived from definitions, provided of course they are consistent extensions of LCF. This correspondence breaks for difficult recursive domains with infinite values. It is not easy to define such domains in HOL and LCF could just axiomatize the domains; still this has its theoretical difficulties in general, but has been automated in certain cases [Pa84a].

However, HOL-CPO is not just another LCF system. Ignoring the problems with recursive domains, we claim it is more powerful and usable than LCF

---

* E-mail: sagerholm@daimi.aau.dk
*** Basic Research in Computer Science, Centre of the Danish National Research Foundation.

since (1) it inherits the underlying logic and proof infrastructure of the HOL system, and (2) it provides direct access to domain theory. These points are the consequences of *embedding* semantics rather than *implementing* logic. One advantage of (1) is that we can exploit the rich collection of built-in types, theorems and tools provided with the HOL system. LCF has almost nothing like that. Another advantage is that we become able to mix domain and set theoretic reasoning in HOL such that reasoning about bottom can be deferred until the late stages of a proof. To support this point, experience shows that the continual fiddling with bottom in LCF is very annoying. Its presence in all types makes LCF clumsy for reasoning about finite-valued types and strict functions [Pa85, Pa84b].

In contrast to (2), domain theory is only present in the logic of LCF through axioms and primitive rules of inference. Therefore fixed point induction is the only way to reason about recursive definitions. Testing that a predicate admits fixed point induction can only be performed in ML by an incomplete syntactic check. By exploiting the semantic definitions of these concepts in domain theory, HOL-CPO does not impose such limitations. Fixed point induction can be derived as a theorem and syntactic checks for admissibility, called inclusiveness, can be implemented, just as in LCF. But using other techniques for recursion or reasoning directly about fixed points allows more theorems to be proved than with just fixed point induction. Inclusive predicates not accepted by the syntactic checks can be proved to be inclusive from the semantic definition.

In this paper we present a number of examples to demonstrate that HOL-CPO supports and extends both the HOL and the LCF worlds. We define non-terminating and arbitrary recursive functions in domain theory and reason about finite-valued types and total functions in set theory (higher order logic) before turning to domain theory. The examples have already been done in LCF by Paulson which makes a comparison of the two systems possible. The first two examples, on natural numbers and lazy sequences, are described in chapter 10 of the LCF book [Pa87] and the third example is based on Paulson's version of a correctness proof of a unification algorithm by Manna and Waldinger [MW81, Pa85]. The unification algorithm is defined as a fixed point and proved total afterwards. Termination is non-trivial and proved by well-founded induction [Ag91].

Before we turn our attention to the examples we give an overview of the formalization of domain theory in section 2 and describe the LCF system in section 3. In section 4 we introduce a cpo of natural numbers and present a few theorems about addition. In section 5 a mapping function for lazy sequences and a generator for infinite sequences are introduced. The correctness proof of the unification algorithm is discussed in section 6. Finally, the conclusions are summarized in section 7.

2 HOL-CPO

In this section we provide an overview of the formalization of domain theory and some of the associated tools [Ag93, Ag94]. This extension of HOL, called

HOL-CPO, constitutes an integrated system where domain theoretic concepts look almost primitive (built-in) to the user. Many facts are proved behind the scenes to support this view. In order to read the paper it is not necessary to know the semantic definitions of the subset of domain theory which is used. Therefore the presentation below shall be very brief. More details can be sought in [Ag94], or in [Wi93] on which the formalization is based.

2.1 Basic Concepts

Domain theory is the study of complete partial orders (cpos) and continuous functions. These notions are introduced as predicates in HOL by their semantic definitions. A *complete partial order* is a set and relation pair which satisfies the predicate `"cpo:(*->bool)#(*->*->bool)->bool"`. If `"(A,R)"` is a cpo then the underlying relation `R` is a partial ordering (reflexive, transitive and antisymmetric) on all elements of the underlying set `A` and there exists a (unique) *least upper bound* (lub) for all non-decreasing chains `"X:num->*"` of elements in `A` (`X` is a chain if `"R(X n)(X(n+1))"` holds for all `n`).

The underlying relation of a cpo `D` is obtained by writing `"rel D"`. If `x` and `y` are elements of `D` , written as `"x ins D"` and `"y ins D"`, then `"rel D x y"` can be read as 'x approximates y' or 'x is less defined than (or equals) y'.

Note that we do not require cpos to have a *least defined* element, also called a *bottom* element, w.r.t. the underlying ordering relation. Cpos which have a bottom are called *pointed* cpos and satisfy the HOL predicate `pcpo` (same type as `cpo`). If `E` is a pointed cpo then the term `"bottom E"` equals the bottom element of `E` . In the following, cpos are usually cpos *without* bottom unless we say explicitly that a cpo is pointed.

A *continuous* function from a cpo `D1` to a cpo `D2` is a HOL function `"f:*1->*2"` such that the term `"cont f(D1,D2)"` is true. It must be monotonic w.r.t. the underlying relations and preserve lubs of chains in `D1` in the sense that `f` applied to the lub of a chain `X` in `D1` is equal to the lub of the chain `"f(X n)"` in `D2` . The results of applying `f` constitute a chain due to monotonicity and therefore have a lub in `D2` . In addition, `f` must be *determined* by its action on elements of the domain cpo `D1` . This means that on elements outside `D1` it should always return a fixed arbitrary value called `ARB` (a predefined HOL constant). The determinedness restriction is necessary to prove that continuous functions constitute a cpo and is induced by the fact that we work with partial HOL functions between subsets of HOL types (corresponding to the underlying sets of cpos). Determinedness occurs everywhere and is the main disadvantage of the formalization. In particular, functions must be written using a dependent lambda abstraction `"lambda D f"` to ensure they are determined. Therefore, many functions become parameterized by cpo variables (corresponding to the free cpo variables of the right-hand side of their definition).

The conditions on cpos and continuous functions ensure the existence of a *fixed point operator*, called `FixI` (the 'I' is explained later), which is useful to define arbitrary recursive functions and other infinite values. Applied to a continuous function `f` on a pointed cpo `E` , it yields a fixed point: `|-f(FixI`

E f) = FixI E f , and in fact the *least* fixed point: |- !x. x ins E ==> (f x = x) ==> rel E(FixI E f)x . The term "FixI E f" equals the least upper bound of the non-decreasing chain $\perp \sqsubseteq f(\perp) \sqsubseteq f(f(\perp))\ldots$ where $\perp$ stands for "bottom E" and $\sqsubseteq$ stands for "rel E".

The proof principle of fixed point induction has been derived as a theorem from the definition of the fixed point operator. It can be used to prove properties of fixed points stated as *inclusive* (or *admissible*) predicates. A predicate is inclusive if it contains lubs of chains of elements in the predicate. Fixed point induction says that "P(FixI E f)" follows from "P(bottom E)" and "!x. P x ==> P(f x)", assuming a pointed cpo E , a continuous function f from E to E and an inclusive predicate P on E . There are a few syntactic-based proof functions to prove these semantic conditions: the cpo prover, the type checker and the inclusive prover (based on the LCF check in [Pa87] on page 199–200), respectively. They only work in certain cases (see below).

2.2 Constructions

There are various standard ways of constructing cpos and continuous functions which allow proofs to be automated in HOL.

The *discrete* construction associates the discrete ordering (identity) with a set and it is therefore useful for making HOL sets into cpos. For instance, the type of natural numbers can be used to define the discrete cpo of natural numbers "discrete(UNIV:num->bool)" using the universal set UNIV (a predicate which is always true, here corresponding to the set of all elements of ":num"). A construction called *lifting* can then be used to extend the cpo with a bottom element as follows "lift(discrete UNIV)". The bottom element of a lifted cpo "lift D" is written as Bt and all other elements are written as "Lft d" for some d in D . It can be proved that "bottom(lift D)" is equal to Bt . The constants Bt and Lft are the constructors of a new datatype in HOL which associates a new element with a type. It is the underlying relation of the lifting construction which makes Bt into a bottom element of "lift D".

There is also a construction for the cpo of continuous functions, relating functions by the pointwise ordering relation. Assuming two cpos D1 and D2 this construction is written as "cf(D1,D2)". Note that the two statements "f ins (cf(D1,D2))" and "cont f(D1,D2)" are equivalent. Finally, we provide a product construction and sum construction written as "prod(D1,D2)" and "sum(D1,D2)", respectively.

A proof function called the *cpo prover* automatically proves any term written using the constructors is a cpo. There is a similar function for pointed cpos.

The constructions on continuous functions include the well-known projection and injection functions associated with the product and sum cpos respectively, and functional composition and currying as well. We also consider the fixed point operator to be a constructor.

In addition, there are two useful constructors associated with the lifting construction on cpos. A determined version of the constant Lft , called LiftI , takes an element d of a cpo D and lifts it to an element of the lifted cpo "lift

D". A construction called *function extension* can be used to extend the domain of a function to the lifted domain in a strict way. It works as follows:

```
|- (ExtI(D,E) f Bt = bottom E) /\
   (!x. x ins D ==> (ExtI(D,E) f (LiftI D x) = f x))
```

where E is pointed cpo and f is a continuous function from D to E.

It is also possible to write continuous functions using the dependent lambda abstraction "lambda D(\x. e[x])" where e[x] must be written using only continuous constructions and variables and constants in appropriate cpos. To prove a function, or more generally any term, is in some cpo (e.g. the continuous function space) a proof function called the *type checker* can be used, provided the term fits within an informal notation. The constructions above and lambda abstraction are part of the notation which can be extended interactively with any terms in cpos (see below). Function application is also part of the notation.

Further, any function between discrete universal cpos as the cpo of natural numbers is trivially continuous. Hence, the cpo of continuous functions between such cpos is itself a discrete universal cpo.

2.3 Adding New Constructions

The collection of constructors for cpos and continuous functions can at any time be extended with user-defined constructors. An ML function is provided to define a new cpo constructor in terms of existing constructors, and similarly new function constructors can be introduced. We make a distinction between new constants that are elements of some cpo and new function constructors of some cpo. The latter are parameterized by cpo variables, like the constructors above. A new constant *to the system* can be any proper left-hand side of a HOL definition which can be proved to belong to some cpo. All of this has been automated such that there are only a few proof functions to use which prove the necessary cpo and membership facts behind the scenes.

However, the constructions do not provide all cpos and continuous functions that we might want. In particular, recursive domains and their associated function constructions must be introduced manually. A fairly tough development gave us lazy sequences and lazy lists using HOL lists ":(*)list" and functions of the form ":num->*" to represent infinite values. These developments are described in [Ag94] which also provides some ideas on how to introduce such cpos more generally using infinite labelled trees. It might be possible to automate these ideas. The cpo of lazy sequences and its associated constructor and eliminator functions are used in section 5.

2.4 Interface

The cpo parameters on function constructions quite quickly become a pain. They make terms difficult to read and write. But fortunately an extension of the built-in HOL parser and pretty-printer can hide the annoying extra information in most cases. This provides two levels of syntax, the internal level of syntax where all parameters occur and the external (interface) level of syntax where the parameters are ignored. Hence, `FixI` , `LiftI` and `ExtI` above are part of the internal level syntax. The last letter 'I' on (internal) names is used to distinguish the constants of the two levels. At the external interface level the terms `"FixI E"`, `"LiftI D"` and `"ExtI(D,E)"` are written as `Fix` , `Lift` and `Ext` , respectively. New function constructors are also introduced in two versions. Furthermore, the interface provides a nicer syntax for the dependent lambda abstraction. The term `"\x::Dom D. e[x]"` can be used for `"lambda D(\x. e[x])"`.

3   The LCF System

The LCF system is very similar to the HOL system (or vice versa, since HOL is a direct descendant of LCF). It has a meta language ML (or Standard ML) in which the logic and theorem proving tools are implemented. Theorems are implemented by an abstract datatype for security and axioms and primitive inference rules are constructors of this datatype. Derived inference rules are ML functions. The subgoal package allows proofs in a backwards fashion using tactics. Constants, axioms, theorems and so on are organized in hierarchies of theories. The main properties of LCF may be summarized as follows:

- LCF supports a first order logic of domain theory.
- The use of LCF to reason about recursive definitions (fixed points) is restricted since only fixed point induction can be used. Besides, fixed point induction is based on an incomplete syntactic check of inclusiveness.
- Extending theories in LCF is done by an axiomatic approach and is therefore unsafe. Checking whether an axiom is safe is difficult since it must be done in domain theory (outside LCF).

Each of these points are discussed below.

The central difference between LCF and HOL lies in their logics. The logic of the HOL system is an implementation of a version of Church's higher order logic. The logic of the LCF system is an implementation of a version of Scott's Logic of Computable Functions, usually abbreviated LCF. In order to be able to distinguish the logic and the system the logic was renamed to PPλ, an acronym of Polymorphic Predicate λ-calculus. PPλ is a first order logic of domain theory, it has a domain theoretic semantics. It differs from higher order logic since it is a first order logic and types denote pointed cpos rather than just sets (cpos can be seen as sets with structure). The function type denotes the cpo of continuous functions whereas HOL functions are total functions of set theory.

Fixed point theory is provided in LCF through axioms and primitive rules of inference. A certain constant of the logic denotes the fixed point operator due to an axiom which states it yields a fixed point and due to the primitive rule of fixed point induction which states it yields the least fixed point. In LCF there is no domain theoretic definition of the fixed point operator. Therefore, fixed point induction is the *only* way to reason about recursive definitions. However, structural induction for many datatypes can be derived from fixed point induction [Pa84a], but well-founded induction cannot. Admissibility of predicates for induction is not defined either; a syntactic check is performed by the rule of fixed point induction. This check is not complete and examples of inclusive predicates exist that are not accepted for fixed point induction in LCF. Paulson gives an example in [Pa84a].

There are quite different traditions of extending theories in LCF and HOL. In HOL there is a sharp distinction between purely definitional extensions and axiomatical extensions. Definitional extensions are conservative (or safe), i.e. they always preserve consistency of the logic. Stating a new axiom is not a conservative extension, it might introduce inconsistency. In LCF there is no such distinction between axioms and definitions. The only way to extend theories with new concepts is by introducing new axioms.

It is not always easy to know whether an LCF axiom is safe or not since this must be justified in domain theory. In particular, an axiom should not violate the continuity of a function. All functions are assumed to be continuous in PPλ since the function type denotes the cpo of continuous functions. Paulson shows how easy it is to go wrong in example 4.11 of his book [Pa87].

4 Natural Numbers

In this section we start the comparison of LCF and HOL-CPO. As a first simple example, we define a cpo of natural numbers and consider a few properties about addition: addition is total, associative and commutative.

In LCF natural numbers are introduced as a recursive datatype where a constant 0 and a strict successor function $SUCC$ are the constructors. Names of constants for the type and for the constructor functions are declared and then axioms about the new constants are postulated. The axioms specify the partial ordering on natural numbers and state strictness and definedness of the constructors. The exhaustion (or cases) axiom is also postulated. It states there are three possible kinds of values of a natural number, namely bottom, zero and the successor of some natural number. Distinctness of the constructors and the structural induction rule are then derived from these axioms and fixed point induction. This is performed automatically by a few ML functions.

It is also easy to define a cpo of natural numbers in HOL, though the method is very different. Instead of introducing a new recursive cpo, we exploit the built-in natural numbers and define |- Nat = discrete(UNIV:num->bool) . Using lifting "lift Nat", we obtain the pointed cpo corresponding to the recursive type of natural numbers in LCF.

The zero element of "lift Nat" is "Lift 0" and a strict successor is obtained from the built-in successor SUC by function extension:

```
|- Suc = Ext(\nn :: Dom Nat. Lift(SUC nn))
|- Suc ins (cf(lift Nat,lift Nat)).
```

Note that SUC is trivially a continuous function from **Nat** to **Nat** since the term "cf(D1,D2)" is a discrete universal cpo when D1 and D2 are.

In LCF, addition is introduced by a recursion equation using an eliminator functional, called *NAT_WHEN*,

$$NAT\_WHEN\ x\ f \perp \equiv \perp$$
$$NAT\_WHEN\ x\ f\ 0 \equiv x$$
$$\forall m.\ m \not\equiv \perp \Rightarrow NAT\_WHEN\ x\ f\ (SUCC\ m) \equiv f\ m$$

which is useful to define continuous functions on natural numbers by cases. From the axiom for addition the usual recursion equations matching the cases above are derived by proof. Note that *NAT_WHEN* must assume the argument of the strict LCF successor is defined, otherwise there would be a conflict with the bottom case. A consequence of this is that most theorems stated about addition inherit this assumption. Definedness assumptions make reasoning about strict functions difficult [Pa85].

In HOL-CPO, a strict addition on "lift Nat" is introduced in the same way as the strict successor, by extending a built-in HOL function $+ :

```
|- Add = Ext(\nn :: Dom Nat. Ext(\mm :: Dom Nat. Lift(nn+mm)))
|- Add ins (cf(lift Nat,cf(lift Nat,lift Nat))).
```

Note, by the way, that neither Suc nor Add are parameterized by any cpo variables since we work with the 'concrete' cpo of (lifted) natural numbers.

In LCF the recursion equations for addition are important in proofs because properties of addition are proved using natural number induction. In HOL we can reuse built-in theorems about addition which probably have been proved by similar inductions once, but without considering the bottom element as in LCF induction. For finite-valued types we can do the set theoretic developments in HOL before adding bottom. It is advantageous to defer reasoning about bottom until as late as possible in a proof, e.g. definedness assumptions tend to accumulate.

The usual recursion equations for addition have been proved in HOL but a reduction theorem is more useful:

```
|- (!n. Add Bt n = Bt) /\
   (!n. Add n Bt = Bt) /\
   (!nn mm. Add(Lift nn)(Lift mm) = Lift(nn+mm)).
```

It states that addition is strict in both arguments and behaves as the built-in addition on lifted arguments.

The next fact we consider states that strict addition is total. That is, provided the arguments of Add are not bottom the result of applying Add will not be bottom. In LCF this fact would be stated by a theorem of the following form:

```
|- !n m. ~(n=Bt) ==> ~(m=Bt) ==> ~(Add n m = Bt).
```

Since we use lifting an equivalent statement in HOL is

```
|- !nn mm. ~(Add(Lift nn)(Lift mm) = Bt).
```

This can be derived immediately from the third clause of the above reduction theorem for addition using the facts that **Bt** and **Lift** are distinct and exhaustive on a lifted cpo.

Finally, let us consider two theorems stating that strict addition is associative and commutative:

```
|- !k m n. Add (Add k m) n = Add k(Add m n)
|- !m n. Add m n = Add n m
```

Their proofs are almost exactly the same in HOL; do a case split on the universally quantified variables (lifted numbers) one by one and reduce using the reduction theorem for addition after each case split. We end up with goals stating that the properties we wish to prove must hold for the built-in addition. So we finish off the proofs by using the desired built-in HOL facts:

```
|- !m n p. m + (n + p) = (m + n) + p
|- !m n. m + n = n + m
```

Such proofs by cases could be automated easily. The LCF proofs require much more thought. They use induction, in fact two nested inductions for commutativity, and rewriting.

5 A Mapping Functional for Lazy Sequences

In this section we define a mapping functional for lazy sequences and an infinite sequence constructor. One theorem is proved by fixed point induction and another is proved by "structural induction" on lazy sequences [Pa84a], i.e. structural induction is used to show the inclusive property holds of all finite sequences; the inclusiveness ensures it holds also of the infinite sequences. The purpose of this section is to show in which way HOL-CPO extends HOL with techniques for reasoning about infinite values, and recursive definitions in general.

The cpo of lazy sequences and its associated constructor and eliminator functions correspond exactly to the LCF type of sequences and its associated functions. The LCF type and constructor functions are introduced automatically by axioms similar to the axioms for natural numbers, using the same ML functions too. Developing the lazy sequences in HOL-CPO was difficult and time-consuming but we reason about sequences using the same techniques as in LCF.

A purely definitional development of a theory of lazy sequences is presented in [Ag94]. It provides a constructor called **seq** for pointed cpos of partial and infinite sequences of data. Hence, if **D** is a cpo then **"seq D"** is a pointed cpo. The bottom sequence is called **Bt_seq** and the lazy constructor function is called **Cons_seq** . These satisfy the following cases theorem

```
|- !D s.
     s ins (seq D) =
     (s = Bt_seq) \/
     (?x s'. x ins D /\ s' ins (seq D) /\ (s = Cons_seq x s'))
```

Further, they are distinct and `Cons_seq` is one-one. There is also an eliminator functional called `Seq_when` which can be used to write continuous functions on sequences by cases. Assuming `"x ins D"`, `"s ins (seq D)"` and `"h
ins (cf(D,cf(seq D,E)))"` for a cpo `D` and a pointed cpo `E`, the following
reduction theorem specifies the behavior of the eliminator:

```
|- (Seq_when h Bt_seq = bottom E) /\
   (Seq_when h(Cons_seq x s) = h x s)
```

The constants `Seq_when` and `Cons_seq` belong to the interface level syntax,
internally they are parameterized by cpo variables (and called `Seq_whenI` and
`Cons_seqI` respectively). In addition, we have derived a theorem for "structural
induction" on lazy sequences from fixed point induction, following Paulson's
approach [Pa84a].

All definitions, theorems and proofs about lazy sequences are very similar
to the ones in LCF. The mapping functional is defined as the fixed point of a
suitable functional as follows:

```
|- !D E.
     Maps =
     Fix
     (\g :: Dom(cf(cf(D,E),cf(seq D,seq E))).
       \f :: Dom(cf(D,E)).
        \s :: Dom(seq D).
         Seq_when
          (\x :: Dom D.\t :: Dom(seq D). Cons_seq(f x)(g f t))s)
|- !D E.
     cpo D ==> cpo E ==> Maps ins (cf(cf(D,E),cf(seq D,seq E)))
```

Internally, the constant `Maps` is parameterized by the cpo variables `D` and `E`
of the definition. Using the reduction theorem for `Seq_when` and the fact that
`Fix` yields a fixed point of a continuous function we can prove the following
reduction equations easily:

```
|- (Maps f Bt_seq = Bt_seq) /\
   (Maps f(Cons_seq x s) = Cons_seq(f x)(Maps f s))
```

where `"x ins D"`, `"s ins (seq D)"` and `"f ins (cf(D,E))"` for cpos `D` and
`E`. A tactic which takes such theorems as arguments can be used to reduce
occurrences of `Maps` and other function constructors using a theorem like this
one and the type checker to prove the assumptions automatically.

We can prove that the mapping functional preserves functional composition,
i.e. assuming `"f ins (cf(D2,D3))"` and `"g ins (cf(D1,D2))"` for cpos `D1`,
`D2` and `D3`, the following equation holds

```
|- Maps(Comp(f,g)) = Comp(Maps f,Maps g)
```

The constant `Comp` is defined as a determined version of the built-in functional
composition (internally it is called `CompI`). The proof is conducted by observing

that the two continuous functions are equal iff they are equal for all sequences of values in **D1** , i.e. iff the following term holds:

```
"!s.
  s ins (seq D1) ==>
  (Maps(Comp(f,g))s = Comp(Maps f,Maps g)s)".
```

Then we use an induction tactic based on the structural induction theorem for lazy sequences. This uses the inclusive prover behind the scenes to prove the equation admits induction. The proof is finished off using reduction tactics for **Maps** and **Comp** .

Finally, we present a functional **Seq_of** which given a continuous function **f** and any starting point value **x** generates an infinite sequence of the form

```
"Cons_seq x(Cons_seq(f x)(Cons_seq(f(f x))...))"
```

or written in a more readable way $[x; f(x); f(f(x)); \ldots]$. The function **Seq_of** is defined as a fixed point as follows:

```
|- !D.
    Seq_of =
    Fix
    (\sf :: Dom(cf(cf(D,D),cf(D,seq D))).
       \f :: Dom(cf(D,D)). \x :: Dom D. Cons_seq x(sf f(f x)))
|- !D. cpo D ==> Seq_of ins (cf(cf(D,D),cf(D,seq D)))
```

The internal version of **Seq_of** is parameterized by a cpo corresponding to the variable **D** in the definition. We have proved the following statement about **Maps** and **Seq_of**

```
|- !x. x ins D ==> (Seq_of f(f x) = Maps f(Seq_of f x))
```

where **D** is a cpo and **f** is a continuous function from **D** to **D** . Informally, the two sequences are equal since they are both equal to a term corresponding to $[f x; f(f x); \ldots]$. The proof of the theorem is conducted by fixed point induction on both occurrences of **Seq_of** ; inclusiveness is proved behind the scenes.

The proofs in LCF and HOL-CPO are based on the same overall idea but tend to be longer in HOL. We must do many simplifications explicitly which are taken care of by LCF rewriting. We must use the reduction tactic to type check arguments of functions before their definitions can be expanded (by applying reduction theorems). LCF rewriting with definitions corresponds to such reductions since it also performs β-conversion.

6 The Unification Algorithm

The problem of finding a common instance of two expressions is called *unification*. The unification algorithm generates a substitution to yield this instance, and returns a failure if a common instance does not exist. Expressions, also called *terms*, can be constants, variables and applications of one expression to another:

```
term = Const name | Var name | Comb term term
```

Variables are regarded as empty slots for which expressions can be substituted. A substitution is a set of pairs of variables and expressions that specifies which expressions should be substituted for which variables in an expression.

Manna and Waldinger synthesized a unification algorithm by hand using their deductive tableau system [MW81] and Paulson made an attempt to translate their proof of correctness to LCF [Pa85]. Paulson did not deduce the algorithm from the proof as Manna and Waldinger did; he stated the algorithm first and then proved it was correct.

A version of Paulson's proof has been conducted in HOL-CPO. In this section we shall not go into the details of this proof but mainly discuss a few points made by Paulson on the LCF proof. The details of the HOL proof are presented in [Ag94]. Although this example is considerably larger than the examples above it does not require deeper insights in domain theory. In fact, domain theory is used very little and only in the last stages of the proof. But the formalization *is* exploited in an essential way. The unification algorithm cannot be defined in pure HOL (at least not directly) since it is not primitive recursive. However, it can be defined as a fixed point easily.

Once we have proved that the unification algorithm defined in domain theory always terminates—this proof is conducted by well-founded induction—we can define a pure set theoretic HOL function. One may therefore argue that this approach provides a method, though probably not the simplest and most direct one, for defining recursive function by well-founded induction in HOL.

Paulson says that LCF does not provide an ideal logic for verifying the unification algorithm since it clutters up everything with the bottom element. For instance, the type of constant and variable names and the syntax type of terms must contain a bottom element, just like all other LCF types. Hence, definedness assertions of the form $t \not\equiv \bot$ occur everywhere because constructor functions for terms are only defined if their arguments are (strictness). To indicate the influence of this problem on the complexity of statements and proofs we show the LCF definitional properties for substitution (derived from a recursion axiom):

$$\bot \, SUBST \, s \equiv \bot$$
$$\forall c. \, c \not\equiv \bot \Rightarrow (CONST \, c) \, SUBST \, s \equiv CONST \, c$$
$$\forall v. \, v \not\equiv \bot \Rightarrow (VAR \, v) \, SUBST \, s \equiv ASSOC \, (VAR \, v) \, v \, s$$
$$\forall t_1 t_2. \, t_1 \not\equiv \bot \Rightarrow t_2 \not\equiv \bot \Rightarrow$$
$$(COMB \, t_1 \, t_2) \, SUBST \, s \equiv COMB(t_1 \, SUBST \, s)(t_2 \, SUBST \, s).$$

In HOL substitution is introduced by a primitive recursive definition:

```
|- (!c s. (Const c) subst s = Const c) /\
   (!v s. (Var v) subst s = assoc(Var v)v s) /\
   (!t1 t2 s.
      (Comb t1 t2) subst s = Comb(t1 subst s)(t2 subst s))
```

Note this is pure HOL, we do not need to use domain theory to define a type of terms and **subst** . Terms and names of constants and variables are represented by HOL types which do not contain bottom, in contrast to the LCF types. All functions on terms used in the proof, except unification itself, can be defined by primitive recursion like **subst** above. Hence, we can do the set theoretic developments first and then turn to domain theory later. We can define discrete cpos of terms and names and lift these to contain a bottom when necessary, just as we did in the natural number example. Besides, we avoid PPλ's explicit statements of totality for functions such as *SUBST* which are obviously total,

$$\forall t\, s.\, t \not\equiv \bot \Rightarrow s \not\equiv \bot \Rightarrow t\, SUBST\, s \not\equiv \bot,$$

since HOL functions are always total.

The unification algorithm is stated as a collection of recursion equations in LCF. In HOL, the unification algorithm is defined as a fixed point of a certain functional, which unfortunately is too large (one page) to be presented here, and the recursion equations are then derived from the fixed point property. It is a continuous partial function as stated by:

```
|- unify ins (cf(term,cf(term,lift attempt)))
```

The cpo of terms is just the discrete universal cpo of all HOL terms of type "**:term**" which can be introduced by the above specification. The cpo of attempts is the sum cpo of a discrete universal cpo with underlying type "**:one**" and a discrete universal cpo with underlying type "**:(name#term)list**", corresponding to the type of substitutions. The first component of the sum can be interpreted as failure and the second as success. The correctness of **unify** is stated as the theorem:

```
|- !t u. ?a. (unify t u = Lift a) /\ best_unify_try(a,t,u)
```

The first conjunct states **unify** is total and the second states it yields the best unifier in a certain sense if a unifier exists, otherwise it yields a failure. The predicate **best_unify_try** is defined in pure HOL (no domain theory).

The unification algorithm is recursive on terms but it is not primitive recursive. In order to unify two combinations "**Comb t1 t2**" and "**Comb u1 u2**" the algorithm first attempts to unify **t1** and **u1** and if it succeeds with the substitution **s** as a result it attempts to unify "**t2 subst s**" and "**u2 subst s**". The latter two terms may be bigger than the original combinations and therefore a primitive recursive definition does not work. However, when this is the case then the total number of variables in the terms are reduced. This argument induces a well-founded relation which can be used to prove termination. It is a kind of lexicographic combination of a proper subset ordering on sets of variables and an 'occurs-in' ordering. A theory of well-founded induction has been developed in HOL [Ag91] but never in LCF, because it is not possible to derive this general kind of induction from fixed point induction. Therefore, well-founded induction is translated to two structural inductions in LCF, one on natural numbers and one on terms. This makes certain statements more complicated than necessary and makes the proof less elegant as well.

Though the unification algorithm is a total function it is not straightforward to define it in 'pure' HOL since it is not primitive recursive. However, going via domain theory and well-founded induction to prove termination it is possible to introduce a pure HOL unification function. We can simply define this function using the choice operator as follows

    |- !t u. Unify t u = (@a. unify t u = Lift a)

Furthermore, we can prove this function yields a best unifier for terms of type ":term".

    |- !t u. best_unify_try(Unify t u,t,u)

From its definition, the recursion equations stating how it behaves on various kinds of arguments can be derived. This approach to derive a pure HOL unification function via domain theory and well-founded induction may be seen as a recursive definition by well-founded induction.

7   Conclusion

A contribution of this work is a comparison of two systems supporting domain theoretic reasoning, namely, LCF and the extension of HOL with domain theory. Using examples we show how HOL-CPO supports a mix of the two different kinds of reasoning provided in HOL and LCF, respectively. In a way, HOL-CPO can be seen as an embedding of the LCF system in HOL which is performed in such a way that the benefits of the HOL world are preserved.

We presented the mechanization of a number of examples in HOL-CPO which have already been done in LCF by Paulson. The natural number example illustrates how we can mix set and domain theoretic reasoning and thereby ease reasoning about finite-valued LCF types and strict functions. The example on lazy sequences gives a definition of an infinite sequence constructor functional as a fixed point and illustrates that we can conduct LCF proofs by fixed point induction and structural induction on infinite-valued recursive domains in HOL-CPO. This kind of reasoning is not possible in 'pure' HOL.

The unification example shows that we can avoid almost all reasoning about bottom that infests the LCF proof since it is an element of the type of expressions. In HOL, bottom is only introduced to allow a fixed point definition of the unification algorithm which is not primitive recursive and therefore cannot be defined in HOL directly. Other recursive functions of the example can be defined by primitive recursion in pure HOL, without using the formalization of domain theory at all.

Further, the example shows that we are not restricted to use fixed point induction for reasoning about recursive functions. The proof of termination of the unification algorithm is conducted by well-founded induction. The LCF proof uses two nested structural inductions to simulate well-founded induction which makes the proof more complicated, and less elegant too. Once it has been shown that the algorithm is total we can be define a total HOL function with the same behavior. Hence, the development can be seen as a way of defining a total HOL unification function by well-founded induction (see the end of section 6).

Some disadvantages of the embedding of domain theory in HOL have also been mentioned. One main problem is that it is time-consuming and not at all straightforward to introduce new recursive domains. Axiomatizing certain recursive types has been automated in LCF. Another problem is that constructors must be parameterized by the domains on which they work. This inconvenience is handled by an interface in most cases but the problem also affects the efficiency of proofs greatly since checking arguments of functions are in the right domains (called type checking) is inefficient.

One may compare the problems in LCF due to bottom to the problems in HOL-CPO due to the parameters on the dependent lambda abstraction and some function constructions. An interface could also be implemented in LCF to hide bottom in many cases but it would always be there in proofs. Often we avoid type checking in HOL-CPO. For instance, in the unification example where the bottom element was a major nuisance in LCF we worked most of the time in set theory where the problem of dependent functions (or bottom) does not exists. Domain theory was only used to define the recursive unification algorithm at a late stage of the proof.

HOL-CPO is a semantic embedding of domain theory in a powerful theorem prover. It was an important goal of this embedding that to preserve a direct correspondence between elements of domains and elements of HOL types. This allows us to exploit the types and tools of HOL directly and hence, to benefit from mixing domain and set theoretic reasoning as discussed above. A semantic embedding does not always have this property. The formalization of $P\omega$ in [Pe93] builds a separate $P\omega$ world inside HOL so there is no direct relationship between, for instance, natural numbers in the $P\omega$ model and in the HOL system. The same thing would be true about a formalization of information systems [Wi93], if it was done. On the other hand, formalizations of $P\omega$ and information systems allow recursive domain equations to be solved fairly easily using the fixed point operator.

Franz Regensburger[4] is working on a very similar project in Isabelle HOL but the formalizations seem to be quite different. Pointed cpos are introduced using type classes and continuous functions constitute a type. Type checking arguments of functions seems not to be necessary but before β-reduction can be performed functions must be shown to be continuous (unlike in our formalization). Recursive domains can be axiomatized in a similar way as in LCF, though this has not been automated as in LCF. He is currently writing a Ph.D. thesis about the work (in German unfortunately). Bernhard Reus[5] works on synthetic domain theory in the LEGO system which implements a strong type theory (ECC) with dependent sums and products. Dependent families can be exploited for the inverse limit construction of solutions to recursive domain equations. This is work in progress for a Ph.D. and the formalization has not been published yet.

---

[4] Technical University, Munich. Email: regensbu@informatik.tu-muenchen.de

[5] Ludwig-Maximilian University, Munich. Email: reus@informatik.uni-muenchen.de

Acknowledgements

This work was supported in part by the DART project funded by the Danish Research Council and in part by BRICS funded by the Danish National Research Foundation. Thanks to Flemming Andersen, Kim Dam Petersen and Glynn Winskel for discussions concerning this work. Glynn made comments on a final draft. I am grateful to Larry Paulson for digging up the LCF proof of correctness of the unification algorithm.

References

[Ag91] S. Agerholm, 'Mechanizing Program Verification in HOL'. In the *Proceedings of the 1991 International Workshop on the HOL Theorem Proving System and Its Applications*, Davis California, August 28–30, 1991 (IEEE Computer Society Press). Also in Report IR-111, M.Sc. Thesis, Aarhus University, Computer Science Department, April 1992.

[Ag93] S. Agerholm, 'Domain Theory in HOL'. In the *Proceedings of the 6th International Workshop on Higher Order Logic Theorem Proving and its Applications*, Jeffrey J. Joyce and Carl-Johan H. Seger (Eds.), Vancouver, B.C., Canada, August 11–13 1993, LNCS 780, 1994.

[Ag94] S. Agerholm, *A HOL Basis for Reasoning about Functional Programs*. Ph.D. Thesis, Aarhus University, Computer Science Department, June 1994.

[GM93] M.J.C. Gordon and T.F. Melham, *Introduction to HOL: A Theorem Proving Environment for Higher Order Logic*. Cambridge University Press, 1993.

[GMW79] M.J.C. Gordon, R. Milner and C.P. Wadsworth, *Edinburgh LCF: A Mechanised Logic of Computation*. Springer-Verlag, LNCS 78, 1979.

[MW81] Z. Manna and R. Waldinger, 'Deductive Synthesis of the Unification Algorithm'. *Science of Computer Programming*, Vol. 1, 1981, pp. 5–48.

[Me89] T.F. Melham, 'Automating Recursive Type Definitions in Higher Order Logic'. In G. Birtwistle and P.A. Subrahmanyam (eds.), *Current Trends in Hardware Verification and Theorem Proving*, Springer-Verlag, 1989.

[Pa84a] L.C. Paulson, 'Structural Induction in LCF'. Springer-Verlag, LNCS 173, 1984. Also in Technical Report No. 44, University of Cambridge, Computer Laboratory, February 1984.

[Pa84b] L.C. Paulson, 'Lessons Learned from LCF'. Technical Report No. 54, University of Cambridge, Computer Laboratory, August 1984.

[Pa85] L.C. Paulson, 'Verifying the Unification Algorithm in LCF'. *Science of Computer Programming*, Vol. 5, 1985, pp. 143–169. Also in Technical Report No. 50, University of Cambridge, Computer Laboratory, March 1984.

[Pa87] L.C. Paulson, *Logic and Computation: Interactive Proof with Cambridge LCF*. Cambridge Tracts in Theoretical Computing 2, Cambridge University Press, 1987.

[Pe93] K.D. Petersen, 'Graph Model of LAMBDA in Higher Order Logic'. In the *Proceedings of the 6th International Workshop on Higher Order Logic Theorem Proving and its Applications*, Jeffrey J. Joyce and Carl-Johan H. Seger (Eds.), Vancouver, B.C., Canada, August 11–13 1993, LNCS 780, 1994.

[Wi93] G. Winskel, *The Formal Semantics of Programming Languages*. The MIT Press, 1993.

A Graphical Tool for Proving UNITY Progress

Flemming Andersen   Kim Dam Petersen   Jimmi S. Pettersson

Tele Danmark Research, Lyngsø Allé 2, DK-2970 Hørsholm

Abstract. A graphical tool for proving **leadsto** progress properties
of UNITY programs is described. The tool allows a user to draw Directed Acyclic Graphs (DAGs) that outlines the proof of UNITY **leadsto**
progress properties. From these DAGs the tool generates proof scripts
that contain proofs of the **leadsto** properties. Edges in the DAGs are
annotated with information that can direct a theorem prover on how to
prove the progress properties that they represent. The proof script generated by the tool can be compiled into another proof script which can
be checked by a theorem prover. Using this graphical tool it is possible,
modulo the strength of the theorem prover, to automatically prove that
a program satisfies a **leadsto** property specified as a DAG which defines
the proof structure.

1   Introduction

The work presented here is part of a project on building a software verification tool for telecommunications industry. The activities in the project are currently based on the UNITY theory [5], which consists of a parallel programming
language and a logic for reasoning about programs. The UNITY theory has
been chosen because it is simple and provides a wide collection of deduction
rules, which makes it very useful in practice. Although simple, it seems powerful
enough to serve as a foundation for other programming languages, such as SDL
[7], C [10] and C++ [17], which are more widely used in the telecommunications
industry. For instance Ulla Binau in her PhD thesis [4] shows how a non-trivial
subset of C++ extended with constructs for parallelism can be translated into
UNITY, and how a subset of UNITY can be translated back to C++.

We have previously implemented a theory for UNITY in the HOL theorem
prover [8, 16] called HOL-UNITY [1] which allows for mechanised reasoning
about UNITY program properties. To automate parts of the verification process
and reduce the need for detailed knowledge about the theorem prover a tactic
[14] has been developed for the theorem prover which attempts automatically to
prove basic UNITY properties. For hiding details concerning the representation
of UNITY in the theorem prover we have also developed a compiler for UNITY
[3, 12]. This compiler translates UNITY programs and properties together with
their proofs into the internal representation used by the HOL-UNITY theorem
prover.

Even with these tools it is still difficult to verify properties. Especially **leadsto**
progress properties, because they are defined inductively and thus must be decomposed to be proved. To prove a progress property in UNITY, or in other

temporal logics, one usually decomposes the property into simpler properties, which can be proved directly from the program text. The desired property can then be constructed by applying a set of inference rules to derive the wanted property from the simpler properties. A mechanised decomposition process involves book-keeping of which properties have actually been verified, and which inference rules to apply to the verified properties in order to derive the desired property.

One way of doing book-keeping is to use graphs for outlining the proof. There are several suggestions in the literature on how to use graphs for outlining proofs. Owicki and Lamport show in [13] how specialised Directed Acyclic Graphs (DAGs), called *proof lattices* can be used to structure the decomposition of a progress property in Temporal Logic. The nodes in the graphs denote assertions and the edges going out from a node denote *strong* progress from this node to the disjunction of the nodes at the end of the edges. A similar approach is found in [9] where Gribomont defines a concept called *proof graphs*. His definition of proof graphs is similar to that of proof lattices even though his graphs are used for specifying safety properties only. The nodes denote assertions, but now the edges denote *weak* progress properties. Finally, Manna and Pnueli [11] introduce a concept called *proof diagrams* for proving progress properties of parallel processes in Temporal Logic. These proof diagrams depict progress by safety edges and progress edges. The safety edges are used to denote program transitions which represent progress or stability, and the progress edges are used to denote the processes that ensures progress.

In the present paper we suggest yet another graphical notation called *annotated proof lattices*, which can be seen as a variant of proof lattices. The main objective of annotated proof lattices is to outline a proof in a way which allows a theorem prover to directly prove progress properties. For this reason it is required that the edges which represents decomposed progress properties are annotated with information that guides the theorem prover in how to prove these properties. The name *annotated proof lattices* is used to distinguish such lattices from those of Owicki and Lamport.

To support mechanised proof of annotated proof lattices we have developed a graphical tool. The tool allows a user to specify a DAG by drawing nodes and annotated edges. For a well-formed DAG the tool is able to generate a proof script for the progress property denoted by the DAG. The generated proof script has two steps. One which proves that each annotated edge is valid, i.e. that the associated progress property is satisfied. The other step derives the overall desired progress property from the edge properties. This means that the graphical tool does not prove anything itself. Rather it generates a detailed proof script from which the property represented by the proof lattice can be proved. In the present tool, the HOL-UNITY compiler translates the proof script into another proof script which can be performed by the HOL-UNITY theorem prover. The graphical tool, the compiler and the theorem prover constitute a system, in which programs and their required properties can be specified.

The paper is organised as follows. In Section 2 a general description of annotated proof lattices is given. The section includes a discussion on how to represent state space restrictions in a proof lattice. In Section 3 it is described how the UNITY progress relation **leadsto** can be used as the basic progress relation of annotated proof lattices. Section 4 presents an example of a proof lattice. Section 5 gives an overview of the implementation of the tool. This section also shows the proof script generated by the tool for the example. Finally, Section 6 contains a discussion on the present work and future directions.

2 Proof Lattices

In [13] Owicki and Lamport describe how a proof lattice for a program can be used to outline the proof structure of a progress formula in temporal logic of form $p \rightsquigarrow q$[1]. A proof lattice is a Directed Acyclic Graph with nodes labeled by temporal formulae. Each graph has a single entry node and a single exit node. For each node labeled r with outgoing edges to nodes $s_1, \ldots, s_n, n \geq 1$, the formula $r \rightsquigarrow s_1 \vee \ldots \vee s_n$ must hold. Owicki and Lamport prove that if there exists a proof lattice for a program with entry node labeled p and exit node labeled q then the formula $p \rightsquigarrow q$ holds. The name *proof lattice theorem* is used for denoting this fact in this paper.

Our aim with proof lattices is to outline progress proofs such that they can be checked mechanically by a theorem prover. To achieve this, a variant of proof lattices called *annotated proof lattices* is defined. In such lattices, edges are annotated with justifications of their associated progress properties. The intention is to pass on the annotations to the theorem prover and thereby direct it in how to verify the lattice property.

Annotated proof lattices differ from those of Owicki and Lamport in that: (1) the nodes are predicates, (2) for each node with a single outgoing edge the edge must be annotated with information by which a theorem prover can prove the associated progress property, and (3) for each node with multiple (> 1) outgoing edges the edges must not be annotated and the start node must imply the disjunction of the nodes at the end of the outgoing edges. In general, annotations have no fixed structure. Since, they are simply considered labels on relations that justifies progress. However, when the UNITY **leadsto** progress relation is considered in the next section, the annotations will represent inference rules that justifies the deduction of **leadsto** properties.

To make annotated proof lattices as general as possible, they are defined for arbitrary progress relations. A progress relation is formally defined as follows:

Definition (Progress Relation) *A* progress relation *is a relation* $\mapsto$ *on predicates that satisfies:*

$$p \mapsto p \tag{1}$$

$$\frac{p \mapsto q, \ q \mapsto r}{p \mapsto r} \tag{2}$$

---

[1] $p \rightsquigarrow q \equiv \Box(p \Rightarrow \Diamond q)$: whenever p is satisfied, then sooner or later q will be satisfied.

$$\frac{[p \Rightarrow (q_1 \vee \ldots \vee q_n), \; q_1 \mapsto r, \; \ldots, \; q_n \mapsto r}{p \mapsto r} \tag{3}$$

The formal definition of annotated proof lattices is given below:

Definition (Annotated Proof Lattice) *An* annotated proof lattice *over a progress relation* $\mapsto$, *a set* A *of annotations, and a set of relations* $\mapsto_a, a \in A$ *that satisfies:*

$$\frac{p \mapsto_a q}{p \mapsto q} \tag{4}$$

is a finite directed acyclic graph with nodes labelled by predicates and edges optionally labelled by annotations such that:

1. *There is a single* entry *node with no incoming edges.*
2. *There is a single* exit *node with no outgoing edges.*
3. *If a node labelled* r *has a single outgoing edge to a node labelled* s, *the edge must be annotated with some* a, *and* $r \mapsto_a s$ *must hold.*
4. *If a node labelled* r *has multiple outgoing edges to nodes labelled* $s_1, \ldots, s_n$, $n > 1$, *none of the edges may be labelled, and* $r \Rightarrow (s_1 \vee \ldots \vee s_n)$ *must hold.*

In the following *annotated relation* is used for any relation $\mapsto_a$.

The main property of annotated proof lattices is presented by the *annotated proof lattice theorem* given below. This theorem states that the property $p \mapsto q$ can be derived from the existence of an arbitrary annotated proof lattice with entry node p and exit node q. To prove this theorem, the assumptions (1)–(4) of the proof lattice relations are used.

Theorem (Annotated Proof Lattice Theorem) *If there exists an annotated proof lattice over a relation* $\mapsto$ *and a set of annotated relations* $\mapsto_a, a \in A$, *with an entry node labeled* p *and an exit node labeled* q, *then* $p \mapsto q$ *holds.*
Proof The theorem is proved by first proving a lemma stating that every node in the proof lattice leads to the exit node. The theorem then follows by specialising this lemma with the entry node.

Lemma *Every node* r *in the proof lattice satisfies* $r \mapsto q$.
The lemma is proved inductively on the length of the longest path from the node to the exit node.
In the base case the length of the longest path is 0; hence $r = q$. In this case $q \mapsto q$ has to be proved. This follows from assumption (1).
Otherwise the length is greater than 0 and r has one or more outgoing edges to nodes, say $s_1, \ldots, s_n, n \geq 1$. If $n = 1$, the node has a single outgoing edge. By definition this edge is annotated, say with a, and $r \mapsto_a s_1$ holds. From assumption (4), $r \mapsto s_1$ can be concluded. From the induction hypothesis $s_1 \mapsto q$ is given; hence by applying assumption (2), $r \mapsto q$ can be derived. If $n > 1$, the node has multiple outgoing edges. From the definition $r \Rightarrow (s_1 \vee \ldots \vee s_n)$ is given, and from the induction hypothesis $s_i \mapsto q, 1 \leq i \leq n$ is given. Applying assumption (3) to this yields $r \mapsto q$. Hence, $r \mapsto q$ has been proved for every node in the lattice. $\square$

The theorem, $p \mapsto q$ now follows by specialising the lemma to the entry node p, which is a node in the proof lattice. □

There are two reasons for presenting the above proof. One, of course, is to prove the theorem. The other, and in this context more important, is to show the constructive nature of the proof. Given a proof lattice, the lemma presents a structured method for deriving that the entry node leads to the exit node. The graphical tool presented later is simply a mechanisation of this method.

Restricted State Space

Owicki and Lamport allow a sub-lattice of a proof lattice to be surrounded by a box annotated with □r to indicate that r holds in conjunction with temporal formulae of all nodes in the sub-lattice. They describe how it is sufficient to prove the property □r at entrance to the sub-lattice and that it propagates through the boxed lattice to the exit node. This is stated by the following *invariant propagation rule* by Owicki and Lamport:

$$\frac{(p \wedge \Box r) \rightsquigarrow q}{(p \wedge \Box r) \rightsquigarrow (q \wedge \Box r)} \tag{5}$$

The purpose of boxing is to restrict the state space by some predicate r. In annotated proof lattices, nodes are labelled by predicates rather than temporal logic formulae. This means that □r cannot be used in annotated proof lattices. Instead the notation $\Box_{\mapsto} r$ is introduced. Informally this notation expresses that the predicate r can always progress from left to right of the relation $\mapsto$. Formally, this is expressed by the following propagation inference rule:

$$\frac{\Box_{\mapsto} r, \ (p \wedge r) \mapsto q}{(p \wedge r) \mapsto (q \wedge r)} \tag{6}$$

Assuming this inference rule holds it is possible to formalise the concept of boxed annotated proof lattices and prove that the annotated proof lattice theorem also holds for such lattices.

Unless otherwise specified the phrase *proof lattice* in the rest of this paper means *annotated proof lattice*.

3 UNITY proof lattices

Annotated proof lattices can be used to outline proofs of the UNITY progress relation **leadsto** such that they can be mechanically verified.

The UNITY **leadsto** progress relation has three parameters, a program and two state dependent predicates. The general properties of **leadsto** are independent of the actual program parameter. For this reason it is in the following implicitly assumed that the program parameter of the **leadsto** relation is bound to some fixed arbitrary program P. Hence, the program parameter is left out in the following.

Since the UNITY **leadsto** relation satisfies the requirements (1)–(3), it is a progress relation. This is implied by the fact that the **leadsto** relation is reflexive, transitive and satisfies the disjunctivity assumption (3) required by the $\mapsto$ relation. As a result, it is valid to specialise $\mapsto$ by **leadsto**.

For these lattices to be of use for mechanical proving, a collection of annotated relations is required. Fortunately, any relation from which **leadsto** can be deduced is a legal annotated relation. This means, all inference rules for deriving **leadsto** can be used for defining an annotated relation.

The **leadsto** inference rules given for logic implication, the basic progress relation **ensures** and the UNITY induction rule are examples of inference rules defining such annotated relations. Such lattices will be called annotated **leadsto** proof lattices.

The ability to mechanically verify a proof lattice depends on the ability to mechanically prove the annotated relations ($\mapsto_a$) and the implications associated with multiple outgoing edges ($p \Rightarrow (q_1 \vee \ldots \vee q_n)$). These proofs can be mechanically composed to deduce the **leadsto** property represented by the proof lattice. As stated by the *annotated proof lattice theorem*, the composition is derived from the structure of the lattice.

Annotated UNITY relations

Presented below is a list of annotated relations valid for **leadsto**. Each relation is justified by an inference rule that derives **leadsto**.

Implication One of the simplest inference rules for deducing **leadsto** is the following, which only requires the corresponding implication to hold.

$$\frac{p \Rightarrow q}{p \text{ leadsto } q}$$

It can be noted that this inference rule is only valid for non-empty programs, i.e. programs with at least one action.

The annotation **implies** will be used to indicate that this inference rule justifies the **leadsto** relation. In Figure 1 an example of how to use the **implies** annotation is given.

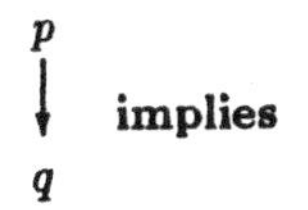

$$p$$
$$\downarrow \quad \text{implies}$$
$$q$$

Fig. 1. Proof lattice with an **implies** edge

Ensures The most important inference rule for deducing **leadsto** is the following, which requires the basic progress inference rule **ensures** to hold:

$$\frac{p \text{ ensures } q}{p \text{ leadsto } q}$$

The annotation **ensures** indicates that the basic UNITY progress relation should be used to prove the property. In Figure 2 below an example of this annotation is showed. The basic progress relation **ensures** expresses that the program has an

Fig. 2. Proof lattice with an **ensures** edge

ensuring action which, when executed in a state that satisfies p, will ensure that q is satisfied after the execution, and furthermore that p holds until q holds. Since this implies that a theorem prover will have to test each action of the program to find the ensuring action, it is possible to specify the ensuring action as parameter to an **ensures**. This makes it possible to optimise the process of proving.

Lattice To avoid large complex lattices, we introduce the annotation **lattice**(l) to denote insertion of another proof lattice. The **lattice** annotated edges play the role of lemma in traditional proving. To avoid mutual dependencies, it is required that the inserted proof lattice has been separately verified. The inference rule that justifies the annotation has the peculiar form:

$$\frac{p \text{ leadsto } q}{p \text{ leadsto } q}$$

The **leadsto** property above the line in the inference rule represents the lattice that has been verified. In Figure 3 an example of a lattice annotation is given.

p

$\downarrow$ **lattice**(l)

q

Fig. 3. Proof lattice with a **lattice** edge

Induction UNITY supports inductive proving of progress described by the inference rule:

$$\frac{\forall m : p \wedge (M = m)\,\mathbf{leadsto}\,(p \wedge (M < m)) \vee q}{p\,\mathbf{leadsto}\,q}$$

This rule defines the annotation **induction**. To use induction requires a *metric function* M, which is a numeric function in the program variables. It is assumed that M is non-negative. The **induction** annotation must be supplied with one parameter: a theorem l of the **leadsto** property above the line in the inference rule. This theorem may be specified by a another proof lattice that has been verified. It is required that the theorem l has the form indicated by the inference rule. An example of a lattice with an **induction** annotation is given in Figure 4.

$$p$$
$$\downarrow \quad \mathbf{induction}(l)$$
$$q$$

Fig. 4. Proof lattice with an **induction** edge

PSP Another deduction rule, which is often used in practice is the **PSP** (Progress-Safety-Progress) rule:

$$\frac{p\,\mathbf{leadsto}\,q,\ r\,\mathbf{unless}\,b}{p \wedge r\,\mathbf{leadsto}\,(q \wedge r) \vee b}$$

The relation **unless** in the inference rule represents a basic safety property. This

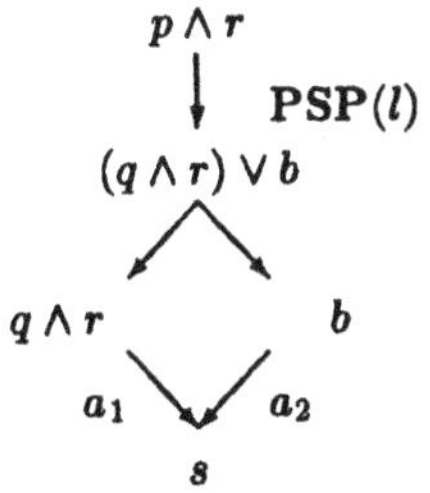

$$p \wedge r$$
$$\downarrow \quad \mathrm{PSP}(l)$$
$$(q \wedge r) \vee b$$

$$q \wedge r \qquad\qquad b$$
$$a_1 \qquad\qquad a_2$$
$$s$$

Fig. 5. Proof lattice with a **PSP** edge

property is supposed to be proved directly by the theorem prover. The **PSP** annotation has one parameter l that represents the assumed **leadsto** relation in the inference rule. It is required that the start and end nodes of a **PSP** annotated edge has the form indicated in the inference rule. An example on how to use the **PSP** rule is sketched in Figure 5.

Cancellation Another inference rule is **cancellation**:

$$\frac{p \text{ leadsto } q \vee b, \; b \text{ leadsto } r}{p \text{ leadsto } q \vee r}$$

The annotation requires two theorems l_1 and l_2 as parameters. These theorems

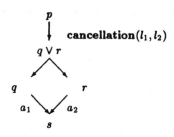

Fig. 6. Proof lattice with a **cancellation** edge

represents the **leadsto** properties assumed in the inference rule and must have the form indicated in the inference rule. An example of how to use the **PSP** rule is sketched in Figure 6.

Infinite branching The only inference rule seen until now which introduces infinite case analysis is the induction rule; but only in a restricted form. The UNITY progress relation **leadsto**, however, allows for a more general infinite branching, as indicated by its inductive definition, given by the three inference rules:

$$\frac{p \text{ ensures } q}{p \text{ leadsto } q}$$

$$\frac{p \text{ leadsto } q \wedge q \text{ leadsto } r}{p \text{ leadsto } r}$$

$$\frac{\forall i : p_i \text{ leadsto } q}{(\bigvee_i p_i) \text{ leadsto } q}$$

where i ranges over some set.

The proof lattices can be extended to describe a wider class of proofs if a notation is introduced for denoting an indexed family of branches. This extension may be achieved, if index parameters are allowed as annotations to the lattice edges. Using this approach, edges can represent infinite collections of branches

Fig. 7. Infinite branching proof lattice

indexed by some binding variable i. Such a variable i is then considered a constant of unknown value in the succeeding part of the lattice, i.e. both when i is used in predicates as well as annotations. In Figure 7 an example of a proof lattice is given. This example has an infinite branching with the edge annotated by λi. This proof lattice is then expected to be proved by the inference rule expressing infinite disjunction of **leadsto** properties.

The relations listed above is by no means complete. As mentioned in the beginning of this section it is possible to use any inference rule from which **leadsto** can be derived.

4 Example

To show what a proof lattice looks like using the tool, one of the proof lattices from a proof of the Sliding Window Protocol [15] is presented in Figure 8. The

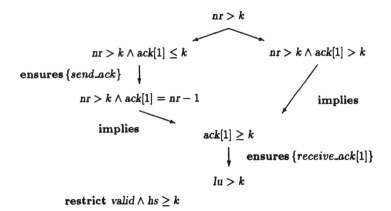

Fig. 8. Example of a proof lattice

lattice demonstrates the steps involved in proving the property:

$$nr > k \ \textbf{leadsto} \ lu > k$$

This **leadsto** property reads: if a message number k has been delivered to the receiver, then sooner or later, the transmitter will receive an acknowledgement for message number k. The split at the top node $nr > k$ into two branches, indicates that the proof has two cases: $ack[1] \leq k$ and $ack[1] > k$ each of which has to be considered separately. In the first case an **ensures** followed by an **implies** proof step leads to $ack[1] \geq k$, a situation, which in the second case is reached by a single **implies** proof step. From this situation, a final **ensures** proof step leads to the goal $lu > k$. In the **ensures** proof steps, additional information on which action actually ensures progress is supplied in the annotation.

The HOL-UNITY compiler performs the translation of the proof script to the notation required by the theorem prover before handing the proof obligations over to the theorem prover. This is discussed in Section 5.

The proof lattice presented in Figure 8 is boxed, i.e. state space restricted with the predicate ($valid \wedge hs \geq k$). This restriction is denoted by the **restrict** annotation at the bottom of the figure. The predicate $valid$ denotes a superset of states reachable by the program. The predicate $hs \geq k$ is required to ensure progress of the lattice.

The generated proof script that derives $nr > k$ **leadsto** $lu > k$ is shown in the next section.

5   Mechanised proof lattice verification

The present HOL-UNITY verification system consists of three tools:

1. a graphical tool which allows the user to draw a DAG. From this it produces a proof script containing the proof obligations for deriving the progress property represented by the proof lattice,
2. a HOL-UNITY compiler which compiles the proof obligations into a notation readable by the HOL-UNITY theorem prover, and finally
3. the HOL-UNITY theorem prover, which takes proof obligations in the generated proof script and performs the actual proof.

The interface between the user and the tools is presented in Figure 9.
The graphical tool is again divided into three parts:

1. a graphical editor, which allows the user to draw a DAG. From the DAG the tool generates a list of text and poly-line objects with their graphical attributes,
2. an analyser that constructs an internal DAG data structure from the graphical objects, and checks that it has the structure of a proof lattice, and
3. a proof generator that produces a proof script, in a notation readable by the HOL-UNITY compiler.

As graphical editor a general purpose public domain image drawing program TGIF [6] is used. This is expected to be replaced in future by a more sophisticated graphical editor, in which the editing modes of TGIF are replaced by UNITY proof lattice boxing and annotation modes.

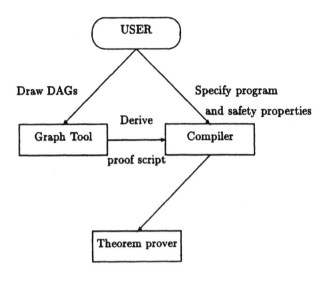

Fig. 9. HOL-UNITY Tools

The analyser and proof generator parts are integrated in the same program. The analyser part scans the output of TGIF for retrieving text and poly-line objects. It then uses a simple coordinate analysis to detect which text objects are nodes respectively annotations of the edges. Following this, it checks that the graph has the structure of a proof lattice, i.e. that it is acyclic, that it has a single entry and a single exit node, and that all edges are properly annotated according to the definition. Finally, it produces an internal data structure of the DAG which allows for forward and backward traversal of edges; and which allows for associating book-keeping information concerning the proof with the individual nodes and edges. This program organisation makes it easy to adapt the proof generator to other theorem provers than HOL-UNITY.

The proof generator part produces a script of proof obligations, which is listed in Figure 10. The listed proof obligations are the result of analysing the proof lattice shown in Figure 8. This proof script can be given as input to the HOL-UNITY compiler.

The presented graphical tool is batch-oriented. A disadvantage of this is that if any part of a proof lattice is changed, then the system will have to regenerate all proof obligations of the lattice; not just those of the changed part. An advantage, however, is that time consuming proofs can be done off-line. The graph tool is able to check that its input is a DAG. However, it cannot prove whether the DAG is a proof lattice; instead sufficient proof obligations are generated to certify that this is the case. Following this, a separate theorem proving session performed will have to verify that the proof obligations actually holds.

```
PROPERTY
LEMMA lem_fig1_0 =
  |= (nr > k) ==>  ((nr > k /\ ack[1]<= k)  \/ (nr > k /\ ack[1] > k))
USING AUTO_TAC END,
LEMMA lem_fig1_1 =
  nr > k /\ ack[1]<= k ENSURES nr > k /\ ack[1] = nr-1
USING AUTO_TAC END,
LEMMA lem_fig1_2 =
  |= (nr > k /\ ack[1] = nr-1) ==>  (ack[1] >= k)
USING AUTO_TAC END,
LEMMA lem_fig1_3 =
  ack[1] >= k ENSURES lu > k VIA receive_ack[1]
USING AUTO_TAC END,
LEMMA lem_fig1_5 =
  |= (nr > k /\ ack[1] > k) ==>  (ack[1] >= k)
USING AUTO_TAC END,
LEMMA fig1 =
  nr > k LEADSTO  lu > k
PROOF
  (d0) nr > k LEADSTO ((nr > k /\ ack[1]<= k) \/ (nr > k /\ ack[1] > k))
       { IMPLIES_LEADSTO lem_fig1_0 }
  (d1) nr > k /\ ack[1]<= k LEADSTO nr > k /\ ack[1] = nr-1
       { ENSURES_LEADSTO lem_fig1_1 }
  (d2) nr > k /\ ack[1] = nr-1 LEADSTO  ack[1] >= k
       { IMPLIES_LEADSTO lem_fig1_2 }
  (d3) ack[1] >= k LEADSTO  lu > k
       { ENSURES_LEADSTO lem_fig1_3 }
  (d5) nr > k /\ ack[1] > k LEADSTO  ack[1] >= k
       { IMPLIES_LEADSTO lem_fig1_5 }
  (11) nr > k /\ ack[1] = nr-1 LEADSTO  lu > k { LEADSTO_TRANS d2 d3 }
  (12) nr > k /\ ack[1]<= k LEADSTO  lu > k { LEADSTO_TRANS d1 11 }
  (13) nr > k /\ ack[1] > k LEADSTO  lu > k { LEADSTO_TRANS d5 d3 }
  (10) nr > k LEADSTO  lu > k { LEADSTO_DISJ [12, 13] }
END
```

Fig. 10. Proof script produced by the graph analyser

The proof script in Figure 10 closely follows the derivation process of the annotated proof lattice theorem. It begins with a collection of lemmas named lem_fig1_x which proves that the DAG is a proof lattice. This implies that an implication property is generated for each node with multiple outgoing edges. Furthermore, if necessary, a property associated with an annotation is generated for each node with a single outgoing edge. These lemmas are followed by a proof of the desired property represented by the lattice. This proof takes the form of a lemma and is divided into two steps: first the corresponding **leadsto** property named **(dx)** is derived for each lemma, next the property named **(lx)** that each node **leadsto** the exit node is proved. The first of these steps **(dx)** derives the **leadsto** property from the annotated property. The next steps **(dx)** applies the derived **leadsto** properties to inference rules according to the structure of the proof lattice. The last **(lx)** property is the desired **leadsto** property stating that the entry node **leadsto** the exit node.

The theorem prover tactic AUTO_TAC [14] is used to verify the properties associated with the annotated edges and implication properties associated with the multiple outgoing edges. The tactic AUTO_TAC is constructed such that it knows how to reduce the various UNITY properties into formulae in first order logic extended with arithmetic. It includes various heuristics for checking the correctness of these. The derivation of **leadsto** properties from the annotated properties is a straightforward process. This mainly consists of an application of the inference rule associated with the annotation. The tactics named IMPLIES_LEADSTO and ENSURES_LEADSTO perform these derivations. Finally, the derivation of the property that every node in the lattice leads to the exit node is performed based on the three requirements (1)–(3) of the progress relation. To perform this derivation the three tactics: LEADSTO_REFL, LEADSTO_TRANS, and LEADSTO_DISJ has been implemented.

6 Discussion

A graphical tool for drawing proof lattices has been presented. From each lattice, the tool produces a script containing proof obligations. The HOL-UNITY compiler [12] translates a proof script into a proof notation readable by the HOL-UNITY theorem prover [1, 8]. The present tool is only a prototype, but it is able to decompose properties specified by a proof lattice. And, it provides the wanted book-keeping on how to derive the property represented by the proof lattice.

The tool has been developed for mechanised verification of the UNITY **leadsto** progress relation using the HOL-UNITY theorem prover. But, as the graphical tool represents the input DAG in a data structure that allows easy forward and backward traversal, the tool is easily adapted to other progress relations or theorem provers.

The HOL-UNITY tools make up a system for specifying programs and properties together with their proofs in an extension of UNITY. Progress properties can be proved by drawing sufficiently detailed proof lattices. The HOL-UNITY compiler translates the output from the graphical tool into HOL notation. Finally, the HOL-UNITY theorem prover attempts to verify the proof obligations generated from the proof lattice.

Presently the system consists of three separate tools each with batch-oriented interfaces. However, the ultimate goal is to integrate the tools into a single tool that allows a user to interactively specify a program and verify its required properties such that errors can be caught as early as possible.

The presented proof lattices have been successfully used for verifying progress properties of two smaller examples [15, 2]. The proof lattices provided a compact and readable view of the required program behaviour and their corresponding proofs. The original proofs were done by a manual translation of these lattices into HOL proof scripts, a tedious job, which inspired the development of the presented tool.

7 Acknowledgements

We would like to thank our colleagues Anders Gammelgaard, Jens Enevold Kristensen, and the anonymous reviewers for helpful comments on improving this paper.

References

1. Flemming Andersen. *A Theorem Prover for UNITY in Higher Order Logic.* PhD thesis, Technical University of Denmark, 1992. Also published as TFL RT 1992-3, Tele Danmark Research, 1992.

2. Flemming Andersen, Kim Dam Petersen, and Jimmi S. Pettersson. *Program Verification using HOL-UNITY.* In *HUG'93: HOL User's Group Workshop, pages 1–17, 1993.*

3. Flemming Andersen, Klaus Elmquist Nielsen, Kim Dam Petersen and Jimmi S. Pettersson. *The HOL-UNITY Language, Reference Manual 1.0.* In preparation.

4. Ulla Binau. *Correct Concurrent Programs: A UNITY design method for Compositional C++ Programs.* PhD thesis, Technical University of Denmark, 1994.

5. K. Mani Chandy and Jayadev Misra. *Parallel Program Design: A Foundation.* Addison–Wesley, 1988.

6. William Chia-Wey Cheng. *TGIF(n) Unix Manual.* Version 2.14, patchlevel 9, 1993.

7. Editors: Ove Færgemand and A. Sarma. *SDL '93 Using Objects: Proceedings of the Sixth SDL Forum, Darmstadt, Germany 1993.* North Holland, 1993.

8. Michael J.C. Gordon. *HOL – A Proof Generating System for Higher Order Logic.* Cambridge University, Computer Laboratory, 1987.

9. E. Pascal Gribomont. *Design, verification and documentation of concurrent systems.* 4th Refinement Workshop, Eds. Joseph M. Morris and Roger C. Shaw, Springer Verlag, 1991.

10. Brian W. Kernighan and Dennis M. Ritchie. *The C Programming Language.* Prentice Hall. 1978, Second Edition 1988.

11. Zohar Manna and Amir Pnueli. *Adequate Proof Principles for Invariance and Liveness Properties of Concurrent Programs.* Science of Computer Programming 4, pp. 257–289, 1984.

12. Klaus Elmquist Nielsen. *The HOL-UNITY compiler.* Technical report in preparation, Tele Danmark Research, 1994.

13. Susan Owicki and Leslie Lamport. *Proving Liveness Properties of Concurrent Programs*. ACM TOPLAS, Vol. 4, No. 3, July 1992. Pages 456–495.

14. Kim Dam Petersen. *HOL-UNITY Tactics – Automatic Proof of Basic Properties*. TFL LD-1994-2, Tele Danmark Research, December 1993.

15. Kim Dam Petersen and Jimmi S. Pettersson. *Proving Protocols Correct – Proving Safety and Progress Properties of the Sliding Window Protocol using HOL-UNITY*. Research Report TFL RR 1993-3, Tele Danmark Research, December 1993.

16. K. Slind. *HOL90 User's Manual*. Technical Report, Technical University of Munich.

17. Bjarne Stroustrup. *The C++ Programming Language*. Addison Wesley. 1991. Second Edition, 1993.

Reasoning About a Class of Linear Systems of Equations in HOL

Catia M. Angelo, Luc Claesen, Hugo De Man

IMEC vzw, Kapeldreef 75, B3001 Leuven, Belgium

Abstract. In this paper we present an algorithm implemented in the HOL system to prove the equivalence of two linear systems of equations (of a certain class) about natural numbers. The procedure has been developed in order to reason about an aspect of programs described in the Silage language. The semantics of Silage has been defined in the HOL system and specifies the class of linear systems of equations that the algorithm copes with.

1 Introduction

The Silage language [Hil85,GHR+90,Nac90a] used as input to the CATHEDRAL [DMRSC86] silicon compilers is a language suitable for describing DSP[1] algorithms represented as a data-flow graph. A basic concept of Silage is the notion of a signal. A signal is an infinite stream of data indexed by discrete and periodic time instants. Signals are defined by function applications. A definition involving signals is indeed a set of equations about the samples of the signals involved. The operations are vectorial and the indexing is implicit. There are operators that can generate signals with periods different from the periods of the signals they are derived from. The periods of the signals are implicit and they are derived by the synthesis tool.

The multi-rate semantics of Silage was defined [ACDM,Ang94] in the HOL system [Gor88] for the verification of the correctness of source-to-source transformations in Silage. These transformations have been applied either manually or automatically in the Silage specification in order to optimize the results of the silicon compilation in terms of area and speed. The multi-rate semantics of Silage has been structured in terms of semantic views. Informally, a semantic view of a Silage program is the meaning of the Silage program focusing on a particular aspect. To prove the equivalence of two Silage programs we prove the equivalence of their meanings with respect to each semantic view. One of the semantic views is concerned with the periods of the signals in Silage and is called period semantics. The period semantics of a Silage function defines a linear system of equations about natural numbers. In a consistent Silage program, there must be a solution for this system and the solution is not unique.

Silage is a synchronous language, i.e. for every consistent Silage program there is an integer number *Silage clock* different from zero such that the period of every signal in the program is a multiple of it. This is because the period of every signal in the program can only be a multiple or a sub-multiple[2] of the period of the input signals of the algorithm specified by the Silage program. There is not necessarily a signal in the program with period equal to the Silage clock but this is possible. The

---

[1] DSP stands for Digital Signal Processing.

[2] If a is a multiple of b then b is a sub-multiple of a.

Silage clock is related to the clock rate of an implementation of the Silage algorithm. Let the *normalized form* of the period semantics of a Silage program be a syntactic representation of its period semantics that explicitly relates the period of each signal in the program to the Silage clock and states that the Silage clock is different from zero. The relation rel_i between the period per_i of a signal i and the Silage clock **SilCk** of a well-formed Silage program is unique and has the following syntactic form:

---

rel_i ::= per_i = SilCk |

                n_i*SilCk ;

where: n_i is a natural number different from zero;

        SilCk is a variable of the Silage program or a new internal variable;

        SilCk does not appear on the left-hand side of any relation rel_i.

For every Silage program well-formed with respect to the periods, there is a *normalized form* of its period semantics. We have developed a proof procedure that automatically *synthesizes* the normalized form of the period semantics of any Silage program well-formed with respect to the periods and that automatically *proves* in the HOL logic that the synthesized representation is equivalent to the period semantics of the Silage program.

To prove in the HOL logic that there are periods such that the period semantics of a program holds, one has to present witnesses for the periods. Since these witnesses are explicit in the normalized form, the proof of the well-formedness of the periods of a Silage program becomes trivial once the normalized form has been proven to be equivalent to the period semantics of the program.

To prove the equivalence between the period semantics of two Silage programs means to prove the equivalence between two systems of equations about natural numbers. We developed a generic proof procedure implemented and tested in the HOL system that proves *automatically* the correctness with respect to the periods of *any* transformation[3]. The core of the proof procedure is the normalization of the period semantics of the Silage programs. In the next sections, we detail this procedure. It implements a backwards proof[4] in HOL with a number of steps that progressively reduce the period semantics of both Silage programs to their normalized forms. If two programs are well-formed with respect to the periods and their normalized forms are equivalent, then the programs have the same behavior with respect to the periods. For both Silage programs, each step synthesizes a new intermediate representation for the period semantics of the program and proves its equivalence to the previous representation. The step is successful if the Silage program is well-formed and the previous steps have been performed. Each step of the procedure is applied to both systems of equations.

---

[3] This procedure is important to verify the correctness of multi-rate transformations that change the periods of the internal signals of a Silage program [Ang94].

[4] The procedure is a tactic implemented in terms of conversions and other tactics [Gor88].

2   A Proof Procedure for any Transformation

The multi-rate semantics of Silage is formalized in terms of constants defined in the HOL logic. The period semantics of a Silage program is a conjunction. Each conjunct constrains the possible periods of the signals involved and corresponds to one or more equations of the linear system.

The first three steps

The first steps of the procedure[5] to prove the equivalence of two Silage programs with respect to the period semantics perform the following tasks:

1. Rewrite the terms in the logic with the definitions of the constants and theorems about them.
2. Then existential quantifications are moved outwards through the conjunction using built-in functions of the HOL system.
3. Reduce a goal about signal variables into a simpler goal where the variables have the type of the periods of the signals, i.e. the naturals [6].

4- Flatten and clean up

This step flattens the conjunction and cleans up the conjuncts. This step is also used between other steps of the proof procedure. The following tasks are performed:

- 4.1- Transform expressions of the form n∗(m∗variable), where n and m are numbers and variable is a variable, into (n'∗variable), where n' is a number and (n' = n∗m). This is done by a conversion that uses the associativity property of multiplication and then performs the multiplication. This conversion was developed using the theorem **MULT_ASSOC** shown below and the conversion of the HOL system[7] **REDUCE_CONV**.

  > **MULT_ASSOC**: ⊢ ∀ m n p. m∗(n∗p) = (m∗n)∗p

- 4.2- Rewrite with the theorems shown below:

  > **MULT_LEFT_SUC_0**: ⊢ ∀ m. (SUC 0)∗m = m
  > **NOT_ZERO_MULTIPLE**: ⊢ ∀ n m. ∼((SUC m)∗n = 0) = ∼(n = 0)
  > **MULT_MONO_EQ**: ⊢ ∀ m i n. ((SUC n)∗m = (SUC n)∗i) = (m = i)

- 4.3- Eliminate duplicate conjuncts flattening the conjunction.
- 4.4- Rewrite with the theorems shown below:

---

[5] These steps are described in detail in [Ang94].

[6] A signal is modeled by a tuple of two components: samples (a function from naturals to polymorphic values), and period (a natural number). Since the two components are independent, a proof about the periods of the signals is equivalent to a proof about natural numbers.

[7] The conversion **REDUCE_CONV** reduces arithmetic and Boolean expressions without variables at all levels possible. For instance, **REDUCE_CONV** "(4∗(6 DIV 2))" returns the theorem ⊢ (4∗(6 DIV 2)) = 12. This conversion has been developed by John Harrison.

```
REFL_CLAUSE: ⊢ ∀ x. (x = x) = T
AND_CLAUSE1: ⊢ ∀ t. t ∧ T = t
AND_CLAUSE2: ⊢ ∀ t. T ∧ t = t
```

5- Eliminate all the divisions

Let var and var$_i$ be variables, and n, n$_i$, m, and m' be natural numbers different from 0. We will use this notation to show how the syntax of the conjuncts evolves in each step of the proof procedure. Let (n DIV k) and n MOD k be[8] respectively the result and the remainder of the integer division of n by k. Consider also the following definition in HOL. Given two natural numbers n and m, the predicate MULTIPLE holds if n is a multiple of m. For every division in the period semantics of Silage there is a MULTIPLE clause.

```
∀ (n:num) (m:num). (MULTIPLE n m) = ((n MOD m) = 0)
```

After the previous steps, the period semantics of each Silage program is reduced to a conjunction (with existentially quantified variables or not) where each conjunct has the following syntax[9]:

```
div_term ::= var DIV n |              /* terminal division */
             div_term DIV n ;         /* non-terminal division /*

term ::= var |                        /* variable */
         n*var |                      /* terminal multiplication */
         div_term |                   /* division */
         n*div_term ;                 /* multiple of a division */

conjunct ::= (term₁ = term₂) |        /* equality */
             (MULTIPLE var n) |       /* terminal MULTIPLE clause */
             (MULTIPLE div_term n) |  /* non-terminal MULTIPLE clause */
             ~(var = 0) ;             /* not-zero clause */
```

The aim of this step is to transform all the divisions into multiplications getting rid of the MULTIPLE clauses. This is performed through the following tasks:

- 5.1- Eliminate non-terminal divisions. Since the multi-rate semantics of Silage guarantees that for every division there is a corresponding MULTIPLE clause, this can be done by searching for non-terminal MULTIPLE clauses. For each non-terminal MULTIPLE clause (MULTIPLE div_term n) introduce a new internal variable defined by (new_var = div_term) and rewrite the other conjuncts with the symmetrical version of the definition as follows (div_term = new_var). This

---

[8] DIV and MOD are built-in constants of the HOL system.

[9] In this paper this can be considered an assumption. The interested reader is referred to [Ang94].

process is recursively repeated until all the non-terminal **MULTIPLE** clauses are eliminated. For instance, this task proved the following theorem[10]:

$\vdash$ ((out = (a DIV 2) DIV 2) $\wedge$ (MULTIPLE (a DIV 2) 2) $\wedge$
  (b DIV 2 = a DIV 2) $\wedge$ (MULTIPLE a 2) $\wedge$
  (c = a) $\wedge$ (MULTIPLE b 2) $\wedge$ (d = b)) =
($\exists$ new_var. (new_var = a DIV 2) $\wedge$
    (out = new_var DIV 2) $\wedge$ (MULTIPLE new_var 2) $\wedge$
    (b DIV 2 = new_var) $\wedge$ (MULTIPLE a 2) $\wedge$
    (c = a) $\wedge$ (MULTIPLE b 2) $\wedge$ (d = b))

– 5.2- Eliminate terms that are multiples of terminal divisions. For each equality where the left-hand side or the right-hand side is a multiple of a terminal division ($n_1 * (var\ \mathbf{DIV}\ n_2)$), introduce a new internal variable defined by (**new_var** = (**var DIV** n_2)) and rewrite the other conjuncts with the symmetrical version of the definition as follows ((**var DIV** n_2) = **new_var**). For instance, this task proved the following theorem:

$\vdash$ ((out1 = 2*(a DIV 2)) $\wedge$ (out2 = 2*(a DIV 2))) =
($\exists$ new_var. (new_var = (a DIV 2)) $\wedge$
    (out1 = 2*new_var) $\wedge$ (out2 = 2*new_var))

– 5.3- Eliminate equalities between terminal divisions. For each equality where both the left and the right-hand side are terminal divisions ((var_1 **DIV** n_1) = (var_2 **DIV** n_2)), introduce a new internal variable defined by (**new_var** = (var_1 **DIV** n_1)) and rewrite the other conjuncts with the symmetrical version of the definition as follows: ((var_1 **DIV** n_1) = **new_var**). For instance, this task proved the following theorem:

$\vdash$ ((b DIV 2 = a DIV 2) $\wedge$ (MULTIPLE a 2) $\wedge$ (MULTIPLE b 2)) =
($\exists$ new_var. (new_var = (b DIV 2)) $\wedge$
    (new_var = (a DIV 2)) $\wedge$ (MULTIPLE a 2) $\wedge$ (MULTIPLE b 2))

– 5.4- Move terminal divisions to the right-hand side of equalities using the theorem below:

DIV_SYM: $\vdash \forall$ x a k. ((a DIV (SUC k)) = x) = (x = (a DIV (SUC k)))

– 5.5- Eliminate all the divisions and **MULTIPLE** clauses. This is done by a procedure that pairs every equality involving a terminal division (x = a **DIV** n) with a **MULTIPLE** clause[11] (MULTIPLE a n) and rewrites with the theorem **FROM_DIV_TO_TIMES** shown below. The multi-rate semantics of Silage guarantees that for every division there is a corresponding **MULTIPLE** clause and for every **MULTIPLE** clause there is a corresponding division.

---

[10] "2" is an abbreviation for the HOL term "SUC(SUC 0)" [Gor88]. Other numeric constants are also abbreviated.

[11] A **MULTIPLE** clause is duplicated if it is to be used to eliminate more than one division.

FROM_DIV_TO_TIMES:
$\forall$ x a k. $((x = a \text{ DIV } (\text{SUC } k)) \wedge (\text{MULTIPLE } a \text{ (SUC } k))) = (a = (\text{SUC } k)*x)$

For instance, this task proved the following theorem:

$\vdash ((a = b \text{ DIV } 2) \wedge (c = b \text{ DIV } 2) \wedge (\text{MULTIPLE } b \text{ } 2)) =$
$((b = 2*a) \wedge (b = 2*c))$

After this step, the period semantics of each Silage program is reduced to a conjunction (with existentially quantified variables or not) where each conjunct has the following syntax:

```
term ::= var |                    /* variable */
         n*var ;                  /* terminal multiplication */

conjunct ::= (term₁ = term₂) |    /* equality */
             ~(var = 0) ;         /* not-zero clause */
```

6- Reorder the conjuncts

At this point of the normalization of the period semantics of Silage programs, the information of each conjunct has to be cross fertilized with the information of all the other conjuncts. In the next steps of the proof procedure we selectively rewrite some conjuncts with the others. To prepare for this step, we first reorder the conjuncts in two phases:

- 6.1- Put conjuncts relating two internal signals at the end of the conjunction. This is a preparation for the next steps. (At the end of the normalization process, the internal variables are eliminated as much as possible.)
- 6.2- Put the not-zero clauses at the end of the conjunction. These clauses will be rewritten in the next step but they will never be used as rewrite rules.

7- Recursively rewrite

Let the (existentially quantified or not) conjunction $(c_1 \wedge c_2 \wedge c_3 \wedge \ldots \wedge c_n)$ be the period semantics of a Silage program that has been syntactically transformed by the previous steps. In this step, the first conjunct c_1 is used to derive a *rewriting rule* c_1'[12] and then c_1' is used to rewrite the rest of the conjuncts obtaining $(c_1' \wedge c_2' \wedge c_3' \wedge \ldots \wedge c_n')$. Then c_2' is used to rewrite the conjuncts $c_3', \ldots, c_n'$ obtaining $(c_1' \wedge c_2'' \wedge c_3'' \ldots \wedge c_n'')$, etc. The process is recursive. At each phase, a rewriting rule is derived from a conjunct which we call *pivot*. For instance, in the first phase c_1 is the pivot and the derived rewriting rule is c_1', then in the second phase c_2' is the pivot and the rewriting rule is c_2'', etc. The process is repeated until the pivot is a not-zero clause. Next we explain how each phase of the recursive rewriting process works.

At this point of the normalization of the period semantics of Silage programs, the conjuncts are either equalities or not-zero clauses. If the pivot is a not-zero clause then the recursive rewriting process is over. Step 6.2 was used to stop this recursion.

---

[12] c_1' might be equal to c_1.

Otherwise a rewriting rule is derived from the pivot. The rewriting rule is always an equality and each of the conjuncts to be rewritten can be either an equality or a not-zero clause. The goal of the rewriting process is to eliminate the variable on the left-hand side of the rewriting rule from the conjuncts to be rewritten[13]. The rewriting rule is derived as follows. If the variable in the term on the left-hand side of the pivot is external and the variable in the term on its right-hand side is internal then the left-hand and the right-hand sides of the pivot are swapped to derive the rewriting rule, otherwise the rewriting rule is equal to the pivot. This favors the elimination of internal variables with respect to the elimination of external variables from the other conjuncts.

If the left-hand side of the rewriting rule is a variable then this variable is replaced by the right-hand side of the rewriting rule in the other conjuncts. Otherwise, the left-hand side of the rewriting rule is a multiple of a variable ($n_1 * var_1$). In this case, if var_1 is present in the conjunct to be rewritten then both of its sides are multiplied[14] by n_1, arithmetical properties are applied if necessary to make the term ($n_1 * var_1$) explicit, and then this term is replaced by the right-hand side of the rewriting rule.

After the rewriting, step 4 is applied in order to clean up the conjuncts. The application of step 4 to conclude each phase of the recursive rewriting process preserves the syntax of the conjuncts for the next phase.

For instance, this step proved the following theorem:

$$\vdash ((in = 3*out1) \wedge (in = 4*out2) \wedge \sim(in = 0) \wedge \sim(out1 = 0) \wedge \sim(out2 = 0)) = ((in = 3*out1) \wedge (3*out1 = 4*out2) \wedge \sim(out2 = 0))$$

After the recursive simplification in a consistent Silage program, the following holds:

- no variable appears more than once in the terms on the left-hand side of equalities[15].
- if we interpret the variable in the term on the left-hand side of each equality as dependent on the variable in the term on its right-hand side, then there is a loop-free graph of dependencies between all the variables[16].

8- Transform the left-hand side of the conjuncts into variables

In this step the left-hand side of each equality is transformed into a variable using the theorem below which transforms a multiplication into a division clause and a **MULTIPLE** clause.

---

[13] If the variable in the term on the left-hand side of the rewriting rule is equal to the variable in the term on its right-hand side then there is an inconsistency in the Silage program because tautological conjuncts such as (**var** = **var**) and (**n*var** = **n*var**) have already been removed at this stage. For instance, the equality (**var** = **2*var**) with the same variable on the left-hand side and on the right-hand side is inconsistent because **var** $\neq$ **0**.

[14] If a factor of n_1 already exists in the coefficient of the variable, this multiplication is unnecessary.

[15] If there were two equalities with the same variable on their left-hand sides, the first equality would not have been used to rewrite the second one.

[16] If there were a conjunction with a loop of dependency such as (**a** = **b**) $\wedge$ (**b** = **c**) $\wedge$ (**c** = **a**), then (**c** = **a**) would not have been rewritten with the previous conjuncts. This kind of reasoning holds for any order of the conjuncts. This can be generalized for any number of conjuncts.

> **FROM_TIMES_TO_DIV**: $\vdash \forall$ k x a. $((\text{SUC k})*x = a) =$
> $((x = a \text{ DIV } (\text{SUC k})) \land (\text{MULTIPLE a } (\text{SUC k})))$

After this step, each conjunct has the following syntax:

> equality ::= $(\text{var}_1 = \text{var}_2)$ |
> $\quad\quad\quad (\text{var}_1 = n_2*\text{var}_2)$ |
> $\quad\quad\quad (\text{var}_1 = (\text{var}_2 \text{ DIV } n_1))$ |
> $\quad\quad\quad (\text{var}_1 = ((n_2*\text{var}_2) \text{ DIV } n_1))$
>
> conjunct ::= equality |
> $\quad\quad\quad (\text{MULTIPLE var}_2 \; n_1)$ |
> $\quad\quad\quad (\text{MULTIPLE } (n_2*\text{var}_2) \; n_1)$ |
> $\quad\quad\quad \sim(\text{var} = 0) \; ;$

9- Unwind, flatten, and clean up

In this step each conjunct is unwound as much as possible using the other conjuncts and then step 4 is applied to simplify the result. The unwinding is performed using the conversion **UNWIND_AUTO_CONV** of the HOL system. **UNWIND_AUTO_CONV** "$\exists$ l1 ... lm. t1 $\land$ ... $\land$ tn" returns a theorem of the form:

> $\vdash (\exists$ l1 ... lm. t1 $\land$ ... $\land$ tn$) = (\exists$ l1 ... lm. t1' $\land$ ... $\land$ tn'$)$
> where tj' is tj rewritten with equations selected from the ti's.

If the Silage program is well-formed then, after this step, the following holds:

- Every variable is related to an expression defined in terms of a unique variable which we call the *reference variable* and denote by **refvar**. The reference variable of a program might depend on the order of its Silage definitions. Two syntactically different but equivalent Silage programs might have different reference variables.
- There is only one not-zero clause and it is about the reference variable.
- If there are **MULTIPLE** clauses then the first arguments of all of them are expressions defined in terms of the reference variable. In this case, each conjunct has the syntax[17] below and steps 10, 11, 12, and 13 will be necessary to find the relation between the reference variable and the Silage clock.

> expression ::= refvar | /* terminal expression */
> $\quad\quad\quad (n*\text{refvar})$ | /* terminal multiplication */
> $\quad\quad\quad (n_1*(\text{expression DIV } n_2))$ | /* non-terminal multiplication */
> $\quad\quad\quad (\text{expression DIV } n) \; ;$ /* non-terminal division */
>
> conjunct ::= $(\text{var} = \text{expression})$ | /* equality */
> $\quad\quad\quad (\text{MULTIPLE expression } n)$ | /* MULTIPLE clause */
> $\quad\quad\quad \sim(\text{refvar} = 0) \; ;$ /* not-zero clause */

---

[17] Expressions of the form $n*(n'*\text{expression})$ are not part of the syntax because they are reduced to $n''*\text{expression}$ by the cleaning task 4.1.

– If there are not **MULTIPLE** clauses then the reference variable is the Silage clock and each conjunct has the syntax below, where **SilCk** denotes the Silage clock of the program.

expression ::= (n∗SilCk) |	/∗ terminal expression ∗/
(n$_1$∗(expression DIV n$_2$)) |	/∗ non-terminal multiplication ∗/
(expression DIV n) ;	/∗ non-terminal division ∗/
conjunct ::= (var = SilCk) |	/∗ terminal equality ∗/
(var = expression) |	/∗ equality ∗/
∼(SilCk = 0) ;	/∗ not-zero clause ∗/
where SilCk = refvar.	

In this case, steps 10, 11, 12, and 13 will have no effect in the normalization of the period semantics of the program. The arithmetic expressions on the right-hand side of each equality can be reduced to a simpler expression of the form n∗**SilCk**, independently of the other conjuncts.

10- Normalize the MULTIPLE clauses

Steps 10, 11, and 12 search for the relation between the reference variable and the Silage clock. This relation has the form **refvar = n∗SilCk**. Using this relation, every period in a consistent Silage program can be related to the Silage clock because every period is already related to the reference variable. This unwinding will be performed in step 13. The information that links the reference variable to the Silage clock is contained in the **MULTIPLE** clauses. Every **MULTIPLE** clause must be prepared to find the relation between the reference variable and the Silage clock in the next steps. This step recursively transforms the **MULTIPLE** clauses without changing the other clauses. The goal of this step is to transform the first argument of each **MULTIPLE** clause into the reference variable. The first argument of a **MULTIPLE** clause is an expression and an expression is a tree of expressions, where each node has at most one child. Since every non-terminal expression is either a multiplication or a division, the goal of this step can be achieved by repeatedly eliminating a multiplication or a division from the root of the tree of expressions until reaching the terminal expression **refvar**. If the first argument of a **MULTIPLE** clause contains divisions, then the proof that its final version is equal to its initial version depends on the first versions of the other **MULTIPLE** clauses. This is explained below. First we show how the recursive process of transformations on each clause works and how the transformations on different conjuncts are combined together to achieve the goal. Then we present the details of each transformation on each kind of conjunct.

– **The recursive transformation process**
Let $c_{i,j}$ be the j-th version of the i-th conjunct and let k be the number of conjuncts in the conjunction, i.e. $(1 \leq i \leq k)$. At the beginning of this step the period semantics of a Silage program is an (existentially quantified or not) conjunction $(c_{1,0} \wedge c_{2,0} \wedge ... \wedge c_{k,0})$. Let $c_{i,j}' \in \{c_{1,0}, c_{2,0}, ..., c_{k,0}\}$ or $c_{i,j}' = \mathbf{T}$. First we generate a new version $c_{i,(j+1)}$ of a conjunct $c_{i,j}$ such that the theorem $\mathbf{thm}_{i,(j+1)}$

below can be proved[18]. How this is done is explained further on.

$$\textbf{thm}_{i,(j+1)}: \vdash c_{i,(j+1)} = c_{i,j} \wedge c_{i,j}{}'$$
where $(0 \leq j \leq (t_i - 1))$, i.e. t_i is the number of transformations applied to the i-th conjunct, and c_{i,t_i} is the final version of the conjunct.

Except for the first version of a conjunct $c_{i,0}$, each version of a conjunct is expressed in terms of the previous versions. By rewriting the last version of each conjunct c_{i,t_i} with the previous versions we prove the following:

$$
\begin{aligned}
\vdash c_{i,t_i} &= c_{i,(t_i-1)} \wedge c_{i,(t_i-1)}{}' = \\
&\quad c_{i,(t_i-2)} \wedge c_{i,(t_i-2)}{}' \wedge c_{i,(t_i-1)}{}' = \\
&\quad \ldots \\
&\quad c_{i,0} \wedge c_{i,0}{}' \wedge c_{i,1}{}' \ldots \wedge c_{i,(t_i-1)}{}'
\end{aligned}
$$

Using this result, we prove the theorem below about the conjunction of the last versions of all the conjuncts.

$$
\begin{aligned}
\vdash (c_{1,t_1} \wedge c_{2,t_2} \wedge \ldots \wedge c_{k,t_k}) &= \\
((c_{1,0} \wedge c_{1,0}{}' \wedge c_{1,1}{}' \ldots \wedge c_{1,(t_1-1)}{}') &\wedge \\
(c_{2,0} \wedge c_{2,0}{}' \wedge c_{2,1}{}' \ldots \wedge c_{2,(t_2-1)}{}') &\wedge \\
\ldots \\
(c_{k,0} \wedge c_{k,0}{}' \wedge c_{k,1}{}' \ldots \wedge c_{k,(t_k-1)}{}')) &
\end{aligned}
$$

If $c_{i,j}{}' \in \{c_{1,0}, c_{2,0}, \ldots, c_{k,0}\}$ or $c_{i,j}{}' = \textbf{T}$ then the following can be proved:

$$\vdash (c_{1,t_1} \wedge c_{2,t_2} \wedge \ldots \wedge c_{k,t_k}) = (c_{1,0} \wedge c_{2,0} \wedge \ldots \wedge c_{k,0})$$

Finally, we use the procedure described in step 4 to clean up $(c_{1,t_1} \wedge c_{2,t_2} \wedge \ldots \wedge c_{k,t_k})$ for step 11.

Next we discuss how $c_{i,(j+1)}$ is generated, how the recursive transformation process stops, and how we prove the theorems $\textbf{thm}_{i,(j+1)}$. If $c_{i,0}$ is not a MULTIPLE clause then:

```
t_i = 1 ;              /* only one trivial transformation */
c_{i,1} = c_{i,0} ;    /* the conjunct is unchanged */
c_{i,1}' = T ;
```

If $c_{i,0}$ is (MULTIPLE expression n) then there are the possibilities below. The recursive process stops when the terminal MULTIPLE clause is reached or the MULTIPLE clause is proven to be a tautology.

$c_{i,j}$	$c_{i,j}{}'$	$c_{i,(j+1)}$
M refvar n	T	M refvar n
M (n*expression) m	T	M expression m'
M (expression DIV n) m	M expression n	M expression (n*m)
where M is an abbreviation of MULTIPLE		
and n, m, and m' are natural numbers different from 0.		

---

[18] The indexed notation used for the conjuncts is not part of the HOL notation for the theorem.

If $c_{i,(j+1)} = c_{i,j}$ then the proof of $thm_{i,(j+1)}$ is trivial. Next we explain how we prove the correctness of the transformations that eliminate a multiplication or a division from the first argument of a **MULTIPLE** clause.

- **The elimination of a multiplication**

Let $c_{i,j}$ be the HOL term (**MULTIPLE** ((SUC n)∗x) (SUC m)). These are the tasks to be performed:

- 1- Rewrite $c_{i,j}$ with the theorem[19] shown below:

> **ELIM_MULTIPLE_PROD:** ⊢ ∀ n x m.
> (MULTIPLE ((SUC n) ∗ x) (SUC m)) =
> (MULTIPLE x ((SUC m) DIV (gcd((SUC n),(SUC m)))))
> where gcd(a,b) is the greatest common divisor of a and b.

- 2- Prove that the term

> MULTIPLE x ((SUC m) DIV (gcd((SUC n),(SUC m))))

is equal to

> MULTIPLE x m'
> where m' is the result of ((SUC m) DIV (gcd((SUC n),(SUC m)))).

This is done by finding the gcd of (**SUC n**) and (**SUC m**) and then performing the division. To find the gcd we use the famous Euclid's algorithm which is based on the fact that if **r** is the remainder when **n** is divided by **m**, then the common divisors of **n** and **m** are the same as the common divisors of **m** and **r**. The algorithm successively reduces the problem of computing a greatest common divisor to the problem of computing the greatest common divisor of smaller and smaller pairs of numbers. It is possible to show that starting with any two positive naturals and performing repeated reductions will always eventually produce a pair where the second number is 0. Then the greatest common divisor is the other number in the pair [ASS85]. The correctness of each step of the algorithm is proven using the theorems below. The conversion of the HOL system **REDUCE_CONV** is used to prove the correctness of the division.

> **GCD_N_M_IS_GCD_M_R:** ⊢ ∀ n m r.
> ((SUC n) MOD (SUC m) = r) ⟹
> (gcd((SUC n),(SUC m)) = gcd((SUC m),r))
>
> **GCD_N_ZERO:** ⊢ ∀ n. gcd(n,0) = n

- 3- If m' is equal to 1 then prove that (**MULTIPLE x m'**) is a tautology. This is done by rewriting this term with the theorem below.

> **MULTIPLE_ONE:** ⊢ ∀ x. (MULTIPLE x (SUC 0)) = T

- 4- Finally, the left-hand side and the right-hand side of the theorem obtained are swapped to obtain $thm_{i,(j+1)}$.

---

[19] The greatest common divisor gcd of two numbers has been defined in the HOL system by John Harrison. We have defined the least common multiple lcm of two numbers based on the definition of gcd. We proved theorems about gcd and lcm for this step of the proof procedure which are based on a theory of basic theorems about gcd developed by John Harrison.

The procedure for the elimination of a multiplication proves, for instance, the following:

$$\vdash (\text{MULTIPLE } (2*x)\ 3) = (\text{MULTIPLE } (5*(2*x))\ 3)$$

- **The elimination of a division**

Let $c_{i,j}$ be the following HOL term $\boxed{\text{MULTIPLE } (e \text{ DIV } (\text{SUC } n))\ (\text{SUC } m)}$.

We generate the HOL term $c_{i,(j+1)}$ as $\boxed{\text{MULTIPLE } e\ ((\text{SUC } n)*(\text{SUC } m))}$ and rewrite it with the theorem below to prove $\text{thm}_{i,(j+1)}$.

ELIM_MULTIPLE_DIV: $\vdash \forall$ e n m.
  MULTIPLE e $((\text{SUC } n)*(\text{SUC } m)) =$
  $((\text{MULTIPLE } (e \text{ DIV } (\text{SUC } n))\ (\text{SUC } m)) \wedge (\text{MULTIPLE } e\ (\text{SUC } n)))$

In this case, $c_{i,j}' = (\text{MULTIPLE } e\ (\text{SUC } n))$. Since the first argument of $c_{i,j}$ is $(e \text{ DIV } (\text{SUC } n))$ and step 8 guarantees that for every division $(e \text{ DIV } (\text{SUC } n))$ there is a clause $(\text{MULTIPLE } e\ (\text{SUC } n))$, we know that $c_{i,j}' \in \{c_{1,0}, c_{2,0}, ..., c_{k,0}\}$. The procedure for the elimination of a division proves, for instance, the following:

$$\vdash (\text{MULTIPLE } e\ 12) = ((\text{MULTIPLE } (e \text{ DIV } 3)\ 4) \wedge (\text{MULTIPLE } e\ 3))$$

At this point, the conjuncts of the period semantics of a consistent Silage program have the following syntax:

```
expression ::= refvar |                      /* terminal expression */
               (n*refvar) |                   /* terminal multiplication */
               (n₁*(expression DIV n₂)) |     /* non-terminal multiplication */
               (expression DIV n) ;           /* division */

conjunct ::= (var = expression) |             /* equality */
             (MULTIPLE refvar n) |            /* MULTIPLE clause */
             ~(refvar = 0) ;                  /* not-zero clause */
```

If this step eliminates all the **MULTIPLE** clauses, then the reference variable is the Silage clock (as discussed at the end of step 9).

11- Merge the MULTIPLE clauses

After the previous steps, the syntactically transformed period semantics of a consistent Silage program might have a number of different **MULTIPLE** clauses and the first argument of these clauses is the reference variable. In this step, the information of these clauses is merged together in one single **MULTIPLE** clause. A number x is a multiple of two numbers $(\text{SUC } n)$ and $(\text{SUC } m)$ if and only if it is a multiple of the least common multiple of $(\text{SUC } n)$ and $(\text{SUC } m)$. The least common multiple of two numbers is their product divided by their greatest common divisor. Two **MULTIPLE** clauses about the same variable x are merged together by using the theorem below:

$$\vdash \forall \text{ m n. } ((\text{MULTIPLE x (SUC n)}) \wedge (\text{MULTIPLE x (SUC m)})) =$$
$$(\text{MULTIPLE x } (((\text{SUC n})*(\text{SUC m})) \text{ DIV } (\gcd((\text{SUC n}),(\text{SUC m})))))$$

Then, as in step 10, the greatest common divisor is found, the result **y** of **((SUC n)∗(SUC m)) DIV (gcd((SUC n),(SUC m)))** is calculated and proven correct by inference in the logic, and the new clause is transformed into **(MULTIPLE x y)**. By repeating this process, any number of **MULTIPLE** clauses about the same variable can be merged. The final **MULTIPLE** clause has the information that relates the reference variable to the Silage clock.

For instance, this step proves that:

$$\vdash ((\text{MULTIPLE x 3}) \wedge (\text{MULTIPLE x 4})) = (\text{MULTIPLE x 12})$$

This step does not change the syntax of the conjuncts.

12- Relate the reference variable to the Silage clock

At this point, if there is a **MULTIPLE** clause, then it is unique and in this step the relation between the reference variable and the Silage clock is made explicit. First the goal is rewritten with the theorem below and then the newly introduced existential quantification is moved outwards through the conjunction.

FROM_MULTIPLE_TO_EXISTS:
$$\vdash \forall \text{ a b. } (\text{MULTIPLE b (SUC a)}) = (\exists \text{ SilCk. b = (SUC a) } * \text{ SilCk})$$

After this step, each conjunct has the syntax below:

expression ::= refvar |	/∗ terminal expression ∗/
(n∗refvar) |	/∗ terminal multiplication ∗/
$(n_1*(\text{expression DIV } n_2))$ |	/∗ non-terminal multiplication ∗/
(expression DIV n) ;	/∗ non-terminal division ∗/
conjunct ::= (var = expression) |	/∗ non-terminal equality ∗/
(refvar = n∗SilCk) |	/∗ terminal equality ∗/
∼(refvar = 0) ;	/∗ not-zero clause ∗/

where SilCk is an internal variable introduced in this step.

13- Unwind and clean up

After these steps are applied in a consistent Silage program, each variable is related to the Silage clock or to the reference variable and the reference variable is related to the Silage clock. To find the relation between each variable and the Silage clock, we repeat step 9 which unwinds and cleans up the conjuncts.

From now on we do not distinguish the reference variable from any other ordinary variable. At this point, each conjunct of the period semantics of a consistent Silage program has the following syntax:

```
expression ::= n*SilCk |                    /* terminal expression */
               (n₁*(expression DIV n₂)) |   /* non-terminal multiplication */
               (expression DIV n) ;         /* non-terminal division */

conjunct ::= (var = expression) |           /* equality */
             ~(SilCk = 0) ;                 /* not-zero clause */

where SilCk is an internal variable introduced in the previous step.
```

The arithmetic expressions on the right-hand side of each equality can be reduced to a simpler expression of the form **n*SilCk**, independently of the other conjuncts.

14- Solve the arithmetic

Consider the syntax that resulted at the end of step 9 for the case when the Silage clock is equal to the reference variable. Combining this syntax with the syntax that resulted in the previous step (when the Silage clock is a new internal variable), we obtain:

```
expression ::= (n*SilCk) |                  /* terminal expression */
               (n₁*(expression DIV n₂)) |   /* non-terminal multiplication */
               (expression DIV n) ;         /* non-terminal division */

conjunct ::= (var = SilCk) |                /* terminal equality */
             (var = expression) |           /* non-terminal equality */
             ~(SilCk = 0) ;                 /* not-zero clause */

where SilCk = refvar or SilCk is the internal variable introduced in step 12.
```

Let e be an abbreviation for **expression**. After the previous steps are applied to a consistent Silage program, every division (e **DIV** n) in the arithmetic expressions on the right-hand side of the equalities is such that n evenly divides e. This is the case for both nested and non-nested divisions.

This step reduces the right-hand side of each equality to an expression with the syntax **n*SilCk** or **SilCk**. An expression is a tree of expressions, where each node has at most one child. The process of simplification of an expression is a bottom-up recursive process. If an expression e is terminal then it is in the form **n*SilCk**. Otherwise e is defined in terms of an expression e_1. If e_1 is equal or proven equivalent to an expression with the syntax n_1***SilCk**, then, by case analysis on the syntax of the non-terminal expressions, and based on the fact that the remainder of each division is zero, we can prove that e is equivalent to an expression with the syntax **n*SilCk** or **SilCk**. This is shown below:

$e = (e_1 \text{ DIV } n_2) =$ /* $e_1 = (n_1 * \text{SilCk})$ */
   $((n_1 * \text{SilCk}) \text{ DIV } n_2) =$ /* Perform the division */
   $((n * \text{SilCk}) \text{ or } \text{SilCk}) \mid$

$e = (n_3 * (e_1 \text{ DIV } n_2)) =$ /* $e_1 = (n_1 * \text{SilCk})$ */
   $(n_3 * ((n_1 * \text{SilCk}) \text{ DIV } n_2)) =$ /* Perform the division */
   $((n_3 * (n_4 * \text{SilCk})) \text{ or } (n_3 * \text{SilCk})) =$ /* Perform the multiplication */
   $(n * \text{SilCk})$

where SilCk is a variable.

Since the leaf of the tree is terminal, applying this process to all nodes of the tree in a bottom-up order, the root is proven equivalent to SilCk or $n * \text{SilCk}$.

In each simplification, the multiplications are proven correct as in step 4 (tasks 4.1 and 4.2). Next we explain how we prove the correctness of the divisions shown below.

$((n_1 * \text{SilCk}) \text{ DIV } n_2) = (n * \text{SilCk})$ or $((n_1 * \text{SilCk}) \text{ DIV } n_2) = \text{SilCk}$

Since $n_i \neq 0$, let (SUC k) and (SUC m) denote n_1 and n_2 in HOL respectively, where k and m are natural numbers. If (SUC k) $\neq$ (SUC m) then the expression ((SUC k)*SilCk) DIV (SUC m) is simplified using the following procedure:

- 14.1- First we calculate the number (SUC k') such that (SUC k) = (SUC m)*(SUC k'). This is proven correct using **REDUCE_CONV**. Then we prove the theorem below.

  ((SUC k)*SilCk) DIV (SUC m) =
  (((SUC m)*(SUC k')) * SilCk) DIV (SUC m)

- 14.2- Using the symmetrical version of the theorem **MULT_ASSOC** introduced in step 4.1 we prove the theorem below.

  (((SUC m)*(SUC k')) * SilCk) DIV (SUC m) =
  ((SUC m) * ((SUC k')*SilCk)) DIV (SUC m)

- 14.3- Using the theorem

  DIV_MULT: $\vdash \forall$ m q. ((SUC m)*q) DIV (SUC m) = q

  we prove the theorem below.

  ((SUC m) * ((SUC k')*SilCk)) DIV (SUC m) = (SUC k')*SilCk

This proves that if (SUC k) $\neq$ (SUC m) then the division is reduced to an expression n*SilCk, where n is denoted in HOL by (SUC k'). Otherwise, if (SUC k) = (SUC m) then, performing only the task 14.3, we prove that ((SUC m)*SilCk) DIV (SUC m) = SilCk.

15- Eliminate unused internal variables

In this step the internal variables that are not used are eliminated.

This concludes the normalization of the period semantics of a Silage program.

Finally, regardless of the order of the conjuncts and regardless whether the variable SilCk is existentially quantified or not, the proof of equivalence between the normalized representations of the period semantics of two behaviorally equivalent Silage programs is trivial.

3 Conclusions

The knowledge about the semantics of a language can be used to define the boundaries of the scope of proofs about aspects of the language. For instance, by knowing that Silage is a synchronous language, i.e. by knowing that the period of each signal in any well-formed Silage program is a multiple or sub-multiple of the period of the input signals of the algorithm specified by the program, we have been able to develop a proof procedure that can automatically verify the correctness of any transformation with respect to the periods.

The search for optimizations in the algorithm towards efficiency is left for future work. One can also investigate extensions and modifications in the algorithm to cope with more generic linear systems of equations.

References

[ACDM] C.M. Angelo, L. Claesen, and H. De Man. "Modeling Multi-rate DSP Specification Semantics for Formal Transformational Design in HOL". To appear in *Formal Methods in System Design: An International Journal*, vol. 5, nos. 1/2, July 1994. Kluwer Academic Publishers.

[Ang94] C.M. Angelo, *Formal Hardware Verification in a Silicon Compilation Environment by means of Theorem Proving*, Ph.D. Thesis, IMEC, Leuven, Belgium, February 1994.

[ASS85] H. Abelson, G.J. Sussman, and J. Sussman. *Structure and Interpretation of Computer Programs*. The MIT Press/McGraw-Hill Book Company, 1985. The MIT Electrical Engineering and Computer Science Series.

[DMRSC86] H. De Man, J. Rabaey, P. Six, and L. Claesen. "Cathedral-II: a Silicon Compiler for Digital Signal Processing". *IEEE Design & Test of Computers*, 3(6):73–85, December 1986.

[GHR+90] D. Genin, P.N. Hilfinger, J. Rabaey, C. Scheers, and H. De Man. "DSP Specification Using the Silage Language". In *IEEE International Conference on Acoustics, Speech and Signal Processing*, pages 1057–1060. Albuquerque, NM, April 1990.

[Gor88] M. Gordon. "HOL: A Proof Generating System for Higher-Order Logic". In G. Birtwistle and P.A. Subrahmanyam, editors, *VLSI Specification, Verification and Synthesis*, pages 73–128. Kluwer Academic Publishers, 1988.

[Hil85] P.N. Hilfinger. "Silage, a High-level Language and Silicon Compiler for Digital Signal Processing". In *IEEE 1985 Custom Integrated Circuits Conference, CICC'85*, pages 213–216. Portland, OR, May 1985.

[Nac90a] L. Nachtergaele. *A Silage Tutorial.* IMEC, Leuven, Belgium, May 1990.

Towards a HOL theory of memory

J-P. Bodeveix, M. Filali and P. Roche

IRIT
Université Paul Sabatier
118 Route de Narbonne
F-31062 Toulouse cédex France
email: bodeveix@irit.fr filali@irit.fr roche@irit.fr

Abstract. This paper introduces a formalization of memory models for multiprocessor architectures based on transition systems. Relations between memory models can be expressed as simulations between the corresponding transition systems. We show how simulation relations are preserved by structuring operators over transition systems. We derive from them proof tactics used to establish simulation relations between basic memory models. These memory models are also proved correct against a formal characterization of memory consistencies.

1 Introduction

Our work deals with the formalization of memory models for shared memory used by concurrent processes. Such processes could be instantiated as hardware components for instance the processors of a multiprocessor architecture or as software components as processes in a distributed architecture accessing a virtual shared memory. The goal of such a formalization is twofold:

first a validation one: the formalization process of a memory model is a first step towards the validation of the implementation of a given memory protocol.

second a software engineering one: we believe that it should be interesting to establish relations between memory models. Such relations could then be used to implement a memory model on another one. More generally, such relations should be essential for the mapping of an application of which semantics relies on a given memory model, on an architecture where another one is available.

Our paper reports on the current state of our work and especially on the HOL [2] formalization of memory models and their relations. In the next section we establish the semantical framework on which our work is based: the *refinement* concept. After presenting and characterizing some memory consistencies, we provide models for each one. The relations between these models are formally established through the refinement relation. Finally, we outline the future of our work and draw some conclusions.

2 Transition systems and refinements

Theoretical aspects about types and refinements have already been dealt with in the literature. For instance [9] reviews two different theories about data abstraction and refinement: one for the functional approach and one for the state based approach. Since our study concerns concurrent systems, as noted by Nipkow [9], we have chosen the state based approach: a transition system provides a natural notion of behaviour.

In this study, we give the semantics of memory and the relations between the different types of memories based on transition systems and simulations between them. This section first outlines the formalization of transition systems and simulations within the HOL system. Then we define composition operations on transition systems and present some theorems about composition simulation.

2.1 Transition systems

We represent a generic transition system as a pair consisting of a set of initial states and a next relation between its states.

Definition 1 (Transition system) *A transition system on an alphabet A is a triplet (Σ, Q_0, T) where*

- *Σ is a state space,*
- *Q_0, the set of initial states, is a subset of Σ,*
- *T, the set of labelled transitions, is a subset of $A \times \Sigma \times \Sigma$.*

Notation 1 *A triplet (l, e, e') of the set of labelled transitions of a system S will be denoted $e \xrightarrow{l}_S e'$.*

Definition 2 (Initial states) *Let $S = (\Sigma, Q_0, T)$ be a transition system. Then* Init$_S$ *characterizes the initial states of the transition system: for any state e of Σ,* Init$_S$ e *is true iff $e \in Q_0$.*

Definition 3 (Next Relation) *Let $S = (\Sigma, Q_0, T)$ be a transition system. Then* Next$_S$ *(also denoted $\rightarrow_S$) is a relation defined over Σ such that $e \rightarrow_S e'$ iff there exits a label l such that $e \xrightarrow{l}_S e'$.*

The transition system is represented in HOL by the following generic type:

```
let ts_rep = ":((*s -> bool) # (*l -> *s  -> *s -> bool))";;
```

2.2 Refinements

We can define several relations between transition systems [4]. We have considered a relation similar to the simulation relation as defined in [5].

Definition 4 (Refinement relation) *Given two transition systems s and s',* *s simulates s' iff there exits a simulation relation such that:*

- *to each state of s at least one state of s' corresponds,*
- *to each initial state of s at least one initial state of s' corresponds,*

- *if from a state e1, s can move to e2, and e1 corresponds to the state e1' of s' then there exists a state e2' such that e2' corresponds to e2 and s' can move from e1' to e2'.*

We formalize such a definition as follows:

```
∀ s s'. s Sim s' = ∃ R.
    ∀ e. ∃ e'. R e e' ∧
    ∀ e. Init s e ⟹ ∃ e'. R e e' ∧ Init s' e' ∧
    ∀ e1 e1' e2. Next_s e1 e2 ∧ R e1 e1' ⟹ ∃ e2'. R e2 e2' ∧ Next_s' e1' e2'
```

It should be noted that our definition is a bit stronger that the one given in [5] since we do not consider only reachable states of **s** but all the states of **s** which are in relation with some state of **s'**.

In the following, we will use the schema(*fig.* 1) to illustrate the simulation relation between two transition systems **s** and **s'**.

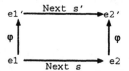

Fig. 1. Refinement relation

2.3 Refinement through operators

In the previous section, the simulation relation is defined in terms of the initial states and the **Next** relation of a transition system. However, refinement is usually proved independently for each label (also called operator in the following).

Definition 5 (Binary relation refinement) *A relation R (denoted $\xrightarrow{R}$) is a refinement of a relation R' through a relation φ (R $\sqsubseteq_\varphi$ R') iff*

$$\forall e1\ e2\ e1'.\ e1 \xrightarrow{R} e2 \wedge \varphi\ e1e1' \Longrightarrow \exists e2'.\ e1' \xrightarrow{R'} e2' \wedge \varphi\ e2e2'$$

We extend this binary refinement relation to transition systems where labels are interpreted as binary relations over the system states. We can now link transition system simulation and refinement between their respective labeled transitions.

Theorem 1 (Refinement through operators) *Given two transition systems $S^1(\Sigma^1, Q_0^1, T^1)$ and $S^2(\Sigma^2, Q_0^2, T^2)$ with the same alphabet A, a sufficient condition for S^2 to simulate S^1 is that there exists a relation φ such that:*

- *for each label $l \in A$, $\xrightarrow{l}_{S^2} \sqsubseteq_\varphi \xrightarrow{l}_{S^1}$*
- *$\forall e \in \Sigma_{S^2}$, $\exists e' \in \Sigma_{S^1}$ such that $\varphi\ e\ e'$*
- *$\forall e \in Q_0^2$, $\exists e' \in Q_0^1$ such that $\varphi\ e\ e'$*

More general theorems concerning preservation of safety properties through refinements could be proved. The automata theory in [4] considers some of them.

2.4   Operations over transition systems

The description of a behaviour through a transition system can be structured through the definition of operator constructors. Thus, a new transition system is derived from existing ones through these operators. We have considered the usual operators defined for regular languages. These operators are the sequence (Seq), the iteration (R_Seq), the loop (Loop), the conjunction (And), the alternative (Alt) and its generalization over a set (I_Alt).

Definition 6 (Relational algebra) *We formalize these operators as follows:*

- And $R\, R' = R \cap R'$ Alt $R\, R' = R \cup R'$
- Seq $R\, R' = \{(e, e') | \exists e''$ *such that* $R\, e\, e'' \wedge R'\, e''\, e'\}$
- R_Seq $n\, R$ *is inductively defined by*
      R_Seq $0\, R = \{(x, x)\}$ $\wedge$ R_Seq $(n + 1)\, R = $ Seq (R_Seq $n\, R$) R
- Loop $R = \bigcup_n$ R_Seq $n\, R$ I_Alt $s\, (R_i)_i = \bigcup_{i \in s} R_i$

These constructors allow structured definitions of operators. Thanks to next refinements laws, the construction of the proof refinement properties between complex operators can be directed by the structure of the operators.

2.5   Refinement laws

Refinement laws concerning Seq, Alt and Loop constructors are very simple: if there is a refinement between basis operators, refinement is transmitted to the constructed operators. For example, the proof of Seq construct refinement law is illustrated by the diagram 2. The refinement relation between op_1 and op'_1 implies the existence of a state e'_2. Then the refinement relation between op_2 and op'_2 implies the existence of a state op'_3 which verifies the desired property.

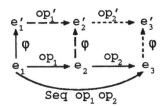

Fig. 2. Refinement of a Sequence

Refinement laws can be defined in HOL as backward chaining proof tactics:

$$\frac{\text{Seq } op_1 op_2 \sqsubseteq_\varphi \text{Seq } op'_1 op'_2}{op_1 \sqsubseteq_\varphi op'_1 \quad op_2 \sqsubseteq_\varphi op'_2} \qquad\qquad \frac{\text{Alt } op_1 op_2 \sqsubseteq_\varphi \text{Alt } op'_1 op'_2}{op_1 \sqsubseteq_\varphi op'_1 \quad op_2 \sqsubseteq_\varphi op'_2}$$

$$\frac{\text{Loop } op \sqsubseteq_\varphi \text{Loop } op'}{op \sqsubseteq_\varphi op'} \qquad\qquad \frac{\text{I\_Alt } s\, op \sqsubseteq_\varphi \text{I\_Alt } s\, op'}{\forall i \in s\, op_i \sqsubseteq_\varphi op'_i}$$

However, the situation is more complex for the And constructor (fig. 3). From $op_1 \sqsubseteq_\varphi op'_1$, and $op_2 \sqsubseteq_\varphi op'_2$, we cannot deduce And $op_1\, op_2 \sqsubseteq_\varphi$ And $op'_1\, op'_2$.

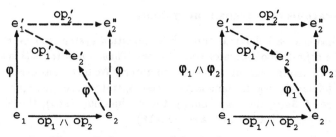

Fig. 3. Refinements of the And constructor

In fact, the refinement relations between op_1 and op_1' (resp. between op_2 and op_2') imply the existence of a satisfying state e_2' (resp. e_2''). However, since the states e_2' and e_2'' are not necessarily equal, we cannot conclude by the existence of a state with the properties of both e_2' and e_2''. A sufficient condition is that the refinement relation φ is functional. Thus, e_2 has only one image through φ, and $e_2' = e_2''$ is the searched state. This hypothesis leads to the following rule:

$$\frac{\text{And } op_1 op_2 \sqsubseteq_\varphi \text{ And } op_1' op_2'}{op_1 \sqsubseteq_\varphi op_1' \quad op_2 \sqsubseteq_\varphi op_2' \quad \text{function } \varphi} \quad \text{FUN\_AND\_TAC}$$

The hypothesis of the functionality of φ is actually too strong. We have investigated another proof rule where the refinement relation φ is split into two conjuncts φ_1 and φ_2 (*fig. 3*). The existence of a state with properties of both e_2' and e_2'' can be derived from some independence hypothesis: constraints imposed on the one hand by op_1' and φ_1 and on the other hand by op_2' and φ_2 must be independent. For this purpose, we introduce the following definition:

Definition 7 (Independent predicates) *We say that two unary predicates P and Q are independent if $\exists x\ P(x) \wedge \exists x\ Q(x) \Longrightarrow \exists x\ P(x) \wedge Q(x)$.*

Then the following theorem can be proved:

Theorem 2 (Independent And Refinement) *Let op_1 be a refinement through φ_1 of op_1' ($op_1 \sqsubseteq_{\varphi_1} op_1'$), and op_2 be a refinement through φ_2 of op_2' ($op_2 \sqsubseteq_{\varphi_2} op_2'$). If for any two elements e_1' and e_2, the predicates $(x \in \varphi_1(e_2) \cap op_1'(e_1'))$ and $(x \in \varphi_2(e_2) \cap op_2'(e_1'))$ are independent, then $op_1 \cap op_2 \sqsubseteq_{\varphi_1 \cap \varphi_2} op_1' \cap op_2'$.*

The last refinement law, illustrated by (*fig. 4*), concerns the stability of the refinement relation by functional composition.

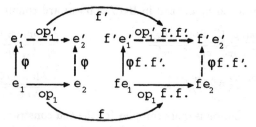

Fig. 4. Functional mapping refinement

Theorem 3 (Functional mapping refinement) *Let op be a refinement through φ of op' (op $\sqsubseteq_\varphi$ op'), and f and f' two functions such that f' is surjective. Then, $\lambda xy.\ op(fx)(fy) \sqsubseteq_{\lambda xy.\ \varphi(fx)(f'y)} \lambda xy.\ op'(f'x)(f'y)$.*

3 Memory consistencies

Several memory coherencies have been defined [8]. The most well known or used ones are the following:

- the *weakest consistency* (although this model is not used) is interesting since it outlines one of the basic idea of memory: every read value must be either the initial value of the register or must have been written before. Such a value can be read at any time and any number of times.
- the *weak consistency*: a written value can be read repeatedly any number of times. An overwritten value cannot be read anymore.
- the *sequential consistency*: if two same values are read by concurrent processes, then they are read in the same order[1].
- the *atomic consistency*: each read value is the last written.

The next example illustrates some of the differences between these consistencies through the sequences processors may read after a given sequence of writes. Let us suppose that the initial value of the variable x is 0 and the processor P_0 has executed the statements $x := 1; x := 2$

Then the sequences (restricted to non zero values) that the processors $P_1, \ldots, P_n$ may read are (we use regular expressions to denote sequences: | is the alternative symbol, * is the repetition symbol):

	Sequences that processors may read			
	P_1	P_2	...	P_n
weakest	$(1\mid 2)^*$	$(1\mid 2)^*$	...	$(1\mid 2)^*$
weak	$(1^*2^*)\mid(2^*1^*)$	$(1^*2^*)\mid(2^*1^*)$	...	$(1^*2^*)\mid(2^*1^*)$
sequential	(1^*2^*)			
	(2^*1^*)			
atomic	1^*2^*			

Let us mention that the weak consistency and the sequential consistency are mostly implemented in distributed environments. Intuitively, we can establish a parallel between the weakest consistency and datagram protocols [12] where messages can be lost or duplicated. Concerning the sequential consistency, we can also make the parallel with atomic broadcast protocols where messages are delivered to every one in the same order. The atomic consistency is generally implemented in centralized environments where, for instance, a hardware bus is used for the implementation. The diagram 5 illustrates a conceptual hierarchy of these different memory models. In section 4, we formalize such a hierarchy.

---

[1] not necessarily the order in which they have been written

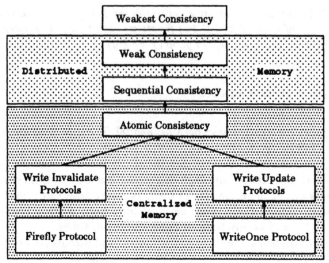

Fig. 5. Consistency protocols hierarchy

3.1 The generic memory transition system

A memory coherency defines the sequences of values, one is allowed to observe after a sequence of read and write operations and an initial value. A read operation is characterized by the issuing process, while a write operation is characterized by the issuing process and the value to be written. We formalize the sequence of read-write operations by the following recursive type:

ReqMem = InitMem num | ReqRead num ReqMem | ReqWrite num num ReqMem

In the preceding type, we have supposed that processes were identified by naturals (type num). The ReqRead operator is parameterized by the issuing process while the ReqWrite operator is parameterized by the issuing process and the value to be written. Moreover, we defined the Read and Write operations on a generic cache *cache as relations with the following signatures:

num -> num -> *cache -> *cache -> bool

For the Read relation, the first parameter represents the issuing process while the second one is the read value. The Write relation has the same parameters: the issuing process then the value to be written. In fact, the preceding types corresponds to the requests with their results superposed.

With respect to transition systems, the state space is the set *cache, the alphabet is the set of labels Read *i v* and Write *i v*. In the HOL formalization, the function **trans_mem** ISt R W builds the transition system associated to its arguments: a characterisation of the initial states, then read and write relations.

It should be noted that our formalization of the sequence of read and write operations presupposes atomic read and write requests. Then, we cannot express within our model the result of simultaneous requests.

3.2 Execution traces

Histories In order to reason about execution traces, we introduce the **History** type as a **List** subtype (where elements are inserted at its tail). An **History**

element H inherits all the operations on List and defines a relation $\prec_H$ on its elements. $v \prec_H v'$ iff the last insertion of v' is more recent than the last insertion of v in H. A basic property of such a relation is its *antisymmetry* when the elements of H are distinct.

Read and write traces In order to superpose the values observed after a sequence of read and writes operations, we use two auxiliary definitions:

- **Exec** which computes the cache obtained after a sequence of requests given an initialization predicate ISt, read R and write W operations. **Exec** is defined inductively as follows:

```
Exec(InitMem i) ISt R W c       = ISt i c
Exec(ReqRead p op) ISt R W c    = ∃ c' v.Exec op ISt R W c' ∧ R p v c' c
Exec(ReqWrite p v op) ISt R W c = ∃ c'.Exec op ISt R W c' ∧ W p v c' c
```

- **ReqReadRel** which tells if the last operation of the sequence may have read a given value **v**. Similarly, **ReqReadRel** is defined as follows:

```
ReqReadRel(InitMem i) ISt R W v      = F
ReqReadRel(ReqRead p op) ISt R W v   = ∃ c c'. Exec op ISt R W c ∧
                                                 R p v c c'
ReqReadRel(ReqWrite p v op) ISt R W v = F
```

We also use the definition of traces: namely the function **RTraces** which gives the set of allowed sequences of read values after a given sequence of requests. The result is defined as a set to take into account the non-determinism of the value returned by a read operation. First, we define an auxiliary function **RD_Traces_aux** returning a set of pairs where:

the first element characterizes the list of read values: each element of this list is a pair: process number, read value.

the second element is the cache obtained after the sequence of requests has been executed.

```
RD_Traces_aux (InitMem i) ISt R W      = { ([], c) | ISt i c }
RD_Traces_aux (ReqRead p op) ISt R W   =
 {(CONS (p,v) l, c)|∃ c1. (l,c1) ∈ RD_Traces op ISt R W ∧ R p v c1 c}
RD_Traces_aux (ReqWrite p v op) ISt R W =
 {(l,c) | ∃ c1. (l,c1) ∈ RD_Traces op ISt R W ∧ W p v c1 c}
```

Then **RD_TRaces** op ISt R W is obtained by taking the first element of each pair contained in **RD_Traces_aux** op ISt R W.

Similarly, we define **WTrace** as follows:

```
WTrace (InitMem i)    = [i]
WTrace (Read p op)    = WTrace op
WTrace (Write p v op) = CONS v (WTrace op)
```

3.3 Consistency formalization

Weakest consistency is simply specified by saying that each value read is a member of the write trace. Then, the following definition[2]:

```
WeakestConsistency ISt R W =
  ∀ op t v. t ∈ RD_TRACES op ISt R W ∧ (∃ p. IS_EL(p,v) t)
    ⟹ IS_EL v (WR_TRACE op)
```

Weak consistency is defined from sequences where each written value is unique. For this purpose, we introduce the list predicate ALL_DIFF [3].

Given a request sequence where each written value is unique, weak consistency is satisfied if on the same processor p we cannot get back to an overwritten value. We express such a property as follows:

```
WeakConsistency ISt R W =
  (WeakestConsistency I_St R W) ∧
  (∀ op t. (ALL_DIFF (WR_TRACE op)) ∧ (t ∈ (RD_TRACES op I_St R W)))
    ⟹ ∀ p v v'. (p,v) ≺ₜ (p,v') ∧ (p,v') ≺ₜ (p,v) ⟹ v = v'
```

Sequential consistency is also defined from sequences where each written value is unique. It is satisfied if two same values are read on the processors p and p' then they are read in the same order. We define it as follows :

```
SeqConsistency I_St R W =
  (WeakestConsistency I_St R W) ∧
  (∀ op t. (ALL_DIFF (WR_TRACE op)) ∧ (t ∈ (RD_TRACES op I_St R W)))
    ⟹ ∀ p q v v'. (p,v) ≺ₜ (p,v') ∧ (q,v') ≺ₜ (q,v) ⟹ v = v'
```

We remark that a similar definition was given in [6]. However, here we have not supposed that values were written in an increasing order, we have only supposed that the values written were different.

Atomic consistency means that a read value is the last written. To formalize it, we first define **LastW op v** which is true iff the last value written is v:

```
LastW (InitMem i) v = (i = v)
LastW (Read p op) v = (LastW op v)   LastW (Write p v op) v' = (v = v')
```

Next, we characterize atomic histories by the predicate **AtomicH**.

```
AtomicH (InitMem i) ISt R W    = T
AtomicH (Read p op) ISt R W    =
  ∀ v. (ReqReadRel (Read p op) ISt R W v) ⟹ (LastW op v)
AtomicH (Write p v op) ISt R W = (AtomicH op ISt R W)
```

---

[2] (IS_EL e l) is true iff e is an element of the list l

[3] In HOL , we have defined it as: ALL_DIFF l = ONE_ONE(λe. EL e l)

At last, a predicate ISt and read/write operations R and W implement an atomic consistency if every history of these operations is an atomic one :

```
Atomic ISt R W = ∀ op. AtomicH op ISt R W
```

3.4 Consistency hierarchy and refinements

The following theorem relates the consistencies defined :

Theorem 4 (Consistency hierarchy) *for any initial state ISt, Read and Write transitions R W, we have : Atomic ISt R W $\implies$ SeqConsistency ISt R W $\implies$ WeakConsistency ISt R W $\implies$ WeakestConsistency ISt R W*

The consistency of new memory models can be established through the refinement of basic memory models of which consistency properties have been already proved. The following theorem validates this methodology :

Theorem 5 (Consistency refinement) *For any consistency predicate $C \in \{WeakestConsistency, WeakConsistency, SeqConsistency, Atomic\}$, we have : $C(I_1, R_1, W_1) \wedge (trans\_mem\ I_2 R_2 W_2) \sqsubseteq (trans\_mem\ I_1 R_1 W_1) \implies C(I_2, R_2, W_2)$*

4 Memory models

In this section, we adopt an operational point of view of memory models. Each basic model is implemented by a canonical transition system specified by a predicate defining its initial states and by read and write transitions. Then the relation between the models is obtained through the refinement relation between the corresponding transition systems. Thanks to theorem 5, a new model satisfies a given consistency if it is a refinement of the corresponding canonical transition system. All these results are summarized by the figure 6.

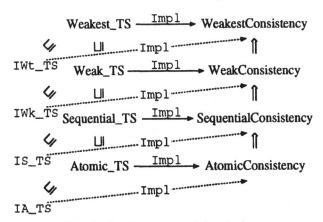

Fig. 6. Consistency models relations

4.1 The weakest memory

The state of the weakest memory is characterized by the set of all written values. Let **WeakestMem** be the associated type with **mk_Weakest** as constructor.

```
type WeakestMem = mk_Weakest (num)set
Weakest_Net (mk_Weakest s) = s
Weakest_Init i c = (c = mk_Weakest {i})
Weakest_Read i v c1 c2 = (c2 = c1) ∧ v ∈ (Weakest_Net c1)
Weakest_Write i v c1 c2 = (c2 = mk_Weakest({v} ∪ (Weakest_Net c1)))
```

4.2 The weak memory

The transition system for the weak memory is more complex since we cannot read an overwritten value. It is represented by the type **WeakMem**, constructed by the operator **mk_Weak**.

```
type WeakMem = mk_Weak num (num -> num) (num -> (num)bag)
Weak_NbProc (mk_Weak s r n) = s
Weak_SetProc c = {p | p < Weak_NbProc c}
Weak_Net (mk_Weak s r n) = n    Weak_Reg (mk_Weak s r n) = r
```

The type **WeakMem** can be seen as a record with fields for: the number of processors (**Weak_NbProc c**: num), the value stored in each cache (**Weak_Reg c**: num->num) and bags containing written values but not yet transferred to the appropriate caches (**Weak_Net c**: num->(num)bag). Thanks to our structuring operators, we have defined the weak memory transition system through three basic transitions:

- an internal read transition **Weak_In_Read**, which consists in reading a value in the local cache,
- an internal write transition **Weak_In_Write**, which consists in writing a value in the local cache and in the bags of the other processors[4].
- and an internal transition **Weak_Loss** which consists in either transferring a value from a bag to a register or in loosing it. We have considered this alternative to represent non reliable distributed systems.

```
Weak_In_Read i v c c' = i ∈ Weak_SetProc c ∧ v = Weak_Reg c i ∧ c' = c
Weak_In_Write i v c c' =
   (i ∈ (Weak_SetProc c) ∧ (Weak_SetProc c' = Weak_SetProc c) ∧
   (v = Weak_Reg c' i) ∧
   (∀ j. j ∈ (Weak_SetProc c) ⟹
        ((j = i) => (Weak_Net c' i = Weak_Net c i)
              | ((Weak_Net c' j = v ⊕ (Weak_Net c j)) ∧
                 (Weak_Reg c' j = Weak_Reg c j))))
Weak_Loss r r' b b' = ∃ v. (b = v ⊕ b') ∧ ((r' = r) ∨ (r' = v))
```

---

[4] ⊕ denotes the insertion of an element in a bag

From the basic **Weak_Loss** transition, we have built a general routing transition obtained through the composition of the **Loop** and **I_Alt** operators.

```
Weak_Loss_i s i r r' b b' =
  (∀ j. j ∈ s ⟹ ((i = j) => Weak_Loss(r i)(r' i)(b i)(b' i)
                          | (r j = r' j) ∧ (b j = b' j)))
Weak_Routing_one s c1 c2 =
  (Weak_SetProc c1 = s) ∧ (Weak_SetProc c2 = s) ∧
  I_Alt s (λ i b1 b2. Weak_Loss_i s i(Weak_Reg b1)(Weak_Reg b2)
                                  (Weak_Net b1)(Weak_Net b2))
        c1 c2
Weak_Routing_s s = Loop (Weak_Routing_one s)
Weak_Routing c1 c2 = Weak_Routing_s (Weak_SetProc c1) c1 c2
```

Then we obtain a general read or write as a sequential composition of the corresponding internal transition and a routing transition:

```
Weak_Init i c = (c ∈ { mk_Weak n (λ p. i) (λ p. ∅)})
Weak_Read i v = Seq (Weak_In_Read i v) Weak_Routing
Weak_Write i v = Seq (Weak_In_Write i v) Weak_Routing
```

We note that the preceding predicate **Weak_Routing**, although rigorously defined, is rather tedious to understand. We can rewrite it (with an OCCAM touch) as follows[5]:

```
LOOP
  ALT i ∈ Weak_SetProc
    ALT v ∈ Weak_Net[i]
      PAR
        Weak_Net[i] := Weak_Net[i] ⊖ v
        ALT
          Weak_Reg[i] := v
          SKIP
```

Weak is weakest In order to establish the simulation relating the corresponding transition systems, we first introduce the morphism $\varphi_{\text{Weak2Weakest}}$ between the two cache specifications **WeakMem** and **WeakestMem** defined as follows:

$$
\text{WeakMem2Set } c = \bigcup_{k \in \text{Weak\_SetProc } c} \{\text{Weak\_Reg } c\ k\} \cup \text{Bag2Set (Weak\_Net } c\ k)
$$

$$
\varphi_{\text{Weak2Weakest}} c\ c0 = (\text{WeakMem2Set } c) \subset (\text{Weakest\_Net } c0)
$$

The basics of the proof that the weak memory transition system simulates the weakest memory transition system consists in proving a simulation relation between **Weak_Routing** and the superset ($\supset$) relation. We obtain the following theorem:

---

[5] the semantics of LOOP and ALT language constructs are given by the already defined Loop and Alt predicates

```
WeakIsWeakest_THM |- (trans_mem Weak_Init Weak_Read Weak_Write)
                      ⊑
                  (trans_mem Weakest_Init Weakest_Read Weakest_Write)
```

4.3 The sequentially consistent memory

As for the weak memory, we define the type SeqMem constructed by mk_Seq. It can be seen as a record with fields for: the number of processors (Seq_NbProc c: num), the value stored in each cache (Seq_Reg c: num->num) and queues containing written values but not yet transferred (Seq_Net c: num->(num)list). Then, as for weak memories, we first define internal read and write transitions and give an OCCAM like specification of an internal routing transition.

```
Seq_In_Read i v c c' = i ∈ Seq_SetProc c ∧ v = Seq_Reg c i ∧ c' = c
Seq_In_Write i v c c' =
     (i ∈ (Seq_SetProc c) ∧ (Seq_SetProc c' = Seq_SetProc c)) ∧
     (!j. j ∈ (Seq_SetProc c) ⟹
       ((j = i) => ((Seq_Net c' j = []) ∧ (Seq_Reg c' j = v))
                 | ((Seq_Net c' j = SNOC v(Seq_Net c j)) ∧
                    (Seq_Reg c' j = Seq_Reg c j))))
Seq_Routing =
  LOOP
    ALT i ∈ Seq_SetProc
    ALT
        Seq_Net[i], Seq_Reg[i] := Rest(Seq_Net[i]), First(Seq_Net[i])
        Seq_Net[i] := Rest(Seq_Net[i])
```

Then we define the initial predicate and the two basic transitions:

```
Seq_Init i c = (c ∈ {mk_Seq n (λ p. i) (λ p. [])})
Seq_Read i v = Seq (Seq_In_Read i v) Seq_Routing
Seq_Write i v = Seq (Seq_In_Write i v) Seq_Routing
```

Sequential is weak The morphism $\varphi_{Seq2Weak}$ establishes the refinement :

```
φ Seq2Weak_s s =
 And (λ c2 c1. (Seq_SetProc c2 = s) ∧ (Weak_SetProc c1 = s))
     (λ c2 c1. (Seq_Reg c2) = (Weak_Reg c1) ∧
               (∀ k. k ∈ s ⟹ listToBag (Seq_Net c2 k) ⊂ Weak_Net c1 k))
 φ Seq2Weak c2 c1 = φ Seq2Weak_s (Seq_SetProc c2) c2 c1
```

The simulation proof is based of the refinement proof of the operators read, write and routing. Then the Seq refinement law establishes the desired result.

```
SeqIsWeak_THM |- (trans_mem Seq_Init Seq_Read Seq_Write)
                    ⊑
                 (trans_mem Weak_Init Weak_Read Weak_Write)
```

4.4 The atomic consistent memory

We define the type `AtomicMem` constructed by `mk_Atomic`:

```
type AtomicMem = mk_Atomic num num
Atomic_NbProc (mk_Atomic s r) = s        Atomic_Reg (mk_Atomic  s r) = r
Atomic_SetProc c = {p | p < (Atomic_NbProc c)}
```

In this model, values are immediately written in the destination caches, so we do not need a routing transition.

```
Atomic_Init i c = (c ∈ {mk_Atomic n i})
Atomic_Read i v c1 c2  =
   (i ∈ Atomic_SetProc c1) ∧ (v = Atomic_Reg c1) ∧ (c2 = c1)
Atomic_Write i v c1 c2 =
   (i ∈ Atomic_SetProc c1) ∧ (c2 = mk_Atomic(Atomic_NbProc c1) v)
```

Atomic is sequential The relation $\varphi_{\text{Atomic2Seq}}$ establishes the refinement :

```
f_φ Atomic2Seq c = mk_Seq (Atomic_NbProc c) (λ i. Atomic_Reg c)(λ i. [])

φ Atomic2Seq c2 c1 = (c1 = f_φ Atomic2Seq c2)
```

The proof of the refinement relation relies on a lemma which states that there is a routing relation between two caches where the second is obtained by transferring *each* queued value to the corresponding register.

```
AtomicIsSeq_THM |- (trans_mem Atomic_Init Atomic_Read Atomic_Write)
                             ⊑
                    (trans_mem Seq_Init Seq_Read Seq_Write)
```

5 Multiprocessor memories

In this section, we are especially interested in shared memory multiprocessors [3] [11]. In such an environment a value is implemented in central memory and for efficiency reasons can be replicated in processor caches. Values are generally grouped into blocks, the block being the unit of transfer between memory and caches. However, in our study, we make a practical simplifying assumption: values are the unit of transfer (we do not consider the problem of mapping a value in a block). Three operations are generally considered: the usual read and write and the *flush* operation due to the limited capacity of a cache.

The coherency that is generally implemented is the strong one (although not explicitly stated). The write invalidate and write update protocols are the two well known coherence schemes. In the following, we only consider write update protocols for lack of space.

Informally, the principles of these protocols are to read a value in the cache if it is locally available, and to update all the cached copies on each write request. As before, the type `Cache_Upd` can be seen as a record with fields for: the number of processors (`NbProc_Upd: num`), the value stored in each cache (`Reg_Upd: num->num`), the state of each cache (`Cached_Upd: num->bool`) and the value stored in memory (`Mem_Upd: num`). We characterize the write update protocols by the following invariant:

```
INV_Upd c =
  (∀ p p'. p ∈ (SetProc_Upd c) ⟹ p' ∈ (SetProc_Upd c) ⟹
    Cached_Upd c p ∧ Cached_Upd c p' ⟹ (Reg_Upd c p = Reg_Upd c p'))
```

The next predicate is associated to the read operation.

```
Read_Upd i v c1 c2 =
  (i ∈ SetProc_Upd c1)∧(SetProc_Upd c1 = SetProc_Upd c2)∧(INV_Upd c2) ∧
  ((Cached_Upd c1 i) ⟹ (c2 = c1)) ∧
  ((∀ p. p ∈ SetProc_Upd c1 ⟹ ¬ Cached_Upd c1 p)⟹(v = Mem_Upd c1)) ∧
  (∀ j. Cached_Upd c2 j = ((j = i) => T | Cached_Upd c1 j)) ∧
  (∀ j. Reg_Upd c2 j = ((j = i) => v | Reg_Upd c1 j))
```

In the last definition, it is interesting to comment the following predicates:

- `INV_Upd c2` whatever we do, the new cache `c2` must respect the invariant.
- `Cached_Upd c1 i ⟹ (c2 = c1)` if a value is locally cached, we do nothing.
- `((∀ p. ¬ Cached_Upd c1 p) ⟹ (v = Mem_Upd c1))` a value not cached is fetched from memory. We remark that if a value is already cached somewhere then we can fetch it from *either* another processor cache *or* memory.
- `Cached_Upd c2 i` the value is eventually cached in `c2`.

The following predicates defines the write, flush operations and init states :

```
Write_Upd i v c1 c2 =
  (i ∈ SetProc_Upd c1)∧(SetProc_Upd c1 = SetProc_Upd c2)∧(INV_Upd c2) ∧
  (v = Reg_Upd c2 i) ∧ Cached_Upd c2 i
Flush_Upd i c1 c2 =
  i ∈ (SetProc_Upd c1)∧(SetProc_Upd c1 = SetProc_Upd c2)∧(INV_Upd c2) ∧
  (Cached_Upd c1 i) ∧ (¬(Cached_Upd c2 i)) ∧
  (∀ p. (p ∈ SetProc_Upd c1) ∧ ¬ (p=i) ⟹
    ((Reg_Upd c2 p =Reg_Upd c1 p)∧(Cached_Upd c2 p =Cached_Upd c1 p))) ∧
  (Cached_Upd c1 i ⟹(Mem_Upd c2 = Reg_Upd c1 i) ∨
              (∃ p. (p ∈ SetProc_Upd c2) ∧ (Cached_Upd c2 p)))
Init_Upd c = INV_Upd c
```

Concerning the `Flush_Upd_DEF` we comment the following predicates:

- `¬ Cached_Upd c2` the value is not anymore cached in the new cache *c2*.
- `(∀ p. p ∈ SetProc_Upd c1 ∧¬ (p = i) ⟹ ...` a flush is non intrusive.
- `(Mem_Upd c2 = Reg_Upd c1 i) ∨...` if the value is not written back to memory, then the value is cached in another processor cache.

6 Conclusion

In this paper, we have formalized in higher order logic, within the HOL system, memory coherencies and memory models. Theorems about the relations between them have been proved.

As mentioned in the introduction, for future work, we plan to study the validation of memory protocols as well as the mapping of memory models in distributed environments [7]. From a theoretical point of view, we hope to enrich the addressed properties and consider liveness ones [1]. We are also interested in the validation of testing methods such as the one proposed in [10].

At last, with respect to methodological aspects, we have dealt mainly with subtyping. Our work can also be seen as a formalization of memory hierarchy through a subtyping relation (*is*). An interesting study, would be to mechanize such a methodology, and allow, as in object oriented programming, to reuse a theorem concerning a type for all its subtypes (a HOL ++ package?).

References

1. A. Arnold. *Systèmes de transitions finis et sémantiques des processus communicants*. Etudes et recherches en informatique. MASSON, 1992.
2. M.J.C. Gordon and T.F. Melham. *Introduction to HOL*. Cambridge University Press, 1994.
3. D. Litaize. Architectures multiprocesseurs à mémoire commune. In *Deuxième symposium architectures nouvelles de machines*, pages 1–40, sep 1990.
4. P. Loewenstein. A formal theory of simulations between infinite automata. *Formal methods in system design*, 3:117–149, 1993.
5. N.A. Lynch and M.R. Tuttle. Hierarchical correctness proofs for distributed algorithms. In *Proceedings of the sixth annual ACM symposium on principles of distributed computing*, pages 137–151, aug 1987.
6. J. Misra. Axioms for memory access in asynchronous hardware systems. *ACM Transactions on Programming Languages and systems*, 3:142–153, 1986.
7. M. Mizuno, M. Raynal, G. Singh, and M.L. Neilsen. An efficient implementation of sequentially consistent distributed shared memories. Technical Report 764, IRISA, oct 1993.
8. D. Mosberger. Memory consistency models. *Operating Systems Review*, 27(1):18–26, jan 1993.
9. T. Nipkow. Formal verification of data type refinement - theory and practice. In *Stepwise refinement of distributed systems*, volume 430 of *LNCS*, pages 561–591. Springer Verlag, 1992.
10. F. Pong and M. Dubois. The verification of cache coherence protocols. Technical Report CENG-92-20, USC, nov 1992.
11. P. Stenstrom. A survey of cache coherence schemes for mutliprocessors. *Computer*, 23(6):11–25, jun 1990.
12. A.S. Tanenbaum. *Computer Networks*. Prentice Hall, 1989.

Providing Tractable Security Analyses in HOL

Stephen H. Brackin* **

Odyssey Research Associates, 301 Dates Drive, Ithaca NY 14850

Abstract. This paper describes tools that let HOL users, even unsophisticated ones, employ sophisticated HOL concepts in conveniently
specifying system designs and proving that these designs have particular nondisclosure security properties. These tools include the Romulus
Interface Process Specification Language (IPSL) and a translator for
translating IPSL specifications into HOL90.

1 Introduction

Designing and evaluating the most trusted computer systems requires formally
specifying these systems and proving that they have particular security properties [6]. The Romulus collection of security tools [13, 14, 15, 16] includes the
means for proving that system designs, formalized as processes, are *restrictive*,
a strong nondisclosure security property. Restrictiveness, a variant of Goguen-
Meseguer noninterference [7], developed by McCullough [9, 11, 10], guarantees
that a system's behavior visible at one security level does not reveal the existence of inputs at higher or incomparable levels. Restrictiveness is of particular
interest because it is a *composable* property, meaning that a system composed
of properly connected restrictive parts is itself restrictive.

Romulus is based on producing proofs in HOL90, so using Romulus requires
knowledge of both computer security and HOL. It is unreasonable, though, to
expect Romulus users to be sophisticated HOL users. The intersection of the
sets of computer security experts and HOL experts has a cardinality of about
six.

This paper describes how Romulus addresses this problem with an Interface
Process Specification Language (IPSL), a translator for translating IPSL specifications into HOL90, and a graphical user interface. These tools not only make
Romulus proofs of restrictiveness available to non-expert users, they also greatly
reduce the effort needed to produce these proofs by even expert users. The techniques and tools used could also be easily generalized to other applications (e.g.,
formalizing computer languages in HOL) that involve defining processes in terms
of process-valued functions.

The work described here extends earlier work by the author and Chin [5].
That earlier work developed a convenient Process Specification Language (PSL)

---

* Supported by Rome Laboratory Contract F30602-90-C-0092
** The author wishes to thank Shiu-Kai Chin, David Rosenthal, Randy Calistri-Yeh,
Doug Long, and the referees for helpful comments.

implemented as a HOL concrete recursive type, an operational semantics for PSL, a collection of easy to understand security properties defined by induction on the structural complexity of PSL objects, and associated tactics. The tactics apply structural-induction rewrite rules until all references to PSL objects disappear.

The computer security properties given in [5] are almost sufficient for guaranteeing restrictiveness in the broad case of *buffered server processes*. A buffered process consists of a FIFO queue and a process being buffered; the buffered process saves its inputs on the queue until the process being buffered is ready to receive them. The process being buffered is a server process if it waits for input in a parameterized state, processes each input by producing zero or more outputs, and then calls itself to again wait for input in a possibly different parameterized state.

The work in [5] followed from earlier work by Sutherland, McCullough, Rosenthal, and others, and continued an alternate approach to restrictiveness from the one taken by Alves-Foss and Levitt [1, 2, 3, 4]. Rosenthal originally identified the security properties sufficient for guaranteeing restrictiveness of buffered server processes [17, 18]. See [5] for history and references.

The work in [5] showed the practicality of reasoning about restrictiveness for server processes at the object-language level, using only conservative extensions of the HOL logic and pure tactics. That paper also demonstrated the practicality and advantage of imposing HOL's strong type checking on message contents and state-machine parameters in process specifications.

Several problems remained, though. This paper discusses these problems in two groups. The first group, of general interest to the HOL community, consists of tractability issues related to

- Providing what are effectively processes defined in terms of arbitrary process-valued functions; and
- Providing strong type checking on the arguments to process-valued functions and the messages in input and output events.

The second group, of interest to those involved with computer security, consists of security issues related to

- Formalizing the use of ranges of possible security levels for input and output events;
- Proving properties that guarantee a slightly stronger form of restrictiveness than that guaranteed in [5];
- Formally checking that the connections between subprocesses are valid in composite processes consisting of connected subprocesses; and
- Formally proving that processes are sometimes secure purely by virtue of their environments.

The remainder of this paper is organized as follows: Section 2 discusses process-specification tractability issues. It describes PSL and its mechanisms for defining input events, output events, and PSL-valued functions. It introduces the parts of IPSL that relate purely to specifying processes and indicates how these parts are translated into HOL90. Section 3 discusses the computer security issues just given and describes how they are addressed in IPSL and the Romulus graphical interface. Finally, Section 4 gives conclusions and suggestions for further work.

The paper uses the notation conventions of Slind's HOL90, Release 6.

2 Process Specification Tractability Issues

This section describes PSL and the techniques it uses both to define processes in terms of process-valued functions and to provide strong typing restrictions on these functions and on events. The section then introduces the parts of IPSL used purely to describe processes, and indicates how the Romulus IPSL translator translates IPSL specifications into the needed HOL constructs.

The mechanism PSL uses to both allow processes to be defined in terms of process-valued functions and to impose strong typing restraints on these functions' arguments is hard to understand. It involves subtle distinctions between functions, names for functions, function values, and names for function values.

Process-valued functions are essential, though. PSL treats parameterized processes as being process-valued functions of these parameters. PSL also treats nonparameterized process' responses to input events as being given by process-valued functions of these input events, and treats parameterized processes' responses to input events as being given by process-valued functions of these parameters and these input events.

The mechanism PSL uses both to distinguish output and input events and to impose strong typing restraints on messages sent or received in these events requires less sophistication, but it still involves a use of `define_type` and type constructors that requires more HOL knowledge than should be expected of typical Romulus users.

Producing specifications using PSL also involves producing a lot of HOL boilerplate, particularly for `define_type` and `new_recursive_definition` uses. Without special tools, this effort is daunting even to experts.

2.1   PSL: Process Specification Language

This subsection describes PSL and how it uses possibly polymorphic concrete recursive types to define processes in terms of process-valued functions and to provide strong typing for the messages in input and output events. The subsection does not repeat the formal definitions of PSL and its semantics given in [5], but it gives detailed informal descriptions, and it introduces an example used repeatedly in the remainder of this paper.

As a specification language, PSL's basic processes are **Skip**, **Send**, **Receive**, and **Call**. **Skip** is the finished process that does nothing. **Send** transmits an output event. **Receive** takes a predicate on input events determining which input events it receives and (effectively) a function determining the process' response to these input events. Call (effectively) makes a particular call to a process-valued function.

The PSL operators ;;, Orselect, If, and Buffered combine other processes. Infix operator ;; is like the sequence operator in CSP [8]. Orselect is like the non-deterministic choice operator + in CCS [12]. If is the if-then-else operator. Buffered takes a predicate on input events, a buffer, and a process to be buffered. It returns the process that puts the input events satisfying the predicate onto the buffer, then passes them on to the process being buffered when that process is ready to receive them.

The type constructor :process('outev,'inev,'invoc) gives PSL as a polymorphic concrete recursive type. The type variables 'outev, 'inev, and 'invoc are themselves intended to be instantiated by (possibly polymorphic) concrete recursive types. If this instantiation is done properly, as described in the remainder of this subsection, PSL processes can be defined in terms of PSL-valued functions, and strong typing restraints can be maintained on the arguments to these functions and on the messages in input and output events.

First consider input and output events. Let an *input port* be an interface through which messages enter a process and an *output port* be an interface through which messages leave a process. These messages can be of any HOL type or any finite combination of HOL types. If the name of each port is taken to be a type constructor for a concrete recursive type of events entering or leaving the process, input events for input ports and output events for output ports, the definitions of these concrete recursive types can be used to impose any desired type constraints on the contents of input or output messages.

For example, suppose a parameterized process accepts natural numbers as inputs though two ports, interpeting the inputs through one port as identification numbers of ships reporting that they are available and interpreting the inputs through the other port as identification numbers of users asking how many ships are available. Further, suppose the process responds to the availability queries by producing pairs of natural numbers, one identifying the user to receive the response and the other giving the requested count.

We can define input and output event types, say exin and exout, for such a process by choosing appropriate port-name type constructors, say reports, queries, and answers:

```
'exin = reports of num | queries of num'
'exout = answers of num => num'
```

When the type variables 'outev and 'inev are instantiated by the types exout and exin, respectively, the PSL objects become arbitrary processes with input ports named reports and queries accepting :num inputs and with an output port named answers producing two :num outputs. Note that answers is Curried, which is normally most convenient.

Now consider defining processes in terms of process-valued functions. Although it is not possible to use define_type to define a concrete recursive type T in terms of T-valued functions, it is possible to use define_type to define T

in terms of *names* for the T-objects returned by particular T-valued functions. The names, which we call *invocations*, are given by a concrete recursive type N whose constructors correspond to the particular T-valued functions used in the definition of T. A separate function, defined with **new_recursive_definition**, maps the invocations to the T-objects they name; the interpretation of T-objects varies with the mapping of N to T. If **foo** is a function mapping :**num** values to T-objects, **FOO** can be taken as the corresponding type constructor mapping :**num** values to invocations. Then the "correct" function mapping invocations to T-objects maps (**FOO x**) to (**foo x**).

Continuing the earlier example, suppose some of the ships, those with identification numbers greater than 99, are submarines whose availability is classified. The process needs to maintain two availability counts, one for non-submarines and the other for submarines. Further, suppose the process includes submarines only in its responses to availability queries from the user whose identification number is 0.

If we choose **exbefore** as the invocation constructor corresponding to the function that gives the process as it waits for the next input, we know that **exbefore** takes a :**num#num** pair (the current counts) as its argument. If we choose **exafter** as the invocation constructor corresponding to the function giving the process after it responds to an input, we know that **exafter** takes a :**num#num** pair (the current counts) and an :**exin** input event (either a report or query number arriving through its appropriate port) as arguments. The invocations are then defined by

`'exinvoc = exbefore of num#num | exafter of num#num => exin'`

When the type variables '**outev** and '**inev** are instantiated as before and the type variable '**invoc** is instantiated by the type **exinvoc**, the PSL objects become arbitrary processes with ports as before containing references to two PSL-valued functions, one on pairs of natural numbers and the other on pairs of natural numbers and the process' possible input events. (A parameterized process must have only one, possibly tupled, parameter in order to define a projection function for it, as explained in Section 3.)

There is no circularity; the appropriate instantiation of the type constructor **process**, :(**exout,exin,exinvoc**)**process**, is defined completely, before functions having values of this type are chosen. The desired process can then be constructed by mapping invocations of the following forms:

1. (**exbefore p**) to

   (**Receive (\ev:exinev. T) (exafter p)**)

   (Since **exafter** is Curried, (**exafter p**) maps input events to invocations.)

2. (**exafter p (reports ship)**) to

   (**Call**
     (**exbefore**
       ((**ship>99**) => ((**FST p**),(**SND p**)+1) | ((**FST p**)+1,(**SND p**)))))

3. `(exafter p (queries user))` to

```
(Send
 (answers
  user
  ((user = 0) => ((FST p)+(SND p)) | (FST p)))) ;;
 (Call (exbefore p))
```

Note that, for parameterized server processes such as the example given, everything except the names of the invocation constructors and the names of the output, input, and invocation types is determined by the name and type of the process parameter, the names of the ports, whether each port is an input or output port, the types of the messages arriving at or leaving each port, and the responses to inputs for each input port. Further, the only name of an invocation constructor that is needed to define these responses to inputs is that of the constructor for the function giving the top-level process as a function of its parameter — in this example, **exbefore**. These observations are the basis for the Interface Process Specification Language (IPSL) described in the next subsection.

2.2 IPSL: Interface Process Specification Language

This subsection gives a description of the parts of IPSL that relate only to defining atomic processes, not their security properties. An *atomic* process is one not decomposed into connected subprocesses; a *composite* process is one decomposed into connected subprocesses. The parts of IPSL relating to security and composite process are described in Section 3. The Romulus IPSL translator translates specifications written in IPSL into HOL90.

IPSL consists of text strings, some giving potentially arbitrary HOL90 or Standard ML input, labeled and delimited by keywords beginning with ?? and ending with :. In IPSL, text strings giving HOL variables, constants, and expressions must not contain explicit calls to the HOL90 term parser (--` and `--); the IPSL translator will supply these calls.

Some IPSL keywords are optional, others are required, and some must be used in combination with others when they are used. Some keywords must have associated values in order for translations into HOL90 to succeed. This paper's descriptions of IPSL take a keyword to mean the keyword with its associated value, if any.

An atomic server process specification consists of the following:

1. **??Process:** — This keyword gives the name of the process. If the process is not named, the IPSL translator creates a name for it. The IPSL translator names the invocation constructor for the top-level process by appending the letters **Top** to the name of the process.

2. **??HOL_functions:** — This optional keyword introduces arbitrary HOL text that defines type or function constants used in the rest of the process' specification.

3. Whether the process is parameterized is optional. If the process is parameterized, the following keywords specify its state parameter.
   - **??StateVar:** — a variable naming the state parameter and giving its type.
   - **??Initial:** — the initial value of the state parameter.

4. **??OutPort:** — This keyword gives the name of an output port and signals the start of an output port specification. If the output port is not named, the IPSL translator creates a name for it. A process specification must include at least one output port for the translator to successfully translate the specification into HOL. The process specification must have an output port specification for each of the process' output ports. An output port specification contains the following:
   - **??MessageVar:** — The port specification must have one or more of these values. The translator assumes that the messages through the port are given by a tuple of variables naming and giving the types of the entries in this tuple. Each **??MessageVar:** value gives one of these variables and its type.

5. **??InPort:** — This keyword gives the name of an input port and signals the start of an input port specification. If the input port is not named, the translator creates a name for it. A process specification must include at least one input port for the translator to successfully translate the specification into HOL. The process specification must have an input port specification for each of the process' input ports. An input port specification contains the following:
   - **??MessageVar:** — The port specification must have one or more of these values. The translator assumes that the messages through the port are given by a tuple of variables naming and giving the types of the entries in this tuple. Each **??MessageVar:** value gives one of these variables and its type.
   - **??Response:** — This keyword gives the PSL process that the process being specified becomes in response to an arbitrary input message received through the port. The port's **??Response:** value is the right-hand side of an equation defining the new process as a function of the port's **??MessageVar:** variables and (optionally) the **??StateVar:** variable of the process being specified.

6. **??EndProcess:** — The optional name value associated with this keyword is ignored by the IPSL translator; the user can include it to create more readable IPSL specifications.

Here is an IPSL specification of the example process in Section 2.1. This specification names this process **counter** and provides security information for it. IPSL intermingles required process and required security specification information, so both must be present for the IPSL translator to produce a HOL90 translation. For the moment, ignore the **??Projection:**, **??LevelFun:**, and **??LevelRange:** values. These are explained in Section 3.

```
??Process: counter

??StateVar: p:num#num
??Initial: (0,0)
??Projection: (standard_dom level secret) => p | ((FST p),0)

??OutPort: answers
??MessageVar: user:num
??MessageVar: count:num
??LevelFun: (user = 0) => secret | unclassified
??LevelRange: unclassified secret

??InPort: reports
??MessageVar: ship:num
??LevelFun: (ship > 99) => secret | unclassified
??LevelRange: unclassified secret
??Response:
(Call
 (counterTop
  ((ship>99) => ((FST p),(SND p)+1) | ((FST p)+1,(SND p)))))

??InPort: queries
??MessageVar: user:num
??LevelFun: (user = 0) => secret | unclassified
??LevelRange: unclassified secret
??Response:
(Send
 (answers
   user
   ((user = 0) => ((FST p)+(SND p)) | (FST p)))) ;;
(Call (counterTop p))

??EndProcess: counter
```

Note the use of counterTop in place of the exbefore used in Section 2.1. Also note that the output event, input event, and invocation types are not specified, and neither is the mapping from invocations to PSL objects. The translator figures out all these things from the ??StateVar:, ??MessageVar:, and ??Response: values, using standard default names formed from the process name counter.

The translator translates these 28 lines of IPSL code into 173 executable lines of HOL and 51 blank or comment lines. This HOL code is readable in a pinch (e.g., when there are subtle syntax errors in the HOL portions of the original IPSL code), but it is not intended for human perusal. Here is a fragment of it, though, slightly edited to fit the page; this fragment shows how the translator solves the problem of defining the interpretations of the invocations. (Note

the use of **counterInEv** instead of **exin**, **counterTop** instead of **exbefore**, and
counterResponse instead of **exafter**.)

```
(* Define the counter functions giving the top-level PSL
process and the response to input events. *)

new_definition(
 "countertop",
 --'^countertop (p:num#num) =
(Receive (\ev:^counterInEv.T) (counterResponse (p:num#num)))
 '--);

new_recursive_definition {
name = "counterresponse",
fixity = Prefix,
rec_axiom = counterInEv_Def,
def =
--'
 (^counterresponse (p:num#num)
   (reports
    (ship:num)
   ) =
((Call
 (counterTop
  ((ship>99) => ((FST p),(SND p)+1) |
               ((FST p)+1,(SND p))))))) /\

 (^counterresponse (p:num#num)
   (queries
    (user:num)
   ) =
((Send
 (answers
   user
   ((user = 0) => ((FST p)+(SND p)) | (FST p)))) ;;
(Call (counterTop p))))
 '--};

(* Define the counter function assigning meanings to
invocations. *)

new_recursive_definition {
 name = "counterInvocVal",
 fixity = Prefix,
 rec_axiom = counterInvoc_Def,
 def =
```

```
__ '
 (^counterInvocVal (counterTop param) =
   (countertop param)) /\
 (^counterInvocVal (counterResponse param inev) =
   (counterresponse param inev))
'--};
```

The IPSL translator is not a complicated program. It is a simple LEX/YACC utility incorporated into the Romulus graphical interface and available as a stand-alone, graphics-free translator called ipsl2hol. The techniques used in it, and even much of its code, could easily be adapted to other applications.

3 Computer Security Issues

This section describes computer security issues addressed by IPSL, the IPSL translator, and the Romulus graphical interface. It briefly describes how IPSL specifies the security-critical aspects of process' environments and how IPSL specifies composite processes. It then describes the HOL security goals that the IPSL translator produces and describes how the security conditions posed by the IPSL translator's goals, if proved, establish a stronger form of restrictiveness than that given in [5]. It gives the proof of the security goal that the IPSL translator produces for the example introduced in Section 2. Finally, it describes some of the additional security specification conveniences provided by the Romulus graphical interface.

3.1 Specifying Security Conditions in IPSL

An IPSL specification defines a *rated* process, one that assigns security levels to each of its input and output events. For both input and output events, the level the IPSL specification assigns to an event is a hypothesis about how the environment created the event or how the environment will be able to use the event. If the specification takes the level of an event receiving or sending a message to be given by a label in that message, the assumption is

- the label level is equal to or higher than the clearance level of whoever or whatever created the event; and
- no one or thing without a clearance level equal to or higher than the label level will ever be able to detect the existence of the event.

An IPSL specification also defines *level ranges* associated with each port. The intended interpretation of an input-port level range is that the process' environment is assumed to never send an input to that port unless the input's level falls in that range. The intended interpretation of an output-port level range is that the process will never produce an output from that port unless the output's level falls in that range.

Finally, for parameterized processes, an IPSL specification also defines an *invariant* and a *projection function*. An invariant is an optional, arbitrary, property that the attainable state parameters will be shown by induction to possess. The projection function is essentially an induction hypothesis about how the process' state parameter will influence the process' future behavior visible at each security level. A projection function returns, for a security level and a parameterized process' state parameter, precisely that information in the state parameter capable of influencing the process' future behavior visible at that level.

The optional IPSL keyword **??LevelTheory**: names a HOL theory defining both a set of security levels and a dominance relation on these levels. The optional keyword **??DomRelation**: gives the name and type of this dominance relation. The optional keyword **??LevelVar**: names and gives the type of an arbitrary security level. The IPSL translator uses standard default values for all these things if an IPSL specification does not give them.

If a process is parameterized, the optional IPSL keyword **??Invariant**: gives a predicate satisfied by all attainable values of the state parameter. The keyword gives the right-hand side of an equation defining the predicate as a function of the **??StateVar**: variable. The IPSL translator takes this predicate to be **T** (i.e., always satisfied), if an IPSL specification does not give it.

If a process is parameterized, its IPSL specification must include the keyword **??Projection**:. This keyword gives the right-hand side of an equation defining the process' projection function as a function of the **??LevelVar**: and **??StateVar**: variables.

Every port specification must include a **??LevelFun**: keyword that gives the right-hand side of an equation defining the security level of a message entering or leaving the port as a function of the **??MessageVar**: variables. It must also include a **??LevelRange**: keyword giving lower and upper bounds on the security levels of the messages passing through the port.

For composite processes, the keyword **??ProcessInFile**: gives the basenames, with the extension .ipsl, of files containing IPSL specifications of child processes. In typical Romulus uses, in which composite processes are created with the Romulus graphical interface, only the graphical interface generates this keyword. Composite process specifications also typically use several instances of the keyword **??Connection**: to identify connections between ports on the composite process and ports on its immediate subprocesses or connections between the ports of immediate subprocesses. In typical cases, this keyword is also generated only by the Romulus graphical interface.

The translator creates HOL90 specification files for atomic and composite processes. It creates **_globals** specification files for composite processes having **??HOL_functions**: entries. The specification file giving the theory of an immediate subprocess makes the **_globals** theory of the composite process, if any, into a parent theory. The specification file giving the theory of a composite process makes the theories of each of the immediate subprocesses into parent theories.

The IPSL translator uses **new_recursive_definition** and the values from the **??LevelVar**:, **??LevelFun**:, and **??LevelRange**: keywords to define functions that assign security levels to input and output events and to define predicates true of input and output events precisely when their levels fall inside the level range of the port where the event arrives or leaves.

See [15] for a complete definition of IPSL, its syntax, and the conditions that must be true for a translation of an IPSL specification into HOL90 to succeed.

3.2 Security Goals

The IPSL translator produces three different forms of security goals. Descriptions of these goals follow.

For a fully specified server process, the goal is a condition guaranteeing restrictiveness — BPSP_restrictive or BNPSP_restrictive for parameterized or non-parameterized processes, respectively. (BPSP stands for "buffered, parameterized server process" and BNPSP stands for "buffered, non-parameterized server process.") These predicates are essentially the same as the corresponding predicates described in [5], with the following exceptions.

All conditions are restated to be conditional on having input events satisfy an arbitrary predicate, a predicate that can be instantiated to assert that the levels of these events are in their appropriate ranges. The conditions on output events are also strengthened to assert that all output events satisfy an arbitrary predicate, a predicate that can be instantiated to assert that the levels of these events are in their appropriate ranges. Note that these input and output predicates can also be instantiated by stronger conditions if the Romulus community ever finds that desirable.

Note that imposing arbitrary conditions on input and output events does not affect the relationship between BPSP_restrictive or BNPSP_restrictive and general restrictiveness. The general theory of restrictiveness is defined for arbitrary *sets* of input and output events [9, 11, 10]; the _restrictive predicates guarantee restrictiveness for inputs and outputs given by those members of arbitrary HOL types that satisfy arbitrary predicates.

BPSP_restrictive includes a new BPSP_samepath condition. This condition checks that if the projections of two state parameters to a level are equal, if this level dominates the level of an input event, and if the process' responses to this input for these state parameters make the same nondeterministic choices, then the choices made in evaluating conditional expressions by these responses are also the same.

Unlike the other BPSP_restrictive subpredicates, BPSP_samepath is not a purely behavioral property. It eliminates some processes that are intuitively insecure on the basis of *probabilistic* reasoning rather than strict deduction. As an example, a process that nondeterministically chooses to give an unclassified user either a random sequence of characters or a top-secret file cannot be *proven* to have ever behaved insecurely, though it is obviously insecure. The BPSP_samepath condition fails for such processes. This extra condition removes the only significant discrepancy between the security conditions taken as being sufficient to guarantee restrictiveness in [5] and those identified by Rosenthal [17, 18].

A proof of the security goal for the example introduced in Section 2 follows. The goal file produced by the IPSL translator contains comments giving the first step in this proof and four lines, not given here, creating a file for communicating the proof result to the Romulus graphical interface.

```
e(BPSP_restrictive_TAC);


e(BPSP_nowritesdown_TAC THEN
  ASM_REWRITE_TAC [definition "romlemmas" "standard_dom"]);


e(BPSP_nolowchange_TAC THEN
  UNDISCH_TAC
  (--'~(standard_dom
        level
        ((ship > 99) => secret | unclassified))'--) THEN
  ASM_CASES_TAC (--'ship > 99'--) THEN
  ASM_REWRITE_TAC
  [theorem "romlemmas" "standard_dom_unclassified"] THEN
  DISCH_THEN (fn th => REWRITE_TAC[th]));


e(BPSP_samepath_TAC);


e(BPSP_lowresponsesame_TAC THEN
  UNDISCH_TAC
  (--'((standard_dom level secret) =>
        param:num#num | (FST param,0)) =
     ((standard_dom level secret) =>
        param' | (FST param',0))'--) THEN
  ASM_CASES_TAC (--'standard_dom level secret'--) THEN
  ASM_REWRITE_TAC[] THEN
  DISCH_THEN (fn th => ASSUME_TAC th THEN REWRITE_TAC[th])
  THENL
  [ASM_CASES_TAC (--'ship > 99'--),
    ASM_CASES_TAC (--'user = 0'--)] THEN
  ASM_REWRITE_TAC[] THEN
  SUBGOAL_THEN
  (--'FST (param:num#num) = FST (param':num#num)'--)
  (fn th => REWRITE_TAC[th]) THEN
  IMP_RES_TAC (theorem "pair" "PAIR_EQ"));
```

The proof shows the following: each **answer** output in response to a **query** input is at a security level dominating the level of the input; the projections to a level of the state parameters before and after a **report** input not visible at that level are equal; the "same path" condition holds; and state parameters whose projections to a level are equal are not distinguished by the responses visible at that level, either in **answer** outputs or projected new state parameters, to inputs not visible at that level. All of the tactics used are described in [15].

For a partially specified server process, the goal is a *manifest security* goal, a goal asserting that a process is secure purely by virtue of its environment. A manifest security goal has the form

```
'!outev inev.
 ((inpred inev) ==>
  (dom (outlev outev) (inlev inev))) /\
 (outpred outev)'
```

where the variables `inpred`, `outpred`, `inlev`, `outlev`, and `dom` are instantiated by predicates on input and output events, functions assigning security levels to input and output events, and a dominance relation on security levels. The goal asserts that every possible output is at a level that dominates the level of every possible input, and every possible output satisfies whatever condition is asserted by the predicate on output events.

For a composite process, the goal is a *hookup validity* goal. This goal is a conjunction of assertions of the following form, one for each connection between the composite process and one of its immediate subprocesses or between two of its immediate subprocesses.

```
(! <variables in a message from sending port>.
 (<sending port predicate on events>
  (<sending port name>
   <variables in sending port message>)) ==>
 ((<receiving port predicate on events>
   (<receiving port name>
    <variables in receiving port message>
   )) /\
  (<level assigned to sent event> =
   <level assigned to received event)))
```

Note that the level ranges on the different ends of a connection between ports do not have to be equal. Romulus treats a level range as an assumption about (for inputs) or a guarantee to (for outputs) a process' environment. There is nothing wrong with having a weaker assumption about a port's inputs than a condition that is actually guaranteed for another connected port's outputs. Also, for connections between processes and their immediate subprocesses, the sending port can be an input port and the receiving port can be an output port. As before, there is nothing wrong with weakening an input assumption when going from a process to one of its subprocesses or with weakening an output guarantee when going from a subprocess to the full process.

3.3   Graphical Interface Conveniences

The Romulus graphical interface is a tool for preparing process specifications, determining which processes are possibly insecure in their assumed environments, translating process specifications into HOL90, preparing formal statements of goals that must be proved to show processes are secure, and checking whether all necessary security goals have in fact been proved. The interface allows the user to do the following: create pictorial representations of processes, ports, connections between ports, and level ranges associated with ports; produce hard copies of these pictures; and complete these pictorial representations into full IPSL specifications. The interface uses IPSL as its basic language for storing information about processes, though it also uses additional files to contain the numbers giving the sizes, shapes, and positions of icons representing processes and ports.

The Romulus graphical interface also provides basic error checking on process specifications as they are being created. It checks that security levels are always chosen from a pre-defined, user-supplied (or default) set of security levels and that level ranges are always non-empty. It also checks that the level range on the receiving end of a connection between ports always includes the level range on the sending end of this connection.

For much, much more information on the Romulus graphical interface, see [15].

4 Conclusions

Although Romulus' developers have had limited opportunity to expose unsophisticated HOL users to the Romulus tools and techniques for specifying processes and their security properties, students in a Romulus course had no difficulty in understanding IPSL or the steps needed to translate IPSL specifications into HOL. This experience indicates that a simple interface language and translator can make tricky, sophisticated, and burdensome HOL specifications available to such users. These tools and techniques could also apparently be easily adapted to help in any areas dealing with processes specifications.

For Romulus, a main point of interest for future work is investigating the use of HOL's abstract theories to make proofs of restrictiveness for generic processes easily reusable in proofs of restrictiveness for composite processes containing instances of these generic processes. Possible future work of theoretical interest also includes producing full formal HOL definitions of composite processes, creating HOL proofs that conditions such as BPSP_restrictive do guarantee restrictiveness, and creating HOL proofs that the composition of properly connected restrictive processes is restrictive.

References

1. J. Alves-Foss and K. Levitt. A Model of Event Systems in Higher Order Logic: Sequence and Event System Theories. Technical Report CSE-90-45, Division of Computer Science, University of California, Davis, November 1990.
2. J. Alves-Foss and K. Levitt. A Security Property in Higher Order Logic: Restrictiveness and Hook-Up Theories. Technical Report CSE-90-46, Division of Computer Science, University of California, Davis, December 1990.

3. J. Alves-Foss and K. Levitt. Mechanical Verification of Secure Distributed Systems in Higher Order Logic. In *Proceedings of the 1991 International Meeting on Higher Order Logic Theorem Proving and its Applications*, pages 263–278, 1991.

4. J. Alves-Foss and K. Levitt. Verification of Secure Distributed Systems in Higher Order Logic: A Modular Approach Using Generic Components. In *Proceedings of the Symposium on Security and Privacy*, pages 122–135, Oakland, CA, 1991. IEEE.

5. S. Brackin and S-K Chin. Server-process restrictiveness in HOL. In *HOL User's Group Workshop*, Vancouver, Canada, August 1993. Springer Verlag.

6. Department of Defense. *Trusted Computer System Evaluation Criteria*, December 1985. DoD-5200.28-STD.

7. Joseph A. Goguen and José Meseguer. Security policy and security models. In *Proceedings of the Symposium on Security and Privacy*, pages 11–20, Oakland, CA, April 1982. IEEE.

8. C. A. R. Hoare. *Communicating Sequential Processes*. Series in Computer Science. Prentice-Hall International, Englewood Cliffs, NJ, 1985.

9. D. McCullough. Specifications for multilevel security and a hook-up property. In *Proceedings of the Symposium on Security and Privacy*, pages 161–166, Oakland, CA, April 1987. IEEE.

10. D. McCullough. Foundations of Ulysses: The theory of security. Technical Report RADC-TR-87-222, Rome Air Development Center, May 1988.

11. D. McCullough. Noninterference and the composability of security properties. In *Proceedings of the Symposium on Security and Privacy*, pages 177–186, Oakland, CA, April 1988. IEEE.

12. R. Milner. Communication and Concurrency. Prentice-Hall, NY, 1990.

13. ORA. *Romulus Theories*. Technical Report TM-94-0016, Odyssey Research Associates, Ithaca, NY, March 1994.

14. ORA. *Romulus Library of Models*. Technical Report TM-94-0017, Odyssey Research Associates, Ithaca, NY, March 1994.

15. ORA. *Romulus User's Manual*. Technical Report TM-94-0018, Odyssey Research Associates, Ithaca, NY, March 1994.

16. ORA. *Romulus Overview*. Technical Report TM-94-0019, Odyssey Research Associates, Ithaca, NY, March 1994.

17. D. Rosenthal. An approach to increasing the automation of the verification of security. In *Proceedings of Computer Security Foundations Workshop*, pages 90–97, Franconia, NH, June 1988. The MITRE Corporation, M88-37.

18. D. Rosenthal. Implementing a verification methodology for McCullough security. In *Proceedings of Computer Security Foundations Workshop II*, pages 133–140, Franconia, NH, June 1989. IEEE Computer Society Press.

Highlighting the Lambda-free Fragment of Automath

N.G. de Bruijn

Eindhoven University of Technology
Eindhoven, The Netherlands

Abstract. PAL is the fragmant of Automath that does not use lambda calculus. It just deals with typed definitions in typed contexts. Its powers and weaknesses are discussed here.

1 Introduction

The ideas of the Automath project started to evolve around the end of 1966 and matured during 1967 and 1968. They made it possible to start a large scale project directed to the verification of mathematics. The soundness of the basic ideas was testified by the fact that it was never necessary to revise the system during the period (running to about 1976) that it dealt with the verification of a considerable amount of mathematical material.

The system was proved to be feasible, in particular in the sense that the amount of time (for humans or machines) and computer memory never showed the slightest indication of running into exponential explosion. Looking back, one should take into account that all this was done by means of the hardware and software technology of the early seventies. What could be called "feasible" with that 1970 technology, can be called "easy" with the present one, of course.

Many of those ideas formed around 1967 Automath were adopted by others, or developed independently by others, and seem to be quite natural today in the world of proof checking systems. In particular this holds for the systems that handle typed lambda calculi and base themselves on the principle of proofs-as-objects. But in 1967 these ideas were quite uncommon or even weird for mathematicians as well as for logicians and computer scientists.

Now, more than 25 years later, it might have some interest to look back in order to see where those ideas came from. As far as historical and philosophical aspects are concerned, this was reported in [12], but the present paper will be slightly more technical. It mainly concentrates on PAL, what can be called the lambda-free fragment of Automath (its name expresses "Primitive Automath Language").

PAL is a mathematical language by itself. All the essentials of the game of mathematics can already be expressed in terms of that fragment. In particular the idea of proofs-as-objects was developed in it. And one can use PAL also for explaining things beyond mathematical objects and proofs. One may think of the description of geometrical constructions, along with their correctness proofs (see [7]).

Comparing Automath to other verification systems based on typed lambda calculus, one can say it is typical for Automath that this fragment PAL was not *replaced* by lambda calculus but rather *enriched* by it. This has the effect that Automath supports a double system for expressing functionality. In a way one can say that it corresponds to the way mathematicians still think. Their idea about functionality seems to be mainly the one of PAL, and lambda calculus is felt as a kind of extra for which proper modes of speech have not yet crystallized in the standard spoken and written mathematical language.

It must be said that upon *replacing* PAL by lambda calculus, the internalization of definitions, such a great thing to have in PAL, cannot be properly maintained, unless one modifies the typed lambda calculi in a way that supports internal definitions (see Section 5). But even in spite of that escape possibility, the author thinks that the structure of PAL should be the core of any verification system. Therefore this paper explains PAL in quite some detail, albeit not very formally.

General references. For information about the Automath project in general the reader might consult [4,6,10,16], for more detailed language information [2,3,16] for language theory to [14].

Much of this material will soon be easier accessible in [1], containing a selection of about 1000 pages from the Automath literature.

2 The structure of PAL

First it should be mentioned that there is an even more primitive language SEMIPAL (see [3]), but we do not separately discuss this here. It suffices to mention that SEMIPAL is obtained from PAL by just omitting the types, both for variables and for expressions, and that it already contains the notion of definitional equivalence handled in PAL.

SEMIPAL can still express new mathematical functions in terms of old ones, but it is unable to represent mathematical reasoning the way PAL can.

2.1 Syntax of expressions in PAL

The expressions in PAL have a syntax that is more or less standard in mathematics. We start from a set of *identifiers*, for which no separate syntactic description will be given here. One might take any string of letters (roman or greek, upper or lower case) and digits. Separation marks inside those strings are not allowed (they might give trouble with parsing). And the words PN, **type**, **prop** are forbidden as identifiers.

Examples of identifiers:

$$x, \ \theta, \ y1, \ sin, \ XYZ.$$

The syntax of expressions in PAL can now be given by means of the following

BNF grammar:

$$< \text{expression} > ::= < \text{identifier} > \mid < \text{identifier} > (< \text{expression string} >)$$

$$< \text{expression string} > ::= < \text{expression} > \mid < \text{expression} >, < \text{expression string} >$$

Examples of expressions: with the identifiers $f, g, x, x3$ one can build in PAL the expressions

$$x, \quad f(g(x3, f(x, g)), g(x)), \quad f(g(f, f)).$$

2.2 Books written in PAL

The idea of a *book* in Automath is something that is to be interpreted as the complete description of a mathematical theory, or even of the whole of mathematics. The designer of the language should not be held responsible for what is written in the book, that is up to the users of the system. The designer should see to it that the users can express everything they want.

The structure of a PAL book will not be described formally, but more or less intuitively by means of the flag-and-flagstaff style shown in figure 1.

A *book* is a sequence of *lines*. There are three kinds of lines:

(i) *Context lines*,
(ii) *Definitional lines*,
(iii) *Primitive lines*.

Lines of type (i) are used for the introduction of a context, and for lines of the type (ii) or (iii) it has to be made clear in what context they are written.

The Ω's are metavariables representing expressions. That is to say, figure 1 becomes a real book only after those Ω's have been replaced by expressions.

The numbers in parentheses on the left do not belong to the official text, but are used to facilitate the present explanation.

The context lines (lines (2), (5), (7), (11), (12), (15), (17), (18)) have a text written on a flag. The text has one of the forms $< \text{identifier} >:< \text{expression} >$ $< \text{identifier} > : \textbf{type}, < \text{identifier} > *$. The identifiers mentioned here are called *variables*. The definitional lines have the form

$$< \text{identifier} > := < \text{expression} > : < \text{expression} >,$$

and the primitive lines

$$< \text{identifier} > := \text{ PN } : < \text{expression} > .$$

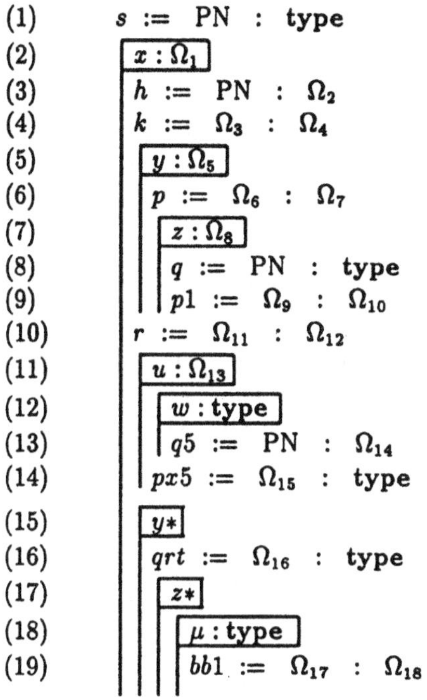

$$(1) \quad s := \text{PN} \ : \ \textbf{type}$$

(2) $\boxed{x : \Omega_1}$

(3) $h := \text{PN} \ : \ \Omega_2$

(4) $k := \Omega_3 \ : \ \Omega_4$

(5) $\boxed{y : \Omega_5}$

(6) $p := \Omega_6 \ : \ \Omega_7$

(7) $\boxed{z : \Omega_8}$

(8) $q := \text{PN} \ : \ \textbf{type}$

(9) $p1 := \Omega_9 \ : \ \Omega_{10}$

(10) $r := \Omega_{11} \ : \ \Omega_{12}$

(11) $\boxed{u : \Omega_{13}}$

(12) $\boxed{w : \textbf{type}}$

(13) $q5 := \text{PN} \ : \ \Omega_{14}$

(14) $px5 := \Omega_{15} \ : \ \textbf{type}$

(15) $\boxed{y*}$

(16) $qrt := \Omega_{16} \ : \ \textbf{type}$

(17) $\boxed{z*}$

(18) $\boxed{\mu : \textbf{type}}$

(19) $bb1 := \Omega_{17} \ : \ \Omega_{18}$

Figure 1. A PAL-book in flag-and-flagstaff style (Ω 's stand for expressions).

Contexts. The *context* of a line is derived from the flags and flagstaffs in the obvious way. Lines (3) and (4) are in the context of the variable x, line (6) in the context of x, y, lines (8) and (9) in the context of x, y, z, etc. The context of line (1) is empty.

The context of a context line does *not* include the variable of that line itself. Example: the context of line (7) is x, y.

It is possible to *reopen* a context that had been closed before. This is indicated by flags with an asterisk. The $y*$ on the flag in line (15) gets us back in the context that was created by the y in line (5). So line (16) has the same context as line (6), and line (19) has the context x, y, z, μ.

Further interpretation. The colon is to be read as "is a". So the $x : \Omega_1$ in line (2) is read as "x is an Ω_1". In other words, Ω_1 is the type of x. But the $\Omega_{16} : \textbf{type}$ in line (16) says that Ω_{16} is a type itself.

The combination $:=$ is read as "is defined by", unless it appears in front of the word PN. So line (6) says that p is defined by Ω_6. This Ω_6 has the type Ω_7, and accordingly we also say that p has the type Ω_7.

In line (6) the new symbol p is introduced, and it is given a "meaning", viz. Ω_6. But line (3) is different. The "PN" stands for "primitive notion", and line

(3) says that h is not defined, but that it will nevertheless be considered to have a fixed meaning. And it will be something of type Ω_2.

Lines (1) and (8) introduce primitives too, but now these primitives are types.

Typed context notation. A context is not just a sequence of variables, but of *typed* variables. We express this by saying, for example, that line (19) has the *typed context*

$$[x : \Omega_1][y : \Omega_5][z : \Omega_8][\mu : \textbf{type}].$$

In an obvious way this is called a context of *length* 4. The parts like $[x : \Omega_1]$ are called *abstractors*, and such a sequence of abstractors is called a *telescope*.

Dependency. We say that the things written to the left of the sign $:=$ *depend* on the sequence of varables of the context. So the q of line (8) depends on x, y, z, and $bb1$ of line (19) depends on x, y, z, μ. Lines (8), (14), (16) give examples of types depending on variables, so-called *dependent types*. Line (8) gives a primitive dependent type.

Degrees. We say that **type** has degree 1. If $\Omega : \textbf{type}$ we say that Ω has degree 2. And if $\Omega_a : \Omega_b$ then the degree of Ω_a is 1 plus the degree of Ω_b.

In practice we only use degrees 1,2,3, but there is no technical reason for such a restriction.

All identifiers of the book distinct. We can make a list of the identifiers of the book: the variables and the identifiers written to the left of the sign $:=$. Variables introduced with an asterisk are not counted. So in the case of figure 1 the identifiers are $s, x, h, k, y, p, z, q, p1, r, u, w, q5, px5, qrt, \mu, bb1$.

We shall require that the list of identifiers of the book is free of repetitions. For the long books we have in practice this is an awkward restriction, so something has to be done about it. But in our present discussion of PAL all identifiers are assumed to be different.

2.3 Validity of books

Validity of a PAL book is described in two rounds. The first round is a matter of well-formedness of expressions relative to the lines in which they are used, but correctness of typings does not come up yet. That first round will be referred to as "well-formedness".

In the second round, to be called "type-correctness", the typings will be taken into account.

Well-formedness of expressions. This is about acceptability of expressions with respect to a particular place in the book. It does not matter whether such expression actually occur in the book or not.

The notion will be introduced recursively.

Let Ω be an expression and let (j) be a line of the book. Ω has the form $\xi(\Psi_1, \ldots, \Psi_k)$ (the case that Ω is an identifier will be treated as the case $k = 0$). The ξ is called the *head* and $\Psi_1, \ldots, \Psi_k$ are called the (first order) *subexpressions*.

We say that Ω is well-formed with respect to line (j) if the following conditions are satisfied:

(i) ξ is either one of the variables of the context of line (j) or it is an *old constant*, i.e., the identifier appearing in front of the := in one of the previous lines (definitional or primitive).

(ii) If ξ is a variable of the context of line (j) then $k = 0$. If ξ is the identifier of a previous definitional or primitive line, then k is equal to the length of the context of that line.

(iii) The subexpressions $\Psi_1, \ldots, \Psi_k$ are well-formed with respect to line (j).

Instantiation. The syntactic process of adding a string of subexpressions to an old constant, i.e., passing from ξ to $\xi(\Psi_1, \ldots, \Psi_k)$, is called *instantiation*. It can be interpreted as supplying expressions for the variables of the context of that old constant.

An abbreviational convention. Some violations of the above rule (ii) can be admitted. If an old constant had been introduced in a context $x_1, \ldots, x_h$, and if $k < h$, then we can still allow $\xi(\Psi_1, \ldots, \Psi_k)$ but then we have to interpret it as $\xi(x_1, \ldots, x_{h-k}, \Psi_1, \ldots, \Psi_k)$. According to rule (i) we then have to require that $x_1, \ldots, x_{h-k}$ are the first $h - k$ variables of the context of line (j).

In particular an old constant can be used without subexpressions as long as our present context is the same as the one of that old constant, or an extension of that context.

This convention is very practical, but by no means essential for understanding the rules of PAL.

Well-formedness of a book. A book is called well-formed if, apart from what has already been said in Section 2.2 about the structure of lines (including all identifiers of the book being distinct), all the expressions of the book are well-formed with respect to the line in which they occur.

So for the book of figure 1 this means that Ω_1 is well-formed with respect to line (2), Ω_2 with respect to line (3), Ω_3 and Ω_4 with respect to line (4), etc.

Delta reduction. Assume that thebook is well-formed. Let some expression $\xi(\Psi_1, \ldots, \Psi_k)$ be well-formed with respect to line (j), of the book, let $k > 0$ and let ξ be the identifier of a definitional line (i). Let that line read $\xi := \Omega_a : \Omega_b$, or $\xi := \Omega_a : \textbf{type}$, and let $x_1, \ldots, x_k$ be its context. The expression Ω_a may contain these variables. Simultaneously substituting $\Psi_1, \ldots, \Psi_k$ for $x_1, \ldots, x_k$, respectively, we get an expression Ω_c.

We use the term "simultaneous substitution" in order to express that only the x's occurring in Ω_b itself are to be replaced and not the x's that are introduced during the substitution process. We have to be aware of the fact that $\Psi_1, \ldots, \Psi_k$ may contain such x's.

Replacing $\xi(\Psi_1, \ldots, \Psi_k)$ by this Ω_c is called *head delta reduction*.

The $\Psi_1, \ldots, \Psi_k$ can be called first-order subexpressions of $\xi(\Psi_1, \ldots, \Psi_k)$, their first order subexpressions can be called second order subexpressions of $\xi(\Psi_1, \ldots, \Psi_k)$, and thus we may get subexpressions of any order.

We can now formulate what delta reduction is: apply head delta reduction either to the expression itself or to any of its subexpressions. Applying reduction to a subexpression means of course that the rest of the expression is left unchanged: If Ψ_2 reduces to Σ then $\xi(\Psi_1, \Psi_2, \Psi_3)$ reduces to $\xi(\Psi_1, \Sigma, \Psi_3)$.

If an expression Ω_a is well-formed with respect to line (j), and if Ω_a turns into Ω_b by delta reduction, then Ω_b is again well-formed with respect to line (j).

Definitional equality. Two expressions, Ω_p and Ω_q are called *definitionally equal* if there is an expression Ω_r such that both Ω_p and Ω_q can be reduced to Ω_r by means of a sequence of delta reductions.

Definitional equality is decidable, since strong normalization can be proved for delta reduction (cf.[14]). Normal forms are, of course, expressions that contain no defined constants: only primitive constants and variables.

Type evaluation. Consider a well-formed book. Let S be the set of all expressions which are well-formed with respect to at least one line of the book. We shall define a mapping, to be called "typ", that attaches to any element of S either an element of S or the word **type**. The mapping is defined recursively.

(i) If θ is a variable, introduced on its flag by $\theta : \Gamma$, we define $\mathrm{typ}(\xi) = \Gamma$. If its flag says $\theta : \textbf{type}$ we take $\mathrm{typ}(\xi) = \textbf{type}$.

(ii) If $\Omega = \xi(\Psi_1, \ldots, \Psi_k)$ where ξ is a constant, we act as follows. ξ is the identifier of a line (i). If that line reads $\xi := \Omega_a : \textbf{type}$ or $\xi := \mathrm{PN} : \textbf{type}$, we take $\mathrm{typ}(\Omega) = \textbf{type}$. If it reads $\xi := \Omega_a : \Omega_b$ or $\xi := \mathrm{PN} : \Omega_c$, then we take $\mathrm{typ}(\Omega) = \Omega_c$, where Ω_c is obtained from Ω_b by simultaneous substitution of $\Psi_1, \ldots, \Psi_k$ for the context variables $x_1, \ldots, x_k$ of line (i).

Type-correct expressions. We assume that our book is well-formed, and want to define type-correctness of an expression Ω with respect to a line (j). It is assumed that this expression is at least well-formed with respect to that line. The notion will be given recursively. As before, we consider two cases.

(i) If Ω is a variable θ, then it is type-correct.

(ii) Let $\Omega = \xi(\Psi_1, \ldots, \Psi_k)$ where ξ is a constant. So ξ is the identifier of a line (i). Let

$$[x_1 : \Theta_1] \cdots [x_k : \Theta_k]$$

be the typed context of that line. The Θ's can be either expressions or the word **type**.

Since dependent types are possible, the Θ's may contain $x_1, \ldots, x_k$ (although, according to the well-formedness, no Θ_h can contain $x_h, \ldots, x_k$).

For for $h = 1, \ldots, k$, let Γ_h be what we get if in Θ_h we simultaneously replace each x_1 by Ψ_1, each x_2 by Ψ_2, etc. We can now finally express the condition of type-correctness for Ω: for each h $(h = 1, \ldots, k)$ $\mathrm{typ}(\Psi_h)$ should be definitionally equivalent to Γ_h.

Type-correctness of a well-formed book. A book is called type-correct if all expressions are type-correct with respect to the lines in which they occur, and all its definitional lines are correctly typed.

The definitional line $\xi := \Omega_a : \textbf{type}$ is called correctly typed if $\text{typ}\Omega_a = \textbf{type}$. The definitional line $\xi := \Omega_a : \Omega_b$ is called correctly typed if $\text{typ}(\Omega_a)$ is definitionally equal to Ω_b.

Since validity is well-formedness plus type-correctness, this completes the definition of the notion of a valid PAL book.

3   Proofs as objects

Expressing functionality in PAL. Before we enter into the representation of reasoning in PAL it should be made clear how functions can be defined and how these definitions can be used later.

Let us consider line (4) in figure 1. According to the rules of the language (assuming the validity of the book), we can have, in some later line, an expression $k(\Sigma)$, where $\text{typ}(\Sigma)$ is definitionally equivalent to Ω_1. This $k(\Sigma)$ is definitionally equivalent to what we get if in Ω_3 we replace each x by Σ. So we can interpret line (4) as the definition of a *function*. We cannot say that k is the *name* of the function, but nevertheless the value of the function for the argument x can be written (at least with respect to line (4)) as $k(x)$, which is definitionally equal to Ω_3. And later, the expression $k(\Sigma)$ is definitionally equivalent to what we expect to be the value of that function for the argument Σ. This is what we mean by saying that line (4) represents a function. One might say that it corresponds to the notion of function of the 18-th century, when functions were still metamathematical things, and no mathematical objects. The functions had no names, but there was a consistent notation for the values of the function.

What was described here for values of a single variable can be repeated for functions of several variables. So in line (9) of figure 1 we can see a function of three variables x, y, z.

A *function call* requires that we provide expressions to be substituted for these three variables.

Structure of theorems and their applications. The flag-and-flagstaff style can do more than building functions. It can also represent the mixture of building and proving that is the essence of the mathematics game. Essentially that is what is often called Fitch-style natural deduction with variables.

If we inspect in this setting how we write a theorem, and how we apply it later, we are more or less forced to discover the principle of "proofs as objects". We need no lambda calculus for that.

In general, a theorem has to be written in a mixed context. On some of the flags we introduce variables, on others we state assumptions.

As an example we take a theorem that makes a statement $T(x, y)$ in the context of a variable x of type P_1, an assumption $A_1(x)$, a variable y of type $P_2(x)$, and an assumption $A_2(x, y)$.

If a later application of this theorem has to be machine-checked, what information do we have to give?

First we have to supply something of type P_1 that can play the role of x, just as if we had been dealing with a function call. We give some expression X. The machine has to check its type. Next we have to convince the machine that $A_1(X)$, the "updated" form of the assumption $A_1(x)$ is satisfied. We have to supply some reference where a proof of this can be found, something like "formula (27)". The machine has to check that this is indeed a proof of $A_1(X)$.

Next we have to supply something for y, and it has to have as its type $P_2(X)$, updated form of $P_2(x)$. We supply Y. And finally we have to supply a reference to a proof of $A_2(X,Y)$, the updated form of $A_2(x,y)$. Such a reference might be "formula (24)". The machine has to check that it is a proof indeed. If it is satisfied, it accepts that there is a proof of $T(x,y)$.

Instead of the answer "formula (24)" the situation is often slightly more complicated. Instead of this reference, it might be a *composite reference*, like a reference to a formula (18) that indicated a statement $Z(u,v)$ given in a declarational context "let $u : U$, $v : V$". And the mathematician gives a pair of expressions K and L with $K : U$ and $L : V$. So instead of the reference "formula (24)" he might have given "formula (18), with u and v replaced by K and L".

Summarizing, the technical work that is expected from the machine is very similar to what has to be done in the case of a function call. Instead of supplying values only, we have to supply both values and proof references. In both cases we have the same updating machinery. And what the machine has to do in order to check that formula (27) is a sufficient argument for $A_1(X)$, has to be organized in the same way as checking that Y has the type $P_2(X)$. Therefore it is very natural to try to put the two things on a common basis, and that is exactly how the idea of "proofs as objects" was discovered in the earliest stages of the Automath project.

It is clear that a reference like "formula (27)" is not a reference to a proposition but to a place where that proposition is proved. One might say that this proof reference is a *name* for a proof. It is of the same kind as a name for a mathematical object and is manipulated similarly. So we get something that was never done in mathematics before: references are not to the names of theorems but to names of their proofs, or to composite expressions that represent such proofs. And similarly, assumptions have to get names that have to be used in expressions. The part of the book where an assumption is supposed to hold is indicated by a flagpole, just like we do for the range of validity of a variable. So instead of "assume $A_1(x)$" we have to think in terms of "let u be a proof for $A_1(x)$" and to see this as a typing, where the type of u is "proof for $A_1(x)$".

So theorems, and lemmas, and all parts of proofs of theorems where some step is proved, get the form of definitional lines of the form "$p := \Sigma$: proof of Γ". And axioms are to be expressed in primitive lines "$p := \text{PN} : \text{proof of } \Gamma$". An axiom is like a theorem, with the difference that we do not *have* a proof, but just *act* as if we had one.

A more extensive account of how functionality in PAL leads to "proofs as objects", and how it has possibilities for other areas as well, can be found in [13].

An example. Figure 2 pesents a little theorem (probably due to Lindenbaum) written in PAL expressing that instead of the usual three axioms for an equivalence relation (reflexivity, symmetry and transitivity) two will do: reflexivity (line (5)) plus an axiom (line (10)) that $x \equiv y$ and $z \equiv y$ together imply $z \equiv x$. From those two axioms first symmetry (line (11)) is proved, and then, after the intermediate steps (14) and (15), transitivity in line (16). The assumptions which are active in line (16) are $x \equiv y$ (line (7)) and $y \equiv z$ (line (13)). Line (16) proves $x \equiv z$.

In line (17) we have a different proof, again based on line (14), but not using line (15). Lines (18) and (20) are proofs obtained from the proof in (16) by delta reductions, and so are (19) and (21) obtained from (17). Actually line (20) can be considered as a condensed one-line proof; it would remain correct if lines (14) to (19) would be skipped. The same thing holds for the one-line proof (21). Since a and b are primitives, the proofs of (20) and (21) are in normal form. It follows that the proofs (16) and (17) are essentially different, in the sense that they are *not* definitionally equal.

(1)	$\boxed{\alpha : \mathbf{type}}$
(2)	$\boxed{x : \alpha}$
(3)	$\boxed{y : \alpha}$
(4)	$EQ := \mathrm{PN} : \mathbf{prop}$
(5)	$a := \mathrm{PN} : EQ(x, x)$
(6)	$\boxed{y*}$
(7)	$\boxed{u \ : \ EQ(x, y)}$
(8)	$\boxed{z : \alpha}$
(9)	$\boxed{v \ : \ EQ(z, y)}$
(10)	$b := \mathrm{PN} \ : \ EQ(z, x)$
(11)	$c := b(y, a(y)) \ : \ EQ(y, x)$
(12)	$\boxed{z*}$
(13)	$\boxed{w \ : \ EQ(y, z)}$
(14)	$d := c(y, z, w) \ : \ EQ(z, y)$
(15)	$e := b(d) \ : \ EQ(z, x)$
(16)	$f := c(z, x, e) \ : \ EQ(x, z)$
(17)	$g := b(z, y, d, x, u) \ : \ EQ(x, z)$
(18)	$h := c(z, x, b(c(y, z, w))) \ : \ EQ(x, z)$
(19)	$k := b(z, y, c(y, z, w), x, u) \ : \ EQ(x, z)$
(20)	$l := b(z, x, b(b(y, z, w, z, a(z))), x, a) \ : \ EQ(x, z)$
(21)	$m := b(z, y, b(y, z, w, z, a(z)), x, u) \ : \ EQ(x, z)$

Figure 2. A PAL-book expressing a theorem with various proofs.

The example has something that had not been mentioned before: it has a second thing of degree 1, the word **prop** in line (4). This is used in many Automath-like languages in order to maintain at least *some* distinction between the object world and the proof world. Nevertheless **prop** works in exactly the same way as **type**.

4   An evaluation of PAL

This section will discuss some of the strong points and some of the weak points of PAL.

It is a natural mathematical language. We can use PAL such that it closely follows the way mathematicians organize what they write and speak. But it is more than that. The relatively simple language rules set the standards for mathematical rigor. On the metalevel it is easy to explain what is a variable, a definition, a theorem, an axiom, a logical derivation rule, a proposition, a proof. We need not warn against circular reasoning or circular definitions: it would just violate the language rules to try such things. And the user of the language is forced to be explicit and precise about the context at any moment.

Internalization of logic. Part of what a mathematician would calls his logic, is nothing but obeying the PAL language rules. But there may be other logical things which we call *derivation rules*. Some of these rules might be taken as primitives and written by means of PN-lines (like the double negation rule), other derivation rules may be derived from them. And such derived derivation rules have exactly the same form as mathematical theorems. Later they can be applied just as mathematical theorems can be applied. We get the feeling that mathematics and logic are being built up together and that we finally got rid of the chicken-and-egg problem.

It can do other things too. The idea of typing can be used for more things than for saying that an object is a well-described thing of a specified type, or saying that a particular construction is a proof for a certain statement. A traditional third possibility lies in the world of Greek constructive geometry, where constructions can be described and their relation to mathematical objects can be studied and provided with proofs. Then we have three worlds we discuss simultaneously with a single substitutional mechanism, with three kinds of context flags, three kinds of definitional lines, three kinds of primitive lines, three kinds of usage of dependent types. More about this can be found in [7,9,13].

PAL may be more than the core of just a language for mathematics. It may be a language for science.

It gives a feeling of independence. Assuming that PAL is a new system, the usual reaction is that it should be established to be in accordance with existing systems. But this is unreasonable, at least if it has to be compared to systems that have not made their rules of operation explicit. Actually PAL is little more than the game of abbreviation and substitution, and it is difficult to see how other systems could live without such a thing.

We should not start from a given logical system and then say that what is written in PAL is in accordance with that system. It might be better to turn the tables around, and to take PAL as a standard format for scientific reasoning.

It is not PAL itself that is to be compared with other systems. What has to be compared is the material that the user writes in PAL, in particular the choice of the primitives. And one should not forget the role of the *interpretations*, i.e., the relations between the formal matter in the books and the more intuitive things in the minds of mathematicians. It is difficult to see how we can formally compare two logical frameworks with different modes of interpretation into a non-formal world.

Weakness of PAL. PAL is on the level of 18-th century mathematics. Untill well in the 19-th century the function *values* were objects, subject to mathematical operations, but the functions themselves were metalinguistic. One could write $\sin(x)$ and x^2, but sin and ^{2} were not considered as objects.

In PAL on can introduce as a primitive a type $M(A, B)$ of all functions from A to B. If f has that type, and if some p has type A, one can introduce something like $val(f, p)$ as a primitive, in order to talk about the function values. If we have such an f we can write later, in a context $[f : M(A, B)][x : A]$, a line $g := val(f, x) : B$, and then our function is expressed by g in the way of PAL functionality. But we can *not* do it the other way around. If in a context $[x : A]$ we have defined some identifier g by means of an expression (depending on x) of type B, and then leave that context, we are unable to indicate an element of $M(A, B)$ whose values are equal to those we get from that g by instantiation.

We have the same weakness in the world of logic: if in the context of an assumption p we have proved q, we are unable to get the implication $p \to q$ outside the context.

In both cases described here, we feel the need for a mechanism to express the *discharging* of a flag.

Replacing primitives by flags. This is a thing we often want, like replacing axioms by assumptions. We can do it in PAL only with axioms written in an empty context. The same thing can be said for primitive objects. If we have introduced, say, the type of natural numbers N in the beginning of our book as a primitive in the empty context, we can at once replace the book by another one that starts with a flag saying "let N be a type", with a flagpole running in front of the whole book. With PN's written in a non-empty context such a thing cannot be done, and yet there is a need for it.

Figure 2 gives an example, with the PN in line (4). If we have these 21 lines (or at least the first 16 of them) in a book, and if later in that book we have a candidate for the application of the Lindenbaum theorem, we cannot *apply* those earlier lines. There is no way to substitute into a PN.

This does not yet mean that the 16 lines were useless. We can use them as a *blueprint*. If later we have a type β and we give a relation on that type by means of an expression Ω, then we can *copy* the blueprint from line 2 onwards, replacing the word PN by that Ω, and replacing α by β throughout.

This technique of copying blueprints may be useful, but it is not what we expect from a mathematical language. Mathematics is the art of abbreviation, where we write things in full when they occur for the first time, but just want to flip our fingers if we need them again.

5 Extending PAL

The syntax of PAL does not have expressions with bound variables, but it is not hard to add such expressions to the syntax, playing around with a bit of lambda calculus inside PAL.

Once we have such lambda expressions, we can set rules for discharging flags, with the effect that the variables on the flags are replaced by bound variables in lambda expressions.

It is here that various kinds of type theory begin to diverge. There are several kinds of flags, and the discharge rules need not be the same for all of them. We have object-oriented and proof-oriented flags, but more important is the difference in degrees. Let us define the *degree of a flag* as the degree of the variable on that flag. So in figure 1 the flag in line (1) has degree 2, and the other flags have degree 3.

Choices made in Automath. In designing the Automath language AUT68 it was felt that, in spite of the fact that there were no *technical* reasons for restrictions, the only thing that could be interpreted in terms of the traditions of standard mathematics is the discharge of flags of degree 3. It was also decided that bound variables in lambda expressions should be limited to degree 3. So for variables of degree 2 the instantiation of PAL remained the only way to express functionality, but for variables of degree 3 Automath got *two* functionality mechanisms: instantiation as well as lambda application.

The use of unrestricted discharging. The practical restriction to degree 3 was felt at once as a toll paid to the habits of contempory mathematics. Having no such restrictions would make the system simpler to describe, and much more powerful. First, contexts would be no longer necessary: everything could be discharged into the empty context. After that, PN lines could be replaced by flags, and those new flags could be removed again by discharging. So we would get get a book consisting entirely of definitional lines in the empty context.

Now suppose we are only interested in the last line of the book (if our interest lies in any other line we just throw the rest of the book away so that this line becomes the last one). The previous lines can be made superfluous by sytematic delta reduction applied to the contents of the last line. This last line becomes a correct book in itself. Its identifier is irrelevant, so all what remains is a typing $\Omega : \Gamma$. These ideas were described at an early stage of Automath (see [5]).

This has the effect that the theory of the language becomes a theory of typed lambda calculus with lambda-structured types, for which R. Nederpelt [15] showed strong normalization.

It must be said that the action of removing all definitions is of theoretical interest only. In practice it gives the exponential explosion that mathematical language just wants to avoid. The same thing can be said about normal forms.

Why remove definitions? The process of removing definitions can be seen as β-reduction. In order to explain this, we take a book of two lines, both in empty context:

$$b := \Omega_1 : \Omega_2$$

$$c := \Omega_3 : \Omega_4$$

The second line may be a statement that contains b, and its validity may depend on the fact that b is defined as Ω_1. If b would have been a variable then the typing $\Omega_3 : \Omega_4$ might be incorrect, and in general Ω_3 cannot even be expected to be type-correct.

In the following we use the Automath notation for the application that leads to the value of a function F at the argument p. It is not written as $F(p)$ (this notation is used for instantiation) or as $F.p$ but as $\langle p \rangle F$.

Eliminating the definition of b means that the typing in the second line is replaced by

$$\langle \Omega_1 \rangle [b : \Omega_2] \Omega_3 \ : \langle \Omega_1 \rangle [b : \Omega_2] \Omega_4$$

Applying β-reduction to the pair $\langle \Omega_1 \rangle [b : \Omega_2]$ on both sides gives exactly what we mean by elimination of the definition of b. But there is trouble. Let us inspect the left-hand side $\langle \Omega_1 \rangle [b : \Omega_2] \Omega_3$. It has the form $\langle p \rangle F$ where F is $[b : \Omega_2] \Omega_3$. This F need not be type-correct, since it corresponds to having the last line with a variable b instead of the properly defined constant b. But usual typed lambda calculi require that, for an application $\langle p \rangle F$ to be correct, *both* parts p and F are correct.

In order to abolish this conflict one does not have to expel the definitions from the book. A much more attractive way is to revise the rules of typed lambda calculi. A very simple kind of typed lambda calculus that works without the requirement that both p and F have to be correct is $\Delta\Lambda$ (see [8,11]). Actually the definition of $\Delta\Lambda$ evolved from streamlining what an efficient (i.e., explosion avoiding) Automath checker has to do when checking a book with definitions.

The conclusion is that if one replaces one's typed lambda calculus by a modification into the direction of $\Delta\Lambda$, one will be able to maintain the internalized definitions structure that PAL has, and one might even tolerate local abbreviations, i.e. abbreviatons introduced inside an expression and used inside that same expression.

References

1. Selected Papers on Automath, edited by R.P. Nederpelt, J.H. Geuvers and R.C. de Vrijer, Studies in Logic, North-Holland Publishing Co., to appear 1994.

2. N.G. de Bruijn: Automath, a language for mathematics. Department of Mathematics, Eindhoven University of Technology, TH-report 68-WSK-05, 47 p., 1968. Reprinted in revised form, with two pages commentary, in: Automation and Reasoning, vol 2, Classical papers on computational logic 1967-1970, Springer Verlag 1983, pp. 159-200.

3. ———: The mathematical language Automath, its usage, and some of its extensions. Symposium on Automatic Demonstration (Versailles December 1968), Lecture Notes in Mathematics vol. 125, Springer Verlag 1970, pp. 29-61. To be reprinted in [1].

4. ———: Automath, a language for mathematics. Séminaire Math. Sup. 1971. Les Presses de l'Université de Montréal 1973, 58 p.

5. ———: AUT-SL, a single line version of Automath. Report, Department of Mathematics, Eindhoven University of Technology, 1971. To be reprinted in [1].

6. ———: A survey of the project Automath. In: To H.B. Curry: Essays in combinatory logic, lambda calculus and formalism, ed. J.P. Seldin and J.R. Hindley, Academic Press 1980, pp. 579-606.

7. ———: Formalization of constructivity in Automath. In: Papers dedicated to J.J. Seidel, ed. P.J. de Doelder, J. de Graaf and J.H. van Lint. EUT-Report 84-WSK-03, ISSN 0167-9708, Department of Mathematics and Computing Science, Eindhoven University of Technology, 1984, pp. 76-101. To be reprinted in [1].

8. ———: Generalizing Automath by means of a lambda-typed lambda calculus. In: Mathematical Logic and Theoretical Computer Science, Lecture Notes in pure and applied mathematics, 106, (ed. D. W. Kueker, E.G.K. Lopez-Escobar, C.H. Smith) pp. 71-92. Marcel Dekker, New York 1987.

9. ———: The use of justification systems for integrated semantics In: Colog-88, International Conference on Computer Logic Tallinn USSR, December 1988, Proceedings. Ed. P. artin-Löf and G. Mints. Lecture Notes in Computer Science Nr. 417, pp. 9-24. Springer-Verlag 1990.

10. ———: Checking mathematics with computer assistance. Notices American Mathematical Society, vol 8(1), Jan. 1991, pp 8-15.

11. ———: Algorithmic definition of lambda-typed lambda calculus. In: Logical Environments. Editors G. Huet and G. Plotkin. Cambridge University Press 1993, pp. 131-146.

12. ———: Reflections on Automath. In: Selected Papers on Automath, edited by R.P. Nederpelt, J.H. Geuvers and R.C. de Vrijer, Studies in Logic, North-Holland Publishing Co., to appear 1994.

13. ——: On the roles of types in mathematics. To be published.

14. D.T. van Daalen: The language theory of Automath. Ph.D. thesis, Eindhoven University of Technology, 1980. Parts of this thesis will be reprinted in [1].

15. R.P. Nederpelt. Strong normalization in a typed lambda calculus with lambda structured types. Ph.D. thesis, Eindhoven University of Technology, 1973. Parts of this thesis will be reprinted in [1].

16. L.S. van Benthem Jutting: Checking Landau's "Grundlagen" in the Automath system. Ph.D. Thesis, Eindhoven University of Technology, 1977. Mathematical Centre Tracts nr. 83, Amsterdam 1979. Parts of this thesis will be reprinted in [1].

First-Order Automation for Higher-Order-Logic Theorem Proving

Holger Busch

SIEMENS AG
Corporate Research and Development
Otto–Hahn–Ring 6, 81739 München, Germany.
E–Mail: busch@zfe.siemens.de

Abstract. In comparison with higher-order-logic proof checkers, automatic theorem provers for first-order logic typically require many fewer basic proof steps for proving comparable goals. Therefore it is vital to increase the automation of higher-order logic systems in order to really benefit from their particular advantages.

Two approaches for incorporating first-order proof automation into the system LAMBDA have been investigated. In the first one, the automatic first-order theorem prover SEDUCT has been integrated. More recent work has resulted in tactics implementing and extending first-order proof techniques directly in LAMBDA.

This paper describes and compares methods, implementation issues, results, and the scope of application of both approaches.

1 Introduction

The expressiveness of higher-order logic allows for more natural specifications than first-order logic, better reusability through parameterization and generalization, formally bridging various levels of abstraction, embedding specific calculi, support for transformational reasoning, and more [3, 6, 14, 1, 19].

Regarding the degree of automation and efficiency for simple routine tasks, which tend to form the majority of machine-supported proving, most higher-order-logic proof checkers perform worse than first-order provers. On the other hand, higher-order logic often allows more compact terms than an equivalent representation in first-order-logic. Furthermore, in practice theoretical decidability problems of higher-order logic play a very minor role. Much can and is being done for curing the inefficiency problems of higher-order-logic proof checkers.

One approach is to delegate appropriate proof goals to an external automatic verification tool. If that one is optimized for its specific proof task, it is likely to perform better than a prover with general-purpose reasoning mechanisms. Promising results have proved that this approach is able to automate a significant portion of the tedious parts of machine proofs [1, 18, 16]. An interface to the Boyer-Moore theorem prover has made available its powerful induction heuristics in HOL [4]. This first approach, however, inherently incurs the risk of flaws through unsafe exchange of proof information in addition to a non-negligible overhead.

A second approach is to make automatic proof procedures part of a monolithic higher-order-logic system, either using safe basic inference mechanisms only, or extending the prover kernel by more efficient direct routines. The system ISABELLE [20] provides first-order sequent calculus. The proof checker PVS [22] is a recent system with built-in decision procedures for automation of low-level deductions. Additionally an external BDD package is used. An intelligent automatic tactic for hardware verification in HOL has been presented in [2].

In this paper, ways of enhancing the higher-order-logic system LAMBDA with more automation are investigated and compared. Following the first approach, an advanced first-order theorem prover SEDUCT has been integrated (cf. Sect. 2.1) with LAMBDA. According to the second approach, we have yielded significant improvements within LAMBDA. The new tactics, which are solely composed from safe functions but nevertheless efficient, considerably increase the degree of automation even without specialization on a particular application area.

This work serves the purpose of creating a reduced instruction set proof environment (Rispe) which combines automated reasoning with the facilities of transformational reasoning in higher-order logic, while requiring minimum learning effort on the part of the user. This is achieved by supplying a few natural proof commands, which internally activate automatic proof routines. Rispe is summarized in a companion paper [8].

The paper is organized as follows. After short outlines of SEDUCT and LAMBDA (Sect. 2), their integration is described (Sect. 3). Section 4 is a presentation of an efficient tactic for first-order reasoning in LAMBDA. A small example of reasoning with Hoare calculus is discussed in the final section.

2 Overview of LAMBDA and SEDUCT

Neither LAMBDA, nor SEDUCT exclusively follows one of the paradigms of an interactive proof checker or a fully automatic theorem prover. The raw LAMBDA-system includes a few automatic proof procedures and SEDUCT provides limited interactive features. The complementary strengths of these systems suggested to investigate the advantages of joining them.

2.1 The First-Order Prover SEDUCT

The automatic theorem prover SEDUCT [23] has inherited concepts of the Larch prover [13], but it extends that system in many respects. SEDUCT supports full many-sorted first-order logic with equality and inequality. It combines an extended first-order sequent calculus with completion-based term rewriting [9] including conditional equations and inequalities. Hardwired routines allow efficient treatment of inductively-defined sorts and linear orders. The system comprises structural and computational induction [5].

Automatic proof routines are likely to fail more often than a step-by-step procedure with an interactive prover. Therefore the analysis of incomplete proofs is important for judiciously modifying proof goal and prover input. SEDUCT encompasses a facility to display proof trees.

The system is implemented and run in PROLOG. In order to avoid introducing its syntax and input, everything is explained in terms of LAMBDA.

Limitations. The system lacks polymorphism and the generalization facilities of higher-order logic. This forces the user to supply specific instances, where a general rule scheme suffices in LAMBDA. Compared with LAMBDA, more detailed external planning of proofs is required. The interactive features of SEDUCT do not support transformational reasoning.

2.2 Theorem Proving in LAMBDA

The core of LAMBDA [11] has conceptual similarities with the systems HOL [15] and ISABELLE [20]. It uses classical higher-order logic with polymorphic types, is implemented in ML, and provides tactics, conversions, and other common proof functions. Its higher-order unification feature enables formal synthesis by resolution-based application of generic rule schemes. Several generalization and instantiation functions are offered. They are important transformation tools in interactive proofs.

An extensive library of system utilities for programming automatic proof routines exists. Various analysis functions particularly help to save CPU time, if a syntactical criterion is formalized for dynamically deciding whether an expensive heuristic is worth being invoked by a tactic.

The user-interface of LAMBDA includes a browser with various menus for selecting and parametrizing proof functions, searching rules, displaying logic definitions and other information. An animator allows the execution of logic specifications. The graphical interface DIALOG [10] addresses formal design of digital circuits.

Notation. Sequents in LAMBDA are constructed from two (implicitly conjuncted) lists of hypotheses and one assertion, hence they are meta-implications. For instance, the sequent

    P1 $ P2 $ G // Q1 /\ Q2 $ H |- R1 \/ R2

which we write in this paper for short as

    P1, P2, Q1 /\ Q2 |- R1 \/ R2

has the three hypotheses P1, P2, Q1 /\ Q2 . The terminator symbols for hypothesis lists, G, H allow unification of two sequents with different numbers of hypotheses in the process of resolution. In LAMBDA, a rule consists of a list of premises and one conclusion, each of which is a sequent. A theorem is a rule with empty premise list. A rule can be considered as a meta-meta-implication.

The syntax of terms in LAMBDA is largely ML. The examples of this paper therefore should be readable without further detailing the LAMBDA syntax.

Criticism. The large number of different commands and features are a severe learning barrier even for mathematically skilled users. Moreover, the low degree of automation forces the user to interactively guide too many trivial operations, such as hypotheses permutations. Being a higher-order-logic system, another seeming disadvantage is the computational inefficiency of LAMBDA compared with automatic first-order provers. The results discussed in Sect. 4, however, are much better than assumed.

3 Integration of SEDUCT

The set of valid first-order formula provable in SEDUCT is a subset of all valid LAMBDA formulas. Therefore a conjecture which can be represented and proven in the first-order logic of SEDUCT is taken for true in LAMBDA, assuming that the SEDUCT prover and the interface between both systems work soundly. There are even classes of goals beyond the expressiveness of the first-order logic of SEDUCT which can be proven after some preprocessing.

The link comprises a tactic-based preprocessor, a syntax translator, a generator for proof scripts, and a UNIX interface to send a proof task to SEDUCT and to collect a message of success or failure. The script generator solely addresses the automatic capabilities of SEDUCT. Along with the goal, declarations of constants, datatypes, and axioms about items occurring in the transferred goal are generated.

All preprocessing is performed within LAMBDA, in order to keep the overall procedure as safe as possible by restricting the kernel translator to straightforward syntax transformations.

From the point of view of the LAMBDA user, the invocation of SEDUCT is just like the call of a tactic, which is optionally given some lemmas and instantiation information, and which either discharges a given LAMBDA top goal completely or fails. In either case, a proof tree generated by SEDUCT is displayed.

3.1 Free Variables

The syntax checker of SEDUCT rejects terms with free variables and with bound higher-order variables. There is a simple evasion of this in the case of free non-polymorphic first-order variables in a LAMBDA goal: they could be transformed into quantified first-order variables within LAMBDA.

The following solution has actually been chosen for better efficiency. As SEDUCT allows used-defined typed constants, free LAMBDA variables are treated as if they were constants.

Although the actual substitution is performed outside LAMBDA, it is mimicked here in terms of the choice-operator (**any**) of LAMBDA. The choice-operator selects an arbitrary member of a datatype satisfying a specified predicate. The constants used in SEDUCT in place of free LAMBDA variables are not restricted at all, which corresponds to setting the choice predicate to TRUE:

```
|- P (xc)
--------------          {xc := any c:xtype.TRUE}
|- P (x:xtype)
```

The substitution corresponds to lifting a successful proof of a property P of any constant xc with type xtype to a proof for a variable x of the same type. Informally this is true, because both proofs are not any different in LAMBDA. The argument is the same for proofs involving free function variables, which are reduced to SEDUCT proofs about functional constants.

Polymorphism is neither available in SEDUCT, but different sorts are accepted. Type variables of LAMBDA terms are similarly treated as if they were fixed datatypes without any assumption upon their generation. Thus, a proof with an anonymous constant type is assumed to be interchangeable with a proof using a type variable.

The following example demonstrates that goals with free polymorphic higher-order variables are provable in SEDUCT.

Example 1.

```
    forall x. Q1(x) ->> lex (m (d11 x)) (m x) == true ,
    forall x. Q1(x) ->> lex (m (d12 x)) (m x) == true ,
    forall x. Q1(x) /\ P(d11 x) /\ P(d12 x) ->> P(x)
|- (forall x. Q1(x) ->>
       lex (m (if (P(d11 x) ->> P(d12 x)) = TRUE then d11 x else d12 x))
           (m x) == true)  /\
    forall x.  Q1(x) /\
      P(if (P(d11 x) ->> P(d12 x)) = TRUE then d11 x else d12 x)
      ->> P(x)
```

The proof in raw LAMBDA is not difficult, but unnecessarily tedious. It does not require the definition of the function lex or instantiations of the free variables Q1, P, $d_{i,j}$, and m.[1]

Almost all semantical differences[2] between terms in LAMBDA and their counterparts used in SEDUCT boil down to the substitution of free variables, with additionally assuming that the basic first-order axiomatization is the same.

3.2 Inclusion of Lemmas

In order to make sure the semantics of corresponding constants is identical in both systems, it is assumed that SEDUCT is ignorant of LAMBDA operators, except for the most basic ones. Therefore, every required axiom or lemma has to be specified by the user and is translated along with the goal.

A particular difficulty arises: the free variables and type-variables of lemmas must be unified in LAMBDA to corresponding items of the goal in order to

---

[1] *Example 1* is also proven fully automatically by the first-order tactic of Sect.4.

[2] The mapping of datatype constructors, in particular polymorphic ones, is more difficult. It is not further discussed in this paper in order to avoid further details about SEDUCT. A flavour of the problems is given in Sect.3.3.

be usable by SEDUCT. Where a user does not specify instantiation, universal quantification is introduced for first-order free variables. The responsibility of determining appropriate instances is this way shifted to the intentionally simple instantiation routines of SEDUCT. The treatment of type variables is discussed in Sect. 3.3.

For example, if the commutativity of the addition of natural numbers

Example 2. |- x + y == y + x

is to be proved, axioms like

    |- a + 0 == a

must be made explicitly known to SEDUCT. In this particular example, arithmetic libraries exist in SEDUCT, but it cannot be decided generally, whether and to what extent predefined theories are available, as long as there is no well-defined common database accessible by both systems.

After introducing universal quantification and enhancing the goal with other basic properties of the addition,

```
forall a. a + 0 == a /\ 0 + a == a,
forall a,b. a + 1'b == 1'(a + b) /\ 1'a + b == 1'(a + b)
|- x + y == y + x
```

it is still not provable. Ignoring the induction capabilities of SEDUCT, this goal is only proved fully automatically by SEDUCT, if an appropriately instantiated induction rule is injected. In LAMBDA, the induction rule

```
|- forall n. P(n) ->> P(1'n)
|- P(0)
-------------------------------
|- forall n. P(n)
```

is instantiated and inserted resulting in the enriched goal

```
(forall y. 0 + y == y + 0) /\
   (forall n.
       (forall y. n + y == y + n) ->> forall y. 1'n + y == y + 1'n)
   ->> (forall n,y. n + y == y + n),
forall a. a + 0 == a /\ 0 + a == a,
forall a,b. a + 1'b == 1'(a + b) /\ 1'a + b == 1'(a + b)
|- x + y == y + x
```

which is apt for SEDUCT.

Although this trivial example is proved automatically in LAMBDA, it illustrates a possible cooperation between LAMBDA and SEDUCT. Rule schemes can be instantiated by means of higher-order facilities in LAMBDA, the result of which is understood by SEDUCT. Thus the limitations of that tool are overcome.

3.3 Polymorphism

There is a subtlety with generic types. In LAMBDA, polymorphic function constants like the length-function for lists may occur in different instances, depending on the type of its arguments. If the argument is a polymorphic constant as well, e.g., the empty list [], appropriate type instances must be generated before corresponding lemmas can be used by SEDUCT. Appropriate instances of polymorphic constants therefore have to be determined. Without this, lemmas like the following could not be used in SEDUCT-proofs of LAMBDA-goals.

```
|- length [] == 0
```

One of the major advantages of polymorphism is the possibility to have generic rules, whose types are instantiated automatically as needed when these rules are applied to a goal. In order to be able to use such generic lemmas in SEDUCT, it is necessary to instantiate the types of polymorphic variables and constants to those of the goal. Otherwise, even a simple goal like

Example 3.   forall x. length [x] == 1 |- length [0] == length [true]

could not be proved, because different constants for the empty lists, the list construction and length functions would be generated. A heuristic has been installed that computes reasonable instantiations of polymorphic variables and different instances of polymorphic constants. Those are treated by the translator as if the following goal were given:

```
forall x. length [x] == 1, forall x. length1 [x] == 1
|- length [0] == length1 [true]
```

This example demonstrates that it would not make sense to simply unify all occurrences of the same polymorphic constant. Note, however, that the soundness is not affected if useless instantiations are determined within LAMBDA, since the worst case is a goal not provable by SEDUCT.

3.4 Higher-Order Goals

Function applications have to be transformed into ones with one tuple argument, for there are no higher-order functions in SEDUCT. According to Sect.3.1, free function variables are mapped onto constants in SEDUCT, thus they cannot be instantiated. There are ways in many cases, though, to enable quantification over functions, which is literally possible in higher-order logic only.

Particular quantifications can be removed by means of rules which introduce restricted variables not occurring in the previous goal:

```
|- P(x')                      P(x')            |- Q
-----------------      ,      --------------------
|- forall x. P(x)             exists x. P(x) |- Q
```

Other goals with higher-order quantification can be represented in first-order logic [17]. Consider the following goal.

Example 4. |- forall f. exists g. f x == g x

A polymorphic apply-function is defined in LAMBDA:

```
fun apply (f,x) = f x;
```

With this auxiliary function, the goal is transformed in LAMBDA into

```
|- forall f. exists g. apply(f,x) == apply(g,x)
```

The functional polymorphic type of *f* and **g** is now abstracted to a first-order polymorphic type variable. The abstracted LAMBDA goal is proven by SEDUCT, and the resulting theorem applied to the previous goal with functional types.

The following goal is a more interesting example.

Example 5. |- forall y,t. exists f. forall x. f x == t x y

SEDUCT proves this goal, if other polymorphic *apply*-functions are defined,

```
fun apply2 (f,x,y) = f x y;
fun swap g y x = g x y;
```

and if the goal is rewritten and extended accordingly within LAMBDA:

```
forall t1,x,y. apply (apply(t1,y),x) == apply2 (t1,y,x),
forall t,x,y. apply2 (swap t,y,x) = apply2(t,x,y))).
|- forall y,t. exists f. forall x. apply(f,x) == apply2(t,x,y)
```

Abstracting the goal to[3]

```
forall t1,x,y. a (a1(t1,y),x) == b1 (t1,y,x)
forall t,x,y. b1 (c t,y,x) == b (t,x,y),
|- forall y,t. exists f. forall x. a(f,x) == b(t,x,y)
```

proving this one by SEDUCT, and applying it to the rewritten goal succeeds. As all generalizations, introduction of auxiliary functions, and other transformations are performed in LAMBDA, the overall soundness essentially depends on the substitution of variables by constants as discussed in Sect. 3.1.

It is doubtful, however, whether these first-order representations are sensible, since they blow up syntactical expressions considerably. In addition, the construction of the different apply-functions and their type-dependent instances is non-trivial (see Sect. 3.3). Therefore, hypotheses with second-order universal quantification are ignored, and goals with assertions with second-order existential quantification even rejected, for it makes much more sense to provide appropriate instantiations in LAMBDA before passing the goal to SEDUCT.

---

[3] Different instances of the polymorphic *apply*-functions are replaced by individual variables (cf. Sect.3.3).

3.5 Discussion

With the extensions regarding the treatment of polymorphic free variables, a surprising number of proofs, such as *Ex.1*, could be completed by SEDUCT, after the first proof steps with induction, instantiation, resolution-based rule application, or others had been carried out in LAMBDA.

There are several disadvantages, though. First, the number of axioms and lemmas which have to be transferred in order to supply SEDUCT with the knowledge to prove more interesting goals is limited by the inefficiency of completion-based rewriting for large rule sets compared with conversion-based rewriting. Second, the overhead for transforming and exchanging a large quantity and variety of proof information may compensate efficiency and automation advantages. A common database accessible by different provers might be a way out this difficulty. Third, in a typical proof, an automatic tactic may well achieve useful simplification without completely discharging a goal. An external tool which just says *proven* or *not proven* does not allow the user to exploit partial results. More sophisticated analyses of the output of an external prover is possible in principle, however, the overhead appears to be considerable. Fourth, although the critical transformations are carried out in LAMBDA rather than as part of the syntax translator, a potential risk of unsoundness incurred through the use of SEDUCT remains.

Linking an external system to a higher-order-logic prover makes sense for testing the effect of foreign proof procedures without reimplementation. Such a heterogeneous proof environment may serve as a good prototype for experimentation. In turn, a later reimplementation can make use of extended capabilities of the more powerful prover, thus yielding an enhanced procedure. If there are classes of subgoals, for which an external prover or verification system is definitely orders of magnitude better then a comparable tactic could ever or only under very high reimplementation efforts be, the final target prover for real applications should exploit such tools. This is the case for BDD-based verifiers.

4 First-Order Sequent Calculus in LAMBDA

The limitations of an interface to an external theorem prover alone would not have justified the creation of LAMBDA-tactics for first-order automation. Historically, the work for automating low level proof activities during the development of Rispe unintentionally yielded a couple of tactics which appeared to be composable and extendible to a powerful first-order tactic. Some of the tactics constituting the complete first-order tactic are described in the following sections.

4.1 Manipulation of Assertions

The internal mechanisms of LAMBDA have not been designed for manipulating disjunctions as conveniently as a list of logically conjunced, but independently treatable hypotheses. Therefore, rather than installing a Gentzen-like sequent calculus[12], a more suitable variant for LAMBDA is set up. An appropriate sequent form is generated from arbitrary sequents in the first step which consists in splitting, simplifying and negating assertions. This step essentially corresponds to backwards application of rules which manipulate the assertion of a sequent:

```
|- P(r')              P |- Q         NOT P |- FALSE      |- P , |- Q
------------------ ,  ---------- ,   --------------- ,   -----------
|- forall x. P(x)     |- P ->> Q     |- P                |- P /\ Q
```

Hence, the original goal is split into several subgoals, all of which have the degenerate assertion FALSE; thus the further proof is achieved by refutation. The resulting sequents contain the negated original subassertions as hypotheses.

4.2 Simplification of Hypotheses

Subsequent simplification steps are performed which may split the goal into various subgoals. For some proof goals a naïve subgoaling strategy could internally produce a large number of parallel subgoals. For this reason a subgoal minimization strategy was implemented.

Rather than using the pure first-order sequent calculus, simplifications are mostly performed through conversions. Their chance of success is greatly improved by a function which permutes hypotheses with a property specified by a predicate to the beginning of the hypothesis list. For instance, modus-ponens like proof steps are efficiently implemented by moving all implicative hypotheses to the end of hypothesis lists, in order to allow the other hypotheses to satisfy the antecedents of the implications.

After having obtained sequents with all information in hypotheses, splitting, simplification, and normalization of hypotheses is performed by means of several rule schemes and conversions such as

```
3: Q, NOT P |- Q1
2: P, NOT Q |- Q1
1: P, Q |- Q1
   --------------
   P \/ Q |- Q1

NOT (exists x. P(x)) == forall x. NOT P(x)

NOT (forall x. P(x)) == exists x. NOT P(x)
```

The effect of the steps summarized so far is illustrated by the next example goal:

Example 6.

```
|- (exists x. x == f(g x) /\ forall x1. x1 == f(g x1) ->> x1 == x) ==
      exists y. y == g(f y) /\ forall y1. y1 == g(f y1) ->> y1 == y
```

which is transformed into two subgoals, the first one of which reads:

```
f (g x') == x',
forall x.
    NOT (g (f x) == x) \/ exists y1. g (f y1) == y1 /\ NOT (y1 == x),
forall x1. f (g x1) == x1 ->> x1 == x'
|- FALSE
```

The equations have been swapped automatically in order to avoid endless loops in rewrites. Other heuristics such as permutation of hypotheses are included in order to increase the chances to discharge many goals just by rewriting, before more expensive heuristics are tried. The tactics up to this point already succeed, if no instantiations are needed.

4.3 Instantiation of Quantified Variables

Since all proof steps concentrate on the left side of the turnstile, just one rule is needed, i.e.,

```
P(y), forall x. P(x) |- Q
--------------------------
forall x. P(x) |- Q
```

which introduces an instantiatable free variable **y**. Much of the success of a first-order tool depends on determining good instantiations. By experience, the overwhelming majority of favourable instances occur as subterms of a proof goal. Hence the most straightforward solution appeared to be the generation of all possible instantiations by collecting all appropriately typed subterms of a goal. Unfortunately, in general too many useless instantiations are produced by this simple heuristic, which could blow up a goal considerably. It turned out that often observing the context in which the instantiation variable is used gives good hints as to which of the possible instantiations make sense. In fact, for the instantiation heuristic of **Rispe** a similarity analysis function has been created. This function generates appropriate generalized contexts[4] of hypotheses subterms containing instantiatable bound variables, and performs comparisons with generalized subterms of the remaining sequent.

In the proof of *Ex.6*, calling this heuristic leads to a new situation, which after simplification and pruning reads:

```
forall x1. f (g x1) == x1 ->> x1 == x',
forall x. NOT (g (f x) == x) \/
          exists y1. g (f y1) == y1 /\ NOT (y1 == x),
f (g x') == x', g (f y1') == y1', NOT (y1' == g x')
|- FALSE
```

Obviously the simplification step has yielded new expressions which are candidates for instantiation of the all-quantified hypotheses. Therefore, the instantiation heuristic is started a second time, yielding a new goal, which, for better readability is just shown with the essential hypotheses:

---

[4] Subroutines for computing generalized goals [6, 11] turned out to be favourable.

```
f (g (f y1')) == f y1' ->> f y1' == x',
f (g x') == x' ->> x' == x',
f (g (f (g x'))) == f (g x') ->> f (g x') == x',
f (g x') == x', g (f y1') == y1', NOT (y1' == g x')
|- FALSE
```

The simplification tactic is now able to prove the remaining goal automatically by a heuristic combining hypothesis permutation and conversion-based rewriting.

From the user's point of view, the first-order routine is a 'push-button'-tactic, which does not need any guidance or input. The complete proof of the initial goal for *Ex.6* is achieved fully automatically in 6 seconds on a Sparc2.

4.4 Limitations

The tactic handles problems of full predicate logic with identity and functions, currently with the limitation that no completion-based rewriting is included, such as required for Pelletier-Problems 63-65 [21]:

Example 7.

```
forall x,y,z. f (f (x,y),z) == f (x,f (y,z))
forall x. f (a,x) == x, forall x. exists y. f (y,x) == a
|- (forall x,y,z. f (x,y) == f (z,y) ->> x == z) /\
   (forall x,y. f (y,x) == a ->> f (x,y) == a)   /\
   ((forall x. f (x,x) == a) ->> forall x,y. f (x,y) == f (y,x))
```

Including Knuth-Bendix completion would be feasible, if it turned out to be worthwhile.[5]

Space problems prohibit full automation for too large problems, such as of Pelletier-Problem 47. Particularly in in this respect a great deal of improvement will arise from heuristic pruning of the solutions found by the instantiation tactic. For example, as the instantiation heuristic is called repeatedly, sometimes too many instances are generated, which affect the speed of other subtactics. The tactic currently does not use Skolem functions, because simplification and instantiation are interleaved and called repeatedly. In some cases, instantiations generated in two different cycles could be determined in one, if Skolem functions were introduced by rewriting with the rule

```
|- (forall x. exists y. P(x,y)) == exists g. forall x. P(x,g x)
```

Even without optimization, however, the proof of Pelletier-Problem 34 [21] is accomplished in 40 seconds:

---

[5] *Example 7* is proved fully automatically in less than 10 seconds by invoking SEDUCT.

Example 8.

```
|- (exists x. forall y. P x == P y)==((exists x. Q x)==(forall y. Q y))
   ==
   ((exists x. forall y. Q x == Q y)==((exists x. P x)==forall y. P y))
```

The current first-order tactic could be extended to handle even second-order instantiations. For instance, the goal

Example 9.   `|- forall y,t. exists f. forall x. f x == t (x,y)`

is transformed by the automatic tactic into

```
NOT (f x' == t' (x',y')) |- FALSE
```

Just a simple instantiation (`INST "f x ~> t' (x,y')";`) is needed. The resulting goal is discharged immediately. Another subtactic could be installed for determining such instantiations automatically. A look-ahead function could restrict the call of this tactic to those cases where an intermediate goal contains second-order meta-variables.

Automatic case analyses for finite datatypes can be added. Case analysis heuristics guided by occurrences of primitive and recursive datatype constructors in proof goals are promising as well. Other heuristics will address automatic expansion of function applications. First experiments have been made already.

4.5 Discussion

The entire first-order tactic could have been implemented just in terms of an extended set of basic sequent calculus rules along with some instantiation. Furnishing the tactic with conditional conversions was a better choice. In addition, the special heuristics for swapping and permuting hypotheses improve the chances of succeeding without completion procedures[9].

Several interesting observations have been made when testing the first-order tactic. Even if the automatic tactic achieves only a partial proof, this result enables subsequent proof steps to continue at that point. Such steps are often case analyses or inductions. Extending the current tactic in LAMBDA by further heuristics which exceed the capabilities of first-order provers is better than defining auxiliary functions that allow a first-order prover to prove higher-order goals (Sect. 3.4). The efficiency is not affected for those examples, where the tactic so far succeeds, because in this case the supplementary heuristics are not invoked. The modularity of the overall tactic allows tuning, exchanging, and adding subtactics as well as individually testing their performance and effects.

There is much potential for further optimization and extension. Nonetheless, the examples done so far demonstrate that the first-order tactic contributes a good deal of automation with acceptable efficiency.

5 Using the First-Order Tactic in Rispe

The first-order tactic is available via the command FOSC within Rispe. The following example has been taken from [14], where a simple imperative programming language has been embedded in higher-order logic and some basic rules of the Hoare calculus have been derived from the definitions. Rather than detailing the formal definitions, the proof steps are shown in terms of the Rispe commands pertaining to that proof. The subject of the example proof is the derivation of an introduction rule for loop invariants.

Example 10. > PROVE "SPEC(p & b,c,p) |- SPEC(p,While b c, p & (Neg b))";

The first steps are expansion (ESIMP, c.f. [8]) and a call of the first-order tactic.

```
> ESIMP "{*}";
1: forall s,t. p s /\ b s /\ c (s,t) ->> p t
   |- forall s,t. p s /\ (exists n. Iter n(b,c)(s,t)) ->> p t /\ NOT(b t)

> FOSC "";
2: forall s,t. p s /\ b s /\ c (s,t) ->> p t,
   p s', Iter n' (b,c) (s',t') |- NOT (b t')
1: forall s,t. p s /\ b s /\ c (s,t) ->> p t,
   p s', Iter n1' (b,c) (s',t'), NOT (b s') \/ NOT (c (s',t')) |- p t'
```

The proof of the resulting two subgoals is the same, therefore only the first one is shown in the sequel. Each subgoal is proved by structural induction[6] over the number of iterations followed by another call of equational expansion and the first-order tactic.

```
> IND    "n1',*s',*t'";
2:    forall s,t. p s /\ b s /\ c (s,t) ->> p t,
      forall s',t'. p s' /\ Iter n1'1' (b,c) (s',t') /\
           (NOT (b s') \/ NOT (c (s',t'))) ->> p t'
   |- forall s',t'.
           p s' /\ Iter (1'n1'1') (b,c) (s',t')
           /\ (NOT (b s') \/ NOT (c (s',t'))) ->> p t'
1: forall s,t. p s /\ b s /\ c (s,t) ->> p t
   |- forall s',t'. p s' /\ Iter 0 (b,c) (s',t') /\
           (NOT (b s') \/ NOT (c (s',t'))) ->> p t'
```

The base case is discharged merely by equational simplification. The step case is not completely proved by simplification, but requires a second call of the first-order tactic, which completely discharges the step case:

```
1: forall s,t. p s /\ b s /\ c (s,t) ->> p t,
   forall s',t'. p s' /\ Iter n1'1' (b,c) (s',t') /\
        (NOT (b s') \/ NOT (c (s',t'))) ->> p t'
```

---

[6] The parameters *s',*t' cause the induction tool to collect all hypotheses containing the variables s',t' and introduce universal quantification.

```
|- forall s',t'.  p s'  /\
      (if b s' = TRUE
       then exists s.  c (s',s) /\ Iter n1'1' (b,c) (s,t')
       else s' == t') /\
      (NOT (b s') \/ NOT (c (s',t')))) ->> p t'

> FOSC"";
```

The example demonstrates that results of the first-order tactic can be useful even if a goal cannot be discharged in the first run. Case analyses and induction yield new subgoals which are again manipulated by the first-order tactic.

6 Conclusion

The recent success in integrating advanced automation techniques in interactive theorem proving environments demonstrates that the disadvantageous complexity of interactive theorem proving which was hitherto caused by too many trivial interactive low-level steps is not inherent in higher-order-logic systems.

Two alternative approaches have been investigated in LAMBDA. We have first discussed how a wide class of formulas with polymorphic and even higher-order variables can be proved by a first-order theorem prover. The second approach precludes soundness problems, avoids the need for translating higher-order logic formulae into first-order logic, exchanging supplementary proof information, and allows proof continuation if an automatic routine succeeds only partially. An extendible tactic for first-order automation within LAMBDA has been presented. Relying on a mixture of conversions, inference rules, instantiation, and other tactics, it is far more efficient than an implementation in terms of basic sequent calculus rules would have been. More automation and acceleration is expected by improved instantiation, expansion, and pruning heuristics, automatic case analyses, and optimized rewrite strategies.

Earlier work has yielded an automatic generator of explicit induction rule schemes in LAMBDA [7]. For instance, they enable automatically proving the termination of most user-defined functions. The automation of transformational proof activities has been obtained with other Rispe routines [8].

LAMBDA turned out to be an excellent platform for discovering and safely experimenting with new heuristics, in particular by its rich tactic-programming facilities, which at the same time limit the access to the safety-critical parts of the kernel inference machinery.

References

1. 'Higher Order Logic Theorem Proving and Its Applications', 6th International Workshop, HUG'93, Vancouver, B.C., Canada, August 11-13 1993, Proceedings, J.J. Joyce and C.-J.H. Seger (Eds.), LNCS 780, Springer, 1994.
2. 'Towards a Super Duper Hardware Tactic' M. Aagard, M. Leeser, and P. Windley, in [1], pp. 399–412.

3. F. Andersen, K.D. Petersen, and J.S. Pettersson, 'Program Verification using HOL-Unity', in [1], pp. 1–15.

4. R. Boulton, 'Boyer-Moore Automation for the HOL System', in 'Higher Order Logic Theorem Proving and its Applications', North-Holland, pp. 133–142, edited by L. Claesen and M. Gordon, 1992.

5. R.S. Boyer and J.S. Moore, 'A Computational Logic', Academic Press, ACM Monograph Series, 1979.

6. H. Busch, 'Transformational Design in a Theorem Prover', in *THEOREM PROVERS IN CIRCUIT DESIGN*, IFIP Transactions A-10, edited by V. Stavridou, T.F. Melham, and R.T. Boute, pp. 175–196, North-Holland, 1992.

7. H. Busch, 'Rule-Based Induction', in *FORMAL METHODS IN SYSTEM DESIGN - Special Issue on HOL'92*, Kluwer, Vol. 5, Issue 1 & 2, July/August 1994.

8. H. Busch, 'A Reduced Instruction Set Proof Environment', in *2nd Int. Conf. on THEOREM PROVERS IN CIRCUIT DESIGN: Theory, Practice, and Experience*, edited by R. Kumar and T. Kropf, Bad Herrenalb, Germany, Sept. 26-29,1994.

9. Dershowitz, N. and J.-P. Jouannaud 1990. Rewrite Systems. In *Handbook of Theoretical Computer Science* (Vol. B: *Formal Models and Semantics*), pp. 243–320. Amsterdam, North–Holland.

10. S. Finn, M. Fourman, M. Francis, and R.Harris, 'Formally Based System Design - Interactive Synthesis Based on Computer-Assisted Formal Reasoning', in *Formal VLSI Specification and Synthesis - VLSI Design Methods-I*, edited by L. Claesen, North-Holland, pp. 139–152, 1990.

11. S. Finn, M. Fourman, M. Francis, B. Harris, R. Hughes, and E. Mayger, Abstract Hardware Limited, LAMBDA Documentation, 1993.

12. J. Gallier, Logic for Computer Science. New York, Harper & Row, 1986.

13. S.J. Garland and J.V. Guttag, 'A Guide to LP - The Larch Prover'. Digital Equipment Corporation Systems Research Center, Report 82, 1991.

14. M.J.C. Gordon, 'Mechanizing Programming Logics in Higher Order Logic', in *Current Trends in Hardware Verification and Automated Theorem Proving*, edited by G. Birtwistle and P.A. Subrahmanyam, Springer,pp. 387–439, 1989.

15. M.J.C. Gordon and T.F. Melham, 'Introduction to HOL: A theorem proving environment for higher order logic', Cambridge University Press, 1993.

16. J.J. Joyce and C.-J. Seger, 'The HOL-Voss System: Model-Checking inside a General-Purpose Theorem Prover', in [1], pp. 1–15.

17. M. Kerber, 'How to Prove Higher Order Theorems in First Order Logic' SEKI Report SR-90-19 (SFB), University of Kaiserslautern, Fachbereich Informatik, 1990.

18. R. Kumar, T. Kropf, and K. Schneider, 'Integrating a First-Order Automatic Prover in the HOL Environment', in *International Tutorial and Workshop on the HOL Theorem Proving System and its Applications*, Davis, pp. 170–176, 1991.

19. T.F. Melham, 'Abstraction Mechanisms for Hardware Verification', in *VLSI Specification, Verification and Synthesis*, edited by G. Birtwistle and P.A. Subrahmanyam, Kluwer, pp. 27–72, 1989.

20. L.C. Paulson, 'Introduction to Isabelle', Computer Laboratory, University of Cambridge, 1992.

21. F.J. Pelletier, 'Seventy-Five Problems for Testing Automatic Theorem Provers', Journal of Automated Reasoning 2, 1986.

22. J. Rushby and M. Srivas, 'Using PVS to Prove Some Theorems of David Parnas', in [1], pp. 162–173.

23. K. Stroetmann and C. Bendix Nielsen, 'A Guide to SEDUCT, Report, Siemens AG, Corporate Research & Development, D-81730 München, August 1993.

Symbolic Animation as a Proof Tool

Juanito Camilleri and Vincent Zammit

Department of Computer Science and A.I.,
University of Malta, Msida, Malta

Abstract. This paper illustrates how animation conversions [14] which
help in preliminary debugging of behavioural definitions, can subsequently
be used as effective proof tools which play an important role in the verification of properties related to the definitions. We illustrate this point
by specifying a simple compiler to map constructs of a toy imperative
programming language into instructions which run on a rudimentary
abstract machine. The same conversions used to symbolically compile
programs of the language and to execute the resulting machine instructions are used in the verification of the compiler. This paper suggests
that conversions provide a sound basis for a proof methodology with
formal animation acting as an integral step in a verification process.

1 Introduction

It is well known that behavioural definitions can be animated through the smart
use of *conversions*: a special class of inference rules in HOL [8] which can map
a term of the logic to a theorem asserting the equality of that term to some
other term. Due to the way in which the logic of the HOL system is represented
in the strongly-typed, general-purpose programming language ML, the user can
write and compose conversions tailored for a specific application, be it a simple
representation of beta-conversion or the more involved animation of the definitions of a compiler, an abstract machine etc. In the case of beta conversion for
example, a term of the form $(\lambda x.\ t_1)t_2$ can be mapped by a conversion to a class
of theorems

$$\vdash\ (\lambda x.\ t_1)t_2\ =\ t_1\,[t_2/x]$$

One can also write a conversion which maps an arbitrary term t of a programming language to a class of theorems expressing that the result of compiling
t in some environment e is equivalent to some list of instructions i

$$\vdash\ \texttt{Compile}\ t\ e\ =\ i$$

or, moreover, that the result of executing a list of instructions i in some initial
state s yields a final state s'

$$\vdash\ \texttt{Execute}\ i\ s\ =\ s'$$

NOP	A skip instruction
HLT	Stops the machine by setting the program counter to 0
POP	Pop the top of the stack
JMP b n	Jump forward by n if b is true, jump backward by n otherwise
JZ b n	Pop stack then, if the result is zero, jump forward by n if b is true, jump backward by n if b is false
JNZ b n	Pop stack then, if the result is non-zero, jump forward by n if b is true, jump backward by n if b is false
OP1 op	Pop one value from stack, perform op, push result
OP2 op	Pop two values from stack, perform op, push result
GET x	Push the contents of memory location x onto the stack
PUSH v	Push value v onto the stack
PUT x	Pop the top of the stack and store the result in memory location x

Figure 1: The instruction set.

There are several reasons for considering the conversion approach to animation of behavioural definitions. First, it is relatively straightforward to write conversions to animate behavioural definitions. These conversions can also be used for symbolic simulation or *partial simulation* when variables are used to represent components of a system while the behaviour of some other component is simulated. Second, because the same definitions are used in both animation as well as verification, we avoid errors resulting from discrepancies between the model of the definitions which is animated and the definitions whose properties are verified. Furthermore, the results of the animation are theorems backed up by formal proof. Hence, assuming that the implementation and underlying logic of the theorem prover are sound, if the animation of a definition conveys unexpected behaviour then this must be due to oversights. Third, when developing a specification for a large system, one can take an incremental and compositional approach to animation. In other words, one can write conversions on-the-fly to animate the definitions of the subsystems and then compose these conversions into a conversion which animates the whole system. For example, when defining a compiler, one may choose to write conversions to animate the compilation of expressions and declarations separately and then use those conversions when constructing a conversion to animate the compilation of commands.

Despite these advantages, the need for such formal symbolic manipulation within the framework of a theorem prover can be questioned by those who view animation as the quick informal way to get an impression that behavioural definitions convey their intuitive intent [3] [1] [10] [16]. If reassurance of correct behaviour is all that is required then formal animation of definitions through conversions usually involves more effort than one is willing to justify. There is, however, one final advantage of using a formal approach to animation which justifies the effort involved—the same conversions used for animation can be used later to facilitate the verification of meta-theorems related to the definitions in question. This is the seed of a methodology which introduces the cost effective use of formal animation of definitions as an integral step towards verification.

This paper illustrates this point by defining the operational semantics of a toy imperative programming language IMP which executes on a machine driven by an assembly language. We define conversions to animate the execution of instructions on the machine as well as the compilation of IMP commands to machine instructions. Of course, the conversions are rather helpful to convey that both the definitions of the compiler and those of the target machine behave as expected in specific cases. The main point, however, is the illustration of the effective use of these same conversions in the correctness proof of the compiler.

2   The target machine and instruction set

Consider a machine whose state is a tuple $(pc, (memory, stack))$ where the program counter pc is of type num, $memory$ is a function which maps machine addresses to values, and $stack$ is a list of values. The instruction set is described in Fig.1 and is defined in HOL using the type definition package [12]. The resulting theorem in higher-order logic, is a complete and abstract characterisation of the data-type instruction and asserts the admissibility of defining functions over instructions by primitive recursion. Primitive recursion over instructions is used when defining a function Step which determines the effect on the state when the instructions are executed (see Fig.2).

Instructions are fetched according to the following definition which ensures that the computation stops once the program counter goes beyond the length of the instruction list.

```
Fetch =
⊢ ∀ins n.
    Fetch ins n =
    (((n = 0) ∨ n > (LENGTH ins)) => HLT | EL(PRE n)ins)
```

Note that EL i $[x_0; x_1; ...; x_n] = x_i$ and LENGTH is a function which determines the length of a list.

Given that PC is a function that extracts the program counter of a state, the behaviour resulting from the execution of one instruction on the machine is defined as follows:

```
⊢ ∀ins ms. MachineStep ins ms = Step(Fetch ins (PC ms))ms
```

```
⊢ (Step NOP = (λ(pc,mry,stk). (SUC pc,mry,stk))) ∧
  (Step HLT = (λ(pc,mry,stk). (0,mry,stk))) ∧
  (∀d n. Step(JMP d n) = (λ(pc,mry,stk). (Jump d n pc,mry,stk))) ∧
  (∀d n.
    Step(JZ d n) =
    (λ(pc,mry,stk).
      (((HD stk = 0) => Jump d n pc | SUC pc),mry,TL stk))) ∧
  (∀d n.
    Step(JNZ d n) =
    (λ(pc,mry,stk).
      (((HD stk = 0) => SUC pc | Jump d n pc),mry,TL stk))) ∧
  (∀a.
    Step(PUT a) = (λ(pc,mry,stk). (SUC pc,Store(HD stk)a mry,TL stk))) ∧
  (∀a. Step(GET a) = (λ(pc,mry,stk). (SUC pc,mry,CONS(mry a)stk))) ∧
  (∀v. Step(PUSH v) = (λ(pc,mry,stk). (SUC pc,mry,CONS v stk))) ∧
  (Step POP = (λ(pc,mry,stk). (SUC pc,mry,TL stk))) ∧
  (∀op1.
    Step(OP1 op1) =
    (λ(pc,mry,stk). (SUC pc,mry,CONS(op1(HD stk))(TL stk)))) ∧
  (∀op2.
    Step(OP2 op2) =
    (λ(pc,mry,stk).
      (SUC pc,mry,CONS(op2(HD(TL stk))(HD stk))(TL(TL stk)))))
```

Figure 2: The definition of the Step function.

The definition of MachineStep can be used to define the symbolic execution of a series of steps by primitive recursion on the number of steps taken.

```
⊢ (∀ins ms. MachineSteps 0 ins ms = ms) ∧
  (∀n ins ms.
    MachineSteps(SUC n)ins ms =
    MachineSteps n ins (MachineStep ins ms))
```

Execution carried out according to this definition can be animated in HOL. One can write a conversion which given an instruction list and an initial state, executes the instructions according to the definition of the machine starting from the initial state. Each step of the animation is done by a systematic specialisation, unfolding and simplification of MachineStep with the instructions i and the current state s of the machine. The result is a theorem of the form:

⊢ **MachineStep** i s = s'

where s' is the state after one step of execution. Of course, one can apply the same procedure to the instructions in state s' to yield

⊢ **MachineStep** i s' = s''

where s'' is the state after two steps of execution. This can be repeated until the machine halts—i.e. either it encounters a HLT instruction or it executes all its instructions. The trail of theorems generated as described above and the definition of MachineSteps can be used to automatically prove a final theorem asserting the result of executing the program on the machine being discussed. For example, the outcome of a conversion implemented to capture the behaviour described above could be:

```
⊢ MachineSteps
  34
  [PUSH 5; PUT 1; GET 1; OP1 PRE; JZ T 5; GET 1; OP1 PRE;
   PUT 1; JMP F 6; HLT]
  (1,(λx. 0),[]) =
  (0, (λx. ((x = 1) => 1 | 0)),[])
```

which in this particular case states that the execution of the list of instructions:

```
[PUSH 5; PUT 1; GET 1; OP1 PRE; JZ T 5; GET 1; OP1 PRE;
 PUT 1; JMP F 6; HLT]
```

in an initial state where the program counter has value 1, all memory locations have value 0, and the stack is empty, terminates in 34 steps leaving the program counter with value 0, and the original stack and memory unaltered except for. the value of the variable stored in the first memory location which is updated to 1.

Such a conversion is very useful in animating any given list of instructions, while producing theorems which can later be used in elaborate proofs of meta-theorems. Such proofs, however, often require lemmas about the execution of partially defined list of instructions. In fact, in order for the conversion described above to be particularly useful in the proof of the compiler described later, it needs to be general enough to handle instruction lists like, for example, the following:

```
[PUSH 5; PUSH 7; JMP T (1 + LENGTH A)] ++ A ++ [OP2 $+; HLT]
```

In other words, the conversion was implemented to handle an instruction list of the general form A_1 ++ A_2 ++ ... ++ A_n, where not all the lists A_i are instantiated. Animation is done step by step as described previously until, either execution terminates naturally, or there is not enough information on how the next execution step will follow. For example, the instruction list above can be animated by the conversion, from state $(1, (m, []))$ until the machine terminates naturally within five steps:

⊢ MachineSteps
   5
   ([PUSH 5;PUSH 7;JMP T(1+(LENGTH A))] ++ (A ++ [OP2 $+;HLT]))
   $(1,m,[])$ =
   $(0,m,[12])$

3   The language IMP

IMP is a simple imperative language with expressions, SKIP, assignment, conditionals, a WHILE construct, and local variable declaration. The syntactic classes of this rudimentary language are represented by the recursive types exp and cmd defined as follows:

exp	::= VAR string	(local variable)
	| CONST num	(constant)
	| UNOP (num->num) exp	(unary operator)
	| BINOP (num->(num->num)) exp exp	(binary operator)
cmd	::= SKIP	(skip)
	| ASSIGN string exp	(assignment)
	| IF exp cmd cmd	(two-armed conditional)
	| SEQ cmd cmd	(sequence)
	| WHILE exp cmd	(while loop)
	| DECL string cmd	(local variable declaration)

The results of defining these recursive types in HOL are theorems which state the admissibility of defining functions over the syntactic structure of IMP programs by primitive recursion [12]. These theorems are of direct utility since this is how the compiler is defined.

4   The IMP compiler

The compilation of IMP commands and expressions proceeds with respect to the naming environments generated by declarations in various blocks. Let

   env:renv # maddress

represent an environment where renv is a function string->num defining the mapping between identifiers and locations in memory, and maddress is the next free location.

In the case of commands, the function

   Comcmd: cmd -> env -> script

is defined by primitive recursion on the type cmd, where script is a list of instructions representing the target machine code:

```
⊢ (∀e. Comcmd Skip e = ComSkip) ∧
  (∀i ex e. Comcmd(Assign i ex)e =
     ComAssign(FST e i)ex(FST e)) ∧
  (∀c1 c2 e. Comcmd(Seq c1 c2)e =
     ComSeq(Comcmd c1 e)(Comcmd c2 e)) ∧
  (∀ex c1 c2 e. Comcmd(If ex c1 c2)e =
     ComIf ex(Comcmd c1 e)(Comcmd c2 e)(FST e)) ∧
  (∀ex c e. Comcmd(While ex c)e =
     ComWhile ex(Comcmd c e)(FST e)) ∧
  (∀i c e. Comcmd(Decl i c)e = Comcmd c(update i e))
```

Details of the compilation algorithm for each of the constructs of IMP are not included here. As an example, however, consider the following definition of the compilation of WHILE constructs:

```
⊢ ∀ex c_prog s.
    ComWhile ex c_prog s =
    (let ex_prog = Comexp ex s
     in
      let c_length = LENGTH c_prog
      in
      ex_prog ++
      ([JZ T(2 + c_length)] ++
       (c_prog ++ [JMP F (1 + ((LENGTH ex_prog) + c_length))])))
```

where Comexp is the expression compiler. The compilation of IMP expressions can be defined similarly, while the compilation of a program can be viewed as the compilation of a command in an initial environment init_env where all variables are mapped to 0 and the next free memory location is 1:

```
⊢ ∀prg. Comprog prg = Comcmd prg init_env
```

4.1 A conversion to animate the compiler

The behavioural definitions of the IMP compiler can be animated by a conversion which maps any term Comcmd c e to a theorem

$$⊢ \text{Comcmd } c \; e = i$$

where i represents the list of instructions resulting from the compilation of command c in environment e. This is done by recursively rewriting with the definition of the compiler until every construct in the command is compiled. For example, the command

```
If (Const 0) C₁ (Decl 'x' (Assign 'x' (Const 1)))
```

where (Const 0) represents false, can be symbolically compiled under the initial environment $(\lambda a.0,1)$:env to yield:

```
⊢ Comcmd (If (Const 0) C₁ (Decl 'x' (Assign 'x' (Const 1))))
  (λa.0,1) =
  [PUSH 0; JZ T (2 + (LENGTH(Comcmd C₁ ((λa.0,1)))))] ++
  ((Comcmd C₁ ((λa.0,1)))) ++ [JMP T 3; PUSH 1; PUT 'x']
```

Note the conversion's ability to work around variables such as C_1 which represents an arbitrary command. The generality of this conversion makes it particularly useful in the proof of the compiler described later.

5   An Operational Semantics for IMP

One can argue that the formal definition of the behaviour of a machine can act as an operational semantics for a high-level programming language such as IMP as long as there is a formal mapping between the constructs of the language and the instructions which drive the machine. In this sense, we have already presented an operational semantics of IMP programs. In Fig.3, however, we define an alternative operational semantics as a transition relation in the style of Plotkin [15].

The automation of a definitional mechanism for inductively defined relations in HOL and the use of this tool to apply the principle of rule induction in proofs of properties of such relations is discussed extensively in [13] and [5]. The definition of the relations EXPEVAL and CMDEVAL in HOL(see Fig.3), yield theorems asserting that the relations are the least relations closed under the rules. Of course, one can write conversions to animate the behaviour of these relations: in this case such conversions would act respectively, as an evaluator of expressions and an interpreter of commands.

6   A correctness proof of the Compiler

The main point of this section is not the fact that our trivial compiler is correct, but, the use of the compilation and execution animation conversions in the correctness proof of the compiler.

Our goal is to prove that according to the definitions of the operational semantics of IMP given that a program *prg* in some initial environment and store yields a final store *s* then, on executing the corresponding compiled code, the target machine terminates in a state which corresponds to *s*. We express this in HOL as follows:

```
∀prg s. (CMDEVAL prg init_env init_store s) ==>
  (∃n. MachineSteps n (Comprog prg) (1, init_store, []) =
  (0,s,[]))
```

where init_env = $((\lambda x.0),1)$:env and init_store = $(\lambda x.0)$:store.

The proof is tackled in two steps. First, the correctness of the compilation of IMP expressions is established, then this is used as a lemma in the correctness proof of the command compiler. The desired goal is a corollary of the latter step.

$$E1 \quad \frac{}{\text{EXPEVAL (Const } v) \ e \ s \ v}$$

$$E2 \quad \frac{}{\text{EXPEVAL (Var } i) \ e \ s \ s((\text{FST } e)i)}$$

$$E3 \quad \frac{\text{EXPEVAL } ex \ e \ s \ v}{\text{EXPEVAL (Unop } op_1 \ ex) \ e \ s \ (op_1 \ v)}$$

$$E4 \quad \frac{\text{EXPEVAL } ex_1 \ e \ s \ v_1 \qquad \text{EXPEVAL } ex_2 \ e \ s \ v_2}{\text{EXPEVAL (Binop } op_2 \ ex_1 \ ex_2) \ e \ s \ (op_2 \ v_1 \ v_2)}$$

$$C1 \quad \frac{}{\text{CMDEVAL Skip } e \ s \ s}$$

$$C2 \quad \frac{}{\text{CMDEVAL (Assign } i \ ex) \ e \ s \ (\text{assign } e \ s \ i \ v)} \qquad \text{EXPEVAL } ex \ e \ s \ v$$

$$C3 \quad \frac{\text{CMDEVAL } c_1 \ e \ s \ s' \qquad \text{CMDEVAL } c_2 \ e \ s' \ s''}{\text{CMDEVAL } (c_1; c_2) \ e \ s \ s''}$$

$$C4 \quad \frac{\text{CMDEVAL } c_1 \ e \ s \ s'}{\text{CMDEVAL (If } ex \ c_1 \ c_2) \ e \ s \ s'} \qquad \text{EXPEVAL } ex \ e \ s \ \text{tt}$$

$$C5 \quad \frac{\text{CMDEVAL } c_2 \ e \ s \ s'}{\text{CMDEVAL (If } ex \ c_1 \ c_2) \ e \ s \ s'} \qquad \text{EXPEVAL } ex \ e \ s \ \text{ff}$$

$$C6 \quad \frac{}{\text{CMDEVAL (While } ex \ c) \ e \ s \ s} \qquad \text{EXPEVAL } ex \ e \ s \ \text{ff}$$

$$C7 \quad \frac{\text{CMDEVAL } c \ e \ s \ s' \qquad \text{CMDEVAL (While } ex \ c) \ e \ s' \ s''}{\text{CMDEVAL (While } ex \ c) \ e \ s \ s''} \qquad \text{EXPEVAL } ex \ e \ s \ \text{tt}$$

$$C8 \quad \frac{\text{CMDEVAL } c \ (\text{update } i \ e) \ s \ s'}{\text{CMDEVAL (Decl } i \ c) \ e \ s \ s'}$$

Figure 3: The definition of the transition relations EXPEVAL and CMDEVAL

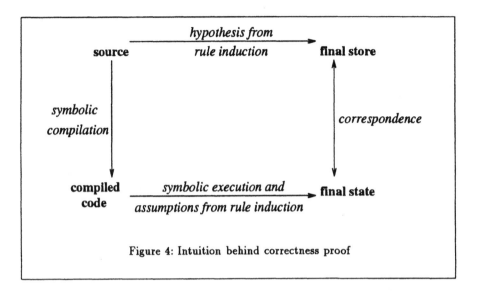

Figure 4: Intuition behind correctness proof

In each case, the proof proceeds by rule induction followed by the use of the compilation and execution animation conversions to obtain a general form of the final state which can be shown to correspond to the final store (see Fig.4).

6.1 Proof of the expression compiler

The aim is to prove that if an expression ex evaluates to v under the environment e and store s, then the execution of the compiled code corresponding to ex would push v to the top of the stack, leaving the initial store unaltered. Formally this is expressed as:

$\forall ex\ e\ s\ v.$ (EXPEVAL ex e s v) ==>
($\forall st.\ \exists n.$ MachineSteps n (Comexp ex (FST e)) $(1,s,st)$ =
(LENGTH(Comexp ex (FST e))+1, s, CONS v st))

Applying the principle of rule induction on EXPEVAL, transforms the term above into four subgoals which have the same format:

$\exists n.$ MachineSteps n (Comexp ex (FST e)) $(1,s,st)$ = $(p,s,$CONS v' $st)$

The subterms (Comexp ex (FST e)) are then animated using the symbolic compilation conversion, producing theorems of the form:

$\vdash$ Comexp ex (FST e) = A_1 ++ A_2 ++ A_3 ++ $\cdots$ ++ A_m

Rewriting with these theorems reduces the subgoals to the form:

$\exists n.$ MachineSteps n $(A_1$ ++ A_2 ++ A_3 ++ $\cdots$ ++ $A_m)$ $(1,s,st)$ = $(p,s,$CONS v' $st)$

The instruction lists $A_1, \ldots, A_m$ can be animated separately using the symbolic execution conversion to yield theorems of the form:

$\vdash$ MachineSteps n_i A_i $(1, s_i, st_i)$ = (p_i, s_{i+1}, st_{i+1})

If some A_i, where $1 \leq i \leq m$, cannot be animated (i.e. if A_i is not instantiated to a specific list of instructions), then the corresponding theorem of the above form can be deduced by a simple manipulation on the assumptions. Furthermore, through the use of basic arithmetic theorems, one can show that

$\vdash$ p_i = LENGTH (A_i) + 1

This is always possible since the animation conversion was programmed to stop execution whenever the program counter exceeds the length of the script.

The final outcome is a series of theorems of the form:

$..\vdash$ MachineSteps n_1 A_1 $(1, s_1, st_1)$ = (LENGTH(A_1)+1, s_2, st_2)

$..\vdash$ MachineSteps n_2 A_2 $(1, s_2, st_2)$ = (LENGTH(A_2)+1, s_3, st_3)

$\vdots$

$..\vdash$ MachineSteps n_m A_m $(1, s_m, st_m)$ = (LENGTH(A_m)+1, s_f, st_f)

These were used in an automated transitive argument to yield the theorem:

$..\vdash$ MachineSteps $(n_1+n_2+ \ldots +n_m)$ $(A_1$ ++ $\cdots$ ++ $A_m)$ $(1, s_1, st_1)$ = (LENGTH$(A_1$ ++ $\cdots$ ++ $A_m)$+1, s_f, st_f)

Finally, showing that $s_f = s$ and $st_f =$ CONS v' st proves the goal.

To illustrate this further, consider the subgoal resulting from the rule induction in the case of unary operators:

$\exists n.$ MachineSteps n (Comexp (Unop op_1 ex) (FST e)) $(1, s, st)$ = (LENGTH(Comexp(Unop op_1 ex)(FST e))+1, s, CONS $(op_1$ $v)$ st)

on the assumption that

$\forall st.$ $\exists n.$ MachineSteps n (Comexp ex (FST e)) $(1, s, st)$ = (LENGTH(Comexp ex (FST e))+1, s, CONS v st)

Symbolically compiling (Unop op_1 ex) under (FST e), produces the theorem:

$\vdash$ Comexp (Unop op_1 ex) (FST e) = (Comexp ex (FST e)) ++ [OP1 op_1]

Specialising the assumption and subsequently choosing a value for n, yields the theorem:

$..\vdash$ MachineSteps n' (Comexp ex (FST e)) $(1, s, st)$ = (LENGTH(Comexp ex (FST e))+1, s, CONS v st)

and, animating [OP1 op_1] with initial state $(1, s, \text{CONS } v \; st)$ produces:

$\vdash$ **MachineSteps** 1 [OP1 op_1] $(1, s, \text{CONS } v \; st) = (2, s, \text{CONS } (op_1 \; v) \; st)$

Combining these two theorems proves the required subgoal:

```
..⊢ MachineSteps (n'+1)
     ((Comexp ex (FST e))++[OP1 op₁]) (1,s,st) =
     (LENGTH(Comexp ex (FST e))+2,s,CONS(op₁ v)st)
```

This procedure can be used whenever execution flows through the sublists of instructions A_1 gradually to A_m with no haphazard jumps from one subscript to another. This is the case for the compiled code generated for expressions, thus we can treat each sublist of instructions separately, producing a set of theorems which can be combined in the proof of the subgoals generated by rule induction. The proof of correctness of the command compiler is slightly more complex.

6.2 The correctness of the command compiler

The correctness of the command compiler can be expressed as the following goal:

```
⊢ ∀c e s s'. CMDEVAL c e s s' ==>
    (∀st. ∃n. MachineSteps n (Comcmd c e)
    (1,s,st) = (LENGTH(Comcmd c e)+1,s',st))
```

As before, the proof proceeds by rule induction which in the case of **CMDEVAL** produces eight subgoals of the general form:

```
∃n. MachineSteps n (Comcmd c e) (1,s,st) = (p,s',st)
```

Applying the symbolic compilation conversion on Comcmd $c \; e$ produces a theorem of the form:

$\vdash$ **Comcmd** $c \; e = A_1 \; \texttt{++} \; \cdots \; \texttt{++} \; A_m$

It is not possible to consider the subscripts A_1 to A_m separately because, in general, execution does not flow gradually from left to right in the compiled code of some constructs. A case in point is the code generated for the looping constructs which includes forward as well as backward jumps. The execution of the whole script $A_1 \; \texttt{++} \; \cdots \; \texttt{++} \; A_m$, however, can be animated to produce theorems of the form:

$\vdash$ **MachineSteps** $n_i \; (A_1 \; \texttt{++} \; \cdots \; \texttt{++} \; A_m) \; (p_i, s_i, st_i) = (p_{i+1}, s_{i+1}, st_{i+1})$

Once again, if animation is not possible in the case of some uninstantiated sublist of instructions A_j, the corresponding theorem, of the form above, can be obtained through manipulation of the assumptions.

Applying the animation conversion and manipulating the assumptions recursively, one can prove a set of theorems:

..⊢ **MachineSteps** n_1 $(A_1$ ++ $\cdots$ ++ $A_m)$ $(1,s,st) = (p_2,s_2,st_2)$

..⊢ **MachineSteps** n_2 $(A_1$ ++ $\cdots$ ++ $A_m)$ $(p_2,s_2,st_2) = (p_3,s_3,st_3)$

$\vdots$

..⊢ **MachineSteps** n_x $(A_1$ ++ $\cdots$ ++ $A_m)$ $(p_x,s_x,st_x) = (p_f,s',st)$

which can be used to prove the theorem:

..⊢ **MachineSteps** $(n_1+n_2+ \cdots +n_x)$ $(A_1$ ++ $\cdots$ ++ $A_m)$ $(1,s,st) =$
(p_f,s',st)

Hence the proof of the subgoal.

To illustrate the above, consider the case of the two-armed conditional when the expression evaluates to tt. The subgoal is:

$\exists n.$ **MachineSteps** n (Comcmd (If ex c_1 c_2) e)
$(1,s,st) =$
$(1+$LENGTH(Comcmd(If ex c_1 c_2)e)$,s',st)$

with assumptions:

1. **EXPEVAL** ex e s tt

2. $\forall st.$ $\exists n.$
MachineSteps n (Comcmd c_1 e) $(1,s,st) =$
$(1 +$ LENGTH(Comcmd c_1 e)$,s',st)$

Applying the compilation conversion on Comcmd(If ex c_1 c_2)e yields:

⊢ Comcmd(If ex c_1 c_2)e = A ++ B ++ C ++ D ++ E

where

A = Comexp ex (FST e)
B = [JZ T (2 + LENGTH C)]
C = Comcmd c_1 e
D = [JMP T (1 + LENGTH E)]
E = Comcmd c_2 e

The following set of theorems can then be obtained:

1. Applying the correctness theorem of the expression compiler on the first assumption and manipulating the resultant theorem further, yields:

..⊢ **MachineSteps** n' $(A$ ++ $\cdots$ ++ $E)$ $(1,s,st) =$
$(1+$LENGTH $A,s,$CONS tt $st)$

2. Animating A ++ $\cdots$ ++ E from initial state (1+LENGTH A,s,CONS tt st),
   yields:

   $\vdash$ MachineSteps 1 $(A$ ++ $\cdots$ ++ $E)$ (1+LENGTH A,s,CONS tt st) =
   (2+LENGTH A,s,st)

3. By manipulating the second assumption one gets:

   ..$\vdash$ MachineSteps n'' $(A$ ++ $\cdots$ ++ $E)$ (2+LENGTH A,s,st) =
   (2+ LENGTH A + LENGTH C,s',st)

4. Animating with initial state $\cdot$(2+(LENGTH A)+(LENGTH C),s',st) yields:

   $\vdash$ MachineSteps 1 $(A$++$\cdots$++$E)$ (2+(LENGTH A)+(LENGTH C),s',st) =
   (3+(LENGTH A)+(LENGTH C)+(LENGTH E),s',st)

Using the four theorems above one can prove the subgoal:

..$\vdash$ MachineSteps (n'+1+n''+1) $(A$ ++ $\cdots$ ++ $E)$ $(1,s,st)$ =
(1+LENGTH$(A$++B++C++D++$E)$,s',st)

7   Conclusion

The concluding remarks of [4] postulate that the same animation conversions used to formally assert and illustrate the behaviour of definitions, can be reused to simplify proofs of theorems related to the definitions. This point has been confirmed and justified by the work presented in this paper through conversions which both animate the compilation and execution of programs of a language as well as play a central role in the proof of its compiler.

One may choose to argue that, for the simple language chosen to illustrate this point, the work dedicated to the implementation of the conversions is overkill. This argument does not hold when one scales up to a realistic programming language [4], with a complex compiler and operational semantics. In such cases, automation of part of the proof process is crucial for clarity and sanity, especially where certain arguments occur persistently in the subgoals.

The construction of generic conversions on-the-fly, to illustrate the behaviour of definitions as they are specified, is almost certainly a good investment if the ultimate intention is to prove some interesting properties about the definitions. In general, animation conversions provide a sound basis for an elegant proof methodology.

8   Acknowledgements

Thanks to Mike Gordon for inspiring our line of thought.

References

1. Aagaard and Leeser: Verifying a logic synthesis tool in Nuprl. 4th Workshop on Computer Aided Verification. (1992)
2. R. Boulton, M. Gordon, J. Herbert, and J. Van Tassel: The HOL Verification of ELLA Designs. Technical Report 199, University of Cambridge Computer Laboratory, August 1990. Revised version in the Proceedings of the International Workshop on Formal Methods in VLSI Design, Miami. (1991)
3. Albert John Camilleri: Executing Behavioural Definitions in Higher Order Logic. Ph.D. thesis, University of Cambridge Computer Laboratory. (1988)
4. Juanito Camilleri: Symbolic Compilation and Execution of Programs by Proof: A case study in HOL. Technical Report 240, University of Cambridge Computer Laboratory. (1991)
5. Juanito Camilleri and T.F. Melham: Reasoning with Inductively Defined Relations in the HOL Theorem Prover. Technical Report 265, University of Cambridge Computer Laboratory. (1992)
6. Paul Curzon: A verified compiler for a structured assembly language. In Proc. 4th International HOL Users Meeting. (1991)
7. Paul Curzon: Of what use is a verified compiler specification? Technical Report 274, University of Cambridge Computer Laboratory. (1992)
8. M.J.C. Gordon: HOL - A Machine Oriented Formulation of Higher Order Logic. Technical Report 68, University of Cambridge Computer Laboratory. (1985)
9. Matthew Hennessy: The Semantics of Programming Languages: An Elementary Introduction using Structural Operational Semantics. Wiley. (1990)
10. Hall and Windley: Simulating microprocessors from formal specifications. Higher Order Logic Theorem Proving and its Applications, Horth Holland. (1992)
11. J.J. Joyce: A verified compiler for a verified microprocessor. Technical Report 167, Computer Laboratory, University of Cambridge. (1989)
12. T.F. Melham: Automating Recursive Type Definitions in Higher Order Logic. In *Current Trends in Hardware Verification and Automated Deduction*, G.Birtwhistle and P.A.Subrahmanyam eds., Springer Verlag. (1988)
13. T.F. Melham: A package for Inductive Relation Definitions in HOL. In Proc. 4th International HOL Users Meeting. (1991)
14. Lawrence Paulson: A higher-order implementation of rewriting. In Science of Computer Programming **3** (1983) 119–149
15. Gordon D. Plotkin: A Structural Approach to Operational Semantics. Technical Report, Department of Computer Science, Aarhus University Denmark. (1981)
16. Rajan: Executing HOL. In Higher Order Logic Theorem Proving and its Applications. Horth Holland. (1992)

Datatypes in L2

Nick Chapman[1], Simon Finn[1], Michael P. Fourman[2]

[1] Abstract Hardware Ltd.
[2] Abstract Hardware Ltd. and Edinburgh University

Abstract. We describe the axiomatisation of a subset of Standard ML's datatypes in L2 (the LAMBDA Logic). The subset includes parameterisation and mutual recursion but has restrictions on the use of function type construction. We sketch a set-theoretic model for these datatypes. Finally, we briefly discuss the relationship between L2's datatypes and datatypes in HOL.

1 Introduction

LAMBDA is a proof assistant designed for the specification and verification of digital systems. User-defined datatypes are an important tool for expressing well-structured specifications.

Early versions of LAMBDA (prior to LAMBDA 4.0) used a 'free' logic, allowing terms that may not denote. This logic could support a rich set of datatypes – essentially[3] the same as Standard ML [8]. The semantics of these datatypes can be described in a standard domain-theoretic way [4]; in fact the presence of the existence predicate, E, means that the information-theoretic domain ordering is actually expressible in the logic (which therefore contains LCF as a sub-logic). In about 1991, we decided to change the logic used within the LAMBDA system. The basic reason for this change is that the old logic appears to be too expressive for the intended usage of the LAMBDA system; hardware designers are rarely impressed by having to consider the subtle distinction between the two functions $\lambda x. \perp$ and $\perp$, for example.

The new LAMBDA logic – now known as L2 – borrows heavily from HOL, but with a concrete syntax based on Standard ML. The philosophy of LAMBDA is somewhat different from that of HOL; rather than trying to reduce every proof to a small number of axioms, we are (relatively) happy to allow the system to construct new axioms from user-supplied definitions. This difference becomes most apparent in the treatment of recursive functions – where LAMBDA doesn't require function definitions to be primitive recursive (see [3] for details) – and in the current work on datatype definition.

SML-style datatype definitions provide a natural way to express specifications, as we had discovered using the 'old' logic, so we wanted to provide them

---

[3] Standard ML allows the definition of datatypes that are too general – in the sense that you can't traverse them with a well-typed recursive function; LAMBDA doesn't support these datatypes.

as part of L2 too. Melham [7] had already shown how to embed a useful subset of the datatype language within HOL. However, it is simply not possible to provide the full generality of SML datatypes within HOL's set-theoretic model – a simple argument about set cardinalities shows this; Gunter [5] provides a constructive proof – in HOL – that it's not possible in any other sort of model for HOL either.

Given the constraint of keeping the logic consistent, what kind of datatypes can we allow? We believe that the version[4] of L2 supported by LAMBDA 4.3 (as described in [2]), which includes parameterisation, mutual recursion and the (limited) use of function space constructors, is pretty close to the maximal datatype language that can be supported by a HOL-like logic. The L2 datatype sublanguage is, in fact, very similar to the 'full class of [datatype] specifications' outlined by Gunter in [6]. The principal difference is that L2 datatype definitions are able to make use of existing type constructors (and we give sufficient conditions for this use to be 'legal') whereas Gunter excludes this, although she adds:

> 'It is also possible to extend the notion of admissibility to include occurrences of certain kinds of type constructors, but the precise definition of this case is quite complicated and we omit it here.'

2 Design Aims

Our design aims for datatypes in L2 are:

1. The syntax should be the same as that used for datatypes in Standard ML.
2. The class of datatypes provided should be as rich as possible within L2's classical, polymorphic, higher-order type-theory.
3. Any restrictions imposed on the ML datatypes should be semantically rather than syntactically based.
4. The induction rules generated by the system should be easy to use within the LAMBDA proof system.

We have made the following restrictions with respect to Standard ML's datatypes:

2.1 Function Space Restriction

Every datatype must be small enough to be modelled as a set. In particular, within the body of a datatype definition, there must be no occurrence of that datatype on the left-hand side of a function arrow. This restriction is treated semantically, so that

---

[4] L2 evolves as our ideas evolve; in particular, the original version of L2 – supported by LAMBDA 4.0 – had much poorer support for datatypes.

```
datatype ('a,'b) arrow = Arrow of 'a -> 'b;
datatype bad = Bad of (bad,bool) arrow;
```

is, of course, illegal.

To enforce the semantic restriction, LAMBDA computes, for each datatype and each type parameter, whether that parameter is 'dirty' (occurs on the left of a function-space arrow or as a subtype parameter) or 'clean'. (Subtype parameters are 'dirty' because L2 subtype construction is not, in general, monotonic; increasing the size of the carrier of the parameter to a polymorphic subtype may *decrease* the size of the carrier of the result. In fact – with a suitable subtype predicate – the size of the subtype can be arbitrarily related to the size of the parameter type.) Recursive instances of the datatype within the body of its definition are legal only if they occur in clean positions.

For simplicity, we make the conservative assumption that a parameterised datatype actually depends on all of its type parameters. This means that LAMBDA may occasionally reject definitions which we could, semantically, allow. For example:

```
datatype 'a ignore = X
datatype funny = Y of funny ignore -> bool
```

LAMBDA will reject this definition of funny, because it assumes that funny ignore – which occurs on the left of a function arrow – actually depends on funny. If this restriction became irksome, we could keep track of which datatypes embed which of their parameters but, for the moment, this seems an unnecessary refinement.

2.2 Non-emptiness Restriction

Every datatype must be non-empty. For example, the definition

```
datatype empty = Empty of empty;
```

is *not* allowed. Note that we impose a semantic restriction rather than saying something syntactic like 'every datatype must contain a nullary constructor'. This means that we can allow useful definitions such as

```
datatype 'a gentree = Tree of 'a gentree list * 'a;
```

LAMBDA enforces the non-emptiness constraint on datatypes by means of an abstract interpretation. Each L2 parameterised datatype is associated with a boolean function; this function has one boolean parameter for each type parameter of the datatype and returns a boolean result. For a non-parameterised datatype, this function degenerates into a single boolean value.

Informally, the interpretation of the boolean value true is that we know that the carrier of the corresponding type is non-empty. (As in HOL, all *legal* L2 types have non-empty carriers. Since we are trying to *establish* that a given datatype is legal, however, we can't make that assumption here. As we will see later, empty sets *do* make an appearance in our model for datatypes; what we have to show is that all *legal* types are modelled by non-empty sets.) The boolean function corresponding to an L2 datatype tells us whether we can construct an element of that datatype, on the assumption that we are given elements of some of the parameter types (those for which the parameter in the abstraction is **true**).

A recursive datatype definition will give rise to a recursive equation for the corresponding boolean function. We solve such recursive equations by taking the least fixed point of the corresponding functional i.e. we assume that the datatype is empty unless we can prove otherwise. (We can only guarantee to find a fixed point because we know that the functional is *monotonic*. This wouldn't necessarily be the case if we allowed the recursively defined datatype to occur on the left-hand side of a function arrow. In practice, this means that we have to check that this doesn't occur *before* we check for non-emptiness.) The datatype definition is legal (or at least, not illegal on the grounds of emptiness) if the boolean function returns **true** when all its parameters are **true**. An example may make this clearer. Suppose we have the L2 definitions:

```
datatype ('a,'b) choice = A of 'a | B of 'b;
datatype 'a list = nil | :: of 'a * 'a list;
datatype 'a tree = ('a, 'a tree list) choice;
```

The corresponding boolean functions would satisfy the following equations:

```
f_choice(a,b) = a \/ b
f_list(a)     = true \/ (a /\ f_list(a))
f_tree(a)     = f_choice(a,f_list(f_tree(a)))
```

which have the least-fixed point solutions:

```
f_choice(a,b) = a \/ b
f_list(a)     = true
f_tree(a)     = true
```

2.3 Parameter Uniformity Restriction

For a parameterised datatype, all instances of the datatype occurring in the body of the declaration must have identical parameters to the defining instance. For example, the following definition is *not* allowed:

```
datatype 'a up = Up of 'a | Down of ('a up) up;
```

This restriction is needed to ensure that the induction rule generated for the datatype (see below) is well-typed.[5]

---

[5] Such datatypes, although legal in Standard ML, are actually useless in practice for just the same reason – the impossibility of writing well-typed recursive functions to traverse them.

For mutually recursive datatypes, we have an obvious[6] generalisation of this restriction. All the datatypes being defined together must have the same number of type parameters and all instances of any of the mutually recursive datatypes occurring in the body of any of the declarations must have the same type parameters as occur in the head of that declaration. For example,

```
datatype 'a gentree    = Tree of 'a gentreelist * 'a
     and 'b gentreelist = List of 'b gentree list
```

is legal, but

```
datatype ('a,'b) swap1 = X | Y of ('a,'b) swap2
     and ('c,'d) swap2 = A | B of ('d,'c) swap1
```

is not, because the occurrence of ('d,'c)swap1 within the definition of ('c,'d)swap2 is illegal – the type parameters don't occur in the same order.

3 Axiomatisation

Given a legal L2 datatype definition, LAMBDA produces a number of rules to axiomatise the properties of that datatype. These rules fall into 3 classes:

1. For each unary constructor, LAMBDA produces a rule stating that it is a 1-1 function i.e. two terms built using the constructor are equal only if they have equal arguments.
2. For each pair of distinct constructors, LAMBDA produces a rule stating that two terms built using different constructors are unequal.
3. For each datatype, LAMBDA produces an induction rule stating that every value in the datatype can be built using one of the constructors.

The first two classes of rules are uninteresting and will not be discussed further. By contrast, constructing appropriate induction rules is somewhat non-trivial and – for parameterised or mutually-recursive datatypes – also requires the axiomatisation of a number of auxiliary functions, as will be described below.

The first of these auxiliary functions is the **extend** function. The **extend** function corresponding to a parameterised datatype takes one parameter – a predicate – for each type parameter of the datatype definition and produces a predicate which operates on the datatype itself. Roughly speaking, the **extend** function applies each predicate to all subterms of the corresponding type and conjoins the results. For example, the L2 datatype definition

```
datatype 'a tree =
  Empty
| Just of 'a
| Pair of bool -> 'a
| Many of 'a tree * 'a tree list;
```

---

[6] This generalisation is so 'obvious', in fact, that we needed 6 months to discover it.

would produce the induction rule:

```
G // H |- forall t,l. Ptree#(t)
                    /\ extend'list (fn x => Ptree#(x)) l
                          ->> Ptree#(Many (t,l))
G // H |- forall f. Ptree#(Pair f)
G // H |- forall x. Ptree#(Just x)
G // H |- Ptree#(Empty)
----------------------------------------------------------
G // H |- forall t. Ptree#(t)
```

which uses the function **extend'list** – previously generated from the definition of the **list** datatype – and would also define the **extend'tree** function:

```
fun extend'tree p Empty       = TRUE
  | extend'tree p (Just x)    = p x
  | extend'tree p (Pair f)    = forall b:bool. p (f b)
  | extend'tree p (Many(t,l)) =
      extend'tree p t /\ extend'list (extend'tree p) l
```

so that **tree** can itself be used in future datatype definitions. In addition to the explicit induction rules, LAMBDA allows the definition of 'primitive recursive' functions that manipulate the newly introduced datatype. For example, LAMBDA would recognise the following function definitions as primitive recursive:

```
fun countItems Empty     = 0
  | countItems (Just x) = 1
  | countItems (Pair f) = 2
  | countItems (Many (t,tl)) =
      countItems t + countItemsInList tl

and countItemsInList []       = 0
  | countItemsInList (t::ts) =
      countItems t + countItemsInList ts
```

We discuss LAMBDA's definition of 'primitive recursive' in more detail later. The combination of the explicit datatype axioms together with the principle of definition of primitive recursive functions is categorical i.e. they determine the structure of the values of the datatype (up to isomorphism).

When we have mutually recursive datatype definitions, expressing the induction rules requires an extra family of auxiliary functions – the **convert** functions. For example the definition

```
datatype 'a T  = Node of 'a * 'a TL
     and 'b TL = Nil | Cons of 'b T * 'b TL
```

generates the following pair of induction rules:

```
G // H |- forall x,tl. convert'T'TL (fn t => PT#(t)) tl
                        ->> PT#(Node(x,tl))
----------------------------------------------------------------
G // H |- forall t. PT#(t)

G // H |- forall t,tl. convert'TL'T (fn tl => PTL#(tl)) t
                       /\ PTL#(tl)
                          ->> PTL#(Cons(t,tl))
G // H |- PTL#(Nil)
----------------------------------------------------------------
G // H |- forall tl. PTL#(tl)
```

Each of these induction rules uses an additional 'convert' auxiliary function. The intuition behind the convert functions is that the predicate convert'X'Y P holds of an object y of type Y precisely if P holds of all the immediate subterms of y which are of type X e.g. convert'T'TL converts an (inductive) predicate on T into a predicate on TL. The definition of these functions is

```
fun convert'T'TL f Nil          = TRUE
  | convert'T'TL f (Cons (x,y)) = f x /\ convert'T'TL f y

fun convert'TL'T f (Node (x,y)) = f y
```

In general, defining n mutually recursive datatypes generates n induction rules and n groups of convert functions, where each group contains $n - 1$ mutually recursive functions.

It would have been possible to define the induction rules *without* introducing the auxiliary convert functions. For example, we could have produced the following, apparently simpler, rules:

```
G // H |- forall x,tl. PTL#(tl) ->> PT#(Node(x,tl))
G // H |- forall t,tl. P#(t) /\ PTL#(tl) ->> PTL#(Cons(t,tl))
G // H |- PTL#(Nil)
----------------------------------------------------------------
G // H |- forall t. PT#(t)

G // H |- forall x,tl. PTL#(tl) ->> PT#(Node (x,tl))
G // H |- forall t,tl. P#(t) /\ PTL#(tl) ->> PTL#(Cons(t,tl))
G // H |- PTL#(Nil)
----------------------------------------------------------------
G // H |- forall tl. PTL#(tl)
```

This – allowing for differences in the logic – is how we treated mutually recursive datatypes in LAMBDA 3.X. The reason that we don't use these seductively simple rules within the current version of LAMBDA is that they are *hard to use*. There are two reasons for this:

1. If we are using an induction rule to perform case analysis rather than full-blown induction, the 'simple' rules force us to consider constructors from all the mutually-recursive datatypes, rather than only the datatype of interest.
2. When we use an induction rule for 'real' induction, we need to instantiate the meta-variables (PT and PTL above) to produce the concrete induction scheme for the particular predicate that we wish to prove. LAMBDA's higher-order unification will instantiate one of these meta-variables for us when we apply the induction rule, but we will then have to instantiate the other(s) by hand. What makes this particularly annoying is that we normally need to define some auxiliary functions in order to perform the instantiation – we need, in fact, to define the **convert** functions by hand.

The apparently more complex induction rules than LAMBDA now generates solve both of these pragmatic problems.

4   Primitive Recursion

LAMBDA will recognise a function definition as primitive recursive if it can show by a simple syntactic check that the corresponding function always is total.[7]The syntactic conditions that a primitive-recursive function must fulfill are as follows. Suppose the function is defined by a series of clauses, each with the function symbol applied to n symbols. For each occurrence of the function symbol in the body of any clause

1. The function must be applied to at least 1 argument.
2. For some i, $0 \leq i < n$, the first i arguments must be identical to the first i patterns at the head of that clause. The $i + 1$'th argument must be strictly smaller than the $i + 1$'th pattern.

For mutually recursive functions, we slightly generalise the above rules. Suppose several mutually-recursive functions are defined by clauses. Then, for each occurrence of any of the mutually-recursive functions in the body of any of the clauses:

1. The function must be applied to at least 1 argument.
2. For some i, $0 \leq i < n$, where n is the number of patterns occurring in that particular clause[8], the first i arguments must be identical to the first i patterns. The $i + 1$'th argument must be strictly smaller than the $i + 1$'th pattern.

What does 'strictly smaller' mean? An expression is *smaller* than a pattern if one of the following holds:

---

[7] LAMBDA also allows the definition of non primitive-recursive functions. To make effective use of such a function the user has to discharge a side condition that says, essentially, that the function 'terminates'. This will be discussed in detail in [3].

[8] For mutually recursive functions, n may vary from clause to clause because different functions may have different numbers of parameters; for each individual function, the number of patterns in each clause should still be constant.

1. The pattern is a variable (N.B. *not* a constructor) and the expression is the same variable or consists of the application of that variable to one or more arguments.
2. The pattern is a nullary constructor and the expression is the same constructor.
3. The pattern and expression each consist of an application of the same unary constructor and the argument in the expression is smaller than the argument in the pattern.
4. The expression and pattern are both labelled records (this includes tuples) with the same labels and each subexpression is smaller than the corresponding subpattern.
5. The expression is smaller than a strict subpattern of the pattern.

An expression is *strictly smaller* than a pattern if it is smaller than the pattern, but not identical to it.

5 Axiomatisation within LAMBDA

In this section we describe the concrete form of the induction rules and auxiliary functions produced by LAMBDA.

5.1 Auxiliary Functions – extend

As noted above, LAMBDA generates higher-order 'extend' functions which take one parameter – a predicate – for each type parameter of the original datatype definition and produce a predicate which operates on the datatype itself. We characterised this function as applying each predicate to all subterms of the corresponding type and conjoining the results. This characterisation of extend is slightly too simple:

1. If the type parameter is embedded in the range of a function type, then the extend function must quantify over the range of the function, as for Pair in the above example. This means that we are interpreting 'subterm' in a semantic rather than a syntactic sense.
2. If the type parameter is ever embedded in the domain of a function type – i.e. the type is 'dirty' – then the corresponding predicate is never applied. This doesn't cause a problem because we define extend functions precisely so that we can use parameterised datatypes in the definition of new, indirectly recursive, datatypes (as we used list in the definition of tree, for example) and our restriction on datatype definitions excludes recursion through such 'dirty' parameters.

In general, the mutually recursive datatype definition

```
datatype ('a₁₁, ..., 'a₁ₙ) D₁ = ...
and ...
and ('aₖ₁, ..., 'aₖₙ) Dₖ = ...| Cₖᵢ | ...| Cₖⱼ of tₖⱼ | ...
```

gives rise to the k functions `extend'D`$_1$ ... `extend'D`$_k$. Conceptually, we define these functions as described below; in practice LAMBDA also performs a 'pattern-lifting' phase (essentially beta-reduction plus simplification of trivial conjuncts) to improve the readability of the generated definitions.

For nullary constructors, the `extend` function always returns TRUE

$$\text{extend'D}_x \ p_1 \ \dots p_n \ C_{xi} = \text{TRUE}$$

For unary constructors, its value depends on the structure of the type of the constructor

$$\text{extend'D}_x \ p_1 \ \dots p_n \ (C_{xi} \ v_{xi}) = [[t_{xi}]] \ v_{xi}$$

where the operation $[[\_]]$ is defined by

$[[t]] = \text{fn } v \Rightarrow \text{TRUE}$,
where t is any type containing no instance of a clean parameter.

$[['a_{xj}]] = p_j$,
where $'a_{xj}$ is the j'th parameter type and $'a_{xj}$ is a clean parameter.

$[[('a_{x1}, \ \dots, \ 'a_{xn})D_y]] = \text{extend'D}_y \ p_1 \ \dots p_n$,
where D_y is one of the mutually recursive datatypes[9] – possibly D_x itself.

$[[(t_1, \ \dots, \ t_l)D]] = \text{extend'D} \ [[t_1]] \ \dots [[t_l]]$,
where D is some other datatype constructor and some t_j contains a clean parameter. Note that this condition logically implies that the j'th parameter position of D must be clean.

$[[\{l_j \ : \ t_j\}]] = \text{fn } \{l_j \ : \ v_j\} \Rightarrow \bigwedge_j ([[t_j]] \ v_j)$,
where $\{l_j \ : \ t_j\}$ is a labelled record type and some t_j contains an instance of a clean parameter.

$[[t_1 \rightarrow t_2]] = \text{fn } f \Rightarrow \text{forall } x \ : \ t_1. \ [[t_2]] \ (f \ x)$,
where t_2 contains an instance of a clean parameter.

The case $[[(t_1, \ \dots, \ t_l)T]]$ where T is a type abbreviation is handled by expanding the abbreviation.

Note that the predicate p_j will never be applied if the corresponding type parameter, $'a_j$, is dirty. We could eliminate these parameters altogether, but we choose not to do so; this means that if, in the future, we change the definition of 'clean' – to take account of datatypes which don't embed their arguments, for example – we won't have to change the type of any existing `extend` function.

---

[9] This rule means that the `extend` functions for mutually recursive datatypes must also be mutually recursive.

5.2 Auxiliary Functions – convert

As noted above, the predicate convert'X'Y P holds of an object y of type Y precisely if P holds of all the immediate subterms of y which are of type X i.e. convert'X'Y converts an (inductive) predicate on X into a predicate on Y. This means that the convert'X'Y function will have type

$$(X \rightarrow om) \rightarrow Y \rightarrow om$$

Suppose we have the mutually recursive datatype definition

```
datatype ('a₁₁, ..., 'a₁ₙ) D₁ = ...
and ...
and ('aₖ₁, ..., 'aₖₙ) Dₖ = ...| Cₖᵢ | ...| Cₖⱼ of tₖⱼ | ...
```

Then, for nullary constructors, the convert function always returns TRUE

$$\text{convert}'D_x'D_y \ P_x \ C_{yi} = \text{TRUE}$$

For unary constructors, convert function depends on the structure of the type of the constructor

$$\text{convert}'D_x'D_y \ P_x \ (C_{yj} \ v_{yj}) = [[t_{yj}]] \ v_{yj}$$

where the compilation operation [[_]] is here defined to be

[[t]] = fn x => TRUE,
where t is any type containing no instance of any of the mutually-recursive datatypes.

$$[[('a_{y1}, \ldots, 'a_{yn})D_x]] = P_x$$

$$[[('a_{y1}, \ldots, 'a_{yn})D_z]] = \text{convert}'D_x'D_z \ P_x,$$
where D_z, distinct from D_x but possibly the same as D_y, is one of the mutually-recursive datatypes.

$$[[(t_1, \ldots, t_l)D]] = \text{extend}'D \ [[t_1]] \ldots [[t_l]],$$
where D is a previously-defined datatype constructor.

$$[[\{l_j : t_j\}]] = \text{fn} \ \{l_j : v_j\} => \bigwedge_j([[t_j]] \ v_j),$$
where $\{l_j : t_j\}$ is a labelled record type.

$$[[t_1 \rightarrow t_2]] = \text{fn} \ f => \text{forall} \ x : t_1. \ [[t_2]] \ (f \ x)$$

As for the extend functions, we handle the case $[[(t_1, \ldots, t_l)T]]$ where T is a type abbreviation by expanding the abbreviation.

As for the extend family of functions, LAMBDA performs pattern-lifting to optimise the definitions produced by the above naive algorithm.

5.3 Induction Rules – Construction

Suppose we have the mutually recursive datatype definition

    datatype ('a_{11}, ..., 'a_{1n}) D_1 = ...
    and ...
    and ('a_{k1}, ..., 'a_{kn}) D_k = ...| C_{ki} | ...| C_{kj} of t_{kj} | ...

LAMBDA will produce k induction rules, one for each datatype. The rule for each datatype consists of a conclusion plus one premiss for each constructor of that datatype. For the datatype D_x, the conclusion will be

    G // H |- forall w : ('a_{x1}, ..., 'a_{xn})D_x. PD_x#(w)

The premiss corresponding to a nullary constructor, C_{xi}, of type D_x will be

    G // H |- PD_x#(C_{xi})

For a unary constructor, C_{xi}, of type t_{xi} -> ('a_{x1}, ..., 'a_{xn})D_x, the premiss will be

    G // H |- forall \bar{v}_{xi}. pre_{xi} → PD_x#(pat_{xi})

where $\langle \bar{v}_{xi}, pat_{xi}, pre_{xi} \rangle$ = $[[t_{xi}]]$ and the compilation operator $[[\_]]$ is defined as follows:

$[[\{1_j : t_j\}]]$ = $\langle @_j\ \bar{v}_j, \{1_j : pat_j\}, \bigwedge_j pre_j \rangle$,
where $\{1_j : t_j\}$ is a labelled record type, $\langle \bar{v}_j, pre_j, pat_j \rangle$ = $[[t_j]]$, and we use the notation '$@_j\ \bar{v}_j$' to represent vector concatenation.

$[[('a_{x1}, ..., 'a_{xn})D_x]]$ = $\langle v, v, PD_x#(v) \rangle$,
where v is a new variable.

$[[('a_{x1}, ..., 'a_{xn})D_y]]$ =
    $\langle v, v, convert'D_x'D_y\ (fn\ z => PD_x#(z))\ v \rangle$,
where D_y is another of the mutually-recursive datatypes and v is a new variable.

$[[(t_1, ..., t_l)D]]$ =
    $\langle v, v, extend'D\ (fn\ pat_1 => pre_1)\ ...\ (fn\ pat_l => pre_l)\ v \rangle$,
where D is a previously-defined datatype constructor, there is an occurrence of one of the mutually recursive datatypes in at least one of the t_j, $\langle \bar{v}_j, pat_j, pre_j \rangle$ = $[[t_j]]$, and v is a new variable.

$[[t_1 \to t_2]]$ = $\langle f, f, forall\ x. pre[v \leftarrow (f\ x)] \rangle$,
where t_2 contains one of the mutually recursive datatypes,
$\langle v, v, pre \rangle$ = $[[t_2]]$, v is a variable, and f and x are new variables.

```
[[t₁ -> t₂]] =
    <f,f,forall x. (fn pat => pre) (f x)>,
```
where t_2 contains one of the mutually recursive datatypes,
`<v̄, pat, pre> = [[t₂]]`, pat is not a variable, and f and x are new variables.

As before, we handle the case `[[(t₁, ..., tₗ)T]]` where T is a type abbreviation by expanding the abbreviation.

`[[t]] = <v,v,TRUE>`,
where v is a new variable and none of the above rules apply.

6   Sketch of Semantics

How do we build a set-theoretic model for L2 datatypes? In general the L2 model would be similar to Pitts' set-theoretic model for HOL [9]. We then have to explain how to add the denotations of recursive datatypes.[10] We then proceed in something like the following stages:

1. We model an L2 datatype as the least fixed point of a monotonic function on a suitable lattice of sets (with a suitable appeal to Tarski's Fixed Point Theorem justify the existence of a fixed point.) The restrictions that we have made on the form of L2 datatypes are just what we need to ensure that such a monotonic function exists and that the resulting fixed point is a *non-empty* set. In particular:

   (a) We made the restriction that all instances of the datatype occurring in the body of the declaration must the same parameters as the defining instance. This means that we can treat the parameter types as fixed when we construct the fixed point and then parameterise the result. (If we didn't have this restriction we would need to find the fixed point of a functional rather than just a function.)

   (b) The restriction that recursive occurrences of the datatype occurring in the body of its definition may only occur in 'clean' positions is precisely what we need to show that the function is monotonic. (Here we need to make the assumption that previously defined parameterised datatypes give rise to functions that are indeed monotonic in their 'clean' parameters. We can justify this by an induction on the number of previously-defined datatypes.)

2. We next need to show that the newly-defined parameterised datatype is a monotonic function of its 'clean' parameters. This should be standard argument involving the least fixed points of monotonic functions.

---

[10] We also have to explain how to handle non-primitive recursive functions; this will be treated in [3] – the techniques used there are remarkably similar to our treatment of datatypes.

3. At this stage in the argument, we have established that the L2 datatype can be represented as a set. We next have to show that the abstract interpretation is correct i.e. that it is conservative in its prediction about whether the datatype is non-empty.

4. Next we have to consider the **extend** functions. If we regard them as functions on sets (represented by their characteristic functions) we can see that we can define the **extend** function for a datatype – as a least fixed point – in much the same way as we defined the datatype itself.

5. Finally the induction rules can be justified by an argument involving least fixed points of monotonic functions. The only complication here is that **extend** functions appear to ignore their 'dirty' arguments i.e. P_i is treated as if it were **fn _ => true** whenever the i'th parameter type is dirty. This isn't actually a problem, because when the i'th parameter type is dirty, P_i actually *is* **fn _ => true** i.e. 'dirty' types are treated as fixed and non-empty throughout the proof. (We could simplify this proof by making the definition of the **extend** function match the datatype definition more exactly, but that wouldn't be very user-friendly.)

7 Relationship of L2 datatypes to HOL datatypes

As we noted in the introduction, the main technical difference between L2 datatypes and Gunter's[6] HOL datatypes is that L2 datatype definitions may make use of existing type constructors. In some respects, this difference is not important because it is always possible to expand out the use of such type constructors by introducing new, mutually-recursive, datatypes. For example, we could treat the definition:

```
datatype 'a gentree = Tree of 'a gentree list * 'a;
```

as if it were:

```
datatype 'a gentree  =
  Tree of 'a gentree_list * 'a

and 'a gentree_list =
  Nil | Cons of 'a gentree * 'a gentree_list
```

If we do this consistently, we can reduce a collection of L2 datatype definitions into a form equivalent to Gunter's[6] 'full class of specifications'. (Our function-space condition is sufficient to show that the expanded form meets Gunter's admissibility conditions.) This is perhaps the simplest way to give a meaning to L2 datatype definitions.

Doing this at the source level would have a distinct price however. The two types **'a gentree list** and **'a gentree_list** are isomorphic but they are not identical. This means that it would not be possible to apply useful general purpose functions such as **map** to an object of type **'a gentree_list** and so it would be necessary to develop a separate theory of lists for each such 'instantiation' of the **list** constructor.

8 Future Work

When we started the first draft of this paper, we believed that our characterisation of L2 datatypes was essentially complete, and that the datatypes we described were in some sense 'maximal' for a HOL-like logic.[11]Since then, we have had a couple of ideas for extensions.

We currently treat all subtyping as 'dirty'. This means, for example, that if we add an integer index to each node of a gentree and specify, using subtyping, that such indices must all be distinct then we can't use the resulting type in any future datatype definition. Given the current HOL (or L2) type scheme, this seems to be unavoidable. The problem is that we can't tell whether or not the subtype predicate makes the subtype non-monotonic in the size of the subtype's parameters, so we have regard *all* the subtype's parameters as potentially nonmonotonic i.e. 'dirty'.

We believe that it may be possible to make a small change to the type scheme to remove this restriction, although we haven't worked out all the details yet. The basic idea is to borrow Standard ML's concept of 'imperative' type variables to keep track of which type parameters are 'clean' and which are 'dirty'. Standard functions have normal 'applicative' types, but (rather ironically) quantifiers get 'imperative' types rather like Standard ML's ref constructor.

Two reviewers pointed out the close relationship between the definition of a datatype and the associated principle of definition for functions on that datatype. Although we have successfully defined induction rules using parameterised datatypes, we have not done so well with the definitional principle. For example, we defined the function countItems as:

```
fun countItems Empty    = 0
  | countItems (Just x) = 1
  | countItems (Pair f) = 2
  | countItems (Many (t,tl)) =
      countItems t + countItemsInList tl

and countItemsInList []      = 0
  | countItemsInList (t::ts) =
      countItems t + countItemsInList ts
```

Here the recursion pattern for countItems, in particular the use of the auxiliary function countItemsInList, is exactly what one would expect if we had defined a local 'a tree_list datatype rather than using 'a tree list in the datatype definition. A more natural definition of countItems would be something like:

```
fun countItems Empty    = 0
  | countItems (Just x) = 1
```

---

[11] With the exception – already noted – that we can define a better function space restriction by keeping track of whether a type constructor actually uses all its type parameters.

```
| countItems (Pair f) = 2
| countItems (Many (t,tl)) =
    countItems t +
      fold'list (0, op +) (map'list countItems tl);
```

Here we are assuming that the **fold'list** and **map'list** functions would be automatically generated from the datatype definition for **list** and, crucially, that we can regard this definition as primitive recursive. There is clearly considerable scope for investigating LAMBDA's definition of 'primitive recursion'.

9 Acknowledgements

Our treatment builds on the work of Matt Fairtlough[1] and we are grateful to Matt for many useful discussions. We would also like the thank the anonymous referees many of whose constructive suggestions have been incorporated into this paper. The work reported in this paper was partially funded by the projects *Formal System Design* (IED/2/1292) and *Synthesis, Optimisation and Analysis* (JESSI AC8).

References

1. Matt Fairtlough, Research into ML Datatypes, in *Formal System Design (IED Project 1292) Deliverable D13*, Edinburgh University, February 1992.
2. Simon Finn, Michael P. Fourman, L2 – The LAMBDA Logic, in *LAMBDA 4.3 Reference Manuals*, Abstract Hardware Limited, September 1993.
3. Simon Finn, Michael Fourman, John Longley, *Partial Functions in a Total Setting*, in preparation.
4. M.P. Fourman and W.K. Phoa, A Proposed Categorical Semantics for Pure ML, in *ICALP '92 International Colloquium on Automata, Languages, and Programming*, *Wien Austria*, Springer-Verlag LNCS, 1993.
5. Elsa L. Gunter, Why We Can't have SML Style datatype Declarations in HOL, in *Higher Order Logic Theorem Proving and its Applications (HOL'92)*, ed. L.J.M. Claessen, M.J.C. Gordon, North-Holland 1993.
6. Elsa L. Gunter, A Broader Class of Trees for Recursive Type definitions in HOL, in *Higher Order Logic Theorem Proving and its Applications (HUG'93)*, ed. Jeffrey J. Joyce, Carl-Johan H. Seger, Lecture Notes in Computer Science 780, Springer-Verlag 1994.
7. Thomas F. Melham, Formalizing Abstraction Mechanisms for Hardware Verification in Higher Order Logic, PhD Thesis and Technical Report 201, University of Cambridge, August 1990.
8. Robin Milner, Mads Tofte and Robert Harper, *The Definition of Standard ML*, MIT Press, 1990.
9. Andy Pitts, Set-Theoretic Semantics, in *The HOL System DESCRIPTION*, HOL88 Documentation, 1991.

A Formal Theory of Undirected Graphs in Higher-Order Logic

Ching-Tsun Chou ⟨chou@cs.ucla.edu⟩

Computer Science Department, University of California at Los Angeles
Los Angeles, CA 90024, U.S.A.

Abstract. This paper describes a formal theory of undirected (labeled) graphs in higher-order logic developed using the mechanical theorem-proving system HOL. It formalizes and proves theorems about such notions as the empty graph, single-node graphs, finite graphs, subgraphs, adjacency relations, walks, paths, cycles, bridges, reachability, connectedness, acyclicity, trees, trees oriented with respect to roots, oriented trees viewed as family trees, top-down and bottom-up inductions in a family tree, distributing associative and commutative operations with identities recursively over subtrees of a family tree, and merging disjoint subgraphs of a graph. The main contribution of this work lies in the precise formalization of these graph-theoretic notions and the rigorous derivation of their properties in higher-order logic. This is significant because there is little tradition of formalization in graph theory due to the concreteness of graphs. A companion paper [2] describes the application of this formal graph theory to the mechanical verification of distributed algorithms.

1 Introduction

In view of the many applications in computer science that graph theory has, the scantiness of work on the formalization of graph theory in HOL—the mechanical theorem-proving system for higher-order logic developed by Gordon *et al.* [5, 6]— is somewhat surprising. The only prior work we are aware of is that of Wong [9], who was concerned mainly with formulating and proving properties of paths in *directed* graphs in order to reason about railway signaling schemes. As we are interested in the mechanical verification in HOL of such distributed algorithms as those in [4, 8], we need a formal theory of *undirected* graphs in HOL that includes at least the notion and important properties of *trees*. Consequently we have gradually formalized a considerable amount of graph theory in HOL over the past two or three years. The purpose of this paper is to describe this formal graph theory in some detail; a companion paper [2] describes its application to the mechanical verification of distributed algorithms.

Our graph theory consists of five segments with the following dependency relation:

$$\text{BASIC} \leftarrow \text{TREE} \leftarrow \text{ORIENT} \leftarrow \text{CONSTR}$$
$$\downarrow$$
$$\text{ACI}$$

where X → Y means that X depends on Y. The segment BASIC (described in Section 5) defines the basic notions of graph theory, including a somewhat unusual definition of undirected *labeled* graphs, and derives their basic properties. The segment TREE (Section 6) is a theory of paths, cycles, bridges, reachability, connectedness, acyclicity, and trees (a tree is defined to be a connected and acyclic graph). After proving many facts about these notions, it culminates with the theorem stating that there is a unique path between any two nodes in a tree. The uniqueness of paths implies that a tree with a distinguished node—called the root—can be oriented and viewed as a family tree by taking the root as the progenitor. The segment ORIENT (Section 7) is a theory of such oriented trees and contains theorems for doing both top-down and bottom-up inductions in a family tree and for distributing recursively an associative and commutative operation with an identity (an ACI operation, for short) over the subtrees of a family tree. The latter task relies on the segment ACI (Section 4), which defines a constant for generalizing ACI operations to finite sets. Finally, the segment CONSTR (Section 8) introduces several constructors of graphs that preserve properties like finiteness, connectedness and acyclicity (and hence "tree-ness" as well). Almost all theorems in our graph theory hold for both finite and infinite graphs. If a theorem holds for finite graphs only, it will be so stated explicitly.

Our graph theory is strictly definitional, meaning that it is developed from the initial theory of HOL without introducing any axioms other than definitions. The definitional approach has two well-known advantages. The first is *consistency*: definitions never introduce any inconsistencies. Since the initial theory of HOL is consistent [6], our graph theory is consistent as well. The second is *eliminability*: definitions can, in principle, be eliminated. So an auxiliary definition not involved in the statement of a result can safely be forgotten as far as that result is concerned, even if it is used in the proof of that result. To be sure, sticking to the definitional approach is sometimes laborious, since even "obvious" propositions must be honestly proved. But the logical security thus gained is well worth the effort.

The main contribution of this work lies *not* in the mathematical sophistication of the graph theory formalized, which is elementary (and all covered in, say, the first two chapters of Even's book [3]), but in the precise formalization of graph-theoretic notions and the rigorous derivation of their properties in higher-order logic. This is significant for two reasons. Firstly, it is not easy to formalize a piece of mathematics in *any* mechanical theorem-proving system. Most proofs and many definitions in typical mathematical textbooks contain gaps that are easy for a human reader to fill (at a subconscious level, in most cases) but too hard for a mechanical prover to overcome. So the human operator of a mechanical prover has to judiciously formulate and arrange definitions and theorems so that the desired results can be proved with as little effort as possible. This involves a lot of planning and experimentation. Secondly, it is particularly difficult to formalize those branches of mathematics, such as graph theory and combinatorics, in which there is little tradition of formalization because of the very concreteness of the objects being studied. In fact, as de Bruijn [1] pointed out, the more abstract a piece of mathematics is, the smaller the gaps usually are and the easier the formalization is.

The rest of this paper is organized as follows. Sections 2 and 3 are preliminaries about higher-order logic and predicates as sets, respectively. The segments of our graph theory are described in Sections 4 through 8. Due to space limitations, only the most significant theorems are listed and no proof is given. Section 9 is the conclusion.

2 Higher-Order Logic

The version of higher-order logic used in this paper is the one supported by the HOL theorem-proving system [6]. In general, the following typographic conventions are followed when writing formulas: lowercase Greek letters stand for type variables, *slanted* font for type constants, *italic* font for term variables, and sans serif font for term constants. Free variables in a definition or theorem are implicitly universally quantified unless otherwise stated. The symbol $\triangleq$ means "equals by definition"; "iff" means "if and only if". Lists are enclosed by square brackets ($[\cdots]$); list concatenation is denoted by $\frown$ (written as an infix).

3 Predicates as Sets

A predicate P is a function whose type is of the form $\alpha \to bool$, where α is called the **domain** of P. In this paper a set is *identified* with its characteristic predicate so that the following holds:

$$\{\, x : \alpha \mid P(x)\,\} \;=\; P$$

With this identification, the usual operations on sets, such as $\subseteq$, $\cap$, $\cup$, and $\setminus$ (set difference), are applicable to predicates as well. In particular, $x \in P$ is now synonymous with $P(x)$ and will be used interchangeably with it. The meaning of the $\{\cdots \mid \cdots\}$ notation for sets is given by:

$$t \in \{x : \alpha \mid Q[x]\} \;=\; Q[t]$$
$$t \in \{x :: P \mid Q[x]\} \;=\; t \in P \land Q[t]$$

where the $Q[x]$ notation stresses that the variable x may (but does not necessarily!) occur free in the term Q and $Q[t]$ is the term obtained by substituting t for the free occurrences of x in Q (with the usual proviso preventing capture of free variables). It should be pointed out that both $\in$ and the $\{\cdots \mid \cdots\}$ notation are used in this paper solely for the sake of readability: they are never used in the actual implementation, in which $x \in P$ is replaced by $P(x)$ and definitions of the forms $S \triangleq \{x \mid Q[x]\}$ and $S \triangleq \{x :: P \mid Q[x]\}$ by $S(x) \triangleq Q[x]$ and $S(x) \triangleq P(x) \land Q[x]$, respectively. (Definitions are the only places where the $\{\cdots \mid \cdots\}$ notation is used.)

Other notions involving sets that are needed in this paper are the following. **Restricted quantifications** over sets are defined by:

$$(\forall x :: P.\ Q[x]) \triangleq (\forall x.\ x \in P \Rightarrow Q[x])$$
$$(\exists x :: P.\ Q[x]) \triangleq (\exists x.\ x \in P \wedge Q[x])$$
$$(\varepsilon x :: P.\ Q[x]) \triangleq (\varepsilon x.\ x \in P \wedge Q[x])$$

The **union** of an indexed family of sets, $F : \iota \to \alpha \to bool$, over a set of indices $I : \iota \to bool$ is:

$$\mathsf{Union}(F)(I) \triangleq \{\, x : \alpha \mid \exists j :: I.\ x \in F(j) \,\}$$

Let $f : \alpha \to \beta$ and $P : \alpha \to bool$. The **image** of P under f is:

$$\mathsf{Image}(f)(P) \triangleq \{\, y : \beta \mid \exists x :: P.\ y = f(x) \,\}$$

For a binary relation $R : \alpha \to \alpha \to bool$, f is **one-one modulo** R over P iff:

$$\mathsf{OneOneMod}(R)(f)(P) \triangleq \forall x\, y :: P.\ (f(x) = f(y)) \Rightarrow R(x)(y)$$

See Section 5.2 for how **OneOneMod** is used. "P is a finite set" is denoted by $\mathsf{Finite}(P)$; see [7] for how to define Finite in HOL.

4 Generalizing ACI Operations to Finite Sets

Let $\mathsf{ACI}(op, id)$ mean that $op : \beta \to \beta \to \beta$ (which will be written as an infix) is an associative and commutative operation and $id : \beta$ is an identity for it:

$$\mathsf{ACI}(op, id) \triangleq$$
$$(\forall x\, y\, z.\ x\ op\ (y\ op\ z)\ =\ (x\ op\ y)\ op\ z\,)\ \wedge$$
$$(\forall x\, y.\ x\ op\ y\ =\ y\ op\ x\,)\ \wedge$$
$$(\forall x.\ x\ op\ id\ =\ x\,)$$

For example, both $\mathsf{ACI}(+, 0)$ and $\mathsf{ACI}(*, 1)$ are true.

Let $f : \alpha \to \beta$, $P : \alpha \to bool$, and $x : \alpha$. Extending the method used in [7] to define the cardinalities of finite sets, a constant Sop (for "set operation") can be introduced, via the *constant specification* mechanism in HOL [6], with the following property:

$$\mathsf{ACI}(op, id) \wedge \mathsf{Finite}(P) \wedge x \notin P \Rightarrow$$
$$(\,\mathsf{Sop}(op, id)(f)(\emptyset)\ =\ id\,)\ \wedge \tag{1}$$
$$(\,\mathsf{Sop}(op, id)(f)(\{x\} \cup P)\ =\ f(x)\ op\ \mathsf{Sop}(op, id)(f)(P)\,)$$

Informally, (1) amounts to saying that:

$$\mathsf{Sop}(op, id)(f)(\{x_1, \ldots, x_n\})\ =\ f(x_1)\ op\ \cdots\ op\ f(x_n)\ op\ id \tag{2}$$

Equation (2) shows clearly why the assumption $\mathsf{ACI}(op, id)$ is necessary: since $\{x_{i_1}, \ldots, x_{i_n}\} = \{x_1, \ldots, x_n\}$ for any permutation $(i_1, \ldots, i_n)$ of $(1, \ldots, n)$, the order in which op is applied to $f(x_i)$'s had better be unimportant; the identity id is needed to take care of the empty set (i.e., when $n = 0$). For example, $\mathsf{Sop}(+, 0)(f)(P)$ and $\mathsf{Sop}(*, 1)(f)(P)$ are what are usually written as $\sum_{x \in P} f(x)$ and $\prod_{x \in P} f(x)$, respectively. In particular, if $f(x) = 1$ for all x, then $\mathsf{Sop}(+, 0)(f)(P)$ is just the cardinality of P.

Let $I : \iota \to bool$ and $F : \iota \to \alpha \to bool$. It can be deduced from (1) that:

$$\mathsf{ACl}(op, id) \wedge \mathsf{Finite}(I) \wedge (\forall j :: I.\ \mathsf{Finite}(F(j))) \wedge$$
$$(\forall j\ k :: I.\ (j \neq k) \Rightarrow (F(j) \cap F(k) = \emptyset)) \Rightarrow \quad (3)$$
$$(\mathsf{Sop}(op, id)(f)(\mathsf{Union}(F)(I)) = \mathsf{Sop}(op, id)(\mathsf{Sop}(op, id)(f) \circ F)(I))$$

where $\circ$ denotes function composition. Informally, (3) says that:

$$\mathsf{Sop}(op, id)(f)(F_1 \cup \cdots \cup F_m) =$$
$$\mathsf{Sop}(op, id)(f)(F_1)\ op\ \cdots\ op\ \mathsf{Sop}(op, id)(f)(F_m) \quad (4)$$

provided that the F_j's are finite and mutually disjoint. For example, if $(op, id) = (+, 0)$, then (4) can be rephrased as:

$$\sum_{x \in F_1 \cup \cdots \cup F_m} f(x) = \sum_{j \in \{1, \ldots, m\}} \left(\sum_{y \in F_j} f(y) \right)$$

5 Basic Notions of Graph Theory

5.1 Nodes, Edges, and Links

Let α and β be the types of **nodes** and **edges**, respectively. Then $\alpha \times \beta \times \alpha$ is the type of **links**. A link (p, e, q) is essentially a directed edge, where $\mathsf{src}(p, e, q) \triangleq p$ is its **source**, $\mathsf{des}(p, e, q) \triangleq q$ is its **destination**, $\mathsf{via}(p, e, q) \triangleq e$ is its **via**, and $\mathsf{rev}(p, e, q) \triangleq (q, e, p)$ is its **reverse**. Two links l and l' are **almost equal** iff they are either identical or reverse of each other: $(l \approx l') \triangleq (l = l') \vee (l = \mathsf{rev}(l'))$.

5.2 What Is a Graph?

An (**undirected, labeled**) **graph** consists of a set $N : \alpha \to bool$ of nodes, a set $E : \beta \to bool$ of edges, and a set $L : \alpha \times \beta \times \alpha \to bool$ of links representing the incidence relation between nodes and edges, such that:

$$\mathsf{Graph}(N, E, L) \triangleq$$
$$(\mathsf{Image}(\mathsf{src})(L) \subseteq N) \wedge (\mathsf{Image}(\mathsf{des})(L) \subseteq N) \wedge$$
$$(\mathsf{Image}(\mathsf{via})(L) = E) \wedge (\mathsf{Image}(\mathsf{rev})(L) = L) \wedge \quad (5)$$
$$\mathsf{OneOneMod}(\approx)(\mathsf{via})(L)$$

Definition (5) is convenient for proving closure properties of constructors of graphs (Section 8), because it contains no quantifiers and many theorems about operations on sets (such as Image, OneOneMod, and $\subseteq$) can be proved separately. But the reader may doubt whether it captures faithfully the notion of undirected labeled graphs. The following theorem should lay such doubts to rest:

$$\text{Graph}(N, E, L) =$$
$$(\forall l :: L.\ \text{src}(l) \in N \land \text{des}(l) \in N \land \text{via}(l) \in E \land \text{rev}(l) \in L)\ \land$$
$$(\forall e :: E.\ \exists l :: L.\ e = \text{via}(l))\ \land \qquad\qquad (6)$$
$$(\forall l\ l' :: L.\ (\text{via}(l) = \text{via}(l')) \Rightarrow (l \approx l'))$$

For a graph G, the first clause on the right-hand side of (6) requires that every link in G has its src, des, via, and rev each belonging to the appropriate component of G, the second requires that every edge in G is the via of some link in G, and the third requires that if two links in G share the same via, then they must be almost equal (i.e., either identical or reverse of each other). Consequently, for any edge e in G, there are exactly two links in G with e being their via and these two links are reverse of each other. (Note that these two links may be identical, which is the case iff e is a self-loop.) This shows that (5) captures the notion of undirectedness correctly. Note also that (5) allows infinite graphs, self-loops, and multiple edges. As the reader will see, much graph theory can be developed without ruling out these possibilities.

The three components of a graph are accessed by: $\text{Node}(N, E, L) \triangleq N$, $\text{Edge}(N, E, L) \triangleq E$, and $\text{Link}(N, E, L) \triangleq L$.

5.3 Finite Graphs

A graph is **finite** iff its components are all finite:

$$\text{GFinite}(N, E, L) \triangleq \text{Finite}(N) \land \text{Finite}(E) \land \text{Finite}(L)$$

Note that, by (5), the finiteness of N and E already implies the finiteness of L.

5.4 Subgraphs

A graph is a **subgraph** of another graph iff the components of the former are subsets of the corresponding components of the latter:

$$(N, E, L) \sqsubseteq (N', E', L') \triangleq (N \subseteq N') \land (E \subseteq E') \land (L \subseteq L')$$

Clearly, the subgraph relation is a partial order, i.e., it is reflexive, antisymmetric, and transitive.

5.5 Adjacent Nodes, Edges, and Links of a Node

Let G be a graph (i.e., $\text{Graph}(G)$) and n be a node in G (i.e., $n \in \text{Node}(G)$). The set of **adjacent edges** of n in G is:

$$\text{AdjEdge}(G)(n) \triangleq \{\, e \mid \exists l :: \text{Link}(G).\ (n = \text{src}(l)) \land (e = \text{via}(l)) \,\} \qquad (7)$$

Note that, by (5), substituting "$n = \mathsf{des}(l)$" for "$n = \mathsf{src}(l)$" in (7) would have resulted in an equivalent definition of AdjEdge.

Let e be an adjacent edge of n in G (i.e., $e \in \mathsf{AdjEdge}(G)(n)$). The **outgoing link**, **incoming link**, and **opposite node** of n over e in G are:

$$\mathsf{out}(G)(n)(e) \triangleq \varepsilon\, l :: \mathsf{Link}(G).\ (\, n = \mathsf{src}(l)\,) \wedge (\, e = \mathsf{via}(l)\,) \tag{8}$$

$$\mathsf{inc}(G)(n)(e) \triangleq \mathsf{rev}(\mathsf{out}(G)(n)(e))$$

$$\mathsf{opp}(G)(n)(e) \triangleq \mathsf{des}(\mathsf{out}(G)(n)(e))$$

To avoid being bogged down in proofs involving ε-terms, a definition like (8) is used solely to prove the *well-definedness* and *uniqueness* theorems for the defined constant, from which all other properties of the defined constant are derived. For example, the well-definedness theorem for out is:

$$\forall G :: \mathsf{Graph}.\ \forall n :: \mathsf{Node}(G).\ \forall e :: \mathsf{AdjEdge}(G)(n).$$
$$\mathsf{Link}(G)(\mathsf{out}(G)(n)(e)) \wedge (\,\mathsf{src}(\mathsf{out}(G)(n)(e)) = n\,)$$
$$\wedge (\,\mathsf{via}(\mathsf{out}(G)(n)(e)) = e\,)$$

and the uniqueness theorem for out is:

$$\forall G :: \mathsf{Graph}.\ \forall n :: \mathsf{Node}(G).\ \forall e :: \mathsf{AdjEdge}(G)(n).$$
$$\forall l :: \mathsf{Link}(G).\ (\,\mathsf{src}(l) = n\,) \wedge (\,\mathsf{via}(l) = e\,) \Rightarrow (\,l = \mathsf{out}(G)(n)(e)\,)$$

Similar well-definedness and uniqueness theorems are proved for all other constants defined using the ε-symbol, but they are not listed due to lack of space.

6 Trees

In this section, G is a graph, l is a link in G, and p and q are nodes in G. Free occurrences of G (respectively, l, p and q) in a theorem are implicitly quantified over by $\forall G :: \mathsf{Graph}$ ($\forall l :: \mathsf{Link}(G)$, $\forall p :: \mathsf{Node}(G)$ and $\forall q :: \mathsf{Node}(G)$).

6.1 Walks, Paths, and Cycles

A **walk** from p to q in G is a list of *links* in G such that the first link starts at p, the last link ends at q, and each link except the last link ends at where the next link starts:

$$(\,\mathsf{Walk}(G)(p)(q)([\ ])\,) \triangleq (p = q)\,) \wedge$$
$$(\,\mathsf{Walk}(G)(p)(q)([h]^\frown t)\,) \triangleq h \in \mathsf{Link}(G) \wedge (\,p = \mathsf{src}(h)\,) \wedge \mathsf{Walk}(G)(\mathsf{des}(h))(q)(t)\,)$$

Walks can also be defined as lists of *edges*, but we have found it more convenient to work with links. A walk is **edge-simple** iff it does not repeat any edge:

$$(\,\mathsf{EdgeSimple}([\ ]) \triangleq \mathsf{T}\,) \wedge$$
$$(\,\mathsf{EdgeSimple}([h]^\frown t) \triangleq \mathsf{Every}(\,\lambda l.\ \mathsf{via}(l) \neq \mathsf{via}(h)\,)(t) \wedge \mathsf{EdgeSimple}(t)\,)$$

where $\mathsf{Every}(P)(t)$ asserts that P holds for every element of t:

$$(\,\mathsf{Every}(P)([\;]) \triangleq \mathsf{T}\,) \wedge$$
$$(\,\mathsf{Every}(P)([h]^\frown t) \triangleq P(h) \wedge \mathsf{Every}(P)(t)\,)$$

A **path** is an edge-simple walk:

$$\mathsf{Path}(G)(p)(q)(w) \triangleq \mathsf{Walk}(G)(p)(q)(w) \wedge \mathsf{EdgeSimple}(w)$$

A **cycle** is a *non-empty* path that starts and ends at the same node:

$$\mathsf{Cycle}(G)(p)(w) \triangleq \mathsf{Path}(G)(p)(p)(w) \wedge (w \neq [\;])$$

It is important that cycles are non-empty, for otherwise no graph (except the empty graph) would be acyclic (Section 6.3).

6.2 Reachability and Connectedness

Two nodes are **reachable** from each other iff there exists a walk between them:

$$\mathsf{Reachable}(G)(p)(q) \triangleq \exists\,w.\ \mathsf{Walk}(G)(p)(q)(w)$$

As a relation on $\mathsf{Node}(G)$, $\mathsf{Reachable}(G)$ is reflexive, symmetric, and transitive. In other words, it is an equivalence relation; the equivalence class containing p is just $\mathsf{Reachable}(G)(p)$ (viewed as a set). Although *not* every walk is a path, the existence of a walk does imply the existence of a path:

$$\mathsf{Reachable}(G)(p)(q) \;=\; \exists\,w.\ \mathsf{Path}(G)(p)(q)(w) \tag{9}$$

Theorem (9) is a good example of those facts about graphs that are obvious intuitively but messy to prove formally. Its proof is an inductive argument that, so to speak, repeatedly removes cycles from a walk until it becomes a path. A graph is **connected** iff any two of its nodes are reachable from each other:

$$\mathsf{Connected}(G) \triangleq \forall p\,q :: \mathsf{Node}(G).\ \mathsf{Reachable}(G)(p)(q)$$

It follows from (9) that a graph is connected iff there is a path between any two of its nodes:

$$\mathsf{Connected}(G) \;=\; \forall p\,q :: \mathsf{Node}(G).\ \exists\,w.\ \mathsf{Path}(G)(p)(q)(w) \tag{10}$$

6.3 Bridges and Acyclicity

A **bridge** is an edge (given as the via of a link) whose removal disconnects its two end points:

$$\mathsf{Bridge}(G)(l) \triangleq \neg\mathsf{Reachable}(\mathsf{DeleteEdge}(l)(G))(\mathsf{src}(l))(\mathsf{des}(l)) \tag{11}$$

where $\mathsf{DeleteEdge}(l)(G)$ is the graph obtained from G by deleting l's via but keeping its src and des:

$$\mathsf{DeleteEdge}(l)(N, E, L) \triangleq (N, E \setminus \{\mathsf{via}(l)\}, L \setminus \{l, \mathsf{rev}(l)\})$$

If l is a bridge in G, then it can be shown by (list) induction on walks that any walk from $\mathsf{src}(l)$ to $\mathsf{des}(l)$ in G must contain l. Conversely, if l is not a bridge, then (11) implies that there is a walk from $\mathsf{src}(l)$ to $\mathsf{des}(l)$ in $\mathsf{DeleteEdge}(l)(G)$, i.e., a walk that does not contain l. In other words, an edge is a bridge iff any walk from one end of it to the other must pass through it:

$$\mathsf{Bridge}(G)(l) =$$
$$\forall\, w :: \mathsf{Walk}(G)(\mathsf{src}(l))(\mathsf{des}(l)). \;\exists\, w'\, w''.\; (w = w'^\frown [l]^\frown w'') \tag{12}$$

If l is a bridge in G, then (12) implies that a cycle w in G containing l must cross l at least twice. But then w is not a path, which is a contradiction. So w cannot contain l and hence must be a cycle in $\mathsf{DeleteEdge}(l)(G)$. Conversely, if l is not a bridge, then (11) implies that there is a path from $\mathsf{src}(l)$ to $\mathsf{des}(l)$ in $\mathsf{DeleteEdge}(l)(G)$, which, together with $\mathsf{rev}(l)$, forms a cycle that is in G but not in $\mathsf{DeleteEdge}(l)(G)$. In other words, an edge is a bridge iff no cycle passes through it:

$$\mathsf{Bridge}(G)(l) =$$
$$\forall\, p :: \mathsf{Node}(G).\;\forall\, w.\; \mathsf{Cycle}(G)(p)(w) \Rightarrow \mathsf{Cycle}(\mathsf{DeleteEdge}(l)(G))(p)(w) \tag{13}$$

A graph is **acyclic** iff it contains no cycle:

$$\mathsf{Acyclic}(G) \triangleq \neg \exists\, p :: \mathsf{Node}(G).\;\exists\, w.\; \mathsf{Cycle}(G)(p)(w)$$

It follows from (13) that a graph is acyclic iff each of its edges is a bridge:

$$\mathsf{Acyclic}(G) = \forall\, l :: \mathsf{Link}(G).\; \mathsf{Bridge}(G)(l) \tag{14}$$

From (12) and (14), it can be shown by (list) induction on paths that a graph is acyclic iff there is at most one path between any two of its nodes:

$$\mathsf{Acyclic}(G) = \forall\, p\, q :: \mathsf{Node}(G).\;\forall\, w\, w' :: \mathsf{Path}(G)(p)(q).\; (w = w') \tag{15}$$

6.4 What Is a Tree?

A tree is a connected and acyclic graph:[1]

$$\mathsf{Tree}(G) \triangleq \mathsf{Graph}(G) \wedge \mathsf{Connected}(G) \wedge \mathsf{Acyclic}(G) \tag{16}$$

It follows immediately from (10) and (15) that a graph is a tree iff there is a unique path between any two of its nodes:

$$\mathsf{Tree}(G) = \forall\, p\, q :: \mathsf{Node}(G).\; \exists!\, w.\; \mathsf{Path}(G)(p)(q)(w) \tag{17}$$

That unique path is picked out by ThePath:

$$\mathsf{ThePath}(G)(p)(q) \triangleq \varepsilon\, w.\; \mathsf{Path}(G)(p)(q)(w) \tag{18}$$

Most theorems about trees below are proved from (17) and (18).

---

[1] In the actual implementation we define $\mathsf{Tree}(G) \triangleq \mathsf{Connected}(G) \wedge \mathsf{Acyclic}(G)$ in keeping with the convention that a predicate on graphs (except Graph of course) presupposes that its argument is a graph. In this paper we use the definition (16) above in order to simplify notations.

7 Trees with Roots

In this section, T is a tree, l is a link in T, and r (the **root**) and n are nodes in T. Free occurrences of T (respectively, l, r and n) in a theorem are implicitly quantified over by $\forall T :: \mathsf{Tree}$ ($\forall l :: \mathsf{Link}(T)$, $\forall r :: \mathsf{Node}(T)$ and $\forall n :: \mathsf{Node}(T)$).

7.1 Orienting a Tree with Respect to a Root

A link is **rootward** iff it is the first link on the path from its src to the root:

$$\mathsf{Rootward}(T)(r)(l) \triangleq \mathsf{Path}(T)(\mathsf{src}(l))(r)(\,[l]^\frown \mathsf{ThePath}(T)(\mathsf{des}(l))(r)\,)$$

Every link in T is oriented by Rootward in a unique way, viz., either it or its rev, but never both, is rootward:

$$
\begin{aligned}
&(\,\mathsf{Rootward}(T)(r)(l) \lor \mathsf{Rootward}(T)(r)(\mathsf{rev}(l))\,) \land \\
&\neg(\,\mathsf{Rootward}(T)(r)(l) \land \mathsf{Rootward}(T)(r)(\mathsf{rev}(l))\,)
\end{aligned}
\tag{19}
$$

7.2 Viewing a Tree with a Root as a Family Tree

By taking the root r as the progenitor, the tree T can be viewed as a family tree. The **parent edge** of a *non-root* node is the (unique!) adjacent edge whose corresponding outgoing link is rootward:

$$\mathsf{ParEdge}(T)(r)(n) \triangleq \varepsilon\, e :: \mathsf{AdjEdge}(T)(n).\ \mathsf{Rootward}(T)(r)(\mathsf{out}(T)(n)(e))$$

Note that $\mathsf{ParEdge}(T)(r)(r)$ is unspecified, as the root has no parent. The **kid edges** of a node are those adjacent edges whose corresponding incoming links are rootward:

$$\mathsf{KidEdge}(T)(r)(n) \triangleq \{\, e :: \mathsf{AdjEdge}(T)(n) \mid \mathsf{Rootward}(T)(r)(\mathsf{inc}(T)(n)(e))\,\}$$

It follows from (19) that every adjacent edge of the root is a kid edge and that an adjacent edge of a non-root node is either the parent edge or a kid edge, but never both:

$$
\begin{aligned}
\mathsf{KidEdge}(T)(r)(n) = {}& \\
\mathbf{if}\ (n = r)\ &\mathbf{then}\ \mathsf{AdjEdge}(T)(n) \\
&\mathbf{else}\ \ \mathsf{AdjEdge}(T)(n) \setminus \{\mathsf{ParEdge}(T)(r)(n)\}
\end{aligned}
$$

If e is the parent edge (respectively, a kid edge) of a node n, then $\mathsf{opp}(T)(n)(e)$ is called the **parent node** (a **kid node**) of n:

$$\mathsf{ParNode}(T)(r)(n) \triangleq \mathsf{opp}(T)(n)(\mathsf{ParEdge}(T)(r)(n))$$

$$\mathsf{KidNode}(T)(r)(n) \triangleq \mathsf{Image}(\mathsf{opp}(T)(n))(\mathsf{KidEdge}(T)(r)(n))$$

The **descendent nodes** (or simply **descendents**) of a node n are those nodes p such that the path from p to the root must pass through n:

$$\mathsf{DescNode}(T)(r)(n) \triangleq$$
$$\{\, p :: \mathsf{Node}(T) \mid \mathsf{Path}(T)(p)(r)(\,\mathsf{ThePath}(T)(p)(n)^\frown \mathsf{ThePath}(T)(n)(r)\,)\,\}$$

It can be shown that every node is a descendent of the root:

$$\mathsf{DescNode}(T)(r)(r) \;=\; \mathsf{Node}(T)$$

In general, the descendents of a node consist of the node itself and the descendents of its kid nodes:

$$\mathsf{DescNode}(T)(r)(n) \;=\; \{n\} \cup$$
$$\mathsf{Union}(\mathsf{DescNode}(T)(r))(\mathsf{KidNode}(T)(r)(n)) \qquad (20)$$

Furthermore, the right-hand side of (20) is a *partition* in the sense that a node is never a descendent of any of its kid nodes:

$$n \notin \mathsf{Union}(\mathsf{DescNode}(T)(r))(\mathsf{KidNode}(T)(r)(n)) \qquad (21)$$

and its kid nodes have mutually disjoint sets of descendents:

$$\forall\, k \; k' \;::\; \mathsf{KidNode}(T)(r)(n).$$
$$(k \neq k') \;\Rightarrow\; (\,\mathsf{DescNode}(T)(r)(k) \cap \mathsf{DescNode}(T)(r)(k') = \emptyset\,) \qquad (22)$$

Theorems (20), (21), and (22) ensure that (1) and (3) can be applied to obtain:

$$\mathsf{ACI}(op, id) \wedge \mathsf{GFinite}(T) \;\Rightarrow$$
$$\mathsf{let}\; S \;\triangleq\; \mathsf{Sop}(op, id)(f) \circ \mathsf{DescNode}(T)(r) \;\mathsf{in} \qquad (23)$$
$$(\, S(n) \;=\; f(n) \; op \; \mathsf{Sop}(op, id)(S)(\mathsf{KidNode}(T)(r)(n))\,)$$

Informally, (20) says that:

$$\mathsf{DescNode}(T)(r)(n) \;=\; \{n\} \cup \bigcup_{k \in \mathsf{KidNode}(T)(r)(n)} \mathsf{DescNode}(T)(r)(k)$$

and, if $(op, id) = (+, 0)$, (23) says that:

$$\sum_{p \in \mathsf{DescNode}(T)(r)(n)} f(p) \;=\; f(n) + \sum_{k \in \mathsf{KidNode}(T)(r)(n)} \left(\sum_{q \in \mathsf{DescNode}(T)(r)(k)} f(q) \right)$$

7.3 Top-Down and Bottom-Up Inductions in a Family Tree

Let P be a predicate on nodes. If P holds at the root and, whenever P holds at a node, P holds at each of its kid nodes, then clearly P holds at every node in the tree:

$$P(r) \wedge (\,\forall\, n \;::\; \mathsf{Node}(T).\; P(n) \Rightarrow (\,\forall\, k \;::\; \mathsf{KidNode}(T)(r)(n).\; P(k)\,)\,)$$
$$\Rightarrow (\,\forall\, n \;::\; \mathsf{Node}(T).\; P(n)\,)$$

This is called **top-down induction** and proved by list induction on paths emanating from the root. The opposite direction, called **bottom-up induction**, says that if P holds at a node whenever P holds at each of its kid nodes, then P holds at every node in the tree, provided that the tree is finite:

$$\mathsf{GFinite}(T) \wedge (\,\forall n :: \mathsf{Node}(T).\ (\,\forall k :: \mathsf{KidNode}(T)(r)(n).\ P(k)\,) \Rightarrow P(n)\,)$$
$$\Rightarrow (\,\forall n :: \mathsf{Node}(T).\ P(n)\,) \tag{24}$$

Bottom-up induction is proved by complete induction on the sizes of subtrees of T. Note that (24) has no special "base case" because its antecedent already implies that P holds at every leaf (a leaf is a node with no kid). Also, the finiteness assumption $\mathsf{GFinite}(T)$ is stronger than absolutely necessary: it is sufficient to assume that there is no infinite path in T. But then the proof would require transfinite induction, since a node can still have infinitely many kid nodes. Theorem (24) seems good enough for most applications.

8 Constructors of Graphs

In this section we introduce several constructors of graphs. We have chosen them *not* because of any theoretical consideration, but with an eye to the practical application of reasoning about such distributed algorithms as those in [4, 8].

8.1 Empty Graph

The **empty graph** is the graph that contains no node, no edge, and no link:

$$\mathsf{GEmpty} \triangleq (\emptyset, \emptyset, \emptyset)$$

Clearly, it is a finite tree:

$$\mathsf{Tree}(\mathsf{GEmpty}) \wedge \mathsf{GFinite}(\mathsf{GEmpty})$$

and a subgraph of any graph:

$$\forall G :: \mathsf{Graph}.\ \mathsf{GEmpty} \sqsubseteq G$$

8.2 Single-Node Graphs

For any node n, the **single-node graph** $\mathsf{SingNode}(n)$ is the graph that contains n as its sole node and no edge and no link:

$$\mathsf{SingNode}(n) \triangleq (\{n\}, \emptyset, \emptyset)$$

Clearly, it is a finite tree:

$$\forall n.\ \mathsf{Tree}(\mathsf{SingNode}(n)) \wedge \mathsf{GFinite}(\mathsf{SingNode}(n))$$

and a subgraph of any graph containing n:

$$\forall G :: \mathsf{Graph}.\ \forall n :: \mathsf{Node}(G).\ \mathsf{SingNode}(n) \sqsubseteq G$$

8.3   Merging Two Disjoint Subgraphs via a Link

In this subsection, H is a graph, G_1 and G_2 are two disjoint subgraphs of H, l is a link in H whose src is in G_1 and whose des is in G_2, and all theorems have the following implicit assumptions:

$$\forall\, H\, G_1\, G_2 :: \mathsf{Graph}.\ G_1 \sqsubseteq H \wedge G_2 \sqsubseteq H \wedge (\mathsf{Node}(G_1) \cap \mathsf{Node}(G_2) = \emptyset)\ \Rightarrow$$
$$\forall\, l :: \mathsf{Link}(H).\ \mathsf{src}(l) \in \mathsf{Node}(G_1) \wedge \mathsf{des}(l) \in \mathsf{Node}(G_2)\ \Rightarrow$$

The **merge** of $G_1 = (N_1, E_1, L_1)$ and $G_2 = (N_2, E_2, L_2)$ via l is defined by:

$$\mathsf{MergeGLG}(G_1)(l)(G_2) \triangleq$$
$$(N_1 \cup N_2, E_1 \cup E_2 \cup \{\mathsf{via}(l)\}, L_1 \cup L_2 \cup \{l, \mathsf{rev}(l)\})$$

Then $\mathsf{MergeGLG}(G_1)(l)(G_2)$ is not only a graph but a subgraph of H:

$$\mathsf{Graph}(\mathsf{MergeGLG}(G_1)(l)(G_2)) \wedge \mathsf{MergeGLG}(G_1)(l)(G_2) \sqsubseteq H \qquad (25)$$

If both G_1 and G_2 are finite, then $\mathsf{MergeGLG}(G_1)(l)(G_2)$ is too:

$$\mathsf{GFinite}(G_1) \wedge \mathsf{GFinite}(G_2) \Rightarrow \mathsf{GFinite}(\mathsf{MergeGLG}(G_1)(l)(G_2)) \qquad (26)$$

If both G_1 and G_2 are trees, then $\mathsf{MergeGLG}(G_1)(l)(G_2)$ is too:

$$\mathsf{Tree}(G_1) \wedge \mathsf{Tree}(G_2) \Rightarrow \mathsf{Tree}(\mathsf{MergeGLG}(G_1)(l)(G_2)) \qquad (27)$$

8.4   Merging a Link and a Subgraph

In this subsection, H is a graph, G is a subgraph of H, l is a link in H whose src is not in G but whose des is in G, and all theorems have the following implicit assumptions:

$$\forall\, H\, G :: \mathsf{Graph}.\ G \sqsubseteq H\ \Rightarrow$$
$$\forall\, l :: \mathsf{Link}(H).\ \mathsf{src}(l) \notin \mathsf{Node}(G) \wedge \mathsf{des}(l) \in \mathsf{Node}(G)\ \Rightarrow$$

The **merge** of l and G is defined by:

$$\mathsf{MergeLG}(l)(G) \triangleq \mathsf{MergeGLG}(\mathsf{SingNode}(\mathsf{src}(l)))(l)(G)$$

Theorems (25)–(27) can be specialized to obtain respectively (28)–(30) below:

$$\mathsf{Graph}(\mathsf{MergeLG}(l)(G)) \wedge \mathsf{MergeLG}(l)(G) \sqsubseteq H \qquad (28)$$

$$\mathsf{GFinite}(G) \Rightarrow \mathsf{GFinite}(\mathsf{MergeLG}(l)(G)) \qquad (29)$$

$$\mathsf{Tree}(G) \Rightarrow \mathsf{Tree}(\mathsf{MergeLG}(l)(G)) \qquad (30)$$

9 Conclusion

In this paper we describe a formal theory of undirected labeled graphs in higher-order logic developed using the theorem-proving system HOL. It formalizes many graph-theoretic notions and rigorously proves many theorems about them. It has been successfully applied to the mechanical verification of simple distributed algorithms [2]. In the future we hope to extend it to include a theory of minimum spanning trees in order to reason about the well-known distributed algorithm of Gallager, Humblet, and Spira [4] for computing minimum spanning trees.

Acknowledgements. The author is grateful to Professors Eli Gafni and David Martin for their guidance, to Peter Homeier and the anonymous referees for their helpful comments, and to HOL hackers around the world for sustaining a lively and friendly user community.

References

1. N.G. de Bruijn, "Checking Mathematics with Computer Assistance", in *Notices of the American Mathematical Society*, Vol. 38, No. 1, pp. 8–15, Jan. 1991.
2. Ching-Tsun Chou, "Mechanical Verification of Distributed Algorithms in Higher-Order Logic", in this Proceedings.
3. Shimon Even, *Graph Algorithms*, Computer Science Press, 1979.
4. R.G. Gallager, P.A. Humblet, and P.M. Spira, "A Distributed Algorithm for Minimum-Weight Spanning Trees", in *ACM Trans. on Programming Languages and Systems*, Vol. 5, No. 1, pp. 66–77, Jan. 1983.
5. Michael J.C. Gordon, "HOL: A Proof Generating System for Higher-Order Logic", pp. 73–128 of G. Birtwistle and P.A. Subrahmanyam (ed.), *VLSI Specification, Verification and Synthesis*, Kluwer Academic Publishers, 1988.
6. Michael J.C. Gordon and Tom F. Melham (ed.), *Introduction to HOL: A Theorem-Proving Environment for Higher-Order Logic*, Cambridge University Press, 1993.
7. Tom F. Melham, *The HOL pred_set Library*, University of Cambridge Computer Laboratory, Feb. 1992.
8. Adrian Segall, "Distributed Network Protocols", in *IEEE Trans. on Information Theory*, Vol. 29, No. 1, pp. 23–35, Jan. 1983.
9. Wai Wong, "A Simple Graph Theory and Its Application in Railway Signalling", pp. 395–409 of M. Archer *et al.* (ed.), *Proc. of 1991 Workshop on the HOL Theorem Proving System and Its Applications*, IEEE Computer Society Press, 1992.

Mechanical Verification of Distributed Algorithms in Higher-Order Logic

Ching-Tsun Chou (chou@cs.ucla.edu)

Computer Science Department, University of California at Los Angeles
Los Angeles, CA 90024, U.S.A.

Abstract. The only way to verify the correctness of a distributed algorithm with a high degree of confidence is to construct a correctness proof that is so formal and rigorous as to be checkable by a machine. The chief aim of this paper is to show, via a simple but typical example, how such proofs can be constructed and checked using HOL—a mechanical theorem-proving system for higher-order logic. The secondary aim is to demonstrate a method for reasoning about distributed algorithms based on the notions of events and causality. The goal is to perform as much reasoning as possible in the events-and-causality view and then translate the results back to the processors-and-messages view, since the former is more abstract and easier to reason about than the latter. The example we use to illustrate our ideas is the verification of a simple distributed summation algorithm for trees. A companion paper [6] describes the formal graph theory needed for this task.

1   Introduction

Distributed algorithms, as exemplified by those in [10, 18], have to operate in the face of total asynchrony: there is no finite bound on how widely component speeds may vary both in space and in time. Testing can do little to catch those subtle timing-dependent bugs with which distributed algorithms are plagued. For, even in the simplest distributed algorithms, the asynchronous interaction of concurrent activities is far too nondeterministic to be adequately tested. The only way to verify the correctness of a distributed algorithm with a high degree of confidence is to construct a correctness proof that is so formal and rigorous as to be checkable by a machine.

The chief aim of this paper is to show, via a simple but typical example, how such proofs can be constructed and checked using HOL—the mechanical theorem-proving system for higher-order logic developed by Gordon *et al.* [12, 13]. We choose to use HOL for several reasons. The first is *security*: HOL reduces every proof to the repeated applications of a small set of axioms and inference rules, which is more trustworthy than a large set of *ad hoc* proof procedures. The second is *expressiveness*: higher-order logic allows natural formalizations of distributed algorithms, the data structures they manipulate, the properties they possess, and the modes of reasoning used to prove these properties. The third is *programmability*: customized proof procedures can be easily written to automate common patterns of reasoning without compromising the security of HOL. We have found that these traits do make HOL a secure and flexible environment in which to construct natural proofs of distributed algorithms.

The secondary aim of this paper is to demonstrate, in as simple a setting as possible, a method for reasoning about distributed algorithms based on the notions of *events* and *causality*. For us, events are (names of) occurrences of atomic actions of a distributed algorithm and causality is the *essential* temporal precedence relation between events that ignores fortuitous orderings of concurrent events. Executions of a distributed algorithm can be visualized as (in general nondeterministic) unfoldings of causal patterns of events. Compared with the possible linear orderings of events, the possible causal patterns are much fewer and provide a more intuitive explanation for *why* the algorithm works. Consequently, we verify a distributed algorithm in four steps: (a) we introduce an *event algorithm* that is an operational representation of the causal patterns of events of the distributed algorithm, (b) we prove properties of the event algorithm by reasoning about events and causality, (c) we show that the event algorithm can *simulate* the distributed algorithm, and (d) we *translate* the properties of the event algorithm into those of the distributed algorithm via the simulation relation. (A theory of simulation and translation has been expounded in an earlier paper of ours [5] and will be reviewed in Section 2.5.) The goal is to perform as much reasoning as possible in terms of the event algorithm, which is more abstract and easier to reason about than the distributed algorithm. For instance, in the example of this paper, all liveness reasoning is performed in terms of the event algorithm.

The example we use to illustrate our ideas is the verification of DSUM—a simple distributed summation algorithm for trees. Let *Net* be a tree (i.e., a connected and acyclic undirected graph) representing a computer network whose nodes are autonomous processors and whose edges are bi-directional communication links over which the nodes can send messages to each other. Messages are not lost, duplicated, or corrupted and, for each direction over each edge, are delivered in the order in which they are sent. There is no finite bound whatsoever on processor and message delays, but we do require that if an action is continuously enabled then it must eventually be executed. There is a distinguished node, called the *root*, and an assignment *val* of (natural) numbers to nodes. The purpose of DSUM is to compute the sum of *val*'s of all nodes in *Net*, $\sum_{n \in \text{Node}(Net)} val(n)$, at the *root*. To achieve this, every node n in *Net* executes the pseudo-PASCAL program shown in Figure 1, which is divided into six blocks numbered $\boxed{1}$ through $\boxed{6}$. Declared in $\boxed{1}$ are the four program variables at each node n: $pc(n)$ is a "program counter" and, if $pc(n) \neq$ Idle, $par(n)$ (for "parent") contains the edge over which a Start message is received, $cnt(n)$ (for "count") contains the number of Report messages yet to be received, and $sum(n)$ contains the sum of $val(n)$ and the values carried by Report messages already received. Initially, all we know is that $pc(n) =$ Idle for each node n and (this is not mentioned in Figure 1) there is no message in transit anywhere in the network. Block $\boxed{2}$ defines a procedure called by $\boxed{3}$ and $\boxed{4}$. The executions of blocks $\boxed{3}$ through $\boxed{6}$ from all nodes are atomic (i.e., a block is executed in

```
1  constant  val(n) : num ;                    2  procedure Start(E)
                                                   begin
    variable   pc(n) : {Idle, Busy, Done}           pc(n) := Busy ;
               par(n) : edge                         sum(n) := val(n) ;
               cnt(n) : num                          cnt(n) := |E| ;
               sum(n) : num ;                        for each e in E do
                                                        send Start on e ;
    initially (pc(n) = Idle) ;                    end ;


3  when (n = root) ∧ (pc(n) = Idle)            4  receive Start on e
    begin                                          begin
      call Start(AdjEdge(n)) ;                       par(n) := e ;
    end ;                                            call Start(AdjEdge(n) \ {e}) ;
                                                   end ;


5  receive Report(v) on e                       6  when (pc(n) = Busy) ∧ (cnt(n) = 0)
    begin                                          begin
      cnt(n) := cnt(n) − 1 ;                         pc(n) := Done ;
      sum(n) := sum(n) + v ;                         if (n ≠ root) then
    end ;                                              send Report(sum(n)) on par(n) ;
                                                   end ;
```

Fig. 1. DSUM at node n

an uninterrupted step), chaotic (i.e., a block can be executed whenver enabled), and nondeterministic (i.e., when more than one block is enabled, any one of them may be executed). The basic idea underlying DSUM is very simple. At the beginning, the *root* spontaneously wakes up and sends a Start message to each of its neighbors (3). The Start messages are propagated down (assuming the *root* is at top) the tree *Net* and wake up other nodes along the way (4). After receiving a Start message, a leaf sends a Report message carrying its *val* upward (6). A non-leaf node collects Report messages from its kids (i.e., those nodes immediately below it) and adds the values they carry to its own *val* (5). After a node has received a Report message from each of its kids, its *sum* equals the sum of its own *val* and the *val*'s of all nodes below it, which it sends to its parent if it is not the *root* (6). After the *root* has executed 6, *sum(root)* equals the desired global sum $\sum_{n \in \text{Node}(Net)} val(n)$.

Reasoning about DSUM requires the formalization of a considerable amount of graph theory in HOL. This has been done and is reported in a companion paper [6]. Note that we are interested in verifying DSUM *not* for any particular tree, but for *all* trees. Note also that although DSUM is formulated as a summation algorithm, the same idea clearly applies to the distributed computation of the product, the maximum, the minimum, and indeed the cumulative result of *any* associative and commutative operation with an identity (an ACI operation, for short) on values residing at nodes. In fact, by taking as parameters both the underlying network *Net* and the ACI operation on node values, we have verified in one stroke all DSUM-like algorithms for all trees and all ACI operations. But, for the sake of concreteness, we shall concentrate on summation in this paper.

Our approach to formalizing mathematics is strictly definitional, meaning that all our theories are developed from the initial theory of HOL without introducing any axioms other than definitions. The definitional approach has two well-known advantages. The first is *consistency*: definitions never introduce any inconsistencies. Since the initial theory of HOL is consistent [13], all our theories are consistent as well. The second is *eliminability*: definitions can, in principle, be eliminated. So an auxiliary definition not involved in the statement of a result can safely be forgotten as far as that result is concerned, even if it is used in the proof of that result. To be sure, sticking to the definitional approach is sometimes laborious, since even "obvious" propositions must be honestly proved. But the logical security thus gained is well worth the effort.

Common patterns of reasoning in program verification can be formulated as special-purpose *programming logics*, so that these patterns can be expressed succinctly and made more easily applicable. In this paper, we make use of some constructs from Lamport's TLA [15] (Section 2.3) and Chandy and Misra's UNITY [3] (Section 2.4), although we follow neither closely. In keeping with the definitional approach, we embed these constructs in HOL by formalizing their semantics as definitions and proving their properties as theorems of HOL. This *semantic embedding* approach allows arbitrary mixing of special-purpose logics and other mathematical theories and hence is very flexible. Furthermore, we need not worry about the completeness of axioms and inference rules of an embedded special-purpose logic, as new axioms and rules can always be derived from the formalized semantics whenever the need arises.

The rest of this paper is organized as follows. Section 2 reviews some prerequisites for verifying distributed algorithms. Section 3 formalizes DSUM in higher-order logic and lists the properties that we want to prove about it. Section 4 describes the event view of DSUM and derives some of its properties. Section 5 relates the distributed and event views of DSUM in order to deduce the desired properties of the former from those of the latter. Section 6 discusses related work. Finally, Section 7 draws conclusions and suggests future work.

2 Prerequisites for Verifying Distributed Algorithms

2.1 Higher-Order Logic

The version of higher-order logic used in this paper is the one supported by the HOL theorem-proving system [13]. In general, the following typographic conventions are followed when writing formulas: lowercase Greek letters stand for type variables, *slanted* font for types, *italic* font for term variables, and sans serif font for term constants except constructors of recursive types [17], which are written in typewriter font. A name may have subscripts or superscripts (or both), which should be considered as part of the name. Free variables in a definition or theorem are implicitly universally quantified. The symbol $\triangleq$ means "equals by definition"; "iff" means "if and only if". Lists are enclosed by square brackets ($[\cdots]$); list concatenation is denoted by $\frown$ (written as an infix).

Lifted boolean connectives:	Lifted restricted quantifications:
$(\neg P)(x) \triangleq \neg P(x)$	$(\mathbb{W}\, i :: I.\ R[\,i\,])(x) \triangleq \forall i :: I.\ (R[\,i\,])(x)$
$(P \mathbb{A} Q)(x) \triangleq P(x) \wedge Q(x)$	$(\mathbb{E}\, i :: I.\ R[\,i\,])(x) \triangleq \exists i :: I.\ (R[\,i\,])(x)$
$(P \mathbb{W} Q)(x) \triangleq P(x) \vee Q(x)$	Lifted validity:
$(P \Rrightarrow Q)(x) \triangleq P(x) \Rightarrow Q(x)$	$\Vdash P \triangleq (\forall x.\ P(x))$

Fig. 2. Lifted logic of predicates

2.2 Predicates

A predicate P is a function whose type is of the form $\alpha \rightarrow bool$, where α is called the **domain** of P. In this paper predicates are used in two ways: as *sets* and as *propositions*.

Predicates as Sets. In this paper a set is *identified* with its characteristic predicate so that the following holds:

$$\{\, x : \alpha \mid P(x)\,\} = P$$

With this identification, the usual operations on sets, such as $\subseteq, \cap, \cup, \setminus$ (set difference), and $|\cdot|$ (cardinality), are applicable to predicates as well. In particular, $x \in P$ is now synonymous with $P(x)$ and will be used interchangeably with it. Also useful are **restricted quantifications** over sets:

$$(\forall x :: P.\ Q[x]) \triangleq (\forall x.\ x \in P \Rightarrow Q[x])$$
$$(\exists x :: P.\ Q[x]) \triangleq (\exists x.\ x \in P \wedge Q[x])$$

where the $Q[x]$ notation stresses that the variable x may (but does not necessarily!) occur free in the term Q.

Predicates as Propositions. Our method for embedding a special-purpose logic in HOL is based on the idea of using predicates over suitable domains in HOL to represent propositions in the embedded logic. As shown in Figure 2, boolean connectives, restricted quantifications, and validity can all be **lifted** to operate on predicates. Note our convention of using the "doubled" version of a symbol for its lifted counterpart. The definitions in Figure 2 provide the logical infrastructure for an embedded logic. For example, the following HOL theorem is the lifted version of *modus ponens*:

$$\forall P\, Q.\ \Vdash(P \Rrightarrow Q) \Rightarrow (\Vdash P \Rightarrow \Vdash Q)$$

Other inference rules of HOL can be similarly lifted to constitute a **lifted logic**. See [4] for a *uniform* method for lifting HOL proof procedures to handle proofs in the lifted logic based on a *sequent* formulation of lifted validity.

Coercion operators:	Basic temporal operators:

$$\Pi_C(C)(b) \triangleq C(b(0))$$

$$\text{future}(i)(b) \triangleq \lambda j.\, b(i+j)$$

$$\Pi_A(A)(b) \triangleq A(b(0), b(1))$$

$$\Box(P)(b) \triangleq \forall i.\, P(\text{future}(i)(b))$$

Action operators:

$$\Diamond(P)(b) \triangleq \exists i.\, P(\text{future}(i)(b))$$

$$\text{En}(A)(s) \triangleq \exists s'.\, A(s, s')$$

Derived temporal operators:

$$[A](s, s') \triangleq A(s, s') \lor (s' = s)$$

$$P \rightsquigarrow Q \triangleq \Box(P \Rightarrow \Diamond Q)$$

$$\langle A \rangle(s, s') \triangleq A(s, s') \land (s' \neq s)$$

$$\text{Fair}(A) \triangleq \Box(\Diamond(\Pi_A(A))) \;\mathbb{W}\; \Box(\Diamond(\Pi_C(\neg\text{En}(A))))$$

Fig. 3. Some notions from TLA

2.3 Temporal Logic of Actions

We have found some notions from Lamport's TLA (temporal logic of actions) [15] to be useful and hence have formalized them in HOL. They are summarized in Figure 3. Much of this subsection has appeared before in [4, 5].

In TLA there are three kinds of objects on which predicates are needed: **states**, **transitions**, and **behaviors**. A state is an assignment of values to program variables. A transition is a pair of states representing a step of program execution. A behavior is an infinite sequence of states representing a complete history of program execution. Predicates on states, transitions, and behaviors are called **conditions**, **actions**, and **temporal properties**, respectively. To summarize, if the type of states is σ, then:

State $: \sigma$	Transition $: \sigma \times \sigma$	Behavior $: num \to \sigma$
Condition $: \sigma \to bool$	Action $: \sigma \times \sigma \to bool$	Temporal $: (num \to \sigma) \to bool$

In Figure 3, C is a condition, A an action, and P and Q temporal properties. A condition C (respectively, an action A) can be **coerced** into a temporal property $\Pi_C(C)$ ($\Pi_A(A)$) by evaluating it at the first state (the first two states) of a behavior. An action A is **enabled**, $\text{En}(A)$, at a state s iff there exists an A-step from s. A may-A, $[A]$, step is either an A-step or a **stuttering** step, viz., a step that leaves the state unchanged. A must-A, $\langle A \rangle$, step is a non-stuttering A-step. The two **temporal modalities** $\Box$ and $\Diamond$ mean respectively "always in the future" and "sometime in the future", where "the future" includes "now". A temporal property P **leads to** another temporal property Q, $P \rightsquigarrow Q$, iff whenever P holds, Q holds then or later. An action A is **fair** iff A is either infinitely often executed or infinitely often disabled. This notion of fairness is called *weak fairness* in [15].

2.4 Fair Transition Systems

Programs will be represented by **fair transition systems** (FTS's). A FTS is a quadruple of the form:

$$(I : \sigma \to bool,\; N : \sigma \times \sigma \to bool,\; X : \chi \to bool,\; A : \chi \to \sigma \times \sigma \to bool)$$

$$\text{Initially}(I, N, X, A)(C) \triangleq (\forall s.\ I(s) \Rightarrow C(s))$$

$$\text{Stable}(I, N, X, A)(C) \triangleq (\forall s\ s'.\ N(s, s') \Rightarrow C(s) \Rightarrow C(s'))$$

$$\text{Invariant}(I, N, X, A)(C) \triangleq \text{Initially}(I, N, X, A)(C) \wedge \text{Stable}(I, N, X, A)(C)$$

$$\text{Ensures}(I, N, X, A)(C_1)(C_2) \triangleq$$
$$(\forall s\ s'.\ N(s, s') \Rightarrow C_1(s) \wedge \neg C_2(s) \Rightarrow C_1(s') \vee C_2(s')) \wedge$$
$$(\exists x :: X.\ (\forall s.\quad C_1(s) \wedge \neg C_2(s) \Rightarrow \text{En}\langle A(x)\rangle(s)) \wedge$$
$$(\forall s\ s'.\ C_1(s) \wedge \neg C_2(s) \Rightarrow \langle A(x)\rangle(s, s') \Rightarrow C_2(s')))$$

Fig. 4. Some notions from UNITY

where σ is the type of states and χ is the type of indices of fair actions. A behavior is **allowed** by (I, N, X, A) iff its initial state satisfies I, each of its transitions satisfies $[N]$, and, for each $x \in X$, $\langle A(x)\rangle$ is fair in it. The **semantics** of (I, N, X, A) is just the set of allowed behaviors:

$$[\![(I, N, X, A)]\!] \triangleq \Pi_C(I) \wedge \Box(\Pi_A[N]) \wedge (\mathbb{W} x :: X.\ \text{Fair}\langle A(x)\rangle) \tag{1}$$

Note that stuttering is always permitted in any allowed behavior, according to the $\Box(\Pi_A[N])$ part of (1). The reason for this is that we want to model both terminating and non-terminating computations by infinite behaviors, where the former is represented by behaviors in which the "terminal state" is repeated forever from some point on. (Another advantage of always permitting stuttering is that the treatment of *refinement* can be simplified [15], but this fact is not needed in this paper.) To ensure that progress is eventually made, (1) includes an explicit fairness assumption $\mathbb{W} x :: X.\ \text{Fair}\langle A(x)\rangle$ to rule out those behaviors that begin to keep stuttering prematurely, since $\langle B \rangle$ cannot stutter for any B.

We have found some notions from Chandy and Misra's UNITY [3] to be useful and hence have formalized them in HOL. They are summarized in Figure 4, in which (I, N, X, A) is a FTS and C, C_1, and C_2 are conditions. With respect to $S = (I, N, X, A)$, C is **initially** true iff C holds at every initial state of S, C is **stable** iff C is preserved by every transition of S, C is **invariant** iff C is both initially true and stable, and C_1 **ensures** C_2 iff (a) if C_1 holds and C_2 doesn't, then every transition of S either keeps C_1 true or makes C_2 true, and (b) there exists a fair action $\langle A(x)\rangle$ such that if C_1 holds and C_2 doesn't, then $\langle A(x)\rangle$ is enabled and executing $\langle A(x)\rangle$ makes C_2 true. Note that every definition in Figure 4 is expressed in terms of statements each involving at most two states.

We say that a FTS S **satisfies** a temporal property P, denoted $S \models P$, iff every behavior allowed by S satisfies P:

$$S \models P \triangleq \vdash ([\![S]\!] \Rightarrow P) \tag{2}$$

Later we will need the following theorems, which follow from Definitions (1), (2), and Figures 2, 3, and 4, but whose proofs are omitted due to space limitations:

$$\vdash (P \leadsto Q) \mathbb{\land} (Q \leadsto R) \Rightarrow (P \leadsto R) \tag{3}$$

i.e., $\leadsto$ is transitive;

$$\text{Stable}(S)(C) \Rightarrow S \models \Diamond(\Pi_C(C)) \Rightarrow \Diamond(\Box(\Pi_C(C))) \tag{4}$$

i.e., if C is stable, then C will stay true once it becomes true;

$$\text{Invariant}(S)(C) \Rightarrow S \models \Box(\Pi_C(C)) \tag{5}$$

i.e., if C is invariant, then C is always true;

$$\text{Ensures}(S)(C_1)(C_2) \Rightarrow S \models \Pi_C(C_1) \leadsto \Pi_C(C_2) \tag{6}$$

i.e., if C_1 ensures C_2, then C_1 leads to C_2;

$$\begin{aligned}
\text{Stable}(S)(C_1) \land S \models \Pi_C(C_0) \leadsto \Pi_C(C_1) &\Rightarrow \\
\text{Stable}(S)(C_2) \land S \models \Pi_C(C_0) \leadsto \Pi_C(C_2) &\Rightarrow \\
S \models \Pi_C(C_0) \leadsto \Pi_C(C_1 \mathbb{\land} C_2)&
\end{aligned} \tag{7}$$

i.e., if C_0 leads to C_1 and C_0 leads to C_2, where both C_1 and C_2 are stable, then C_0 leads to $C_1 \mathbb{\land} C_2$.

2.5 Simulation and Translation

This subsection reviews some results from [5]. Consider the following two FTS's:

$$S^\flat = (I^\flat : \sigma^\flat \to bool, N^\flat : \sigma^\flat \times \sigma^\flat \to bool, X^\flat : \chi \to bool, A^\flat : \chi \to \sigma^\flat \times \sigma^\flat \to bool)$$
$$S^\natural = (I^\natural : \sigma^\natural \to bool, N^\natural : \sigma^\natural \times \sigma^\natural \to bool, X^\natural : \chi \to bool, A^\natural : \chi \to \sigma^\natural \times \sigma^\natural \to bool)$$

where $S^\flat$ represents a "concrete" program with a "concrete" state space $\sigma^\flat$ and $S^\natural$ represents an "abstract" program with an "abstract" state space $\sigma^\natural$. We say that $S^\flat$ is **simulated** by $S^\natural$ via a **joint invariant** $J : \sigma^\flat \times \sigma^\natural \to bool$ and a **joint action** $M : (\sigma^\flat \times \sigma^\natural) \times (\sigma^\flat \times \sigma^\natural) \to bool$, denoted $\text{Sim}(S^\flat)(S^\natural)(J)(M)$, iff all of the following statements are true:

S1 For any initial state of $S^\flat$, there exists an initial state of $S^\natural$ such that they jointly satisfy J:
$$\forall s^\flat. \; I^\flat(s^\flat) \Rightarrow$$
$$\exists s^\natural. \; I^\natural(s^\natural) \land J(s^\flat, s^\natural)$$

S2 From any joint state satisfying J, if $S^\flat$ can take a step, then $S^\natural$ can also take a step such that the new joint state satisfies J and the joint step satisfies M:
$$\forall s^\flat \; s^\natural. \; J(s^\flat, s^\natural) \Rightarrow \forall t^\flat. \; N^\flat(s^\flat, t^\flat) \Rightarrow$$
$$\exists t^\natural. \; N^\natural(s^\natural, t^\natural) \land J(t^\flat, t^\natural) \land M((s^\flat, s^\natural), (t^\flat, t^\natural))$$

S3 $X^\flat = X^\natural$ and, for each $x \in X^\flat$, the following two statements are true:

S3a $\forall s^\flat \; s^\natural. \; J(s^\flat, s^\natural) \Rightarrow \text{En}\langle A^\natural(x)\rangle(s^\natural) \Rightarrow \text{En}\langle A^\flat(x)\rangle(s^\flat)$

S3b $\forall s^\flat \; s^\natural \; t^\flat \; t^\natural. \; M((s^\flat, s^\natural), (t^\flat, t^\natural)) \Rightarrow \langle A^\flat(x)\rangle(s^\flat, t^\flat) \Rightarrow \langle A^\natural(x)\rangle(s^\natural, t^\natural)$

If $\text{Sim}(S^{\flat})(S^{\natural})(J)(M)$, then the following four kinds of temporal properties of $S^{\natural}$ can be **translated** into those of $S^{\flat}$:

$$S^{\natural} \models \Box(\Pi_C(C^{\natural})) \;\Rightarrow\; S^{\flat} \models \Box(\Pi_C(\text{Trans}(J)(C^{\natural}))) \tag{8}$$
$$S^{\natural} \models \Diamond(\Pi_C(C^{\natural})) \;\Rightarrow\; S^{\flat} \models \Diamond(\Pi_C(\text{Trans}(J)(C^{\natural})))$$
$$S^{\natural} \models \Diamond(\Box(\Pi_C(C^{\natural}))) \;\Rightarrow\; S^{\flat} \models \Diamond(\Box(\Pi_C(\text{Trans}(J)(C^{\natural})))) \tag{9}$$
$$S^{\natural} \models \Box(\Diamond(\Pi_C(C^{\natural}))) \;\Rightarrow\; S^{\flat} \models \Box(\Diamond(\Pi_C(\text{Trans}(J)(C^{\natural}))))$$

where the **translation function** $\text{Trans}(J)$ transforms a condition on "abstract" states into one on "concrete" states via the joint invariant J:

$$\text{Trans}(J)(C^{\natural}) \triangleq \lambda s^{\flat}.\; \exists s^{\natural}.\; C^{\natural}(s^{\natural}) \wedge J(s^{\flat}, s^{\natural})$$

Note that all temporal properties built up from a condition by repeated applications of $\Box$ and $\Diamond$ are translatable. In this paper, only (8) and (9) are needed.

2.6 Graph Theory

To reason about distributed algorithms, some elementary graph theory is indispensable. Below is a self-contained summary of the graph-theoretic notions needed in this paper, which have all been formalized in HOL. The work involved in the formalization is considerable; see [6] for details.

In this paper all graphs are *undirected* and *finite*. A graph G is specified by the following: two types *node* and *edge*, which are respectively the types of node and edge names; two sets $\text{Node}(G) : node \rightarrow bool$ and $\text{Edge}(G) : edge \rightarrow bool$, which are respectively the sets of nodes and edges of G; and a function $\text{AdjEdge}(G) : node \rightarrow edge \rightarrow bool$, which maps each node n in G to the set $\text{AdjEdge}(G)(n)$ of edges in G that are adjacent to n. A **link** of G is a triple of the form (p, e, q), where p and q are nodes in G and e is an edge in G connecting p and q. A link can be viewed as a directed edge. The type $node \times edge \times node$ is abbreviated as *link* and the set of links of G is denoted by $\text{Link}(G) : link \rightarrow bool$. Given a link $l = (p, e, q)$, we define $\text{src}(l) \triangleq p$, $\text{des}(l) \triangleq q$, $\text{via}(l) \triangleq e$, and $\text{rev}(l) \triangleq (q, e, p)$ to be the **source**, **destination**, **via**, and **reverse** of l, respectively. Note that since G is undirected, if l is a link in G, so is $\text{rev}(l)$. For n in $\text{Node}(G)$ and e in $\text{AdjEdge}(G)(n)$, $\text{inc}(G)(n)(e)$ (**incoming link**) denotes the link in G whose des is n and whose via is e, $\text{out}(G)(n)(e)$ (**outgoing link**) denotes the link in G whose src is n and whose via is e, and $\text{opp}(G)(n)(e)$ (**opposite node**) denotes the src of $\text{inc}(G)(n)(e)$ (or, equivalently, the des of $\text{out}(G)(n)(e)$).

A **walk** from p to q in G is a list of links $[l_1, l_2, \ldots, l_n]$ such that $p = \text{src}(l_1)$, $\text{des}(l_1) = \text{src}(l_2)$, $\text{des}(l_2) = \text{src}(l_3)$, $\ldots$, $\text{des}(l_n) = q$. A walk can be empty, in which case $p = q$. A **path** is a walk that does not repeat any edge (but it can repeat nodes). A **cycle** is a *non-empty* path that starts and ends at the same node. A graph G is **connected** iff there exists a walk between any two nodes in G, G is **acyclic** iff there is no cycle in G, and G is a **tree** iff G is both connected and acyclic. The crucial property of a tree is that there exists a *unique* path between any two of its nodes.

Given a tree T and a node r (the *root*) of T, T can be *oriented* with respect to r as follows. Let n be any node in T. The **parent edge** of n, $\mathsf{ParEdge}(T)(r)(n)$, is the first edge on the path from n to r in T, if $n \neq r$; $\mathsf{ParEdge}(T)(r)(r)$ is undefined. The set of **kid edges** of n, $\mathsf{KidEdge}(T)(r)(n)$, is the set of adjacent edges of n which are not $\mathsf{ParEdge}(T)(r)(n)$; in particular, $\mathsf{KidEdge}(T)(r)(r) = \mathsf{AdjEdge}(T)(r)$. The set of **descendent nodes** of n, $\mathsf{DescNode}(T)(r)(n)$, is the set of nodes in T reachable from n (including n itself) without crossing $\mathsf{ParEdge}(T)(r)(n)$; in particular, $\mathsf{DescNode}(T)(r)(r) = \mathsf{Node}(T)$. We further define $\mathsf{ParNode}(T)(r)(n) \triangleq \mathsf{opp}(T)(n)(\mathsf{ParEdge}(T)(r)(n))$ and $\mathsf{KidNode}(T)(r)(n) \triangleq \{\, \mathsf{opp}(T)(n)(e) \mid e \in \mathsf{KidEdge}(T)(r)(n)\,\}$.

3 Formalizing DSUM in Higher-Order Logic

3.1 The Underlying Network

Recall that *Net* is a tree representing the underlying network, *root* is a node in *Net*, and *val* : *node* → *num* is an assignment of (natural) numbers to nodes. Every constant defined in the sequel has (*Net, root, val*) as a parameter, which will be omitted to simplify notations. Likewise omitted are the first parameter of Node, Edge, Link, $\mathsf{AdjEdge}$, inc, out, and opp, which will always be *Net*, and the first two parameters of $\mathsf{ParNode}$, $\mathsf{ParEdge}$, $\mathsf{KidNode}$, $\mathsf{KidEdge}$, and $\mathsf{DescNode}$, which will always be *Net* and *root*.

3.2 DSUM as a Fair Transition System

We shall formalize DSUM as the following FTS:

$$\mathsf{Prog}^D \triangleq (\mathsf{Init}^D, \mathsf{Next}^D, \mathsf{FEvent}, \mathsf{Act}^D)$$

where the superscript D indicates that this is the *distributed view* of DSUM. Later (Section 4) the *event view* of DSUM will be given the superscript E.

A state of Prog^D consists of five functions, the first four of which represent the values of program variables at nodes and the last of which represents the contents of message queues over links:

$$
\begin{aligned}
pc &: node \to pc & par &: node \to edge \\
cnt &: node \to num & sum &: node \to num & mq &: link \to (msg)list
\end{aligned}
$$

The type $state^D$ of states of Prog^D is the cartesian product of the types of the five functions above, where the types pc (program counter values) and msg (messages) are given by the following (degenerate) recursive type definitions [17]:

$$pc ::= \texttt{Idle} \mid \texttt{Busy} \mid \texttt{Done}$$

$$msg ::= \texttt{Start} \mid \texttt{Report}(num)$$

Note that though the function pc assigns a value to each $n : node$, we are only interested in the values of those n's that are in the set Node; similar remarks apply to par, cnt, sum, and mq. To simplify notations, we will abbreviate (pc, par, cnt, sum, mq) as ds and $(pc', par', cnt', sum', mq')$ as ds'.

$Update(f)(x)(v) \triangleq$
$\quad \lambda y.$ if $(y = x)$ then v else $f(y)$

$SendOne(mq)(n)(e)(m) \triangleq$
$\quad$ let $l = out(n)(e)$ in
$\quad \lambda k. mq(k)^\frown(\text{if } (k = l) \text{ then } [m] \text{ else } [\ ])$

$SendMany(mq)(n)(E)(m) \triangleq$
$\quad$ let $L = \{ out(n)(e) \mid e \in E \}$ in
$\quad \lambda k. mq(k)^\frown(\text{if } (k \in L) \text{ then } [m] \text{ else } [\ ])$

$Act^D(rootStart(n))(ds, ds') \triangleq$
$\quad (n = root) \wedge (pc(n) = \text{Idle}) \wedge$
$\quad$ let $E = AdjEdge(n)$ in
$\quad (pc' = Update(pc)(n)(\text{Busy})) \wedge$
$\quad (par' = par) \wedge$
$\quad (cnt' = Update(cnt)(n)(|E|)) \wedge$
$\quad (sum' = Update(sum)(n)(val(n))) \wedge$
$\quad (mq' = SendMany(mq)(n)(E)(\text{Start}))$

$Act^D(recvStart(n)(e))(ds, ds') \triangleq$
$\exists t. (mq(inc(n)(e)) = [\text{Start}]^\frown t) \wedge$
$\quad$ let $E = AdjEdge(n) \setminus \{e\}$ in
$\quad (pc' = Update(pc)(n)(\text{Busy})) \wedge$
$\quad (par' = Update(par)(n)(e)) \wedge$
$\quad (cnt' = Update(cnt)(n)(|E|)) \wedge$
$\quad (sum' = Update(sum)(n)(val(n))) \wedge$
$\quad (mq' = SendMany(Update(mq)(inc(n)(e))(t))(n)(E)(\text{Start}))$

$Act^D(recvReport(n)(e))(ds, ds') \triangleq$
$\exists t v. (mq(inc(n)(e)) = [\text{Report}(v)]^\frown t) \wedge$
$\quad (pc' = pc) \wedge$
$\quad (par' = par) \wedge$
$\quad (cnt' = Update(cnt)(n)(cnt(n) - 1)) \wedge$
$\quad (sum' = Update(sum)(n)(sum(n) + v)) \wedge$
$\quad (mq' = Update(mq)(inc(n)(e))(t))$

$Act^D(sendReport(n)(e))(ds, ds') \triangleq$
$\quad (pc(n) = \text{Busy}) \wedge (cnt(n) = 0) \wedge$
$\quad (n \neq root) \wedge (e = par(n)) \wedge$
$\quad (pc' = Update(pc)(n)(\text{Done})) \wedge$
$\quad (par' = par) \wedge$
$\quad (cnt' = cnt) \wedge$
$\quad (sum' = sum) \wedge$
$\quad (mq' = SendOne(mq)(n)(e)$
$\quad\quad\quad\quad\quad (\text{Report}(sum(n))))$

$Act^D(rootReport(n))(ds, ds') \triangleq$
$\quad (pc(n) = \text{Busy}) \wedge (cnt(n) = 0) \wedge$
$\quad (n = root) \wedge$
$\quad (pc' = Update(pc)(n)(\text{Done})) \wedge$
$\quad (par' = par) \wedge$
$\quad (cnt' = cnt) \wedge$
$\quad (sum' = sum) \wedge$
$\quad (mq' = mq)$

Fig. 5. Fair actions of DSUM

The initial condition of $\mathbf{Prog}^D$, $\mathsf{Init}^D : state^D \to bool$, asserts that initially every node is $\mathbf{Idle}$ and every message queue is empty:

$$\mathsf{Init}^D(ds) \triangleq (\forall n :: \text{Node}. \ pc(n) = \mathbf{Idle}) \wedge (\forall l :: \text{Link}. \ mq(l) = [\ ])$$

The family of fair actions of $\mathbf{Prog}^D$, $\mathbf{Act}^D : event \to state^D \times state^D \to bool$, plus three auxiliary functions, are defined in Figure 5. The type $event$ of $\mathbf{events}$ is:

$\quad event ::=$
$\quad\quad \mathbf{rootStart}(node) \mid \mathbf{recvStart}(node)(edge) \mid$
$\quad\quad \mathbf{recvReport}(node)(edge) \mid \mathbf{sendReport}(node)(edge) \mid \mathbf{rootReport}(node)$

Note that here we are calling the names of actions "events", whereas in Section 1 we meant by "events" the names of *occurrences* of actions. We may confuse the two notions because each action of $\mathbf{Prog}^D$ can occur at most once. The correspondence between Figures 1 and 5 is as follows:

$\boxed{3}$ is represented by $\mathsf{Act}^D(\mathbf{rootStart}(n))$

$\boxed{4}$ is represented by $\mathsf{Act}^D(\mathbf{recvStart}(n)(e))$

$\boxed{5}$ is represented by $\mathsf{Act}^D(\mathbf{recvReport}(n)(e))$

$\boxed{6}$ is represented by $\begin{cases} \mathsf{Act}^D(\mathbf{rootStart}(n)) & \text{if } (n = root) \\ \mathsf{Act}^D(\mathbf{sendStart}(n)(e)) & \text{if } (n \neq root) \wedge (e = par(n)) \end{cases}$

Clearly, not every value of type *event* is a meaningful event; we define the set of **fair events**, $\mathsf{FEvent} : event \rightarrow bool$, to be:

$\mathsf{FEvent}(\mathbf{xxx}(n)) \quad \triangleq\ n \in \mathsf{Node}$

$\mathsf{FEvent}(\mathbf{yyy}(n)(e)) \triangleq\ n \in \mathsf{Node} \wedge e \in \mathsf{AdjEdge}(n)$

where $\mathbf{xxx}$ is $\mathbf{rootStart}$ or $\mathbf{rootReport}$ and $\mathbf{yyy}$ is $\mathbf{recvStart}$, $\mathbf{recvReport}$, or $\mathbf{sendReport}$. Finally, the action of $\mathbf{Prog}^D$, $\mathsf{Next}^D : state^D \times state^D \rightarrow bool$, is:

$$\mathsf{Next}^D(ds, ds') \triangleq \exists\, ev :: \mathsf{FEvent}.\ \mathsf{Act}^D(ev)(ds, ds')$$

Note that $\mathbf{Prog}^D$ can take no step that does not belong to some fair action of $\mathbf{Prog}^D$. This property is *not* true for all FTS's.

3.3   What Do We Want To Prove about DSUM?

In general, there are two classes of properties that one wants to prove about a distributed algorithm [15]: *safety*, which says that nothing bad will ever happen, and *liveness*, which says that something good will eventually happen. In the case of DSUM, the safety property we want to prove is *partial correctness*, which says that whenever DSUM terminates, the correct outcome has been computed:

$$\mathbf{Prog}^D \models \Box(\Pi_C(\mathsf{Done}^D \Rightarrow \mathsf{Outcome}^D)) \tag{10}$$

and the liveness property we want to prove is the *eventual persistence* of termination, which says that DSUM eventually terminates and stays that way forever:

$$\mathbf{Prog}^D \models \Diamond(\Box(\Pi_C(\mathsf{Done}^D))) \tag{11}$$

where the condition Done^D says that the *root* is **Done**:

$$\mathsf{Done}^D(ds) \triangleq (pc(root) = \mathbf{Done})$$

and the condition $\mathsf{Outcome}^D$ says that the *root* has the desired global sum:

$$\mathsf{Outcome}^D(ds) \triangleq (sum(root) = \sum_{n \in \mathsf{Node}} val(n))$$

$$\text{Cause}(\textbf{rootStart}(n))(occ) \triangleq$$
$$(n = root)$$

$$\text{Cause}(\textbf{recvStart}(n)(e))(occ) \triangleq$$
$$(n \neq root) \wedge (e = \mathsf{ParEdge}(n)) \wedge$$
$$\textbf{nodeStart}(\mathsf{opp}(n)(e))(occ)$$

$$\text{Cause}(\textbf{recvReport}(n)(e))(occ) \triangleq$$
$$e \in \mathsf{KidEdge}(n) \wedge$$
$$\textbf{sendReport}(\mathsf{opp}(n)(e))(e) \in occ$$

$$\text{Cause}(\textbf{sendReport}(n)(e))(occ) \triangleq$$
$$(n \neq root) \wedge (e = \mathsf{ParEdge}(n)) \wedge$$
$$\textbf{recvStart}(n)(e) \in occ \wedge$$
$$\forall k :: \mathsf{KidEdge}(n).\ \textbf{recvReport}(n)(k) \in occ$$

$$\text{Cause}(\textbf{rootReport}(n))(occ) \triangleq$$
$$(n = root) \wedge$$
$$\textbf{rootStart}(n) \in occ \wedge$$
$$\forall k :: \mathsf{KidEdge}(n).\ \textbf{recvReport}(n)(k) \in occ$$

Fig. 6. Causes of events of DSUM

4 Event View of DSUM

The *event view* of DSUM is an alternative, abstract description of DSUM's behaviors in terms of events and the causality relation between them. In the sequel, let *occ* be the set of occurred events (i.e., $ev \in occ$ means that the event ev has occurred) and:

$$\textbf{nodeStart}(n)(occ) \triangleq ((n = root) \Rightarrow \textbf{rootStart}(n) \in occ) \wedge$$
$$((n \neq root) \Rightarrow \textbf{recvStart}(n)(\mathsf{ParEdge}(n)) \in occ)$$
$$\textbf{nodeReport}(n)(occ) \triangleq ((n = root) \Rightarrow \textbf{rootReport}(n) \in occ) \wedge$$
$$((n \neq root) \Rightarrow \textbf{sendReport}(n)(\mathsf{ParEdge}(n)) \in occ)$$

4.1 Causes of Events

Figure 6 lists the (immediate) **cause** of each event of DSUM. The cause of, say, **sendReport**$(n)(e)$ is: (a) n is not the *root* and e is the parent edge of n, (b) n has received a **Start** message over e, and (c) n has received a **Report** message over each of its kid edges. The important thing is that writing down the causes of events involves only local reasoning and hence is quite easy.

4.2 Event Algorithm

Now we can construct the following FTS as an operational representation of the causal patterns of events of DSUM:

$$\mathsf{Prog}^E \triangleq (\mathsf{Init}^E, \mathsf{Next}^E, \mathsf{FEvent}, \mathsf{Act}^E)$$

The state of Prog^E simply records what events have occurred. So the type $state^E$ of states of Prog^E is $event \rightarrow bool$. Since no event has occurred at the beginning, we define:

$$\mathsf{Init}^E(occ) \triangleq (occ = \emptyset)$$

An event ev is enabled when its cause, $\mathsf{Cause}(ev)(occ)$, is true but ev itself has not occurred. The effect of executing ev is to add ev to the set of occurred events. So we define:

$$\mathsf{Act}^E(ev)(occ, occ') \triangleq \mathsf{Cause}(ev)(occ) \wedge ev \notin occ \wedge (occ' = occ \cup \{ev\})$$

Finally, since only fair events can occur, we define:

$$\mathsf{Next}^E(occ, occ') \triangleq \exists\, ev :: \mathsf{FEvent}.\ \mathsf{Act}^E(ev)(occ, occ')$$

4.3 Safety Property of Event Algorithm

The event algorithm Prog^E has a simple invariant:

$$\mathsf{Inv}^E(occ) \triangleq \forall\, ev.\ ev \in occ \Rightarrow \mathsf{FEvent}(ev) \wedge \mathsf{Cause}(ev)(occ)$$

which says that every occurred event is a fair event and has its cause true. It is easy to verify that Inv^E is indeed an invariant of Prog^E:

$$\mathsf{Invariant}(\mathsf{Prog}^E)(\mathsf{Inv}^E)$$

from which and Theorem (5) it follows that:

$$\mathsf{Prog}^E \models \Box(\Pi_C(\mathsf{Inv}^E)) \tag{12}$$

Theorem (12) is the counterpart of (10) for the event algorithm.

4.4 Liveness Property of Event Algorithm

The invariant Inv^E uses the causality relation in the backward direction: it says that if an event has occurred, then its cause must be true. The key liveness property of Prog^E uses the causality relation in the forward direction:

$$\forall\, ev :: \mathsf{FEvent}.\ \mathsf{Prog}^E \models \Pi_C(\mathsf{Cause}(ev)) \rightsquigarrow \Pi_C(\mathsf{Occ}(ev)) \tag{13}$$

which says that if the cause of an event ev is true, then ev itself must eventually occur, where $\mathsf{Occ}(ev)(occ) \triangleq ev \in occ$. By Theorem (6), (13) follows from:

$$\forall\, ev :: \mathsf{FEvent}.\ \mathsf{Ensures}(\mathsf{Prog}^E)(\mathsf{Cause}(ev))(\mathsf{Occ}(ev))$$

which has an easy proof in which the existential witness for the variable x in the definition of $\mathsf{Ensures}$ (Figure 4) is just the event ev.

The next step is to prove:

$$\mathsf{Prog}^E \models \Pi_C(\mathsf{Cause}(\mathtt{rootStart}(root))) \rightsquigarrow \Pi_C(\mathsf{Occ}(\mathtt{rootReport}(root))) \tag{14}$$

whose proof uses the definition of Cause and repeated applications of Theorems (13), (7), and (3) first from $root$ to leaves and then from leaves back to $root$ (see Section 7.3 of [6]). In addition, the applications of Theorem (7) need:

$$\forall\, ev :: \mathsf{FEvent}.\ \mathsf{Stable}(\mathsf{Prog}^E)(\mathsf{Occ}(ev)) \tag{15}$$

which is trivial to prove.

Finally, note that $\mathsf{Cause}(\mathtt{rootStart}(root))$ is identically true, so (14) can be simplified and weakened into:

$$\mathsf{Prog}^E \models \Diamond(\Pi_C(\mathsf{Occ}(\mathtt{rootReport}(root))))$$

from which and Theorems (4) and (15), it follows that:

$$\mathsf{Prog}^E \models \Diamond(\Box(\Pi_C(\mathsf{Occ}(\mathtt{rootReport}(root))))) \tag{16}$$

Theorem (16) is the counterpart of (11) for the event algorithm.

5 Relating Distributed and Event Views of DSUM

We now apply the theory of simulation and translation reviewed in Section 2.5. The "concrete" program will be Prog^D and the "abstract" program Prog^E.

5.1 Simulation of Distributed View by Event View of DSUM

To show that Prog^D is simulated by Prog^E, we need to construct a joint invariant $\mathsf{Inv}^{D,E}$ and a joint action $\mathsf{Next}^{D,E}$ so that $\mathsf{Sim}(\mathsf{Prog}^D)(\mathsf{Prog}^E)(\mathsf{Inv}^{D,E})(\mathsf{Next}^{D,E})$.

The joint invariant $\mathsf{Inv}^{D,E}$ essentially expresses the state of Prog^D in terms of that of Prog^E. It consists of the invariant Inv^E of Prog^E, a local invariant $\mathsf{Inv}_N^{D,E}(n)$ for each node n, and a local invariant $\mathsf{Inv}_L^{D,E}(l)$ for each link l:

$$\mathsf{Inv}^{D,E}(ds, occ) \triangleq \mathsf{Inv}^E(occ) \wedge (\forall n :: \mathsf{Node}.\ \mathsf{Inv}_N^{D,E}(n)(ds, occ))$$
$$\wedge (\forall l :: \mathsf{Link}.\ \mathsf{Inv}_L^{D,E}(l)\ (ds, occ))$$

where:

$\mathsf{Inv}_N^{D,E}(n)(ds, occ) \triangleq$
  $(\ pc(n) = \mathbf{if}\ \mathtt{nodeReport}(n)(occ)\ \mathbf{then}\ \mathtt{Done}\ \mathbf{else}$
  $\qquad\qquad\ \mathbf{if}\ \ \mathtt{nodeStart}(n)(occ)\ \mathbf{then}\ \mathtt{Busy}\ \mathbf{else}\ \mathtt{Idle}\ )$
$\wedge (\ \mathtt{nodeStart}(n)(occ) \Rightarrow$
  $\mathbf{let}\ R = \{\ k \in \mathsf{KidEdge}(n) \mid \mathtt{recvReport}(n)(k) \in occ\ \}\ \mathbf{in}$
  $((n \neq root) \Rightarrow (par(n) = \mathsf{ParEdge}(n)))\ \wedge$
  $(\ cnt(n) = |\,\mathsf{KidEdge}(n) \setminus R\,|\ )\ \wedge$
  $(\ sum(n) = val(n) + \sum_{k \in R} \sum_{d \in \mathsf{DescNode}(opp(n)(k))} val(d)\ )\ )$
$\wedge (\ \mathtt{nodeReport}(n)(occ) \Rightarrow$
  $(\ sum(n) = \sum_{d \in \mathsf{DescNode}(n)} val(d)\ )\ )$

$\mathsf{Inv}_L^{D,E}(p, e, q)(ds, occ) \triangleq$
  $(\ mq(l) = \mathbf{if}\ \mathtt{nodeStart}(p)(occ) \wedge e \in \mathsf{KidEdge}(p) \wedge \mathtt{recvStart}(q)(e) \notin occ$
  $\qquad\ \ \mathbf{then}\ [\mathtt{Start}]\ \mathbf{else}$
  $\qquad\ \ \mathbf{if}\ \mathtt{sendReport}(p)(e) \in occ \wedge \mathtt{recvReport}(q)(e) \notin occ$
  $\qquad\ \ \mathbf{then}\ [\mathtt{Report}(\sum_{d \in \mathsf{DescNode}(p)} val(d))]$
  $\qquad\ \ \mathbf{else}\ \ [\ ]\ )$

The joint action $\mathsf{Next}^{D,E}$ can be viewed as the "parallel composition" of Act^D and Act^E, synchronized through the fair events:

$$\mathsf{Next}^{D,E}((ds, occ), (ds', occ')) \triangleq$$
$$\exists\, ev :: \mathsf{FEvent}.\ \mathsf{Act}^D(ev)(ds, ds') \wedge \mathsf{Act}^E(ev)(occ, occ')$$

Due to space limitations, we are not able to show the proofs of S1–S3 here. They are tedious but not difficult, and none of them involves any temporal reasoning.

5.2 Translation of Safety and Liveness Properties of DSUM

Given that Prog^D is simulated by Prog^E via $\mathsf{Inv}^{D,E}$ and $\mathsf{Next}^{D,E}$, we can use (8) and (9) to translate, respectively, (12) and (16) to obtain:

$$\mathsf{Prog}^D \models \square(\Pi_C(\mathsf{Trans}(\mathsf{Inv}^{D,E})(\mathsf{Inv}^E))) \tag{17}$$
$$\mathsf{Prog}^D \models \Diamond(\square(\Pi_C(\mathsf{Trans}(\mathsf{Inv}^{D,E})(\mathsf{Occ}(\mathtt{rootReport}(root)))))) \tag{18}$$

It is easy to see, from the definition of $\mathsf{Inv}_N^{D,E}$, that:

$$\vdash \mathsf{Trans}(\mathsf{Inv}^{D,E})(\mathsf{Inv}^E) \Rightarrow (\mathsf{Done}^D \Rightarrow \mathsf{Outcome}^D) \tag{19}$$

Clearly, (10) follows from (17) and (19). Similarly, it is easy to see that:

$$\vdash \mathsf{Trans}(\mathsf{Inv}^{D,E})(\mathsf{Occ}(\mathtt{rootReport}(root))) \Rightarrow \mathsf{Done}^D \tag{20}$$

and that (11) follows from (18) and (20). The correctness proof of DSUM is now complete.

6   Related Work

Due to space limitations, we review only those papers on mechanical verification of distributed or concurrent algorithms based on theorem proving (as opposed to *model checking* [16]).

Goldschlag [11] embeds UNITY in the Boyer-Moore prover and uses it to verify a distributed minimum-finding algorithm for trees, based on a detailed hand proof by Lamport. Although the algorithm he verifies is very similar to our DSUM, the two proofs differ in several important ways. Firstly, he uses the conventional invariant method [15] for reasoning about distributed algorithms, while we use a method based on the notions of events and causality. We believe that our method can be scaled up more easily to verify more complex algorithms [7, 8]. Secondly, he uses an *ad hoc* encoding of trees as nested lists, while we have a theory of general graphs [6]. So we will be able to reason about distributed algorithms for general networks [10, 18]. Thirdly, since higher-order logic is more expressive than the logic of the Boyer-Moore prover, the former permits naive and direct formalizations whose counterparts in the latter are subtle and indirect. This is important because the ability to have naive and direct formalizations not only reduces the risk of having *wrong* formalizations, but also helps to communicate the results of proofs to the uninitiated.

Engberg, Grønning, and Lamport [9] implement a translator of TLA into LP (the Larch Prover) and use the combination to verify a distributed algorithm for computing the distances of all nodes to a distinguished node in a graph. Consequently they must face the difficult problem of how to interface two different logics in a meaningful and consistent way, which we avoid by working in a single logic. Unlike us, they do not prove the axioms and inference rules of TLA and the properties of data structures used in their proofs, which they either keep as assumptions or assert as axioms. Indeed, there does not seem to be a clear distinction in LP between definitions and arbitrary axioms.

Von Wright [19] and von Wright and Långbacka [20] embed TLA in HOL and use it to reason about concurrent algorithms. They embed more features of TLA than we do, but we use a sequent-based embedding that allows a *uniform lifting* of HOL proof procedures to handle proofs in an embedded logic (see [4] for details). Also, DSUM is more complex than the algorithms they verify.

Andersen, Petersen, and Pettersson [1] embed UNITY in HOL and use it to verify a lift-control program. They follow UNITY very closely, while we only borrow some concepts from UNITY which we formalize in our own style. Also, there is a considerable amount of automation in their work.

7 Conclusion and Future Work

In this work we have made several "design decisions" that deserve further comment. Firstly, when applied to an algorithm as simple as DSUM, the advantage of our events-and-causality-based method may not be manifest. But it will become so when we verify more complex distributed algorithms (more on this in the next paragraph). This work is therefore a trial run for the larger undertaking. Secondly, if trees are all we need, it is simpler to use recursive types [17] to model trees than to develop a theory of general graphs as we did in [6]. Again, since we aim to verify more complex distributed algorithms that operate in general networks, we decide to develop a theory of general graphs and try it out on a simple example. Thirdly, our approach to embedding programming logics is eclectic (i.e., we embed any features from any programming logics that we find useful) and minimalist (i.e., we embed only those features that we find useful). Thus we follow neither TLA nor UNITY completely.

In the future we hope to extend this work in two directions. Firstly, we hope to verify more complex distributed algorithms. Note that DSUM possesses a *unique* causal pattern of events (described in Figure 6). In contrast, Segall's PIF (propagation of information with feedback) algorithm [18], which can be viewed as a slight modification of DSUM that works for all connected graphs, can generate many causal patterns (in fact, one for each spanning tree of the underlying network). Despite the enormous increase of "true nondeterminism", PIF can also be verified using our events-and-causality-based method [7]. The only extension needed is that the causality relation of PIF is no longer predetermined, but dynamically generated and hence has to be recorded as part of the state of the event algorithm. In fact, we [8] believe that our method can be scaled up to verify algorithms as complex as the distributed minimum spanning tree algorithm of Gallager, Humblet, and Spira [10]. But only experience can tell whether this belief is correct.

Secondly, we hope to automate as large a part of our proofs as possible. Although it is unlikely that the correctness proofs of distributed algorithms can ever be completely automated (because these algorithms manipulate such general data structures as numbers, lists, and graphs), the programmability of HOL allows us to write customized proof procedures to automate many tasks in the proofs. At present we already have procedures for the uniform lifting of HOL proof procedures to handle lifted logic [4], for the automatic derivation of the enabling conditions of actions like those in Figure 5, and for the chasing of causality chains using definitions like those in Figure 6. We expect that more automation be achieved as more experience is gained and more common patterns in reasoning are observed.

Acknowledgements. The author is grateful to Professors Eli Gafni and David Martin for their guidance, to Michael Gordon for his encouragement, to Flemming Andersen, Peter Homeier, Sara Kalvala, Stott Parker, Konrad Slind, and the anonymous referees for their helpful comments, and to HOL hackers around the world for sustaining a lively and friendly user community.

References

1. Flemming Andersen, Kim Dam Petersen, and Jimmi S. Pettersson, "Program Verification Using HOL-UNITY", in [14].
2. G.v. Bochmann and D.K. Probst (ed.), *Computer-Aided Verification, 4th International Workshop*, LNCS 663, Springer-Verlag, 1992.
3. K. Mani Chandy and Jayadev Misra, *Parallel Program Design: A Foundation*, Addison-Wesley, 1988.
4. Ching-Tsun Chou, "A Sequent Formulation of a Logic of Predicates in HOL", pp. 71–80 of L.J.M. Claesen and M.J.C. Gordon (ed.), *Higher Order Logic Theorem Proving and Its Applications, 5th International Workshop*, IFIP Transactions A-20, North-Holland, 1992.
5. Ching-Tsun Chou, "Predicates, Temporal Logic, and Simulations", in [14].
6. Ching-Tsun Chou, "A Formal Theory of Undirected Graphs in Higher-Order Logic", in this Proceedings.
7. Ching-Tsun Chou, "Practical Use of the Notions of Events and Causality in Reasoning about Distributed Algorithms", work in progress.
8. Ching-Tsun Chou and Eli Gafni, "Understanding and Verifying Distributed Algorithms Using Stratified Decomposition", *Proc. of the 7th ACM Symp. on Principles of Distributed Computing*, pp. 44–65, Aug. 1988.
9. Urban Engberg, Peter Grønning, and Leslie Lamport, "Mechanical Verification of Concurrent Systems with TLA", pp. 44–55 of [2].
10. R.G. Gallager, P.A. Humblet, and P.M. Spira, "A Distributed Algorithm for Minimum-Weight Spanning Trees", *ACM Trans. on Programming Languages and Systems*, Vol. 5, No. 1, pp. 66–77, Jan. 1983.

11. David M. Goldschlag, "Mechanically Verifying Concurrent Programs with the Boyer-Moore Prover", *IEEE Trans. on Software Engineering*, Vol. 16, No. 9, pp. 1005–1023, Sep. 1990.

12. Michael J.C. Gordon, "HOL: A Proof Generating System for Higher-Order Logic", pp. 73–128 of G. Birtwistle and P.A. Subrahmanyam (ed.), *VLSI Specification, Verification and Synthesis*, Kluwer Academic Publishers, 1988.

13. Michael J.C. Gordon and Thomas F. Melham (ed.), *Introduction to HOL: A Theorem-Proving Environment for Higher-Order Logic*, Cambridge University Press, 1993.

14. J.J. Joyce and C.-J.H. Seger (ed.), *Higher Order Logic Theorem Proving and Its Applications, 6th International Workshop*, LNCS 780, Springer-Verlag, 1993.

15. Leslie Lamport, "The Temporal Logic of Actions", DEC SRC technical report #79, Dec. 1991. (To appear in *ACM Trans. on Programming Languages and Systems*.)

16. Kenneth L. McMillan, *Symbolic Model Checking*, Kluwer Academic Publishers, 1993.

17. Thomas F. Melham, "Automating Recursive Type Definitions in Higher-Order Logic", pp. 341–386 of G. Birtwistle and P.A. Subrahmanyam (ed.), *Current Trends in Hardware Verification and Automated Theorem Proving*, Springer-Verlag, 1989.

18. Adrian Segall, "Distributed Network Protocols", *IEEE Trans. on Information Theory*, Vol. 29, No. 1, pp. 23–35, Jan. 1983.

19. Joakim von Wright, "Mechanising the Temporal Logic of Actions in HOL", pp. 155–159 of M. Archer et al. (ed.), *Proc. of 1991 Workshop on the HOL Theorem Proving System and Its Applications*, IEEE Computer Society Press, 1992.

20. Joakim von Wright and Thomas Långbacka, "Using a Theorem Prover for Reasoning about Concurrent Algorithms", pp. 56–68 of [2].

Tracking Design Changes with Formal Verification

Paul Curzon

University of Cambridge Computer Laboratory, United Kingdom.
Email: pc@cl.cam.ac.uk

Abstract. Designs are often modified for use in new circumstances. If formal proof is to be an acceptable verification methodology for industry, it must be capable of tracking design changes quickly. We describe our experiences formally verifying an implementation of an ATM network component, and on our subsequent verification of modified designs. Three of the designs verified are in use in a working network. They were designed and implemented with no consideration for formal methods. This case study gives an indication of the difficulties in formally verifying a real design and of subsequently tracking design changes.

1 Introduction

Designs are often modified as requirements change. Such modifications often take a fraction of the original design time to complete. Even if a design can initially be validated in an acceptable time scale, formal verification is unlikely to be accepted if a similar amount of time is required to validate subsequent modified designs. It has been suggested that this is one of the main problems of using formal proof in industry [2]. We describe our experiences verifying a real hardware design using formal proof and of our attempts to subsequently track changes that were made to the implementation.

The device we have considered is a component of an Asynchronous Transfer Mode (ATM) switch. This work is part of a larger project to investigate the formal verification of an ATM network as a whole. We are investigating a working network which carries real user data: the Fairisle ATM network [10], designed at the University of Cambridge. It provides a realistic formal verification case study. The network component we have considered is the Fairisle 4 by 4 switching element [7]. We verified several versions of the element with slightly different designs. Four of these designs have been fabricated of which three are in use in the Fairisle network.

The formal specification and verification work was carried out using the HOL90 theorem prover. It was performed on completed implementations. This is harder than if formal methods are integrated into the design process for several reasons. (i) Why the design was believed to be correct must be at least partially rediscovered. (ii) If the implementors work to a formal specification, the specification and implementation are more likely to agree and less time will be wasted correcting specification errors. (iii) Since it involves analysing the design in great detail, the specification and verification process can identify changes to the implementation that simplify the verification task. Such changes do not necessarily need to compromise other design aims such as efficiency. (iv) Reverse engineering a specification from the implementation is a very time consuming process. (v) Finally, a hardware description language amenable to verification can be used, thus avoiding translation problems. The above problems were exacerbated further for this case study by the fact that little documentation of the implementations had been produced. A significant amount of reverse engineering was required. The behavioural specifications were largely deduced by examining the implementations.

The full details of the specifications of the first element verified are given in a literate document [4] derived from the HOL source files using the HOL mweb tool. An overview is given separately [5].

2   The Fairisle Switch

The Fairisle switch consists of three types of component: input port controllers, output port controllers and a switching fabric. Each port controller is connected to either an input or output transmission line of the switch, and to the switching fabric. The port controllers process incoming or outgoing cells of data. A cell consists of a fixed number of bytes. Some of the bytes contain control information giving, for example, an indication of the ultimate destination of the cell. The port controllers use this information to determine the outgoing transmission link the cell should be transmitted on. This information is placed in a *routeing byte* appended to the front of the cell.

The fabric switches cells from input port controllers to output ones. The switching fabric is the place where cells contend for access to the outgoing transmission lines. If different port controllers inject cells destined for the same output port controller (as indicated by the routeing byte) into the fabric at the same time, then only one will succeed. The others must retry later. The routeing byte also includes one bit of priority information which is used by the fabric when arbitrating clashes. Arbitration takes place in two stages. Firstly, high priority cells are given precedence over low priority ones. Of the remaining cells, the choice is made on a round-robin basis. The input port controllers are informed of whether their cell was successful using acknowledgement lines. The fabric sends a negative acknowledgement to the unsuccessful input ports, and passes the acknowledgement from the requested output port to the successful input ports. This means the output port controllers may reject cells even if they successfully pass through the fabric.

The port controllers and fabric all use the same clock so bytes are read in on each link synchronously. They also use a higher level cell frame clock—the *frame start* signal. It ensures that the port controllers inject data cells into the fabric synchronously so that the routeing bytes arrive at the same time. The behaviour of the switching element is cyclic. In each cycle or *frame*, the element waits for cells to arrive, reads them in, processes them, sends successful ones to the appropriate output ports and sends acknowledgements. It then waits for the next round of cells to arrive. The cells from all the input ports start when a particular bit of any one of them goes high.

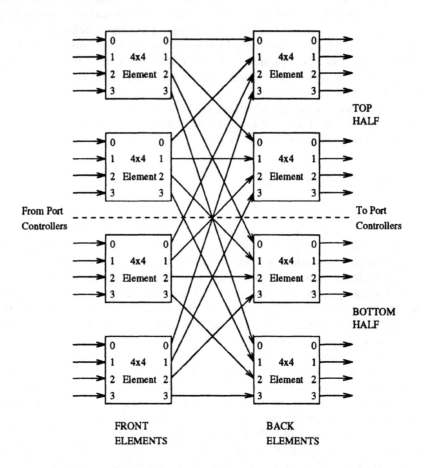

Fig. 1. The 16x16 Switching Fabric

We have been concerned with the verification of the fabric, not the port controllers. In particular, we have verified several versions of the switching elements from which the fabric is built. The switching fabric consists of a series of switching elements connected in a regular array. The simplest (4 by 4) switching fabric consists of a single element which connects 4 input ports to 4 output ports. It is this element that we initially formally verified. To make larger fabrics, several elements can be connected together in a regular array. A 16 by 16 fabric can be made from 8 elements in two rows as shown in Figure 1. Such a fabric has been fabricated. However, the design of the elements used differed from the original.

We modified the specifications and proofs for the fabricated designs used in the 16 by 16 fabric, including one design which was fabricated but found to be unsuitable when used. We also verified four other versions of the element which were not fabricated. These were intermediary designs consisting of some but not all of the differences between the 4 by 4 fabric element and the 16 by 16 fabric elements. Verifying these designs allowed us to evaluate the time taken to make smaller changes than for the fabricated designs.

The elements were designed using a Hardware Description Language: Qudos HDL [6]. This is a simple HDL which allows the structure of hardware to be specified. It does not allow behaviour to be specified directly. The fabricated elements were implemented on 4200 gate-equivalent Xilinx programmable gate arrays.

3 The 4 by 4 Fabric Switching Element

The first element we verified was that for the 4 by 4 fabric. We manually translated the HDL descriptions into higher-order logic. The translation could have been done mechanically, involving only a change of syntax. However we made some changes to the description to aid the verification. It was not intended that these changes alter the design, only the description of it. Both descriptions should describe the same collection of logic gates. In fact several errors were introduced which were then discovered during the verification. The most serious was that two wires were swapped. This was discovered due to a subgoal of the form [T,F] = [F,T] being generated. One side of this equality originated from the specification and the other from the implementation. It was thus obvious that two wires had been crossed, and also which they were from the context of the proof. It was not immediately clear whether the mistake was in the implementation or specification. More details of the errors discovered are given elsewhere [3]. To ensure that changes to the design had not inadvertently been made, the netlists from the two descriptions could have been compared.

Two kinds of changes were made: adding extra layers to the hierarchy and simplifying the description using features of HOL which are not available in Qudos HDL. Multi-level words (that is, words of words) were used to replace single level ones where appropriate. Several of the descriptions were made generic. In the HOL word library, the bits of words are numbered from the opposite end to that used in Qudos HDL so modifications were made to the description to overcome this. The resulting descriptions were fairly standard HOL structural hardware descriptions [8]. The final hierarchy is shown in Figure 2.

The design consists of three main modules, as shown in the circuit diagram for the module FAB4B4 given in Figure 3. The ARBITRATION module reads the routeing bytes of the cells, makes arbitration decisions and passes the results to the other modules with the grant signal. It controls the timing of the decision with respect to the frameStart signal and the time the routeing byte arrives. The outputDisable signal indicates to the other units when a new arbitration decision has been made. The PAUSE_DATASWITCH module switches the cell bytes

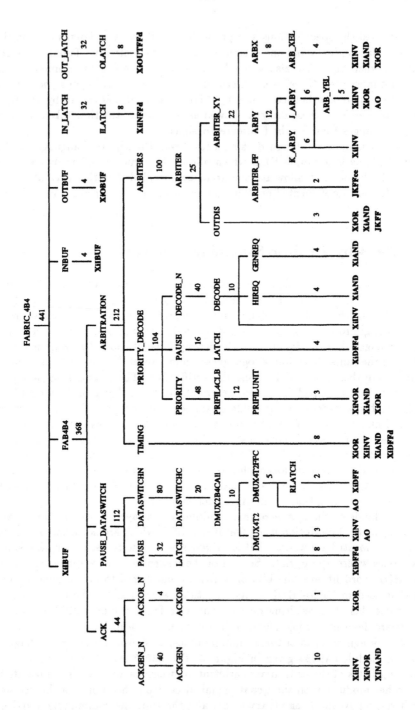

Fig. 2. The design hierarchy of the original element showing the total number of primitive components in each module

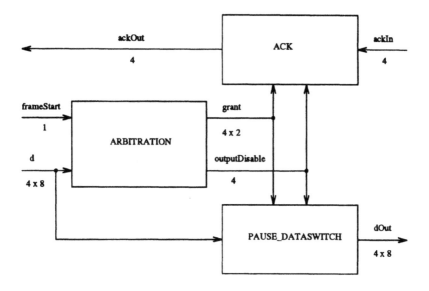

Fig. 3. The Implementation of FAB4B4

from inputs to outputs according to the most recent arbitration decision. If no cell is to be transmitted to an output port zeroed bytes are output. The cell bytes are delayed just long enough in the dataswitch before being output for an arbitration decision to be made on their routeing byte. The acknowledgement unit **ACK** passes acknowledgement signals in the reverse direction. Negative acknowledgements are sent to the input ports until a decision is made. Ports that did not make a request or whose request was turned down continue to receive a negative acknowledgement until the end of the frame.

To formally verify a hardware design, formal behavioural specifications of the design and of all its modules are needed. The switching element had been designed and implemented without any form of formal behavioural specification being created. There was only sketchy, informal documentation. Therefore the formal specifications were reverse-engineered from the HDL description. This was a very time-consuming process and the early versions of the specification contained many errors. The specifications for each of the 43 modules (both behavioural and structural) were written prior to any proof. This took between one and two man-months. No detailed breakdown of this time has been kept, though much of the time was spent attempting to understand the design.

The formal behavioural specifications describe timing diagrams associated with each output signal. As the behaviour of the element is cyclic, the specifications of most modules are based on the frame cycles. The behaviour of each output signal over the period of a frame is described in terms of the state and values on the inputs during that frame. Frames can either be inactive or active. During the former no cells arrive, so no arbitration need be performed. During the latter at least one input port injects a cell into the fabric. A cell being injected is indicated by bit zero of the data signal (the active signal) going high. An inactive frame could be thought of as just a degenerate case of an active one. However, the specification is cleaner if the two cases are treated separately.

A relation **AFRAME** was defined which determines whether a triple of times constitutes an active frame for given frame start and active signals. The first and third time of the triple must correspond to consecutive times that the frame start signal is high. The second time must correspond to the first time within this interval that the active signal of any input is high. A similar relation **IFRAME** defines an inactive frame for a pair of times.

For inactive frames, the behaviour of the element over the whole frame can be treated uniformly. For example, the specification of the data output signal of the element in an inactive frame has the following form.

```
IFRAME ts te ... ⊃
    STABLE (ts + 3) (te + 3) dOut default_dOut
```

Under the assumption that times **ts** and **te** represent an inactive frame, this states that the signal **dOut** is stable, outputing the constant value **default_dOut** in the interval from **ts+3** to **te+3**.

For an active frame, the behaviour is typically split into two parts: that up to some fixed time after the active signal arrives; and that from this point until the end of the frame. The specification of the data output signal in an active frame has the following form.

```
AFRAME ts ta te ... ⊃
    STABLE (ts + 3) (ta + 5) dOut default_dOut ∧
    DURING (ta + 5) (te + 3) dOut (λt. ...(dIn ta)...(dIn (t - 2))...)
```

Under the assumption that times **ts**, **ta** and **te** represent an active frame, this states that the signal **dOut** is stable and outputs the constant value **default_dOut** in the interval from **ts+3** to **ta+5**. In the interval **ta+5** to **te+3**, however, the value output is dependent on the data input at time **ta** (that is, the routeing bytes) and the data input two cycles earlier than the time under consideration (that is the next cell bytes to be input).

The functional behaviour of the element is dependent on the arbitration process. We specified this process as a function on the cell routeing bytes. Given the routeing bytes, the function describes the arbitration decision which will be made. The acknowledgement signals, data to be output and new state are specified in terms of this function. The successful input for each output is specified independently of the other outputs. This is done in several stages. First the inputs requesting the output under consideration are determined. They are then filtered on the basis of their priorities. Finally round-robin arbitration is performed on the resulting set.

Each module in the design was formally verified separately. The proofs were roughly of two kinds. The simplest were for the low level modules. Their specifications were in the form of an equation stating that the outputs at some time

were a function of the inputs at earlier times. The specification for the module
ARB_XEL given below is a simple example of this kind.

```
ARB_XEL_SPEC ((reqA, y, reqB, reqC), xjk) =
    ∀t. xjk t = ((y t) ∨ (reqA t)) ∧ ((reqB t) ∨ (reqC t))
```

The proofs for such modules were generally straightforward, and could be done
largely automatically. Invariably, manual intervention was required to complete
the proofs, calling on specific lemmas about words, for example.

A second kind of proof involved modules with specifications based on timing
diagrams over the period of a frame such as that for the full element outlined
earlier. A separate proof was performed for each output and state variable. Each
of these proofs was split into separate proofs for inactive frames and for each
interval within active frames. The proofs for inactive frames were virtually identical to the proofs for the start of the interval of an active frame. These proofs
normally involved considering a typical point within the interval, and showing
that the output in question had the value specified at that time. The start of an
interval was normally treated separately, because at that point a change of behaviour occurred. For example, the frame start signal might trigger a change in
behaviour at the start of the interval putting the module into a different state.
Once in that state it might remain there as long as no active signal occurs.
The reasoning required in the two situations is different so they were proved
separately.

As with proofs about equational specifications, the proofs of interval specifications involved an initial, largely automatic part concerned with expanding
definitions and rewriting the values of output signals using the specifications
of the sub-modules. This was followed by a user-guided part, which involved
proving that two expressions on the input values were the same. These proofs
were more tricky, due to the more complex notions being reasoned about. For
example, modules involved with arbitration require reasoning about round-robin
arbitration.

Many errors were found in the specifications by the proof process. The timing
of many modules was incorrectly specified. Such errors resulted in goals such
as ts=ts+1 so were generally easy to detect and correct. Word lengths were
frequently specified incorrectly. This resulted in goals of the form PWORDLEN
m w, requiring that a word w has size m, when the only assumption about the
length of w was PWORDLEN n w, stating that it had size n. A more significant error
concerned the grant signal which encodes the successful input for an output as
a two bit word. It was assumed that the two bits were sampled at the same
time. However, they are actually sampled on consecutive cycles. This resulted in
a goal in which the value of the grant signal at one time was needed, but it was
only known to have the required value at a different time.

Approximately two man-months were spent performing the verification. Figure 4 shows the cumulative time taken as each module was verified. The time
(one week) spent proving general purpose theorems about machine words and
signals is not included. Approximately half the time was spent verifying the upper half dozen modules of the element. The proofs of these modules were more

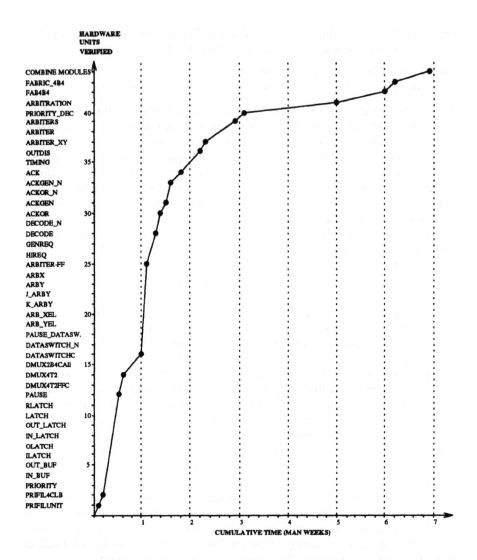

Fig. 4. Time taken to verify the switching element

time-consuming for several reasons. (i) There were several intervals to consider. (ii) There were several outputs to consider. (iii) The behaviours were defined in terms of complex notions. (iv) There were more errors in the formal specifications of these modules due to their increased complexity: understanding and correcting the errors was very time consuming.

No detailed record of the time spent designing the element was recorded. However, the designer estimated that had it been designed from scratch the initial design time would have been in the order of several months. The time spent testing would have been in the order of several weeks. An error was discovered after the testing process had been completed when the fabric was in use. It is likely that this error would have been discovered by the formal verification if it had been performed prior to fabrication. We did not attempt to verify the faulty design, however. Thus the time spent to formally specify and verify the design was not unreasonable. Had it been performed as an integral part of the design, it is unlikely that it would have unduly slowed the design cycle. The formal specification and verification would have been much quicker if done as the element was designed, for the reasons given earlier. Much of the time was spent attempting to understand the design.

4 Variations on the Design

Several modifications were made to the element for the 16 by 16 fabric. This resulted in two new designs, both of which were fabricated: one for the front elements and one for the back elements.

An extra 5 cycle delay was placed on the frame start line on entry to the front elements and an 8 cycle delay was used for the back elements. This was to allow the port controllers more time to process the data. The longer delay was required for the back element to allow the header to pass through the front element. The data was delayed by 5 cycles in the port controllers. Thus the change ensured that the data and frame start signal still arrived at the same time relative to each other. A faulty version of the back element which had only a 5 cycle delay was originally fabricated. The error was only discovered when the fabricated fabric was tested. This was prior to the verification attempt. Initially the verifier was both unaware that the 5 cycle delay design was faulty and that the 8 cycle delay design existed.

For the 16 by 16 fabric the routeing byte is split in two. The front elements read only the first nibble of the routeing byte. This is the same as for the 4 by 4 fabric. However, the back elements read the second nibble, so their design was changed accordingly. This involved changing the wiring in the routeing byte decoding circuitry.

The original element strips off the routeing byte as it passes through the element since it is not needed by the output port controllers. The back elements of the 16 by 16 fabric must still do this. However, the front elements must not as the back element must receive the routeing byte. Thus, in the front element an extra delay was placed on the main data path so that the outputs would be enabled as the routeing byte arrived, rather than after it had been discarded.

One further change was made to the front elements. The full fabric had to be implemented on two boards. It was desired that an identical board be used to implement the top and bottom halves of the fabric (each holding two front and two back elements). This was to "keep the design simple" [7]. However, the two halves are mirror images of each other. Different outputs from the front element needed to go off the board in the two halves. For the top half, the data bytes routed to outputs 0 and 1 needed to go to the back elements on the same board, whereas those for outputs 2 and 3 needed to go to the elements on the other board. The opposite is required for the bottom half. This can be seen from Figure 1. To overcome this an extra boolean input was added to the element which, when set, made the decoder swap the meanings of the route field in the route byte. Requests for outputs 0 and 1 were sent to outputs 2 and 3 respectively, and vice versa.

We adapted the implementation, specification and proofs of the original element to the new designs, including the erroneous back element. In the process, we also verified four further versions, in which only some of the changes had been made. The front element effectively consists of two different designs one for the top element and one for the bottom element with a flag to switch between them. We thus verified these two designs (with no flag input) as designs in their own right. Two other designs verified were versions of the front fabric which removed the routeing byte as the cell passed through. Verifying the designs separately allowed us to evaluate the time taken to make each change. In fact the two designs which removed the byte resulted from a mistake: the verifier initially did not realize that the extra delay had been inserted into the design.

The first new design verified was the erroneous back element. This differs from the original in that the second nibble of the routeing byte is accessed rather than the first. Also the 5 cycle delay occurs on the frame start data path. This design was successfully verified! This was because the specification was reverse engineered from the implementation. Thus the specification erroneously required a 5 cycle delay. This highlights how bugs can become features if the specification is produced from the implementation rather than the other way round. The error *was* discovered when an attempt was made to verify the full 16 by 16 fabric constructed from the faulty elements. It was found because the active signal needed to be defined over one interval for the back elements, but the front elements specified its value over a different interval.

The faulty element took three man-days to verify. This was longer than expected because the new design invalidated an assumption which had been made in the original. It had been assumed that a frame start signal did not occur on the very first cycle after a restart. This was to ensure that the element reset properly. However, in the new design, the signal of interest was on an internal signal—the output of the new delay. Thus its value could not be guaranteed. It had not occurred to the designer that this might cause a problem. On closer inspection, it was found that the assumption was actually stronger than required, and that on the first real frame start signal from outside the chip, the element did reset itself. The element was well-behaved provided the behaviour until after this occurred was ignored. Previously we had required that it be well-behaved from the start. Changing this assumption involved changing the specifications of modules other than those whose implementation had been changed, notably the timing circuitry. In particular the modules, TIMING, OUTDIS and all modules dependent on them had to be reverified to account for the new assumption. Despite this it still only took a few days to complete the verification. We reverified a total of seven modules, and added two new modules.

The second design verified was the version of the front fabric that strips off the routeing byte, delays the frame start signal, but interprets the route information normally. It is the same as the original element except for the delay on the frame start signal. The proof of the back element was modified to obtain the new proof. Changes were made to three modules. This took about a quarter of a man-day. Had the problem with the assumption not occurred, the verification of the back element would have taken little more than this.

The third design verified was the version of the front element that strips off the routeing byte, delays the frame start signal and interprets the route information differently. The previous proof was modified. This involved adding an extra module above DECODE in the hierarchy and changing the proofs of all modules dependent on it. Six modules were modified and one new module added. This took three-quarters of a man-day.

The fourth version verified was the front element that does not strip off the routeing byte but delays the frame start. One new module was added above PAUSE_DATASWITCH to include the extra delay. The proofs for three others were modified. This took half a man-day.

The fifth version verified was the front element that does not strip off the routeing byte, delays the frame start and reinterprets the route information. Three modules were reverified. This took a quarter of a man-day.

Next, the fabricated front element was verified. It has an input which switches its behaviour between that of the previous two versions. Seven modules were reverified. This took a day and a half. This was longer than for the previous versions of the front element, because rather than just making small changes to the earlier specifications, their structure was radically changed. The new specification consisted of a case split between two specifications of the previous form. This meant that for the main lemmas new proofs had to be created. The new proofs were far simpler than those they replaced, but creating them took more time than making modifications to an existing proof.

Finally, after the verification attempt on the 16 by 16 fabric had caused the error in the original back element to be discovered, the correct version of the back element was verified. We reverified a total of 8 modules. This took less than two hours to complete. It involved adding 3 to various time offsets in the specifications, goals and proofs of the original back element proof to account for the change in the delay from 5 to 8 cycles. It would have been better in the long term to make the proof generic with respect to this delay. Then a single proof would cover both designs as well as any future designs with different delays. This would have taken slightly longer however, so was not done since our aim was to complete the verification as quickly as possible.

The cumulative time taken to verify the new designs is show in Figure 5. The modules that were reverified were the top-level ones which had originally taken the most time to verify. However, they were reverified at a rate corresponding to the lowest level modules in the original verification. The original fabricated back element was verified in three days, despite the new design invalidating an assumption in the original proof. The fabricated front element was verified in

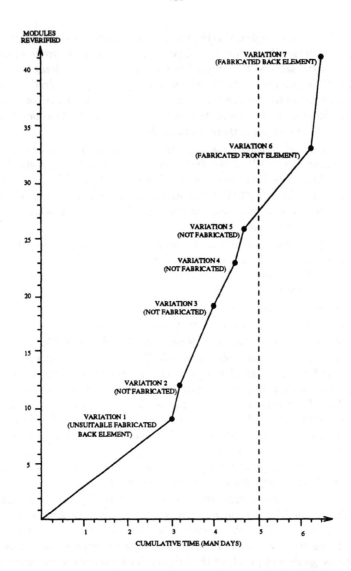

Fig. 5. Time taken to verify the switching element variations

just over three days, despite several other designs being verified unnecessarily on the way. It should also be stressed that these times include the time taken to understand, formally specify and verify the new designs. The original times in Figure 4 only include the verification time and should be roughly doubled to give a comparable time.

The reverification was quicker for several reasons. The most obvious reason was that the design was modular. Only modules that were actually affected needed to be reverified. This does not account for all the improvement, however, as the affected modules were verified far more quickly than originally. One reason these proofs were quicker was because they were split into a series of lemmas. Many lemmas did not need to be changed and so their proofs were not affected. Furthermore, most proofs that did need to be changed only needed to be changed in small ways. Thus the scripts could be rerun with only a few modifications. Also, the reason why the design was correct was well-understood by the verifier at this point. Finally, the new specifications did not contain errors because they were produced by making small changes to a correct specification. Working with correct specifications can save a significant amount of time, because understanding and correcting errors and redoing proofs accordingly can be very time consuming, especially when multiple errors interact. It accounted for much of the time spent verifying the top-level modules of the original element.

5 Conclusions

We have demonstrated that a fully machine-checked formal verification of a real piece of communications hardware can be completed in a time scale roughly comparable to that required for its original design, implementation and informal testing. This was despite the formal specification and verification only starting after the design had been implemented, the documentation being sketchy, and the verifier having no previous knowledge of the design or its application. More significantly, we have demonstrated that real changes made to a real design can be quickly tracked. It took several months to specify and formally verify the original design. Modified designs could be specified and verified in a matter of hours or days. This was despite the fact that the modules of the design that needed to be reverified were the upper levels which took the most time to verify originally. Thus for the reasons noted in the previous section the effort invested in verifying a design is not wasted if the design is subsequently changed.

It would be possible to verify a generic n by n element. The implementations we considered were specifically for a 4 by 4 design, some modules of which could not trivially be treated generically. The modules that could easily by treated in this way were done so. Thus, if the problematic modules were suitably reimplemented we would be able to quickly make the full proofs generic. The designs we verified differ in only small ways, so it might be possible for a generic version to be verified which covers all of them. However, each modification would have required the design to be made generic in a different way. The verifier would therefore have needed to anticipate future changes. It is thus important that whenever possible the verifier is aware of the kind of design changes that are likely. It is easy to see that the delay on the frame start signal could be made generic. In the long term this would have saved time. However, in the short term it would have been slower. In a commercial setting, this is a problem that is likely to arise frequently. If a design is being verified under time pressure, there will be a strong temptation to do the minimum required to verify that design, even if it means modifications will take longer to verify.

The study highlighted a problem when modifying proofs to track design changes. The two versions of the front fabric that removed the routeing byte resulted from a mistake. The verifier forgot that the extra delay had been added. It was only after the verification of the two versions was completed that this was noticed. This problem arose because two separate descriptions of the implementation were being used: Qudos HDL and HOL. The structural specification of the original design was obtained by editing the actual HDL from which the design was implemented. Thus, there was little opportunity for errors of this kind to be introduced. However, in the modified designs, the formal structural specifications were obtained by editing the formal descriptions of the original design rather than the sources since this was quicker. Thus it was easy to miss some of the changes. The verification was still "successfully" completed because the behavioural specifications were written from the implementation. The errors became "features". The faulty design of the back element was initially verified rather than the correct one for similar reasons. It was only when the specifications were used in the proof of a higher level of the hierarchical design that the errors were highlighted by the proof attempt. Using specifications for proofs of higher levels of the design is a good way to validate them. However, we are always dependent on the top level specification being correct. The problem would not have arisen if the original designs had been written in HOL rather than Qudos HDL, or if the designs had been implemented from formal behavioural specifications. Formal methods should be an integral part of the design process, even if the verification is not attempted until after the designs have been completed. Tacking formal methods on afterwards can lead to a false sense of security.

If errors are present in a formal specification or implementation, the verification time will be longer. One of the reasons the modified switching element designs were verified more quickly than the original 4 by 4 fabric was that the specifications were correct before the verification commenced. Formal proof should only be used to find the most obscure bugs. Faster techniques such as simulation and animation of specifications should be used to find as many errors as possible prior to formal proof. Model checking techniques [11] could also be used for this purpose as they can quickly find counter-examples for incorrect designs.

It is possible that the switching elements are simple enough to be verified using model checking tools alone without the need for formal proof. The lower level modules could certainly have been validated in this way. A system such as HOL-Voss [9] in which model checking and formal proof are combined could be used for this purpose. If simplified versions of designs are used to combat the state explosion problem encountered by model checking, guarantees of correctness are weaker since errors may have been simplified away. Furthermore, model checking tools are large, complex programs and as such are likely to contain bugs. This is less of a problem for HOL due to the use of an abstract theorem type which ensures only valid results are produced. Thus, for safety-critical applications formal proof may still be preferable. Ultimately we are interested in performing a hierarchical verification of the implementation of the complete Fairisle switch running a signalling protocol. This is almost certainly beyond the capabilities of current model checking technology.

The reverification work involved modifying existing definitions and rerunning the proofs. Better facilities for doing this are desirable. For example, a debugging tool that could be used to single step through proofs would have been useful. Much time was spent replaying old scripts, waiting for them to fail. This would be a good application for some form of Lazy proof [1] in which fast but less secure versions of tactics and inference rules are used whilst the verifier is present, with the secure versions used to run the proofs securely overnight.

Acknowledgements

This work was supported by SERC grant GR/J11133. I am grateful to Mike Gordon, Brian Graham, Ian Leslie, Malcolm Newey, Wai Wong and the members of the Automated Reasoning Group in Cambridge for their help and advice.

References

1. Richard Boulton. Lazy techniques for fully-expansive theorem proving. *Formal Methods in System Design*, 3(1/2), August 1993.
2. Albert J. Camilleri. Formal methods in industry. Tutorial at the 1991 International Workshop on the HOL Theorem Proving System and its Applications.
3. Paul Curzon. Experiences formally verifying a network component. In *Proceedings of the 9th Annual IEEE Conference on Computer Assurance*. IEEE Press, 1994.
4. Paul Curzon. The formal verification of the Fairisle ATM switching element. Technical Report 329, University of Cambridge Computer Laboratory, 1994.
5. Paul Curzon. The formal verification of the Fairisle ATM switching element: an overview. Technical Report 328, University of Cambridge Computer Laboratory, 1994.
6. K. Edgcombe. The Qudos quick chip user guide. Qudos Limited.
7. Daniel Gordon. The 4 by 4 and 16 by 16 switch fabrics: Design and implementation. In *ATM Document Collection 2*. University of Cambridge, Computer Laboratory, February 1993.
8. M.J.C. Gordon and T.F. Melham. *Introduction to HOL: a theorem proving environment for higher order logic*. Cambridge University Press, 1993.
9. Jeffrey Joyce and Carl Seger. The HOL-Voss system: Model-checking inside a general-purpose theorem prover. In Jeffrey J. Joyce and Carl-Johan H. Seger, editors, *Higher Order Logic Theorem Proving and Its Applications, 6th International Workshop, HUG'93*, number 780 in Lecture Notes in Computer Science. Springer-Verlag, 1994.
10. I.M. Leslie and D.R. McAuley. Fairisle: An ATM network for the local area. *ACM Communication Review*, 19(4), September 1991.
11. K. L. McMillan. *Symbolic Model Checking*. Kluwer, 1993.

Weak Systems of Set Theory Related to HOL

Thomas Forster

D.P.M.M.S., University of Cambridge

This is an early version of a survey article designed for the interested non-specialist, and it does not contain any proofs of novel results, though it does contain announcements (of novel unpublished results) and proofs (of frequently underregarded trivialities). For the reader who wishes to take this material further, the chief advantage of this essay will be the bibliography, which would be quite hard for a naïve reader to assemble from scratch. I would like to thank my friend and colleague Juanito Camilleri for the invitation which led to me writing this.

HOL is one of a cluster of systems bounded in strength by Zermelo's set theory. Probably the most popular of the weak set theories is Kripke-Platek. It is much favoured by set theorists studying recursion and also, curiously, by people who do situation semantics. It doesn't fit well with the systems related to HOL and I shall say little about it. The best reference is Barwise.

Why should people be interested in studying systems weaker than ZF? (By the standards of much modern work in set theory, ZF is weak enough already!) Weak systems lay bare connections between things. If you have a system in which you can do A but not B, then A and B are not the same thing. Sometimes discovering facts like this can be very enlightening.

I don't believe this is the motivation behind designing HOL to be no stronger than it is. I think the primary motivation was to have a typed theory that was a strong as it conveniently could be: the advantages of having a typed system being felt to be more important than any strength advantages that might accrue in a trade off. There *is* such a trade off, as we shall see. In general, typed theories are weaker than the untyped theories one might naturally consider them with.

For some of its uses one would like HOL to be as strong as possible. Granted, the proof of correctness of circuits is not *mathematically* onerous (though the time consumption can be enormous) but people naturally want to use HOL for proofs of correctness for programs as well, and proofs of total correctness involve proof of termination. It is by now a commonplace that stronger systems can prove termination of more and more functions. For more on this see Buchholz and Wainer, *op. cit.*. A jolly good read.

With a few peripheral exceptions all of the systems that concern us here are subsystems of Zermelo set theory, or can be interpreted in it. Zermelo set theory is a convenient and natural place to start. It was the first one-sorted axiomatic set theory. It is a theory in a one sorted language with two primitives, $\in$ and $=$. It has the following axioms

1. Extensionality: $(\forall xy)(x = y \longleftrightarrow (\forall z)(z \in x \longleftrightarrow z \in y))$
2. Pairing: $(\forall xy)(\exists z)(\forall w)(w \in z \longleftrightarrow (w = x \vee w = y))$

3. Union $(\forall x)(\exists y)(\forall z)(z \in y \longleftrightarrow (\exists w)(z \in w \wedge w \in x))$
4. Powerset: $(\forall x)(\exists y)(\forall z)(z \in y \longleftrightarrow z \subseteq x)$
5. Infinity: *There is an infinite set.*
6. Separation (Axiom scheme: one instance for each ϕ): $(\forall x)(\exists y)(\forall z)(z \in y \longleftrightarrow (z \in x \wedge \phi(z)))$
7. Empty set: $(\exists x)(\forall y)(y \notin x)$

The difference between Zermelo set theory and Zermelo-Fränkel set theory is the axiom scheme of replacement. This axiom scheme says that anything that is the same size as a set is a set. The motivation behind this is a belief that paradoxes are connected with big collections and so the way to avoid paradox is to ensure that large collections do not turn out to be sets. If this is true, then it should certainly be safe to suppose that anything the same size as a set is a set. It turns out that the axiom scheme of replacement has very strong consequences. In particular it has strong consequences of the kind mentioned above: in *ZF* we can prove termination of functions whose termination cannot be proved in Zermelo set theory.

1 HOL and *TST*

There are of course many type systems one could consider, but the obvious one for ease of illustration is a theory known as *TST*. Traditionally the initials spell Theory of Simple Types but "Typed Set Theory" is better. *TST* is expressed in a language with a type for each nonnegative integer, an equality relation at each type, and between each pair of consecutive types n and $n + 1$ a relation $\in$. The axioms are an axiom of extensionality at each type

$$\forall x_{n+1} \forall y_{n+1} (x_{n+1} = y_{n+1} \longleftrightarrow \forall z_n (z_n \in x_{n+1} \longleftrightarrow z_n \in y_{n+1}))$$

and (at each type) an axiom scheme of comprehension

$$\forall x_1 \ldots x_n \exists y_{n+1} \forall z_n (z_n \in y_{n+1} \longleftrightarrow \phi(x_1 \ldots x_n, z_n))$$

with 'y_{n+1}' not free in 'ϕ'.

TST_k is like *TST* except that there are only k types, labelled $0, \ldots, k-1$. The "theory of negative types"[1] *TNT* (Wang [1952]) and its language are defined analogously, except that the types are indexed by $\mathbf{Z}$. For the purposes of this paper all these theories will be assumed to have the axiom of infinity

$\langle\langle x \rangle\rangle$ is the structure $(x, \mathcal{P}^{\ell}x, \mathcal{P}^{2\ell}x, \ldots)$ thought of as a model of *TST*.

This is clearly a much simpler syntax that HOL. Its relevance here is the following:

THEOREM 1.1 *HOL and TST are each interpretable in the other.*

---

[1] This is actually a misnomer: strictly he should have called it the theory of *positive* and negative types.

It is easy to interpret *TST* in HOL because we can interpret type 0 as being any type we please and thereafter type $n+1$ is interpreted as $\alpha \rightarrow$: bool (where α is the type that interprets type n. The other direction is equally elementary but much more complicated. First we must note that any level of a model of *TST* can be embedded in all higher levels by means of an iteration of the singleton function. For notational convenience we will write singletons functionally, so that $\iota'x =_{df} \{x\}$ and $\iota^2{}'x = \{\{x\}\}$ and $\iota^2{}''x = \{\{\{y\}\} : y \in x\}$ and so on. Then one can embed type m in type $m+n$ by the function $\lambda x.\iota^n{}''x$. We have to reinterpret the membership relation as well. If we let $RUSC(R)$ be $\{\langle\{x\}, \{y\}\rangle : \langle x, y\rangle \in R\}^2$ we interpret $\in$ by $RUSC^n(\in)$, or $RUSC^n(\in)$ composed with some suitable iterate of ι.

To interpret HOL in *TST* we embed the type of individuals and the booleans in type 0. We embed : num in the lowest type containing an interpretation of the naturals. Now suppose we have interpreted two HOL types α and β into two *TST* types m and n, with $m < n$, say. We compose this interpretation for α with the iterated singleton map to get interpretations of both α and β in type n. The, by means of Wiener-Kuratowski ordered pairs we can interpret $\alpha \times \beta$ in type $n+2$. Function types are now no problem because once we have interpreted both α and β in type n every function in $\alpha \rightarrow \beta$ is an element of type $n+3$.

It is possible to do this in more detail!

2 Predicativity, truth definitions and consistency proofs

What do we mean by strength of a system? Because of the second incompleteness theorem no theory can prove the consistency of itself, and if $T \vdash Con(S)$ then $S \nvdash Con(T)$, so "$T \vdash Con(S)$" is clearly a relation we will need to consider. Another relation to consider is "there is an interpretation of S in T" which, in contrast to the last, is not asymmetrical.

Much of the early work on interpreting theories in each other goes back to the time when the canonical work on set theory was Whitehead and Russell, and is therefore informed by a type-theoretical intuition that is now, once again, the theme of the hour.

If we can interpret T_1 in T_2 in such a way that we can prove in T_2 that every theorem of T_1 is true, then clearly $T_2 \vdash Con(T_1)$. What sort of conditions on T_2 are sufficient for us to do this? Well, we must be able to define in T_2 a truth-predicate for the expressions in the range of the interpretation. Now a truth-predicate is an inductively defined set of ordered pairs, and we can illustrate the complications which are relevant here by reference to the simplest case, namely the natural numbers $\mathbb{N}$. If we are working in a theory where we have a concept of cardinal number, and we know what 0 is, and we know what $Succ$ is, we can say: x a natural number iff

$$(\forall Y)(0 \in Y \land (\forall x \in Y)(Succ(x) \in Y))$$

---

[2] This notation is Rosser's

This molecular formula capturing a property of a cardinal number contains a quantifier over—not *cardinals* but—*sets* of cardinals. This is an example of an impredicative definition.[3] This particular example is not *terribly* impredicative: we are defining a set of cardinals by quantifying over sets of cardinal not sets-of-sets-of cardinals, which one could.

2.1 Truth definitions

I shall assume we are dealing with a one-sorted language. This is to keep things simple by preventing proliferation of indices and subscripts. To be specific, I shall spell out the details only for the simplest case where there is one binary relation ("$\in$") in addition to equality.

We will assume that our variables, rather than being x, y, z etc, are all x's with numerical subscripts.[4] This clearly makes no difference to us, qua language users, since it is a trivial relettering, but it does make life a lot easier for us qua students of the language. The subscripts are quite important. We call them *indices*. The purpose of this change in notation is to make visible to the naked eye the fact that we can enumerate the variables: it is much clearer that this is the case if they are written as "$x_1, x_2 \ldots$" than if they are written as "$x, y \ldots$" In fact I think we will also have to assume that no variable is bound more than once in any formula, and that there are no occurrences of any variable outside the scope of any quantifier that binds some other occurrence of that variable. Thus we will outlaw $((\forall x)F(x)) \vee ((\forall x)G(x))$ and $F(x) \vee (\forall x)(Gx)$ even though they are perfectly good wffs. It will make life easier later.

Naturally you expect that a notion of interpretation will crop up if we are trying to define what it is for a sentence to be true in a structure.

A *finite assignment function* is a function that assigns elements of M to finitely many indices. Computer scientists will recognise this immediately as the logician's version of their concept of *state*. They will also recognise that partial (though not finite) assignment functions form a CPO. I have (see above) carefully arranged that all our variables are orthographically of the form x_i for some index i, so we can safely and conveniently think of our assignment function f as being defined on *indices*. [5]

Next we define what it is for a finite assignment function to satisfy a sentence p, (written "f sat p"). We define *sat* first of all on atomic sentences. First a word

---

[3] "Impredicative" is a word due to Russell which he used to describe definitions of properties-of-widgets which make reference to objects of higher type—such as: sets of widgets, or functions from widgets to widgets. The word is unevocative to modern ears, and to understand why he used it we need to know that—at the time—Russell thought that such definitions were not legitimate, did not properly define predicates and so were *impredicative*.

[4] This is why I wanted no type subscripts!

[5] It is better practice to think of the assignment functions as assigning elements of M to the *indices* and write "$f'i = \ldots$", since any notation that involved the actual *variables* would invite confusion with the much more familiar "$f'x_i = \ldots$" where f would have to be a function defined on the things the variables range over.

on use and mention. Notice that in

$$f \; sat \; x_i = x_j$$

we have a relation between a function and an expression, not a relation between f and x_i and x_j. This is usually made clear by putting quotation marks of some kind round the expressions to make it clear that we are mentioning them not using them. Now precisely what *kind* of quotation mark is a good question. Our first clause will in fact be something like

$$f \; sat \; 'x_i = x_j ' \; \text{iff}_{\text{df}} \; f(i) = f(j)$$

But how like? Notice that, as it stands, it contains a name of the expression which follows the next colon:

$$x_i = x_j$$

Once we have put quotation marks round this, the i and j have ceased to behave like variables (they were variables taking indices as values) because quotation is a referentially opaque context. But we still want them to be variables, because we want the content of this clause to read, in English, something like: "for any variables i and j, we will say that f *sat* the expression whose first and fourth letters are 'x', whose third and fifth are i and j respectively (whatever i and j are in this case) and whose middle letter is '$=$', iff $f(i) = f(j)$". Notice (and this is absolutely crucial) that in the piece of quoted English text 'x' and '$=$' appear with single quotes round them and 'i' and 'j' do not. Now to achieve this, ordinary single quotes will not do. Quine invented a new notational device in [1951], which he modestly calls "corners" and which are nowadays known more usually as "Quine quotes" (or "quasi-quotes") which operate as follows: The expression after the next colon:

$$\ulcorner x_i = x_j \urcorner$$

being an occurrence of '$x_i = x_j$' enclosed in quine quotes is an expression which does not, as it stands, name anything. However, i and j are variables taking integers as values, so that whenever we put constants (numerals) in place of i and j it turns into an expression which will name the result of deleting the quasi-quotes. This could also be put by calling it a variable name.

> *Putting quine-quotes round a compound of names of wffs gives you a name of the compound of the wffs named.*

A good way to think of quine quotes is not as a funny kind of quotation mark, for quotation is referentially opaque and quine quotation referentially transparent, but rather as a kind of diacritic, not unlike the LaTeXcommands I am using to write this paper. Within a body of text enclosed by a pair of quine quotes, the symbols '∧' '∨' etc. do not have their normal function of composing *expressions* but instead compose *names of expressions*. This also means that Greek letters within the scope of quine quotes are being used to range over expressions (not sets, or integers). Otherwise, if we think of them as a kind of funny quotation mark, it is a bit disconcerting to find that, as Quine points out, $\ulcorner\mu\urcorner$ is just μ. The reader is advised to read pages 33-37 of Quine [1951] where this gadget is introduced. We say

f sat α iff$_{df}$

f R α for every R satisfying (i) - (vii)

(i) $f\ R^{\ulcorner}\ x_i = x_j^{\urcorner}$ iff$_{df}$ $f'i = f'j$

(ii) $f\ R^{\ulcorner}\ x_i \in x_j^{\urcorner}$ iff$_{df}$ $f'i \in f'j$

(iii) if $f\ R\ \alpha$ and $f\ R\ \beta$ then $f\ R^{\ulcorner}\alpha \wedge \beta^{\urcorner}$

(iv) if $f\ R\ \alpha$ or $f\ R\ \beta$ then $f\ R^{\ulcorner}\alpha \vee \beta^{\urcorner}$

(v) if for no g extending f does $g\ R^{\ulcorner}\alpha^{\urcorner}$ hold then $f\ R^{\ulcorner}\neg\alpha^{\urcorner}$

(vi) if there is some g extending f such that $g\ R^{\ulcorner}F(x_i)^{\urcorner}$ then $f\ R^{\ulcorner}(\exists x_i)(F(x_i))^{\urcorner}$

(vii) if for every g extending f with $i \in dom(g)$, $g\ R^{\ulcorner}F(x_i)^{\urcorner}$ then $f\ R^{\ulcorner}(\forall x_i)(F(x_i))^{\urcorner}$

(Then we say ϕ is true in $\mathcal{M}$ iff the empty partial assignment function *sat* ϕ.)

The point here is that the relation *sat* is an inductively defined set of ordered pairs and as such has a higher-order quantifier in its definition. What this ought to suggest to us immediately is that something like the following is true. Let us suppose we have a theory T in a typed language of set theory, which contains an axiom of extensionality at each type ("distinct sets have distinct members"). Suppose T has an axiom scheme that says that the set of all things at type α that satisfy condition ψ is a set, *as long as the condition ψ does not contain bound variables of type higher than α*. TST of course lacks the italicised restriction. Then in TST we can find an interpretation of T and an inductive truth-definition in such a way that we can prove that every theorem of T is true. This is illustrated beautifully by the results in McNaughton and the general treatment in Wang [1952]. Another standard result was proved by Shoenfield and by Novak. The set theory GB is obtained from ZF by adding a scheme of class existence (so that the class of all x such that Φ exists as long as Φ contains no bound class variables) and substituting for the replacement scheme an axiom that says that the image of a set in a class is a set. GB is consistent if ZF is. However, if we allow bound class variables to appear in the class existence scheme we obtain a new theory, nowadays commonly called "Morse-Kelly", which is stronger than ZF. Quine [1966] is good on this point. See also Wang [1949].

Before we leave the subject of truth-definitions altogether we should mention Levy [1965]. In this beautiful monograph Levy makes *inter alia* the point that if we are trying to define a satisfaction relation on a set of formulæ that is not itself a recursive datatype, then the satisfaction relation we are trying to define is not itself an inductively defined set of ordered pairs, and so can be defined without appeal to a comprehension principle using quantification over objects of higher type. The particular relevance of this is that it enables us to produce satisfaction relations for classes of formulæ that contain, for example, no more than two quantifiers. This will become important later when we consider and compare Zermelo set theory and the theories Mac and KF.

One consequence of this is that, for any sensible system of type theory, one is liable to find that one can prove the consistency of any proper initial segment of it in some larger initial segment. (cf, reflection principles in ZF). Let us go into a little detail on this, and take the example of the construction in *TST* of a truth-definition for the theory of the bottom three types of a model of *TST*. By means of iterating the singleton relation (as in section 1) we can represent the first three types all as sets of the same level (probably level 5), and the satisfaction predicate will be a set of ordered pairs a few levels higher.

Thus we have the theorem

THEOREM 2.1 *For all k and all sufficiently large n, $TST_{k+n} \vdash Con(TST_k)$*

This gives rise to interesting complications when we introduce *polymorphism*, to which we now turn.

3 Polymorphism

There are several forms polymorphism can take. People who can read German should read Specker [1958], which is the best introduction to this topic.[6] His example is duality between points and lines in projective geometry. There is an automorphism (in fact an involution) of the language of projective geometry that swaps quantifiers over points with quantifiers over lines, and swaps "x and y intersect at z" with "z goes through x and y". Let us write this automorphism as Specker does, with an asterisk. Clearly if ϕ is an axiom of projective geometry, so is ϕ^*. Indeed * extends to a endomorphism defined on the recursive datatype of proofs, and this enables us to prove by induction on that datatype that ψ is a theorem of projective geometry iff ψ^* is. The important point is that this is **not** the same as saying that $\psi \longleftrightarrow \psi^*$ is a theorem. Specker says that this scheme of biconditionals is actually the same as adopting Desargues' theorem as an axiom).[7]

Elegant though this example is, it is a little remote from our concerns here. Closer to HOL is the theory *TNT*, which is defined above. The language in which it is expressed has an automorphism too, like the language of projective geometry. In fact it has an infinite group of them, all generated by one which we will notate with an asterisk and which arises as follows. Simply raise every type index attached to a variable in a formula ϕ by one to obtain a new formula ϕ^*. For example, asterisk of

$$(\forall x_2 y_2)(x_2 = y_2 \longleftrightarrow (\forall z_1)(z_1 \in x_2 \longleftrightarrow z_1 \in y_2))$$

is

$$(\forall x_3 y_3)(x_3 = y_3 \longleftrightarrow (\forall z_2)(z_2 \in x_3 \longleftrightarrow z_2 \in y_3))$$

(The reason for working with TNT rather than *TST* at this point is to ensure that * is not an endomorphism but an automorphism, as with projective geometry)

---

[6] Those who can't could consult Scott's review of it in *Mathematical Reviews*.

[7] Specker even shows that the conjunction of finitely many expressions of the form $\phi \longleftrightarrow \phi^*$ is another expression of that form. This depends on * being an involution and doesn't apply in the cases below.

As with projective geometry we notice that ϕ^* is a theorem of TNT iff ϕ is. As before we prove this by induction of the recursive datatype of proofs. (indeed * gives rise to an automorphism of this datatype, though this automorphism is of infinite order and is not an involution). This form of polymorphism, which is the kind we find in HOL and in the type theory of Russell and Whitehead was called by them *typical ambiguity*[8]: since the axioms (and therefore the theorems) are the same at each type, there is no need to put in the type indices. And as before there is no reason to suppose that $\phi \longleftrightarrow \phi^*$ is ever going to be provable.

If Γ is a class of formulæ, we call the scheme $\phi \longleftrightarrow \phi^*$, for $\phi \in \Gamma$, "Γ-ambiguity" or $Amb(\Gamma)$. Ambiguity for all formulæ is just Amb. The full scheme Amb is strong, and is not known to be consistent. We can prove $Amb(\Gamma)$ consistent for various natural classes Γ. For example when Γ is the class of all formulæ that mention only two types. The background to this is that all one can say in a typed set theory with two types can be said in the first-order theory of infinite atomic boolean algebras, and this is known to be a complete theory. Another example of a Γ for which $Amb(\Gamma)$ is known to be consistent relative to TNT is the class of all formulæ of the form $(\forall x_1 \ldots x_n)\Phi$ where ϕ is built up from atomics by the usual boolean operations and restricted quantifiers in the style of Levy (*op. cit.*) **and** quantifiers $\exists x \in \mathcal{P}'y$ and $\forall x \in \mathcal{P}'y$ ($\mathcal{P}'x$ is the power set of x). This is in Kaye-Forster (*op. cit.*). In the terminology of that paper, Γ is Σ_1^P.

Interestingly, in view of the way in which extensionality is proof-theoretically problematic, one can show that if the axiom of extensionality is weakened to allow lots of empty sets (or *urelemente*) but retained for nonempty sets (so that distinct nonempty sets have distinct members) to obtain a system which we call *NFU*, then the axiom scheme $\phi \longleftrightarrow \phi^*$ can be added without any extra consistency strength being gained.

3.1 Automorphisms of type algebras

The idea of polymorphism or typical ambiguity for a type theory is of course tied up with the idea of an automorphism of what one might call the *type algebra* of the theory under consideration. The most straightforward case is *TNT*. Its types are indexed by $\mathbf{Z}$ not by $\mathbb{N}$, so that the type algebra is the monad $\mathbf{Z}$. Asserting the biconditional $\phi \longleftrightarrow \phi^*$ for all ϕ has the same effect as asserting the biconditional $\phi \longleftrightarrow \phi^n$ (where ϕ^n is the result of applying n asterisks to ϕ) for all ϕ and all n. This second, more inclusive scheme is probably what one would naturally think of as an axiom scheme of polymorphism but it follows from the weaker version because the automorphism group of the type algebra $(\mathbf{Z})^9$ is cyclic. The type algebras of even quite simple elaborations of *TST*—consider for example the Church-style type theory with only one type constructor (namely function

---

[8] I know of no good reason for this term to have been replaced by 'polymorphism': people who study *TST* continue to use the old word. I assume this is another example of a neologism arising because people are unfamiliar with the literature.

[9] Both the type algebra and its automorphism group are naturally called $\mathbf{Z}$!

types) and where every type is a function type—get complicated. However it is known that the automorphism group of the type algebra of this last theory is a finitely generated simple group.[10]

4 TNT

Theorem 2.1 tells us that *TST* proves the consistency of all its proper initial segments. This means that *TNT* proves the consistency of TST_k, for each k. Now we were able to infer the consistency of *TNT* from the consistency of *TST* by a simple compactness argument—any proof of an inconsistency in *TNT* can be reproduced inside TST_k for some k—so we know that $TNT \nvdash Con(TST)$. If it did, we could reproduce the compactness argument inside *TNT* and *TNT* would prove its own consistency. Therefore $TNT \nvdash (\forall k)Con(TST_k)$. Therefore *TNT* is ω-incomplete. Although any consistent system extending arithmetic is incomplete, the ω-incompleteness is not always this transparent. It is an open question whether or not $TNT + Amb \vdash (\forall k)Con(TST_k)$. *TNT* is a bit odd in other ways. Although—as we have seen—its consistency follows by a compactness argument, it has no standard model. See Forster [1989].

5 Untyped theories: Z, Mac and KF

Recently Richard Kaye has proved a very useful theorem which enables us to infer the consistency of one sorted theories from typed theories with ambiguity schemes to which they are related. This is a strengthening of a theorem in Specker [1962].

LEMMA 5.1 Kaye, [1991].

Suppose that $M = \langle M_0, M_1, M_2 \ldots \rangle$ is a structure for the language of TST and that Σ is the class of formulae of the form "$\exists x_1 \ldots x_n \Phi(x_1 \ldots x_n, y_1 \ldots y_n)$" for Φ in some class Δ which contains all atomic formulae and is closed under conjunction and substitution of variables and contains $\psi^*(y_1 \ldots y_n)$ whenever it contains $\psi(x_1 \ldots x_n)$. Suppose further that $M \models Amb(\Sigma)$. Then there is a structure for the signature $\langle \in, = \rangle$ that satisfies any σ of the form $\forall y_1 \ldots y_n \Phi(y_1 \ldots y_n)$, where the result of adding suitable type indices to Φ is true in M and the $\mathcal{L}_{TST}$ formula corresponding to Φ is in Σ.

This is a very important result. It means that whenever we can prove the consistency of $Amb(\Gamma)$ relative to *TNT*, we get a consistency result for a one-sorted theory. One particular instance of this is in Forster-Kaye [1991]. Starting from a model of $TST + Amb(\Sigma_1^P)$ (and we know there are such) we can obtain a model of the theory that Kaye and I immodestly called KF. KF has the same axioms as Zermelo except for a restriction applied to the separation scheme (axiom scheme 6). In effect axiom scheme 6 is replaced by a scheme that says that

---

[10] Conway told me this was proved by a Californian by the name of R. J. Thompson, but I have never seen it published.

for any x, the substructure $\langle\langle x\rangle\rangle$ is a model of TST. The more usual description of the version of separation that holds in KF would be "separation for stratified Δ_0 formulæ, and this obliges us to define two new terms. "Δ_0" means "contains no unrestricted quantifiers".[11] A formula of the language of set theory is *stratified* if it can become a formula of the language of TST by adding type indices consistently to the variables in it. So where Zermelo has axiom scheme 6, KF has

6: Stratified Δ_0 separation (Axiom scheme: one instance for each stratified Δ_0 ϕ): $(\forall x)(\exists y)(\forall z)(z \in y \longleftrightarrow (z \in x \wedge \phi(z)))$

In fact a refinement (Kaye-Forster [2???]) extends this to a relative consistency proof of a theory trading under various names in the literature, but which I was brought up by my *Doktorvater* Adrian Mathias to call 'Mac' after Saunders MacLane, who advocated it as an adequate basis for all of mathematics. Mac is like KF except in not having the restiction to stratified ϕ in axiom scheme 6. The ϕ still have to be Δ_0 though. This was first proved by Jensen [1969] and clarified in Lake [1975]. There is also a proof in Mathias [2???]. In fact Mathias shows that all the axioms of Kripke-Platek set theory can be added to Mac without gaining extra consistency strength.

Mac is a very significant weakening of Zermelo. We have just seen that Mac is precisely as strong as TST and HOL, and it is an old result of Kemeny's [1949] that we can prove the consistency of TST in Z.

The principle of replacement is what one adds to Zermelo to obtain ZF. It has been known for many years that ZF proves the consistency of Zermelo set theory. Nowadays some quite refined information is coming to light about the precise strengths of different kinds of replacement. A variant of replacement is the axiom scheme of collection:

$$(\forall x \in X)(\exists y)(\psi(x, y)) \to (\exists Y)(\forall x \in X)(\exists y \in Y)(\psi(x, y))$$

It is easy to show that collection implies replacement. To show that replacement implies collection assume replacement and the antecedent of collection, and derive the conclusion. Thus

$$(\forall x \in X)(\exists y)(\psi(x, y))$$

Let $\phi(x, y)$ say that y is the set of all z such that $\psi(x, z)$ and z is of minimal rank. Clearly ϕ is single-valued so we can invoke replacement. The Y we want as witness to the "$\exists Y$" in collection is the sumset of the Y given us by replacement. Notice the use of the axiom of foundation here. We use it to get a *set* of z which are ψ-related to x. This obstructs the proof of this for stratified formulæ: it is *not* the case that stratified replacement implies stratified collection. The following counterexample is due to Mathias. Consider the assertion: for every natural number n there is a set of size n consisting of infinite sets all of different sizes. This is provable in Zermelo set theory. However in, say, $Z + V = L$ we can show

---

[11] Restricted quantifiers are quantifiers in the style "$(\forall x \in y)(\dots)$" and "$(\exists x \in y)(\dots)$".

that there is no set which collects all these together, because the sumset of such a set would be an infinite set of infinite sets of infinitely many different sizes, and we knwo that Zermelo set theory does not prove the existence of such a set, since in $Z + V = L$ all sets of infinite cardinals are finite. So stratified collection is not provable in Zermelo. However Coret [1970] has shown that every stratified instance of the axiom scheme of replacement is provable in Zermelo set theory. Therefore stratified replacement does not imply stratified collection.

6   Current developments and open problems

Holmes [1995] considers variants of TST where the types are partially ordered and whenever $\alpha \leq \beta$ there is a membership relation defined between objects of type α and objects of type β (not just when $\beta = \alpha + 1$!).

By Kaye's lemma the full scheme of typical ambiguity is equivalent to the consistency of Quine's NF. It is open whether this theory is consistent. It is also open whether or not the result of adding full ambiguity to TNT is a theory that proves $(\forall k)(Con(TST_k))$. It is also open whether or not TNT has an ω-model; and open whether or not TNT has a model in which every set is definable.

7   References

Barwise, J. [1975] Admissible sets and structures, an approach to definablity theory. Springer-Verlag 1975.

Buchholz, W. and Wainer S. [1985] Provably computable functions and the fast-growing hierarchy. Logic and Combinatorics, AMS Contemporary Mathematics v 65 pp 179-198.

Church, A. [1940] A formulation of the simple theory of types. *Journal of Symbolic Logic* 5 pp. 56−68.

Coret, J. [1970] Sur les cas stratifiés du schema de remplacement. *Comptes Rendues hebdomadaires des séances de l'Académie des Sciences de Paris série A* 271 pp. 57−60.

Forster, T.E. [1989] A second-order theory without a (second-order) model. *Zeitschrift für mathematische Logik und Grundlagen der Mathematik* 35 pp. 285−6

Forster, T.E. and Kaye, R.W. [1991] End-extensions preserving power set. *Journal of Symbolic Logic* 56 pp. 323−28.[12]

Forster, T.E. and Kaye, R.W. [2???] More on the set theory KF. unpublished typescript 24pp.

Holmes, M.R. [1995] The equivalence of NF-style set theories with "tangled" type theories; the construction of ω-models of predicative NF (and more) Journal of Symbolic Logic *to appear*

---

[12] Errata. p 327. Line 11 should read 'and $a \in M$ such that $M \models \overline{\overline{\pi`a}} = \overline{\overline{\mathcal{P}`a}}$'. Line 13 the expression following '$M \models$' should be '$\overline{\overline{\pi`a}} = \overline{\overline{\mathcal{P}`a}}$'. Line 26 '(not just $\pi`a = \mathcal{P}`a$)' should read '(not just $\overline{\overline{\pi`a}} = \overline{\overline{\mathcal{P}`a}}$)'. Line 28 '$\pi`a$' should read '$\mathcal{P}`\pi`a$'.

Jensen, R.B. [1969] On the consistency of a slight(?) modification of Quine's NF. *Synthese* **19** pp. 250—63.

Kaye, R.W. [1991] A generalisation of Specker's theorem on typical ambiguity. *Journal of Symbolic Logic* **56** pp 458-466

Kemeny, J. [1949] Type theory vs. Set theory. Ph.D.Thesis, Princeton 1949

Lake, J. [1975] Comparing Type theory and Set theory. *Zeitschrift für Matematischer Logik* **21** pp 355-6.

Levy, A. [1965] A hierarchy of formulæ in set theory. *Memoirs of the American Mathematical Society* no 57, 1965.

Mathias, A.R.D. [2???] Notes on MacLane Set Theory. unpublished typescript.

McNaughton, R. [1953] Some formal relative consistency proofs. *Journal of Symbolic Logic* **18** pp. 136—44.

Mostowski, A. [1950] Some impredicative definitions in the axiomatic set theory. *Fundamenta Mathematicæ* v 37 pp 111-124.

Novak, I.L. [1950] A construction of models for consistent systems. *Fundamenta Mathematicæ* **37** pp 87-110

Quine, W.v.O. [1951] Mathematical Logic. (2nd ed.) Harvard.

Quine, W.v.O [1966] On a application of Tarski's definition of Truth. in Selected Logic Papers pp 141-5

Rosser, J.B. and Wang, H. [1950] Non-standard models for formal logics JSL 15 pp 113-129

Scott, D. S. [1960] Review of Specker [1958]. *Mathematical Reviews* **21** p. 1026.

Shoenfield, J. R. [1954] A relative consistency proof *Journal of Symbolic Logic* **19** pp 21-28

Specker, E. P. [1958] Dualität. *Dialectica* **12** pp. 451—465.

Specker, E. P. [1962] Typical ambiguity. In *Logic, methodology and philosophy of science.* Ed E. Nagel, Stanford.

Wang, H. [1949] On Zermelo's and Von Neumann's axioms for set theory. Proc. N. A. S. **35** pp 150-155

Wang, H. [1952] Truth definitions and consistency proofs. *Transactions of the American Mathematical Society* **72** pp. 243—75. reprinted in Wang: Survey of Mathematical Logic as ch 18

Wang, H. [1952a] Negative types. *MIND* **61** pp. 366—8.

Interval-Semantic Component Models and the Efficient Verification of Transaction-Level Circuit Behavior

David A. Fura[1,2] and Arun K. Somani[2,3]

[1] Flight Critical Info. Processing, Boeing Defense & Space Group, Seattle, WA, 98124, USA
[2] Electrical Engineering Department, University of Washington, Seattle, WA, 98195, USA
[3] Comp. Science & Engineering Department, University of Washington, Seattle, WA, 98195, USA

Abstract. Verifying abstract operations, such as bus transactions, is a hard problem that is not well served by existing theorem proving methods. This problem has aspects in common with the familiar model checking problem of verifying temporal logic formulas against a state machine implementation. The contribution of this paper is a new approach to efficiently handle such proofs for circuits that are built using standard library components. The fundamental idea is to prove and then store away temporal properties for the individual components of the library. Our initial results show that this approach can result in an order of magnitude execution speedup for circuit proofs that use these components. This is combined with a drastic reduction in the level of user interaction needed to construct the proofs. This work provides a promising direction for future research into effective proof automation.

1 Introduction

Proponents of formal methods are fond of pointing out the infeasibility of exhaustive simulation for modern VLSI circuits. Of course, the increasing complexity of hardware presents nontrivial challenges for the formal approaches as well. However, a significant advantage of formal methods based on *theorem proving* is an unmatched support for defining and utilizing abstraction. When high-level abstraction can be performed within subsystems *before* they are composed, theorem proving offers a realistic potential for controlling the verification complexity of real-world hardware systems.

A well known disadvantage of theorem proving is the excessive amount of time and tenacious effort required for some implementation proofs. This is particularly true when the abstraction linking a concrete implementation to the specified behavior is a complicated one, often the case for hardware systems in industry. One approach to address these proofs is to link theorem provers to automated verifiers specialized to handle the difficult cases (e.g., [Hun93],[Joy93],[Kur93]). These 'hybrid' approaches offer considerable promise, however, tools implementing them have yet to reach the level of maturity where they can be fully evaluated. In addition, considerations of proof security argue for approaches that can remain entirely within a theorem proving environment.

This research was partially funded by NASA-Langley Research Center under contract NAS1-18586, Task 10. The NASA technical monitor was Sally Johnson.

As part of a research project to inject theorem proving methods into an industrial ASIC (application-specific integrated circuit) development, we have been investigating techniques to specify and verify interface chips for embedded systems. In contrast to much of the previous work in hardware verification, our work is largely concerned with the *interactions* between hardware subsystems, and much of our focus has been on expressing and verifying these interactions at high levels of abstraction. In [Fur93a] we explained our methods for constructing abstract-level ASIC specifications at the *transaction level* of abstraction.[1] In a future paper we will describe techniques for composing these types of abstract-level models into larger systems. The work described in this paper is concerned with the intermediate step of verifying the implementation correctness of abstract hardware models. All of the definitions and theorems presented in this paper have been formalized within the HOL theorem proving system [Gor93].

The application for our work is a processor interface unit (or PIU) for a commercially-developed fault-tolerant computer system, called the Boeing Fault-Tolerant Embedded Processor (or FTEP). Our initial focus is on the PIU subsystem interface to the local processor – the 'processor port' (or P-Port). Several NASA reports document the PIU and its partial specification and verification ([Fur92], [Fur93b], [Fur93c]). In this paper we describe work done primarily since the publication of these documents.

1.1 Contributions of this Paper

Our initial approach to verify the P-Port required huge amounts of painstaking effort and time. This work employed formal models for logic gates and memory elements derived from the standard models in widespread use then and today. These models can be characterized as defining behavior at individual points in time. In contrast, the circuit properties we were verifying described signal behavior over *intervals* of time (Figure 1). It has become clear that this temporal mismatch is largely responsible for the difficulty of our transaction-level proofs.

In this paper we introduce new behavioral models for clock-level logic components and demonstate how they significantly simplify the problem of verifying transaction-level specifications. The models define component behavior over intervals of time, providing a seamless interface to behavioral targets employing interval semantics. We demonstrate how these new component models simplify transaction-level theorem proving for circuits built using standard library components.

Interval-semantic components exploit one of the important strengths of theorem proving: the ability to *reuse* prior work stored away into theorems (Figure 1). By driving difficult interval proof obligations down into the hardware components themselves, the degree of theorem reuse is increased considerably. These component proofs are amortized over all future system designs that incorporate the components. In contrast, when traditional point-semantics are used, difficult interval proofs must be repeated for every relevant state and output signal of a circuit, even when these signals are produced by common component types.

---

[1] At the clock level of abstraction a time unit represents the passage of a single cycle of the system clock; transaction-level time counts the passage of multi-clock (bus) transactions. As explained later, the data types for the two levels are quite different as well.

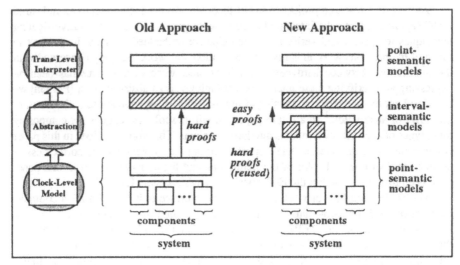

Figure 1: Specification Hierarchy Showing Reuse of Component Interval Proofs.

1.2 Related Work

Interval-semantic component modeling is not new and, to our knowledge, is due to Hanna and Daeche [Han86]. Herbert also describes similar ideas in [Her88]. The intervals used by both groups were at a detailed 'timing level,' where a time unit is a real-time segment much smaller than a system clock cycle duration. The objectives of both efforts were to verify component behavior at a level of abstraction comparable to our clock level.

Of the numerous applications of temporal logic to hardware, the interval temporal logic (or ITL) of Moszkowski [Mos85] comes closest to our work. In contrast to standard temporal logics, ITL temporal operators act over intervals of time rather than over all time. For example, the interval version of the standard operator 'henceforth' acts like the 'during' operator of [Han86] and [Her88], which we also use. It doesn't appear, however, that ITL has been used for component modeling as a means to facilitate circuit-wide theorem proving as we have done.

Our work differs from the component models of [Han86] and [Her88] in two ways:
(a)   We extend the reach of interval semantics to verify higher levels of abstraction, namely the transaction level.
(b)   We extend the application of interval semantics to additional component types, (e.g., counters) using more-powerful operators (e.g., 'fewer than').

The following sections contain additional details of these topics.

1.3 Paper Outline

The rest of this paper is divided into four sections. Section 2 demonstrates interval-semantic models for some representative component types that are used in an example circuit described in Section 3. Section 4 explains how the interval behavior described in Section 3 is verified using these component models. Finally, Section 5 finishes the paper with a concluding discussion.

2 Interval Component Models

This section demonstrates the point-semantic and interval-semantic component models shown at the right hand side of Figure 1. We start with combinational-logic gates, followed by latches and, finally, counters. The component models of this section are based on the actual component library of an ASIC design tool used in the PIU design.

2.1 Logic Gates

Combinational-logic gates have a straightforward interval representation. The following HOL segment illustrates the story for 2-input And gates.[2] The standard (asserted) point behavior is followed by an important type of interval behavior expressed using the **During** construct. Briefly, the form **f During (t1,t2)** is true precisely when the signal **f** is true at every point in the closed interval [t1, t2].

Point-Semantic Definition: And2_Gate

$\vdash_{def}$ And2_Gate a b z = ∀ t. (ASel (a t) ∧ ASel (b t)) , (BSel (a t) ∧ BSel (b t))

Interval Semantic Theorem: AND2_GATE_DURING_A

$\vdash$ And2_Gate a b z ⊃
   ((∀ t1 t2. ((AOf a) Is T) During (t1, t2) ∧ ((AOf b) Is T) During (t1, t2)
        = ((AOf z) Is T) During (t1, t2)) ∧
   (∀ t1 t2. (((AOf a) Is F) Or ((AOf b) Is F)) During (t1, t2)
        = ((AOf z) Is F) During (t1, t2)) ∧
   (∀ t1 t2. ((AOf a) Is F) During (t1, t2) ∨ ((AOf b) Is F) During (t1, t2)
        ⊃ ((AOf z) Is F) During (t1, t2)))

We model all physical, non-state, signals as *2-tuples* to accommodate flexible (clock level) component compositions in a 2-phase clocking discipline. The first value of a sampled signal, accessed using **ASel** (or **FST**), represents the phase-A version; the second is the phase-B version. The operator **AOf** creates a signal carrying the phase-A portion of a 2-phase signal. The infix operator **Is** creates a new signal carrying true precisely when the value carried by the original signal equals the specified constant. **During, Is,** and some of the other operators used in this paper are similar to those found in [Han86], [Her88], [Mel90], [Joy93], and elsewhere. More details of these operators can be found in [Fur94].

Theorem **AND2_GATE_DURING_A** derives And-gate **During** semantics for clock phase A, with a corresponding theorem proven for phase B. The first conjunct says that the output is high throughout an interval precisely when both inputs are. The second conjunct covers the corresponding case for a low output.

The first conjunct is in a very useful form because each input signal resides in its own **During** construct. In circuit proofs where interval properties for gate inputs are used to establish output properties, it is typically the case that the input properties are available

---

[2] In addition to the standard logical connectives ∀ (universal quantification), ∃ (existential quantification), ⊃ (implication), ∧ (conjunction), ∨ (disjunction), and ¬ (negation), the HOL code in this paper uses T and F to represent true and false; the form •1 ⇒ •2 | •3 to represent "if •1 then •2 else •3;" the form λ v. u to represent a function such that (λ v. u) w = u w (juxtaposition represents function application in HOL); and the form ε v. P to represent an arbitrary value of v's type satisfying predicate P.

separately, since the inputs are usually driven by separate components. The left-hand side for the first conjunct is in the proper form to 'match up' with such properties.

The second conjunct says that the output is low over an interval precisely when one of the inputs is low at each time in the interval. Note that the stable-low output does not require either of the inputs, by itself, to be low throughout the interval – only their combination must contain one low value at each time.

The third conjunct expresses the low-output behavior that we have found to be more useful in practice. An implication rather than an equivalence, it says that one of the inputs being low over an interval is enough to ensure that the output is low as well. The left hand side here matches up in (typical) scenarios where individual control signals are used to turn off And gates. Including this conjunct in the theorem therefore helps improve the efficiency of circuit proofs involving And gates, our primary goal.

2.2 Latches

The main benefits of interval component semantics come not from combinational gates, but from state-holding components. The following HOL segment shows the point and **During** semantics for a simple latch that we use to illustrate this. It is a standard D-type latch, clocked on phase A, containing an enable.

Point-Semantic Definition: DELatA_Gate

$\vdash_{def}$ DELatA_Gate d e s q =

    $\forall$ t. (s (t+1) = ASel (e t) $\Rightarrow$ ASel (d t) | s t) $\land$
    (q t = s (t+1) , s (t+1))

Interval-Semantic Theorem: DELATA_GATE_DURING_B

$\vdash$ DELatA_Gate d e s q $\supset$

    (($\forall$ t1 t2 c. ((AOf e) Is T) During (t1, t2) $\land$
        ((AOf d) Is c) During (t1, t2)
        $\supset$ ((BOf q) Is c) During (t1, t2)) $\land$
    ($\forall$ t1 t2. ((AOf e) Is T) At t1 $\land$
        ((AOf e) Is F) During (t1+1, t2)
        $\supset$ ((BOf q) Is ((AOf d) At t1)) During (t1, t2)))

The point definition expresses the expected behavior that the next state (s (t+1)) is either the phase-A input (d t) or the current state, depending on the value of the phase-A enable (e t). The current output is simply the updated state for this flow-through device.

The interval semantics is divided into two parts: (a) flow-through behavior and (b) latching behavior. The first part says that if the enable is high during an interval then the output follows the input during the interval. The second parts says that if the enable is high for a cycle, followed by an interval for which it is low, then the output during this interval is equal to the value of the input at the cycle that the enable was high.

This second behavior is surprisingly time-consuming to prove. It requires: (a) an induction on the *offset* into the interval to establish the correct latch state at these offsets, followed by (b) a specialization of this lemma with t – t1 to obtain the desired result "for all t." This second step involves tedious numerical manipulations. We don't list the actual proof for this theorem because of its length. In Section 4 we contrast it with the simple proof steps that such pre-proved theorems afford circuit verifications.

2.3 Counters

Counters are important components in many control applications. In their familiar role as timers or other 'event counters,' their most important output signal is not the count value, per se, but rather the 'zero' output, which indicates whether or not the count is zero. We focus on this particular output for a rising-edge triggered, n-bit down counter whose point semantics is given by the following HOL segment.

Point-Semantic Definition: DownCntA_Gate

$\vdash_{def}$ DownCntA_Gate n d ld dn s z =

$\forall$ t. (s (t+1) = (BSel (ld t)) $\Rightarrow$ VAL n (BSel (d t)) |
  (BSel (dn t)) $\Rightarrow$ numDECN n (s t) | s t) $\wedge$
  (z t = ((ASel (dn t)) $\Rightarrow$ ((numDECN n (s t)) = 0) | ((s t) = 0)) ,
  ((BSel (dn t)) $\Rightarrow$ ((numDECN n (s t)) = 0) | ((s t) = 0)))

This definition shows the next state (s (t+1)) being set to the data input (d t) if the 'load' input (ld t) is high, otherwise if the 'down' input is high then it equals the decremented version of the current state, otherwise it is the unchanged current state value itself. Type conversion (using VAL) is needed because the state is of type ":num," while the data input is a bit vector – type ":wordn." The function numDECN performs a modulo-2^{n+1} decrement on natural numbers. The counter is parameterized by n, the word size.

The counter z output is complicated by the decrementer's position on the output side of the counter. This output reflects whether or not the 'updated' counter value is 0.

The interval semantics for this counter is shown in the next HOL segment. The counter behavior is divided into three parts, each of which describes During-type counter behavior. The second and third parts also make use of new interval constructs. The form (f FewerThan n CountsSince t1) At t2 says that f is true fewer than n counts since t1 at time t2. The Exactly construct has a corresponding interpretation.[3]

The first part of this definition describes the output behavior when a 0 is initially loaded into the counter. In this case the zero output is active-high until after the first active-high down input is received.

The second and third parts define the counter behavior when a non-0 data value is loaded. The second part covers the interval of time where the counter state has yet to reach 0. This is the period where *fewer than* bs–1 counts of active-high down have been received, and for which the zero output is low.

The third part covers the period of time after an active-high down has arrived for bs–1 counts, but before the bs'th count. In this interval the counter value is 0, as reflected in the active-high zero output.

The proof for this theorem is quite lengthy; much more difficult than the latch proof, for example. However, the effort spent here is especially well rewarded. The behavior it describes is applicable to a wide range of applications and the theorem is valid for an entire class of components, since the counter is parameterized by the size n.

---

[3] In this paper *counts* start at zero.

Interval Semantic Theorem: DOWNCNTA_DURING_B

⊢ **DownCntA_Gate n d ld dn s q z** ⊃

    let bs = VAL n (((BOf d) At t1) in

    ((∀ t1 t2. (bs = 0) ∧
          ((BOf ld) Is T) At t1 ∧
          ((BOf ld) Is F) During (t1+1, t2) ∧
          ((BOf dn) Is F) During (t1+1, t2)
          ⊃ **((BOf z) Is T) During (t1+1, t2)) ∧**

     (∀ t1 t2. (bs > 0) ∧
          ((BOf ld) Is T) At t1 ∧
          ((BOf ld) Is F) During (t1+1, t2) ∧
          (((BOf dn) Is T) FewerThan (bs − 1) CountsSince (t1+1)) At t2)
          ⊃ **((BOf z) Is F) During (t1+1, t2)) ∧**

     (∀ t1 t2 t. (bs > 0) ∧
          ((BOf ld) Is T) At t1 ∧
          ((BOf ld) Is F) During (t1+1, t2) ∧
          (((BOf dn) Is T) Exactly (bs − 1) CountsSince (t1+1)) At t) ∧
          ((BOf dn) Is F) During (t+1, t2)
          ⊃ **((BOf z) Is T) During (t+1, t2)))**

3 Specification for a P-Port Subcircuit

For the purpose of demonstrating our interval components on a realistic verification problem, in this section we briefly describe elements of a small industry-designed circuit, part of the P-Port of the FTEP PIU (Figure 2).

The behavior of the P-Port is fairly straightforward; it receives memory-access requests sourced by a local processor and passes them on to other parts of the PIU. In the figure, the requests are received, on the left, over an Intel 80960MC L-Bus [Int89] and relayed, on the right, onto the internal bus (or I-Bus) of the PIU. Aside from implement-

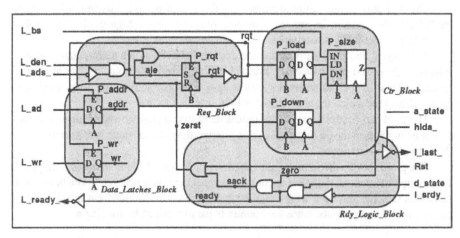

Figure 2: Gate-Level Schematic of Example P-Port Subcircuit.

ing these two bus transfer protocols, the P-Port supports delayed access to the multi-master I-Bus by latching the address and control information received from the L-Bus.

The input and output signals not prefixed by an "L_" or an "I_" are local to the P-Port, except for Rst, which is the port reset. The variables prefixed by "P_" are clock-level state variables and correspond to the shaded state-holding devices in the figure.

Aside from the logic gates of Figure 2, the only other components there are latches and a single 2-bit down-counter. The latches are all D-type except for one SR latch. The clock signals A and B indicate whether the latches are clocked on phase A or B, respectively; the E input is the enable. The counter input IN is a 2-bit data input, while LD and DN are the 'load' and 'down' inputs, respectively, of the last section. The And gates and D-type latches (with enables) were described in the last section as well.

The transaction-level specification for the P-Port [Fur93a] expresses behavior in terms of abstract packets that are quite different from the signals of Figure 2. The fields of these packets include various 'opcodes,' the address, the data array, and the block size. In the course of verifying the transaction level, these packet elements are mapped into their clock-level counterparts according to the definitions in the abstraction predicate. Referring back to Figure 1, this mapping is represented by the vertical line connecting the top two levels shown on the right hand side. The following two subsections describe this mapping for two transaction-level elements: the address and the block size. In Section 4 we explain how the proof obligations described here are verified using the interval-semantic components derived in the last section.

3.1 Transaction-Address Derived Interval Specifications

Figure 3 describes the specification for the transaction-level address, and the mapping of the transaction address to the clock level signals of Figure 2. The specification is a simple one – the L-Bus address (PB_Addr) is transmitted, unchanged, as the I-Bus address (IB_Addr). The common time index (t) indicates that this transmission occurs within the same transaction-level cycle.

The abstraction for the address is straightforward. The L-Bus address is the phase-A value of clock-level signal L_ad on the cycle (at tp') marking the beginning of the t'th L-Bus transaction, defined in terms of the signals L_ads_ and L_den_ . The I-Bus address is defined here in terms of the signal addr. It is sampled on the occurrence of the t'th I-Bus transaction start (at ti'), defined with respect to the clock-level signals rqt and a_state. The

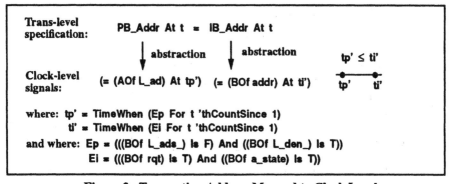

Figure 3: Transaction Address Mapped to Clock Level.

inequality within the relationship **tp'** ≤ **ti'** results because the P-Port may have to wait on a busy I-Bus. The form **TimeWhen** f shown in the figure returns the time for which f is true; it is defined using the Hilbert choice operator (ε). The form **f For n 'thCountSince** t is similar to the **istimeof** operator of [Mel90]. It creates a signal carrying true for all times at which f is true for the **n'th** count since time **t**.

There are a large number of steps involved in the verification of the P-Port transaction address. However, for the sake of our simple example we focus on the single circuit-level specification shown next. This specification describes an important stability property for the signal **addr**, ensuring that its value is the correct one at the beginning of the I-Bus transaction (again at **ti'**).

Transaction Address Proof Obligation:

(P_rqt **Is** F) At tp' ∧
(((BOf L_ads_) **Is** F) And ((BOf L_den_) **Is** T)) At tp' ∧
(((BOf L_ads_) **Is** T) Or ((BOf L_den_) **Is** F)) During (tp' + 1, ti' − 1) ∧
1 ≤ tp' ∧
((BOf zerst) **Is** F) During (tp', ti' − 1)
⊃ ((BOf addr) **Is** ((AOf L_ad) At tp')) During (tp', ti'))

From Figure 3 it is clear that this stability property is an important element of the overall transaction address correctness. The first four preconditions shown here are themselves mapped from transaction-level entities, the precondition and excondition described in [Fur93a]. They provide important facts about the initial value of the **P_rqt** latch (of *Req_Block*) and of the transaction-request signals. The fifth precondition is derived using the other components of the P-Port.

3.2 Transaction-Block-Size Derived Interval Specifications

The transaction-level specification for the block size, shown in Figure 4, is as concise as the address specification. The I-Bus output (**IB_BS**) is simply the block-size input received from the L-Bus (**PB_BS**). The transmission is again single-cycle flow through.

The L-Bus block-size abstraction is also similar to that of the address. In this case it is the sampled clock-level signal **L_bs** that maps to the transaction level. As before, the clock-level time **tp'** is the sampling point.

The abstraction for the I-Bus block size presents an interesting contrast to the previous abstractions. Here, there is no direct I-Bus equivalent to **L_bs**; instead, the block size is defined by the temporal behavior of the output signal **I_last_**. If **I_last_** is brought low immediately after the I-Bus transaction start (again at **ti'**), and remains low until **I_srdy_** is brought low, then the block size is **0** (or one word). If **I_last_** is instead high during this time, remaining high until **I_srdy_** is low for the **0'th** count, and then remains low from then until the return of **I_srdy_** low, then the block size is **1**, and so on.

We note that the slave port participating in the I-Bus transaction uses **I_srdy_** to indicate its successful handling of the current data word. For writes to the slave, this means that the slave has finished storing the word, while for reads it means that the slave is currently driving the data word onto the **I_ad** bus not shown here.

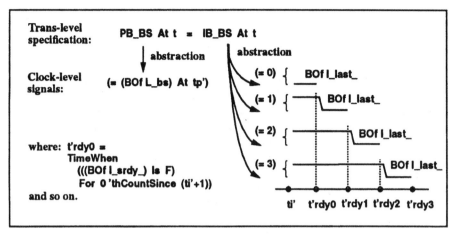

Figure 4: Transaction Block Size Mapped to Clock Level.

The following HOL segment contains a specification for the I_last_ signal consistent with Figure 4. The complexity of this specification is quite representative of specifications involving temporal properties of clock-level signals. On the other hand, this specification could have been even more complex. For one thing, we have left out a few details irrelevant to the following discussion; therefore, this specification is not complete. Secondly, the FewerThan operator used here encompasses more complex behavior than provided by individual standard temporal-logic operators such as 'henceforth' and 'until.' Had we relied on primitive operators such as these, the specification would have been longer. Finally, while the number of block-size cases could be as high as four here (bs = 0, 1, 2, 3), we have reduced it to two (bs = 0, bs > 0).

Transaction Block Size Proof Obligation:

let bs = VAL 1 ((BOf L_bs) At tp') in
let ti'sack = TimeWhen (((BOf I_srdy_) Is F) For bs 'thCountSince (ti' + 1)) in
let ti'psack = TimeWhen (((BOf I_srdy_) Is F) For (bs – 1) 'thCountSince (ti' + 1)) in

((P_rqt Is F) At tp' ∧
 ((((BOf L_ads_) Is F) And ((BOf L_den_) Is T)) At tp' ∧
 (∀ t'. (((BOf I_srdy_) Is F) FewerThan (bs–1) CountsSince (ti'+1)) At (t'–1)
        ⊃ (((BOf L_ads_) Is T) Or ((BOf L_den_) Is F)) During (tp' + 1, t' – 1) ∧
1 ≤ tp' ∧
 (∀ t'. (((BOf I_srdy_) Is F) FewerThan (bs–1) CountsSince (ti'+1)) At (t'–1)
        ⊃ ((BOf Rst) Is F) During (tp', t')) ∧

(Eventually ((BOf I_srdy_) Is F)) Since (ti'+1) ∧
(∀ t'. (((BOf I_srdy_) Is F) ∧ ((BOf I_last_) Is HI)) At t' ⊃
       (Eventually ((BOf I_srdy_) Is F) Since (t'+1)))

⊃ (((bs = 0) ⊃ ((BOf I_last_) Is LO) During (ti' + 1, ti'sack)) ∧
   ((bs > 0) ⊃ (((BOf I_last_) Is HI) During (ti' + 1, ti'psack)) ∧
              ((BOf I_last_) Is LO) During ((ti'psack + 1), ti'sack))))

The variable ti'sack defined at the top of this specification is the time that the I-Bus slave transmits its bs'th active-low l_srdy_. This event marks the end of the active portion of the transaction. The variable ti'psack is the time of the prior active-low l_srdy_ transmission.

Immediately below the let constructs are the preconditions corresponding to those shown in the last section. Again, these define the initial value for the P_rqt latch and the assumed behavior of the L-Bus transaction request signals (L_ads_ and L_den_) and the reset signal (Rst). The form of these preconditions differs from before, being more faithful to the actual PIU specification. For example, here the L-Bus master (the local processor) is assumed to maintain the transaction request lines in an inactive condition only while fewer than bs-1 counts of l_srdy_ are received. This is identical to the actual PIU specification except that here the signal l_srdy_ plays the role of L_ready_.

For the block size, the most interesting precondition defines the assumed behavior of the l_srdy_ signal. This precondition contains two parts. The first says that after the I-Bus transaction is begun (at ti') there will eventually be an active-low l_srdy_ transmitted by the I-Bus slave. The second part says that if the slave transmits an active l_srdy_ while the P-Port is transmitting an inactive-HI l_last_, then the slave will transmit another active l_srdy_ at some point in the future. The value HI is the 'high' value of our 4-valued logic type ":wire", used here because l_last_ is driven by a tri-state buffer (Figure 2).

Finally, at the bottom of the specification is the required behavior for the l_last_ signal. Both of the cases shown there are easily seen to match the behavior expressed in Figure 4, but with far less ambiguity.

4 Verification Using Interval-Semantic Components

The complex specifications of the last section provide a good indication of the non-trivial nature of transaction-level verifications. Fortunately, these verifications can benefit greatly from the interval-semantic component models described in Section 2. In this section we demonstrate some 'easy proofs' (see the right hand side of Figure 1) using these models. We begin with the address stability obligation described in the last section, followed by the obligation for the block size.

4.1 Address Stability

The verification of address stability illustrates the advantages of interval-semantic components particularly well. Recalling Section 3.1, the objective is to show that the address latch output (signal addr) maintains the value loaded into the latch at transaction start. The circuit blocks of Figure 2 playing roles in this proof are *Data_Latches_Block* and *Req_Block*. The signal zerst, input by *Req_Block*, is also important here but, since it has an even bigger role in the block-size verification, we wait to cover it in the next section. As before we simply precondition the results on a suitably-behaved zerst input.

Breaking down the P-Port into a collection of blocks, as evidenced in Figure 2, has been quite helpful in managing the overall verification effort of the P-Port. Each of these blocks has an interval behavior proved for it from the individual component behaviors. For three of the four blocks in Figure 2 the verification proof resembles that shown in the next HOL segment (for *Data_Latches_Block*). The proof for *Req_Block* is somewhat more complex because of a need to adjust certain interval boundaries to match up, however, the additional complexity is low.

'Data Latches Block' Theorem: DATA_LATCHES_BLOCK_THM

```
⊢  Data_Latches_Block  L_wr  L_addr  rqt_  P_wr  P_addr  wr  addr  ∧
   ((AOf rqt_) Is T) At tp'  ∧
   ((AOf rqt_) Is F) During (tp' + 1, t'sack)
   ⊃
   ((((BOf wr) Is ((AOf L_wr) At tp')) During (tp', t'sack) ∧
    ((BOf addr) Is ((AOf L_addr) At tp')) During (tp', t'sack)))
```

Proof:

```
REWRITE_TAC [Data_Latches_Block; AT_EQ_DURING]
THEN REPEAT STRIP_TAC
THEN IMP_RES_TAC DELATA_GATE_DURING_B
```

This theorem statement says that if the request signal, **rqt_**, is high for one cycle, followed by a low interval, then the values held by both the address and read/write latches remain stable, and with the 'appropriate' values – the input values present at the time the request was high. The proof for this theorem is quite trivial. We rewrite the block definition to get to the components, move the components onto the assumption list, and finish up by resolving with the component interval behavior captured in the latch theorem. The theorem **AT_EQ_DURING** is used to convert 'at' behavior, **f At t**, into its interval counterpart, **f During (t, t)**. The simplicity of this proof stands in sharp contrast to the nontrivial proof of the **DELATA_GATE_DURING_B** theorem used here. Furthermore, both of the address and read/write values are handled by this single proof.

We now consider the case for multiple blocks. As an example, the following HOL segment shows the desired address and read/write stability for the subsystem containing the two blocks, *Data_Latches_Block* and *Req_Block*. This theorem statement differs from before by defining the stability in terms of the **P_rqt** precondition, the L-Bus obligation for **L_ads_** and **L_den_**, as well as the internal signal **zerst**, mentioned earlier. Of signifi-

---

'Stable Latches' Theorem: STABLE_LATCHES

```
⊢  Req_Block L_ads_ L_den_ zerst P_rqt rqt rqt_  ∧
   Data_Latches_Block L_wr L_addr rqt_ P_wr P_addr wr addr  ∧
   (P_rqt Is F) At tp'  ∧
   ((((BOf L_ads_) Is F) And ((BOf L_den_) Is T)) At tp'  ∧
   ((((BOf L_ads_) Is T) Or ((BOf L_den_) Is F)) During (tp' + 1, ti' – 1)  ∧
   ((BOf zerst) Is F) During (tp', ti' – 1)  ∧
   1 ≤ tp'
   ⊃
   ((((BOf wr) Is ((AOf L_wr) At tp')) During (tp', ti')  ∧
    ((BOf addr) Is ((AOf L_addr) At tp')) During (tp', ti'))
```

Proof:

```
EXPAND_LET_TAC
THEN REPEAT STRIP_TAC
THEN IMP_RES_TAC REQ_BLOCK_THM
THEN IMP_RES_TAC DATA_LATCHES_BLOCK_THM
```

---

cance is, again, the very short proof required to verify the behavior. Our experience on the P-Port verification is that most of the theorem proofs resemble the two shown in this section.

4.2 The Block Size

The specification for the I_last_ signal, shown in Section 3.2, makes the block-size verification more difficult than it needs to be for our particular circuit. For example, if instead of the two cases (bs = 0, bs > 0) we were to use the four explicit cases (bs = 0, 1, 2, 3), then the proof for I_last_ is similar to the proof for addr of the last section. However, rather than repeating the same basic proof strategy in this section, we instead continue working with the dual-case specification, a specification that works for arbitrary block sizes. Although the interval-semantic components make this proof easier than would be the case otherwise, they cannot prevent a form of complexity associated with the need to perform an induction on the block size.

In some sense the address-stability theorem of the last section is not a complete statement of required behavior since it is preconditioned on an *internal* P-Port signal. However, the behavior of the zerst signal is intimately tied to the actions of *Ctr_Block* and *Rdy_Logic_Block*, combined with I-Bus slave actions, so we cover it here.

A careful look at Figure 2 reveals a circular chain of control that presents us a major problem. For example, the counter in *Ctr_Block*, based on the counter semantics presented earlier, has an indeterminate interval behavior unless the counter LD input is low during the interval under consideration. To provide this, the P_rqt latch in *Req_Block* must not be reset by an active-high zerst signal during the interval. Now, in the event that an active-low I_srdy_ does arrive, it must be the case that the counter output, zero, is low (d_state is assumed to be high). But this returns us back to the counter where we started from. Because there will, in general, be occurrences of I_srdy_ going low before we want zerst to go high, a way has got to be found to break this circular chain of control.

Fortunately, the structural partitioning used in our new approach provides some insight into this problem. The key to breaking the circular chain of control is a fact about the signal zerst expressed in the following theorem. We don't include the proof because of its length.

Ignoring all of the normal preconditioning, what this theorem says is that if I_srdy_ is low fewer than bs–1 counts within an interval then zerst is low throughout the same interval, with some interval spillover irrelevant to the immediate discussion. The proof for this theorem is achieved by contradicting the converse statement that a time does exist

---

Theorem: ZERST_FALSE_BEFORE_ISRDY_BS_TIMES

⊢ let bs = VAL n ((BOf L_bs) At tp') in
  (Req_Block L_ads_ L_den_ zerst P_rqt rqt rqt_ ∧
   Ctr_Block n L_bs rqt_ ready P_size P_load P_loadA P_down P_downA zero ∧
   Rdy_Logic_Block Rst zero d_state hlda_ I_srdy_ sack zerst ready I_last_ ∧
   (standard preconditions) ∧
   ((BOf I_srdy_) Is F) FewerThan (bs − 1) CountsWithin (tp' + 1, t'psack − 1)
   ⊃
   ((BOf zerst) Is F) During (tp', t'psack)

for which zerst goes high. If such an event did occur then there would exist a *first* time, such that for all prior times zerst was low. However, at this transition point the precondition is met for the second part of the interval semantics for the counter[4], which means that the output, zero, is in fact low at the transition point – zerst cannot be high here (see Figure 2).

With this theorem in hand we can successfully prove that when fewer than bs-1 counts of low l_srdy_ occur that the counter output zero is low, hence that l_last_ is high. This is combined with the I-Bus slave obligation (that a high l_last_ leads to an eventual next occurrence of low l_srdy_) to obtain the key fact that l_srdy_ goes active low for bs counts. This fact is needed to process the Hilbert choice operator contained within the TimeWhen operator of the block size obligation (Section 3). More details for all of these steps, in the context of the proof for the P-Port, are contained in [Fur94].

Before leaving this section we point out that the proofs we have just described are not ones that we would want to execute as part of a 'normal' circuit verification procedure; they are too hard. For a standalone verification effort it would probably be better to instead perform a 4-way case split on the block size, and then use the simpler proof techniques described in the last section. The approach that we have adopted is to consider a modified version of the 4-block circuit of Figure 2 as a new 'standard component' that implements 'block-size translations.' When we finish our changes to this model and our proofs, this subcircuit will be verified to work for an arbitrary block size, and will then be added as a component to our design library.

5 Conclusions

Transaction-level verification is a hard problem, and one that is poorly served by existing theorem proving methods. When employing standard hardware component models, implementation proofs are difficult to construct and require large amounts of CPU processing to execute. In this paper we have described an approach to transaction-level verification that is superior to our old theorem proving methods in terms of both human interaction and CPU processing needs. The key aspect of our new approach is the use of hardware components with behavior expressed over temporal *intervals* rather than individual points of time.

Our approach exploits a major strength of theorem proving by emphasizing the *reuse* of difficult proofs stored away as theorems. We drive the interval proof obligations down into the hardware components themselves, obtaining verified component behavioral models expressed over intervals. These proofs are reused, in general, every time the components are brought into a new design. In contrast, when traditional point semantics are used for components, the interval proofs must be repeated for every relevant state and output signal of a circuit, even when these signals are produced by common component types.

The subcircuit contained in the data latches block of Figure 2 illustrates the differences between the two approaches quite well. The simple 3-line tactic shown in Section 4.1 clearly benefits from the nontrivial component interval proof of the theorem DELAT-A_GATE_DURING_B. Had this proof not existed for the latch component, then one similar

---

[4]Here's an instance where the delay provided by the P_load latch combination is needed.

to it would have been required for *each* of the two latch variables in the data latches block.

The simple proof for the data latches block demonstrates a clear reduction in required human workload for our new approach. In addition, CPU processing needs are reduced in some cases by an order of magnitude. For example, the execution time for the latch component proof (**DELATA_GATE_DURING_B**) is roughly 25 seconds, whereas the execution time for the data latches block proof (**DATA_LATCHES_BLOCK_THM**) is roughly 2.5 seconds.[5] The other blocks of the P-Port show comparable speedups.

The use of interval-semantic components is fully consistent with our philosophy of raising the level of behavior within components *before* composing them. Defining interval behavior for the components themselves allows us to completely bypass the construction and verification of a (point-semantic) behavioral clock level for the P-Port. By appropriately partitioning the design into major blocks, one can quickly generate interval proofs with low CPU processing requirements.

We believe that the work presented in this paper demonstrates the feasibility of using theorem proving directly on a specific, but important, class of circuit verification problems that we, at least, had found beyond the practical reach of previously-published approaches. A number of improvements can be made to our methods however. Developing a good scheme for automating interval-semantic proofs is a high priority. It appears that the work described by Hanna and Daeche in [Han86] has progressed much farther along this path than we have.

A second important area of future work is the development of languages for representing abstraction. Such languages must satisfy a number of constraints to be effective. For one, they should be easy for people to understand and use. Standard temporal logics, for example, are too austere and lead to huge descriptions that strain our ability to comprehend specifications. Ideally, abstraction languages should operate over multiple domains, including theorem proving (as we have shown in this paper), model checking, and simulation. As industry-designed hardware systems are represented at higher levels of abstraction we expect that this issue will take on greater importance.

Finally, we point out that the major limitation of our approach is its restricted applicability to circuits built using standard library components. Most systems contain application-specific state machines that clearly do not fall within this category. State-machine verification can be approached as a separate problem to be handed off to an external model checker, a problem for which they are especially well suited. Alternatively, other theorem proving approaches more suitable for state machines can be pursued, such as the proof procedures described in [Sch93].

6 Acknowledgements

Thanks to the reviewers for poring over the details in this paper and suggesting ways to improve our presentation.

---

[5] Each of these times is for HOL88 v2.01 with AKCL 1-530, running on a SPARC 2 with 256 MB of memory, and includes garbage collection. The 2.5 seconds includes the time needed to load the pre-proved **DELATA_- GATE_DURING_B** theorem.

7 References

[Fur92] D. Fura, P. Windley, and G. Cohen, "Formal Design Specification of a Processor Interface Unit," *NASA Contractor Report 189698*, November 1992.

[Fur93a] D.A. Fura, P.J.Windley, and A.K. Somani, "Abstraction Techniques for Modeling Real-World Interface Chips," in J. Joyce and C. Seger (eds.), *Higher-Order Logic Theorem Proving and its Applications*, Lecture Notes in Computer Science 780, Springer-Verlag, 1994.

[Fur93b] D. Fura, P. Windley, and G. Cohen, "Towards the Formal Specification of a Processor Interface Unit," *NASA Contractor Report 4521*, December 1993.

[Fur93c] D. Fura, P. Windley, and G. Cohen, "Towards the Formal Verification of a Processor Interface Unit," *NASA Contractor Report 4522*, December 1993.

[Fur94] D.A. Fura, *Abstract Interpreter Modeling and Verification Methods for Embedded Hardware and Fault-Tolerant Systems*, Ph.D. thesis, Electrical Engineering Department, University of Washington, 1994.

[Gor93] M.J.C. Gordon and T.F. Melham, *Introduction to HOL: A Theorem Proving Environment for Higher Order Logic*, Cambridge University Press, 1993.

[Han86] F.K. Hanna and N. Daeche, "Specification and Verification using Higher-Order Logic: A Case Study," in G.J. Milne and P.A. Subrahmanyam (eds.), *Formal Aspects of VLSI Design*, Elsevier Science Publishers, 1986, pp. 179–213.

[Her88] J. Herbert, "Formal Verification of Basic Memory Devices," Technical Report No. 124, Computer Laboratory, University of Cambridge, February 1988.

[Hun93] H. Hungar, "Combining Model Checking and Theorem Proving to Verify Parallel Processes," in C. Courcoubetis (ed.), *Fifth Conference on Computer Aided Verification*, Lecture Notes in Computer Science 697, Springer-Verlag, 1993, pp. 154–165.

[Int89] Intel Corporation, *80960MC Hardware Designer's Reference Manual*, June 1989.

[Joy93] J.J. Joyce and C.H. Seger, "Linking BDD-Based Symbolic Evaluation to Interactive Theorem-Proving," in *Proceedings of the 30th Design Automation Conference*, IEEE Computer Society Press, June 1993.

[Kur93] R.P. Kurshan and L. Lamport, "Verification of a Multiplier: 64 Bits and Beyond," in C. Courcoubetis (ed.), *Fifth Conference on Computer Aided Verification*, Lecture Notes in Computer Science 697, Springer-Verlag, 1993, pp. 166–179.

[Mos85] B. Moszkowski, "A Temporal Logic for Multilevel Reasoning about Hardware," *IEEE Computer*, Vol. 18, No. 2, February 1985, pp. 10–19.

[Mel90] T.F. Melham, *Formalizing Abstraction Mechanisms for Hardware Verification in Higher Order Logic*, Ph.D. thesis and Technical Report No. 201, Computer Laboratory, University of Cambridge, August 1990.

[Sch93] K. Schneider, R. Kumar, and T. Kropf, "Alternative Proof Procedures for Finite-State Machines in Higher-Order Logic," in J. Joyce and C. Seger (eds.), *1993 International Workshop on Higher Order Logic Theorem Proving and its Applications*, Vancouver, Canada, August 1993, pp. 215–228.

An Interpretation of NODEN in HOL

Brian T. Graham

University of Cambridge Computer Laboratory,
New Museums Site, Pembroke Street, Cambridge, CB2 3QG, England.

Abstract. We describe the use of the HOL system to corroborate the logical consistency of Noden, an integrated HDL and proof system. The Noden logic is interpreted by representing terms of the logic (which are terms of the Noden HDL) as logical statements in the HOL logic. We describe the datatypes representing types and values, and sample the representation of built-in operations, functions and macros as HOL function specifications. An interpretation of Noden truth-valued statements is presented as a translation of these to HOL sequents, and the representation of Noden proof operations as HOL conversions. Results of the work are summarised, including exposed errors and ambiguities of the Noden logic and implementation. We conclude that this approach to providing assurance of soundness and consistency of a less secure proof system is not only useful, but is a practical method of prototyping and an aid to specifying such a system.

1 Introduction

The Noden system [Pyg92, Pyg93] incorporates a hardware description language, the Noden HDL, and a logic and proof environment for reasoning about terms of the HDL. The purpose of the project is to derive results about the soundness and consistency (or otherwise) of the Noden logic. The methodology has been to interpret the Noden logic in the HOL [GM93] logic, and rely upon the consistency of the latter in arguing for the consistency of the former.

Noden began as a subset of the Ella language, extended with a logic and proof assistant. It is intended as a low level proof tool simple enough to be used by design engineers with no particular knowledge of logic. The intention is manifest in the close coupling of the logic and the HDL, the limited number of relatively powerful proof operations[1] available to the user, and a user-friendly, window-based interface. Unlike the HOL system, the logic is not implemented from a small core of axioms and inference rules. Instead it has high-level proof operations which attempt to capture the intended semantics of the HDL. The designers of the system have deliberately chosen to trade off security of the logic implementation for efficiency and ease of use of the prover.

To agree on the meaning of relevant terms, we take the following definitions from [And86].

---

[1] We use the terminology "proof operation" to avoid distinguishing tactics from inference rules.

We say that a wff of a logistic system is *valid* with respect to an interpretation iff the value of that wff is truth (under that interpretation) for all assignments of values to its (free) variables. ... An interpretation of a logistic system is *sound* iff, under that interpretation, the axioms are all valid and the rules of inference preserve validity.

A logistic system is *absolutely consistent* iff there is a wff which is not a theorem. A logistic system is *consistent with respect to negation* iff there is no wff A such that both A and the negation of A are theorems.

In an inconsistent logistic system every wff is a theorem, and thus being a theorem holds no significance. The two forms of definition of consistency are equivalent in the logistic systems with which we are concerned.[2] We define an interpretation by embedding the Noden logic in HOL. Soundness is demonstrated by showing that all of the logistic system's axioms are valid and that the rules of inference preserve validity. Once soundness is demonstrated, we can get consistency by use of the relative consistency theorem for theory interpretation [Far94, Far93]: if T is interpretable in U and U is consistent, then T is also consistent. This assumes, of course, that the HOL logic is consistent (see [Pit93]), and that the HOL system is a sound interpretation of the HOL logic.

We begin with a brief look at the Noden HDL and logic. Next we describe the shallow embedding of the HDL in HOL, including the representation of types, values, and examples of the language's operators. The interpretation of the logic presents both the representation of truth-valued statements and coding of proof operations. We review some results of the work, including errors discovered, questions raised, and problems with discrepancies between the language implementation and the model in HOL, and conclude with an assessment of the contributions of the project.

2 The NODEN System

The Noden HDL is a typed, first-order language, intended for specifying synchronous circuits. Types include _integer, finite integer ranges, _boolean (members _true and _false), bool (members t, f and others), compound types including structures and vectors, and user-defined finite types. The latter include **wire** types, intended for representing the values on a single wire, and **enumeration** types, intended for representing values held on collections of (bool) wires. All types except _boolean include two distinguished members, referred to as illegal and don't care values. These constants are written as the type name prefixed with ! or ? respectively. The compound types differ in the uniformity of component types: vector component types are uniform, while structures are not. Lastly, the type _time has no literals, but can have _integer values added or subtracted; and the pseudo type _const which applies to _integer (other than the illegal and don't-care) values which do not vary over time. Uses for the latter include recursive arguments to macro definitions, and indices into compound type values.

---

[2] Provably so if wff's of the form $A \supset (\sim A \supset B)$ are valid under all assignments of values. See [And86].

Built-in Types

type	members	don't care & illegal values
_integer	...,-2,-1,0,1,2,...	?_integer, !_integer
_time		?_time, !_time
bool	t, f	?bool, !bool
_boolean	_true, _false	

User-definable Types

class	examples
constrained integer	(0..15), (-(2^{31}) .. (2^{31}-1))
wire	bool, tri = (lo | mid | hi)
enumerated	counter = (reset=#0x | inc=#10 | load=#11)
vector	[6]bool, [3]_integer, (bool,bool)
structure	(bool, (tri,[6]bool), _integer)

Values of all types, with the exception of _time, are considered sequences through time, and thus may be sampled at a specific time using the infix operator @, returning again a sequence of (a constant) value through time.

Noden *expressions* include all of the following:

- variables
- constants (including illegal and don't care values)
- IF expressions
- CASE expressions
- LOOP expressions
- function applications

IF expressions can have either _boolean or bool conditions, and the last two arguments must have the same type. If the condition is the illegal or don't care bool value, the illegal value of the same type as the second argument is returned, as is also the case if the ELSE branch is omitted and the condition is _false or f. The CASE expression similarly returns an illegal value if either the value expression is an illegal or don't care value, or it fails to match any limb selector and the optional ELSE branch is omitted. The test for matching can ignore parts of a compound value if don't care values are used in the limb selectors. The LOOP expression takes an initial value, an increment, and a terminating value (all _integer) and an expression. The value returned is a vector of the type of the expression, with each element having the loop counter value substituted.

Built-in operators include three different equality operators, the tautological =, the bool == operator which returns ?bool if either argument is illegal or don't care, and ~= which is similar but returns t if the left operand is a don't care value. All the standard integer arithmetic operators and relations are included, as well as transfer function between bool vectors and _integer's. Logical operators are provided for both _boolean and bool values, as well as bool logical operators for negation, conjunction, disjunction, nand, and nor over all the elements of a bool vector. Additional operators include _DELAY which models a fixed delay element of any given type, CONC which concatenates vectors (or single elements of the component type of the vector), and postfix index and slice operators on structures and vectors.

Many operators have actual Noden_HDL definitions, and users may also define functions and macros. Macros differ from functions in permitting recursion on (_const) _integer arguments. The simplest form of function or macro is composed of a signature and a body which is a Noden expression, as illustrated in the example definition of the macro AND_N.

```
MAC AND_N{n} = ([n]bool: a) → bool:
    IF n = 1 THEN a[1]
            ELSE  a[1] AND (AND_N{n-1}(a[2..n])) FI.
```

The signature indicates that AND_N takes one argument, "a" of vector type [n]bool, and returns a bool result. The expression comprising the body tests for the terminating value 1, and if the test fails a recursive call is made.

The other general form of user-definable "function" is intended for representing circuits, potentially with internal feedback and delay elements. In the following definition, DELAY_BOOL is a delay element of 1 unit time for the type bool. Because there is feedback with the signal mem, it is declared with a MAKE statement as the output of the DELAY_BOOL component. The JOIN statement effects the connection of the inputs to that component. The OUTPUT statement determines which values are accessible from outside the component.

```
FN B2 = (bool: a b) → (bool,bool):
        BEGIN    MAKE DELAY_BOOL: mem.
                 JOIN b OR mem → mem.
                 OUTPUT ((a AND mem), mem)
        END.
```

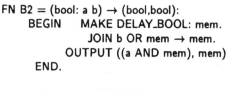
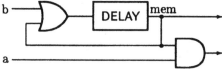

2.1   The Noden Logic and Implementation

The Noden logic extends the HDL with "proof statements" which enable the expression of properties of objects described in the HDL. In describing the Noden logic, it is difficult to separate entirely the logic from its implementation. The existing documentation and definition of the proof operations (ie. rules of inference) is merely a description of the implementation of the same in the system. The lack of such specification is one of the prime motivations of this work.

In addition to the normal logic operators, there is a polymorphic predicate, _LIFT, which holds of a value that is not a don't care or illegal value. Universal and existential quantification are included, and universal quantification of types and classes of types is permitted, using keywords _any, _scalar, _vector, _struct, _wire, _enum, and _ranged to introduce type variables.

Conjectures represent wff's of the logic, and have a restricted form, described by the following BNF-like grammar (where [ ] encloses optional elements).

```
conjecture ::= [ FORALL vars_list ] [ WHERE ⟨bexp⟩ ] conjecture
            | [ EXISTS vars_list ] [ WHERE ⟨bexp⟩ ] conjecture
            | ⊢ ⟨bexp⟩
  vars_list ::= ⟨type⟩ : var_list [ , vars_list ]
   var_list ::= ⟨var⟩ [ , var_list ]
```

The scope of the FORALL and EXISTS quantifiers extends to the far right of the conjecture in every case, and all variables must be bound by an enclosing quantifier. ⟨bexp⟩ represents a term of the HDL of type _boolean. Example conjectures follow.

FORALL _any: ty, ty: a b ⊢ (a == b) = (b == a).

FORALL bool: x y WHERE _LIFT(x) ∧ _LIFT(y) ⊢ _LIFT(x AND y) ∧ _LIFT(x OR y).

Note that the ⊢ symbol is not used to identify proved theorems, but rather demarks the "hypothesis"[3] of the conjecture from the preceding "predicates" and quantifiers. Conjectures are considered "valid" in Noden (ie. provable theorems) iff whenever the predicates hold then the hypothesis is necessarily _true.

Axioms and theorems may be supplied directly by the user, with all the attendant risk of introducing inconsistency. Preferably, conjectures may be proved and become theorems. Both backward and forward proofs may be performed. The latter can only begin with a conjecture in which the hypothesis is a reflexive equation, and thus is necessarily valid. The same proof operations can be used for both proof methods, both because they mainly effect equational transformations such as rewriting, and also because of the restricted form of initial theorem in forward proof.

The logic implementation has only nine proof operations at present. We describe these in the following as suggestive of the rules of the logic.

- **Apply** uses an existing theorem or axiom to rewrite the hypothesis or some predicate. A matching procedure instantiates variables in the theorem, and predicates of the theorem must be satisfied by predicates of the goal, or become added proof obligations. Only single point rewriting is implemented.
- **Use predicate** uses one of the predicates to substitute in either the hypothesis or another predicate.
- **Case split**, as its name implies, considers all possible values an expression of a finite type may take.
- **Open** is used to rewrite with the definition of a function or macro, and also to unfold LET and MAKE bound variables.
- **Split hypothesis** divides a conjunctive hypothesis into separate subgoals, and splits a hypothesis expressing the equality of two structure values into subgoals expressing the equality of corresponding elements of the structures.

---

[3] We use the terms "hypothesis" and "predicate" as used by the authors of Noden. The hypothesis is the expression to the right of ⊢, and a predicate is an expression to the left of ⊢ following WHERE, or a single conjunct of such an expression.

- **Assert** and **Use existential** are used together to supply a witness for an existentially quantified variable, then substitute it in the goal. All predicates in which the variable occur become proof obligations. Any proof in which witnesses have been supplied will substitute the witness (or a Skolem function as the witness) in the delivered theorem when the proof is complete, replacing existential with universal quantifiers.
- **Push time** moves the application of the temporal sampling function applied to a function application inwards to apply it to the arguments of the function.
- Last and most powerful is **Simplify**, which acts mainly on in-built functions, macros, and other operators, effecting transformations reflecting the semantics of these. In the absence of any formal definition of the semantics of the HDL, the simplifier defines the semantics by default.

More proof operations will likely be added in the future, including one for induction proofs.

3 Embedding Noden: the HDL

The wff's of the Noden logic are terms of the HDL, hence we will represent these terms within HOL. We have defined a "shallow" embedding for this purpose, similar in some ways to work of the HOL/Ella project [Bou92, Har93]. The immediate difference is that our model does not identify distinct Noden types with HOL types, but instead with HOL terms. Two aspects of the language make this change necessary. First, quantification over Noden types is allowed but this is not allowed with HOL types, and second, Noden operators such as index and slice are not expressible as well-typed functions in HOL, since the type of the value returned depends on the value of the numeric argument.

Two principles underlie the embedding of the HDL in HOL. First, equivalent constants always share a unique representation. This will make some Noden theorems vacuous in HOL, since syntactically distinct Noden expressions, such as ?(bool,bool) and (?bool,?bool), can have a common HOL representation. Second, we represent Noden types and Noden values by elements of defined concrete HOL datatypes, and use HOL functions to represent operators, functions and macros, thus avoiding the need to define notions of function application, abstraction, term substitution, etc.

We define a hierarchy of HOL types to represent classes of Noden types, as illustrated in the following grammar. This hierarchy allows operations to be defined over classes of types. The type UDty includes wire and enumerated classes of types. The built-in type bool is included in the wire class. The next level, Gty, includes the infinite _integer type as well as UDty types. The top level, Nty, includes a representation for the built-in _boolean type (B_ty), and the class of structure types (S_ty), in addition to Gty types.

```
UDty ::= Wire string | Enum string
Gty  ::= G_integer | G_UD UD_ty
Nty  ::= G_ty Gty | B_ty | S_ty ((Nty)list)
```

The representation of built-in types as values of HOL type Nty is as follows: bool as G_ty(UD_ty(Wire "bool")), _boolean as B_ty, and _integer as G_ty G_integer. The structure-forming operation is S_ty, which takes a list argument, and is used for both vector and structure types. Vectors are simply structures in which the types of components are uniform.

Noden values represent members of defined Noden types, including illegal and don't care values for all types (excepting _boolean). The following grammar for Noden values has distinct constant constructors for _integer (N_iconst), _boolean (N_bconst), and all other nonstructure constants (N_const). Illegal and don't care value constructors take Gty arguments. Thus it excludes _boolean, and identifies illegal and don't care values of structure types with the structures consisting entirely of illegal and don't care values respectively.

```
Nvl ::= N_const UDty string
      | N_iconst integer
      | N_bconst bool
      | N_illeg Gty
      | N_dcare Gty
      | N_struct (Nvl)list
```

The function Nty_of, which relates Nvl values to their associated Nty's, is defined recursively.

```
Nty_of_DEF =
  ⊢ (∀ s u. Nty_of (N_const u s) = G_ty (G_UD u)) ∧
    (∀ i.   Nty_of (N_iconst i)  = G_ty G_integer) ∧
    (∀ b.   Nty_of (N_bconst b)  = B_ty) ∧
    (∀ g.   Nty_of (N_illeg g)   = G_ty g) ∧
    (∀ g.   Nty_of (N_dcare g)   = G_ty g) ∧
    (∀ nl.  Nty_of (N_struct nl) = S_ty (MAP Nty_of nl))
```

Noden's view of constants as sequences of values over time leads to the use of "raised" constant values: functions from a time domain to Nvl. The function Raise_vl:Nvl→(time→Nvl) is defined recursively over the Nvl datatype. The definition of the temporal sampling operator _@ follows; note that it always returns a raised value.

```
Raise_vl_DEF =
  ⊢ (∀ s u. Raise_vl (N_const u s) = (λ t. N_const u s)) ∧
    (∀ i.   Raise_vl (N_iconst i)  = (λ t. N_iconst i)) ∧
    (∀ b.   Raise_vl (N_bconst b)  = (λ t. N_bconst b)) ∧
    (∀ g.   Raise_vl (N_illeg g)   = (λ t. N_illeg g)) ∧
    (∀ g.   Raise_vl (N_dcare g)   = (λ t. N_dcare g)) ∧
    (∀ vl.  Raise_vl (N_struct vl) =
                   (λ t. N_struct (MAP (λ v. Raise_vl v t) vl)))

at_DEF = ⊢ ∀ (rv:^Rvl) (t0:^time). rv _@ t0 = (λ t:^time. rv t0)
```

"Raised" structure values pass the sampling time to each element in the structure. We abbreviate the raised type as Rvl. We have defined a suite of functions to transform functions in the Nvl domain to functions in the Rvl domain, typified by Raise_fn2.

```
Raise_fn2_DEF =
  ⊢ ∀ fn. Raise_fn2 (fn : Nvl -> Nvl -> Nvl) =
           λ(v1:^Rvl) (v2:^Rvl) (t:^time). fn (v1 t) (v2 t)
```

This approach allows us to define most operators in the simpler Nvl domain, and produce a raised version afterwards.

Functions are defined for all operators including IF, CASE and LOOP, functions and macros. Preconditions on the Nty of arguments are included where required. Illegal and don't care values must also be considered, such as in the N_IF function.

```
N_IF_DEF =
  ⊢ ∀ nty:Nty.
      ∀ a::N_bool_or_boolean.
       ∀ b c::(N_ty nty).
       N_IF a b c =
          (((a = N_t) ∨ (a = N_true))  => b
         | (((a = N_f) ∨ (a = N_false)) => c
         | (N_mk_illeg_or_F (Nty_of b))))

  ⊢ R_IF_DEF = Raise_fn3 N_IF
```

Predicates used in the restricted quantifier syntax (eg. ∀ b c::(N_ty nty)) hold when the variables have the specified Nty. Note that IF can have either bool or _boolean conditions, and the types of the other arguments must match. If nty is _boolean, then the returned value is _false rather than an illegal value. The "raised" version of the function is also defined[4].

Every built-in operator, function, and macro has been represented. Those with a Noden HDL definition are translated directly, with only minor syntactic adjustments. Macro definitions which use 1 as the base case have a definition clause for the 0 case in addition, returning an illegal value of the appropriate type, to ensure that the typing of all expressions are provably correct. Otherwise, the expression in the definition would have an undefined recursive call in the base case of the definition.

---

[4] The naming scheme for HOL constants prefixes alphabetic Noden names with R_ for "raised" versions and N_ for non-raised. Symbolic constants are prefixed with "_" for the raised version, and if necessary to differentiate from an existing HOL constant, both versions have "_" suffixed.

```
N_AND_N_DEF =
  ⊢ (∀ a::(N_ty (0 Nty_vec Nty_bool)). N_AND_N 0 a = N_ibool) ∧
    (∀ n.
      ∀ a::(N_ty (SUC n Nty_vec Nty_bool)).
        N_AND_N (SUC n) a =
        N_IF (N_bconst (SUC n = 1)) (a N_EL 1)
          (a N_EL 1 N_AND N_AND_N n (a N_SLICE (2,SUC n))))
```

A scheme for representing user-defined functions follows the work of the HOL/Ella project, but is beyond the scope of this report.

To summarise, we have a representation for all classes of Noden expressions. Variables are represented by HOL variables of HOL type Rvl, and constants by HOL terms also of HOL type Rvl. IF, CASE, and LOOP expressions are represented by an application of the respective operator applied to representations of the arguments. Similarly, function applications are represented by applications of the function representation to the representation of the arguments.

4 Embedding Noden: the Logic

Interpreting the logic requires representing conjectures, theorems and axioms, and modelling all proof operations. Our pragmatic approach seeks to identify Noden theorems with HOL theorems, and represent proof operations as HOL inference rules. Every introduced axiom, such as those in case studies with the Noden prover, should be derivable as a theorem in our model, and validity of Noden proofs must be preserved. We present a translation scheme for conjectures to HOL sequents which retains the Noden syntax remarkably well.

We wish to represent conjectures as HOL sequents (with an empty assumption list). However, conjectures are _boolean-valued expressions, so must be converted to a HOL :bool expression. This is done by equating the term to the _boolean value _true. Quantifiers are represented by HOL quantifiers, and typing information is represented by type predicates using the restricted quantifier syntax. Converting the type of expressions is uniformly handled by two introduced constants, |= and WHERE, defined as follows.

```
TS_DEF = ⊢ ∀ a. |= a = (a = R_true)

WHERE_DEF = ⊢ ∀ a b. WHERE a b = ((a = R_true) ==> b)
```

Thus the translation is exceedingly regular, merely substituting the representing HOL term for each operator, using HOL quantifiers, and representing types as predicates in the restricted quantifier syntax. The translations of two sample conjectures from page 5 follow (empty assumption lists are omitted).

```
∀ ty:Nty. ∀ a b::(R_ty ty).  (|= ((a _== b) _=_ (b _== a)))

∀ x y::R_bool. WHERE(R_LIFT x _∧_ R_LIFT y)
               (|= (R_LIFT(x R_AND y) _∧_ R_LIFT(x R_OR y)))
```

The final and most vital part of this work is the representation of the logic's proof functions in HOL. One immediate difference between the approach to proof in HOL and Noden is the use of the same proof operations for both tactical (backward) and "correct-by-construction" (forward) proofs in Noden. In HOL there are clear differences between the management of these two different proof methods. However, by representing proof operations as conversions, we may apply them to either style of HOL proof.

Type checking must be performed explicitly in all proofs. A pair of conversions have been written. The first can determine the Nty of most well-typed expressions, and the second can prove that type constraints hold for most such expressions. These conversions use a list of records which hold information (name and argument types) and theorems about every operator and constant defined in the model. Two theorems are included in each record: the definition of the constant and a theorem giving the type returned by an application of the constant, under the same type restrictions on arguments as in the definition. New operators can also be handled by adding the appropriate records, including the theorem about the type of value returned. The same data structure will be used in a conversion to represent the **Open** proof operation, which rewrites the definition of known constants.

The **Use Predicate** command is used to substitute a value in the hypothesis or another predicate based on some predicate, which may be either an equation or other _boolean value. First we note the following theorem, which permits us to derive the HOL equality from the form of Noden equality found in conjectures.[5]

```
val R_EQ_equals =
 ⊢ ∀ nty. ∀ a b::(R_ty nty). ((a _=_ b) = R_true) = (a = b)
```

The conversion ANTE_SUBST_CONV effects an approximation of the **Use Predicate** command, using the antecedent of an implication to substitute for a term in the consequent. It permits rewriting a term as T, or substituting for the lhs or rhs of the predicate equation. Rather than effecting a single substitution it presently substitutes for all instances of the term. This can readily be modified to capture, using the above theorem as well, the behaviour of **Use Predicate**.

The **Push time** operation is supported by a set of theorems which justify moving the the temporal sampling function inwards to the arguments of a function.[6]

---

[5] It should be noted that a similar theorem applies for all the _boolean operators defined using HOL :bool operations.

[6] Some operators such as CASE have a different argument structure, hence are treated individually in deriving a result of this sort.

```
val at_1_thm = ⊢ ∀ N a t0. (Raise_fn N a) _@ t0 = Raise_fn N (a _@ t0)

val at_2_thm =
  ⊢ ∀ N a b t0.
  (Raise_fn2 N a b) _@ t0 = Raise_fn2 N (a _@ t0) (b _@ t0)

val at_3_thm =
  ⊢ ∀ N a b c t0.
  (Raise_fn3 N a b c) _@ t0 = Raise_fn3 N (a _@ t0) (b _@ t0) (c _@ t0)
```

The representation of the remaining proof operations is incomplete at the conclusion of the project. The representation of each operation was reviewed, considering specifically its validity for both forward and backward proofs, and its behaviour in the presence of universal quantifiers. Equational transformations were considered which can produce a standard form of term in many cases, to which HOL primitive inference rules and tactics can be applied. These transformations, for example specialising and generalising variables, and discharghing assumptions and undischarging antecedents, are not expressible in Noden but can be done in the HOL logic. The presence of universal quantifiers will interfere with such manipulations, since the introduction and elimination rules for existential quantifiers are not symmetric. This analysis also noted that the model requires an extension to express the context of defined types, in order to justify case analysis over values of such types.

5 Results

We have created an interpretation of the Noden logic which models all operators of the HDL and logic, expresses conjectures and theorems in a syntax very close to the original Noden form, and which can be used to model the behaviour and limitations of proof operattions. This interpretation can provide for the Noden logic the sound logical basis which has been missing until now. Furthermore, it provides a context in which decisions about the design of the language can be made, and in which they must be justified.

One specific contribution of this work has been a means of proving axioms introduced by users to overcome limitations of the available proof operations. The introduction of such axioms is common in the examples of use of the prover. Providing a means of assuring that they are consistent is immediately useful.

Rules for forming proof obligations for induction proofs have been presented in an earlier project report [Gra93]. This has included a proposal to extend the Noden Logic to permit quantifiers within predicates in order to express inductive subgoals. The effects of this change will be far-reaching, as it is likely that new types of subformulae will be added to the logic, such as the negation of quantified expressions, currently prohibited by the restricted syntax of conjectures. The syntax may need to be reconsidered and rationalised. We expect this extension will make the logic more regular, treating quantified expressions as ordinary _boolean terms, to which other operators may be applied. From a purely logical viewpoint this is desirable, in the end making the structure of the logic simpler. The impact of these changes on the functioning of all proof operations needs close consideration. The model can handle such a change immediately, and can guide the revision of the Noden system.

Several implementation errors have been discovered in the current Noden system, often from attempts to precisely determine the semantics of some operator or other to correctly represent it. Examples include the reduction of terms of the form a == a to t, when the result could be !bool, and an incorrect matching test in CASE statement branches. Both errors permit inconsistent theorems (specifically ⊢ ˜ _true) to be derived. Other errors include the inability to parse well-formed terms, and failure to correctly recognise instances of vector constants as vectors.

Another error was discovered when trying to model the CONC operator. This error is more serious because the documentation concurred with the implementation, hence is not dismissable as an implementation error. An ambiguity exists when the arguments have identical vector types, as either a longer vector or a vector of length 2 of the type of the arguments can be returned. The simplifier supports both reductions, and thus permits the derivation of a theorem with a badly typed hypothesis, although a failure is trapped at the last step.

The HOL model deviates from the language definition by using :num rather than :integer values in certain critical areas, including the recursive parameter of macro functions and arguments to slice and element operators. This limits the form of recursion in macros (which need not terminate in Noden), and it impacts arithmetic reductions, prohibiting negative values in places where their meaning is questionable. Noden does accept and reason about negative vector sizes, though the intended interpretation is unclear. This has raised a major area of concern about the semantic definition of Noden.

Omissions rather than errors often limit what can be proved by Noden. Aside from the (temporary) absence of induction proof, many aspects of the informally defined semantics of the HDL cannot be reasoned about in Noden without introduced axioms. These include such fundamental cases as equality between structure objects and reducing conditional expressions with illegal condition values. This is partly attributable to the developmental status of Noden, but it reinforces the need for a full specification of the system. Additional proof operations have been recommended.

6 Conclusions

The use of the HOL system to define an interpretation of the Noden logic has shown that a high security system such as HOL can be used to corroborate the consistency of the logic of a less secure system. Although this work is incomplete, and a definitive answer about changes needed for Noden to be consistent is not yet possible, the feasibility of the method has been shown.

The best result this work could hope to achieve is a complete model of the Noden system with full functionality of proof operations. Not only would this strongly support the consistency of the Noden logic, but it would also provide a formal specification for the implementation of the prover. This latter point is not insignificant. Such a specification could guide the design of the implementation, and hopefully result in the elimination of the sort of errors discovered in the course of this work. It also serves to distill the relevant factors in deciding the semantics of the HDL. With the interpretation as defined, there should be no need for unproved axioms, as seems common in the use of Noden. Every introduced axiom should be derivable by formal proof in this model.

The approach taken in this work is a minimal embedding of the logic within HOL. Deeper embeddings are also possible, where for example, datatypes can be defined to represent theorems and inference rules can be defined inductively over terms and theorems, such as in [vW93]. The limited objectives of, and time available for, this work meant that embedding the Noden HDL in HOL and using the HOL logic infrastructure for modelling the Noden logic met the requirements. This pragmatic approach lends itself to experimentation with added proof operations and other changes to the logic, without costly reimplementation for each change.

By creating a formal model for the Noden logic, we have provided a reference specification for the implementors of the Noden system, against which the implementation may be appraised. Prototyping complex proof tools, and indeed entire logics, in a tool with an expressive logic and large theory base such as HOL, may be a practical and efficient methodology, with a high assurance that the end product will be correct. The authors of Noden [Pyg94] see a significant role for the Noden approach of application domain specific tools, provided they can be supported by more formal models such as outlined herein. The same arguments which support the use of proof tools to design complex circuits surely apply even more strongly to the design of the complex proof tools themselves.

There will doubtless be some adjustment to the Noden HDL and logic in the future. Some of these changes will be in response to issues raised here, and others will be from attempting to include more features of the Ella language. The latter changes could have a great impact on the soundness of the interpretation, and will determine whether continued work on this particular model will be useful. It is our most strongly held opinion that consistency of the Noden Logic cannot be assured without the definition of an interpretation in a well understood logic. Without such assurance of consistency, the system can never hope to be considered trustworthy.

7   Acknowledgements

This research is supported by DRA Malvern under research Agreement No 2029/282. The author would like to thank Paul Curzon, Mike Gordon, Malcolm Newey, Wai Wong, and other members of the Automated Research Group at the University of Cambridge Computer Laboratory, David Tombs and Clive

Pygott of DRA Malvern, and Konrad Slind for their help and advice. The work was carried out using the HOL90 system, and uses the res_quan, integer, string, and arith libraries, as well as the mutrec contrib package.

References

[And86] P.B. Andrews. *An Introduction to Mathematical Logic and Type Theory: to Truth through Proof.* Computer Science and Applied Mathematics Series. Academic Press, 1986.

[Bou92] Richard J. Boulton. A HOL Semantics for a Subset of ELLA. Technical Report No. 254, University of Cambridge, Computer Laboratory, 1992.

[Far93] William M. Farmer. Theory interpretation in simple type theory. In Karl Meinke, editor, *Proceedings of the International Workshop on Higher Order Algebra, Logic and Term Rewriting*, 1993.

[Far94] William M. Farmer, March 1994. Personal communication.

[GM93] M. J. C. Gordon and T. F. Melham. *Introduction to HOL: A theorem proving environment for higher order logic.* Cambridge University Press, 1993.

[Gra93] B. Graham. Soundness of theorem proving methods for hardware verification Report 1. Unpublished progress report, December 1993.

[Har93] John Harrison. Hardware Verification with the HOL-ELLA system. Unpublished draft report on the HOL-ELLA system, 1993.

[Pit93] A.M. Pitts. The HOL logic. In M.J.C. Gordon and T.F. Melham, editors, *Introduction to HOL: A theorem proving environment for higher order logic*, chapter 15–16, pages 191–232. Cambridge University Press, 1993.

[Pyg92] Clive Pygott. A prover for NODEN. Technical report, CSE3, DRA Malvern, August 1992.

[Pyg93] Clive Pygott. A window based prover for a hadware description language. Mod Report Number DRA/CIS/CSE3/B36AK/93002/1.0, CSE3, DRA Malvern, November 1993.

[Pyg94] Clive Pygott. Final report for "formal mehtods for hardware design". Mod Report Number DRA/CIS/CSE3/B36AK/94001/1.0, CSE3, DRA Malvern, April 1994.

[vW93] J. von Wright. Representing higher-order logic proofs in hol. Draft report, September 1993.

Reasoning about Real Circuits

Keith Hanna

University of Kent, Canterbury, Kent, CT2 7NT, UK

Abstract. This paper describes how to specify and reason about the properties of non-ideal logical circuitry in the analogue domain. Device behaviours are characterised by predicates over the voltage and current waveforms present at their ports. In many cases it suffices to formulate these predicates simply in terms of linear inequalities. The behavioural predicate for an overall circuit is obtained by taking the conjunction of the propositions satisfied by the individual components along with the constraints imposed by Kirchhoff's current law. As an illustration, the verification of a circuit for a TTL NOT gate is outlined.

1 Introduction

Digital circuitry comes in two forms: *ideal* and *non-ideal*. In the case of ideal digital circuitry, devices have a well-defined sense of directionality (they have inputs and outputs) and the overall circuit conforms to a given set of design rules. These rules require, for example, that each input is connected to exactly one output, and that each output does not drive more than a certain number of inputs. The behaviour of such idealised logic circuitry can be understood purely in terms of the electrical potentials appearing at its nodes and there is a direct and simple correspondence between the form of the circuit and the syntactic form of the predicate that describes its behaviour.

In practice, however, most digital circuitry does *not* conform to the above ideal. In particular, outputs may be connected together (as they are in *tristate* logic) and some or all of the devices may be adirectional (as, for example, with *pass-transistor* logic). By way of illustration, a typical instance of non-ideal digital circuitry is shown in Fig 1; it is a well-known CMOS implementation [WE85] of an EXCLUSIVE-OR gate making use of pass-transistor circuitry.

In contrast to the case with ideal logic, the behaviour of non-ideal circuitry cannot be understood purely in terms of potentials. Instead, it is necessary to take account of both potentials *and* currents. This consideration means that the principles of specification and formal verification for non-ideal logic are a superset of those for ideal logic.

1.1 Approach

This paper describes an approach to specifying and reasoning about non-ideal digital circuitry within a formal framework. It differs from earlier papers that have addressed this general theme in that it starts from a formalisation of the

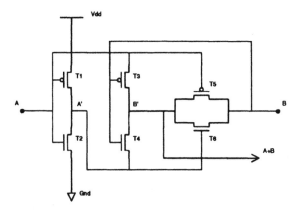

Fig. 1. Circuit diagram of an EXCLUSIVE-OR gate. It operates in the following way:
Case (1) When signal A is high, A' is low. Transistor pair T3 and T4 thus acts
as an inverter, with B' appearing at the output. The transmission gate formed by
transistor pair T5 and T6 is open. **Case (2)** When signal A is low, A' is high. The
transmission gate formed by T5 and T6 is now closed, passing B to the output. The
inverter formed by T3 and T4 is disabled.

laws that pertain to analogue electronics and it treats digital devices as being
analogue in nature. Thus, subject only to the laws adopted for describing analogue electronics offering an accurate description of physical reality, any conclusions drawn about digital devices are necessarily sound. (Alternative approaches,
which begin by formalising the informal, quasi-digital rules used intuitively by
design engineers, can be adequate in certain circumstances but, in many others,
they can all too easily lead to false conclusions.)

In the following sections, we shall be characterising the behaviour of devices
(resistors, diodes, transistors, logic gates, etc.) in terms of the relations they imply between the potentials and currents at their ports and we shall be reasoning
about the behaviour of interconnected devices by using Kirchhoff's Laws. Since
our motivation is primarily to reason about the behaviours of *digital* circuitry,
we shall not, in general, be seeking to use the strongest possible predicates to describe device behaviour, but rather to use weaker ones which, whilst adequately
strong to capture the essential digital properties of a device's behaviour, are
considerably simpler to manipulate.

2 Behavioural Specifications

Our starting point is that, in the analogue electronics domain, a *behavioural
specification* for a device is a relation over both the voltage *and* the current
waveforms at the ports of the device that describes the way that these waveforms
are known to be, or required to be, related. A behavioural specification is thus

an invariant that characterises the useful properties of the device — properties that hold of the signals at its ports under any circumstances.

Treating behavioural specifications as relations, rather than as functions, has several advantages in the present context, including:

- It avoids any implication of causality. Causality is always a slippery concept; at the analogue level it is best avoided completely: potential differences do not "cause" currents to flow any more than currents "cause" potential differences to arise. In particular, notions of *input* and *output* are seldom valid concepts at the analogue level of abstraction.
- Using predicates allows one *to place bounds* on the characteristics of devices whereas the use of functions would require one to describe them *exactly*. Given that the real world is indeterminate and our knowledge of it is only partial, any attempt at an *exact* description of a device's behaviour *would necessarily be a wrong one.*

Steady-state behaviour In general, behavioural specifications are relations over the *waveforms* at a device's ports and thus they characterise its dynamic behaviour. Sometimes, however, we may wish to focus attention on only the *steady-state behaviour* of a device. In such a case, a behavioural specification will then be a relation on the steady-state (that is, constant) potentials and currents at its ports. We caution, though, that a given device may, or may not, exhibit a well-defined steady-state behaviour and so this concept has to be deployed with great care.

Conventions Given an *n*-ported component (see Fig. 2(a)), there are two conventions we shall use for describing the signals at its ports; we name these the *symmetric* and the *asymmetric* conventions. Both are useful.

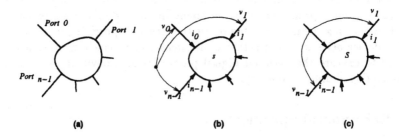

(a)   (b)   (c)

Fig. 2. (a) An *n*-ported component; (b) Voltages and currents in accordance with the symmetric convention; (c) Voltages and currents in accordance with the asymmetric convention

Symmetric convention With the *symmetric* convention (see Fig. 2(b)), we treat all ports of a component in the same way and thus a behavioural specification, *spec*, for an *n*-ported component is a predicate over the *n*-tuple of electrical signals (that is, the voltage-current pairs) at the device's ports. This convention finds most use in formulating generic results — the symmetric treatment of all ports tends to lead to simpler, more concise theorems. It does, however, mean that there is a two-fold redundancy in specifications since the choice of the voltage reference is arbitrary, and since the sum of the currents into a device is known to be zero.

Asymmetric convention The *asymmetric* convention (see Fig. 2(c)) eliminates this redundancy by identifying one of the ports as a *reference port*. With this convention, a behavioural specification, *Spec*, for an *n*-ported component is a predicate over the $(n-1)$-tuple of electrical signals of the remaining ports. The asymmetric convention finds most use for describing the characteristics of specific devices — especially of 2-port ones.

Notice that, in either the symmetric or the asymmetric convention, the reference orientation adopted for describing currents is always *into* the device.

2.1 Some examples: idealised components

We have emphasised throughout that, although our motivation is primarily directed towards specifications of digital components, the methods we use are applicable to analogue components in general. By way of illustration, we provide specifications for two typical, analogue components. (We note that, in contrast to the remaining specifications in this paper, these specifications describe the behaviour of idealised, perfectly-known components).

The first example is of the steady-state behaviour of an ideal resistor. We *assume* — without proof — that this component does indeed have a steady-state behaviour. The specification is thus over the tuple of steady-state potentials and currents that exist at the ports of the device. Using the symmetric convention, the specification (which is parametrized by the resistance, r) is

$$resistor_s \ r \ ((v_0, i_0), (v_1, i_1)) \triangleq \\ ((v_0 - v_1) = i_0 * r) \wedge (i_0 + i_1 = 0))$$

Alternatively, using the (redundancy-free) asymmetric convention, it is

$$resistor_a \ r \ (v, i) \triangleq (v = i * r)$$

The second example is of the dynamic behaviour of an ideal transformer. This time, the specification is over the tuple of potential and current *waveforms* at the ports of the device. In order to express the specification, it is assumed that the waveforms are differentiable functions and that a combinator, *diff*, is available that yields the derivative of a function. (The presentation of the Reals described in [Har92] would provide a suitable theory in which to reason about such specifications). The specification is parametrized by the primary, secondary and mutual inductances of the transformer. Using the symmetric convention (with the potentials of the primary and secondary ports being denoted by v_p, v_p', v_s and v_s'), the specification is

$$trans\ (L_p, L_s, M)\ ((v_p, i_p), (v_p', i_p'), (v_s, i_s), (v_s', i_s')) \triangleq$$
$$\forall t : time.\ ($$
$$(v_p\ t - v_p'\ t) = L_p * (diff\ i_p)\ t + M * (diff\ i_s)\ t\ \ \wedge$$
$$(v_s\ t - v_s'\ t) = L_s * (diff\ i_s)\ t + M * (diff\ i_p)\ t\ \ \wedge$$
$$(i_p\ t + i_p'\ t = 0) \wedge (i_s\ t + i_s'\ t) = 0))$$

2.2 Rectilinear specifications

In general, a behavioural specification is an entirely arbitrary predicate on the tuple of potentials and currents present at the ports of a device. Experience shows, however, that allowing complete generality results in specifications that can be difficult to validate (that is, to confirm that the specification is, in fact, expressing the *intended* property) and quite impossible to manipulate automatically.

Often, therefore, it is preferable to limit attention to specifications that are structured in particular ways — provided, of course, that the desired properties can still be adequately captured. We have found that, for describing and reasoning about the behaviour of digital components, a restricted class of specification that we term *rectilinear* specifications suffices for most purposes. By using rectilinear specifications one achieves a disciplined approach to the development and validation of specifications.

A *rectilinear* behavioural specification, *spec* $((v_0, i_0),\ \dots, (v_{n-1}, i_{n-1}))$, is a predicate made up from the conjunction and disjunction of simple inequalities on the individual voltage and current parameters. We call it 'rectilinear' since its graph consists of regions bounded by hyperplanes parallel to the axes. We note that many operations on specifications preserve the property of rectilinearity (that is, given rectilinear specifications they yield rectilinear ones).

2.3 Examples of rectilinear specifications

Specification of a Diode Components used in logic circuits are often highly non-linear. For example, a junction diode (Fig. 3(a)) is described by the *ideal diode equation*[HJ83]:

$$I = I_s(e^{V/V_T} - 1.0)$$

In Fig. 3(b) a fairly strong specification for the behaviour of the diode is shown; this specification conveys much of the information that a designer of analogue electronic circuits would require to know about the device. On the other hand, Fig. 3(c) shows a weaker, but rectilinear, version of the same specification. We have found, in exploring the analogue behaviour of digital circuitry, that this second form, evidently much easier to describe and to reason about, is still adequately strong to infer the properties of interest to the digital designer.

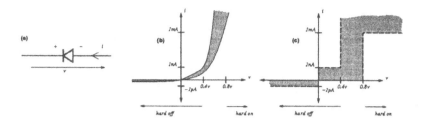

Fig. 3. (a) A diode. (b) A strong specification. (c) A weaker, rectilinear specification.

Specification of a Power Supply As a second example of a rectilinear specification, consider a power supply (Fig. 4(a)) having the following characteristic: it is guaranteed to maintain an output potential of 5V ± 0.5V provided that a current of not more than 2A is drawn or a current of not more than 1A is supplied. This specification can be described by the predicate shown in Fig. 4(b). This asserts:

- The only circumstance under which the potential will be less than 4.5 V is when there is a current of at least 2 A flowing.
- The only circumstances under which the potential will be greater than 5.5 V is when there is a current of at least 1 A flowing with reverse polarity.

As we shall see, specifications of this general form tend to occur frequently in many contexts; they can be thought of as a generalised form of a *Thevenin* voltage source.

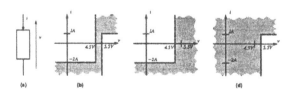

Fig. 4. (a) A power supply; (b) A rectilinear specification of a power supply; (c) and (d) Components of the *conjunctive normal form* of (b).

2.4 Standard Forms

There are a variety of standard forms in which specifications can be expressed. Adoption of one or other of these forms is often advantageous in allowing the systematic development of a specification — a benefit not to be underestimated for some of the more complex specifications, especially those of devices with more than two ports.

Conjunctive Implicative Form A standard form that is particularly useful is one that we term the *conjunctive implicative form* (CIF); this takes the form of a conjunction of a set of implications and is similar, in many respects, to a **case** expression.

A good example of the use of the CIF is provided by a specification for a junction transistor. In linear circuits, junction transistors tend to be operated only in the so-called *forward active mode* (in which their base-emitter junctions are forward biassed and their base-collector junctions are reverse biassed). In digital circuits, however, junction transistors are used in all four possible bias modes [HJ83]. The CIF provides a natural way to structure a suitable specification. One can associate one implication with each of the bias modes, with the antecedent of each implication delimiting an appropriate region of parameter space and the consequent describing the device's behaviour in that region. Thus, the specification (a predicate over the tuple of voltages and currents at the emitter, base and collector ports of the transistor) can take the form:

$$
\begin{aligned}
&junction\_transistor\_spec\ ((v_e, i_e),\ (v_b, i_b),\ (v_c, i_c)) \stackrel{\wedge}{=} \\
&\quad forward\_active\_mode\ (v_e, v_b, v_c) \Rightarrow \\
&\qquad forward\_active\_spec\ ((v_e, i_e),\ (v_b, i_b),\ (v_c, i_c)) \qquad \wedge \\
&\quad reverse\_active\_mode\ (v_e, v_b, v_c) \Rightarrow \\
&\qquad reverse\_active\_spec\ ((v_e, i_e),\ (v_b, i_b),\ (v_c, i_c)) \qquad \wedge \\
&\quad cutoff\_mode\ (v_e, v_b, v_c) \Rightarrow \\
&\qquad cutoff\_spec\ ((v_e, i_e),\ (v_b, i_b),\ (v_c, i_c)) \qquad \wedge \\
&\quad saturation\_mode\ (v_e, v_b, v_c) \Rightarrow \\
&\qquad saturation\_spec\ ((v_e, i_e),\ (v_b, i_b),\ (v_c, i_c))
\end{aligned}
$$

One particular advantage with using the CIF for a specification is that omitting (either by intent or accidentally) one or more of the conjuncts is safe — it merely weakens the specification; it does not render it invalid.

Normal Forms For some purposes it is useful to express a specification in a *normal* (or *canonical*) form. One such form is the *conjunctive normal form* (CNF). A behavioural specification in CNF consists of the conjunction of a series of disjunctions. That is, it is of the form

$$ spec\ ((v_0, i_0),\ \dots ,\ (v_{n-1}, i_{n-1})) \stackrel{\wedge}{=} P_1 \wedge \cdots \wedge P_m $$

where each P_i is of the form $Q_{i,1} \vee \cdots \vee Q_{i,\ell_i}$. If the original specification is a rectilinear one, then each of the disjunctions will consist of a simple inequality.

As an example, the power supply specification shown in Fig. 4(b) may be split up into the two components shown in Figs. 4(c) and (d).

3 Operations on specifications

We now look at the definition of some typical operations on specifications. Some of these operations are relevant to specifications of any kind. For example, we have the usual lattice operators $\sqcap$ and $\sqcup$, defined by

$$p \sqcup q \triangleq \lambda x.\, p\ x \wedge q\ x \qquad \text{and} \qquad p \sqcap q \triangleq \lambda x.\, p\ x \vee q\ x$$

and we have the *unrealisable* specification, $\top$, and the *don't care* specification, $\bot$, defined by

$$\top \triangleq \lambda x.\, false \qquad \text{and} \qquad \bot \triangleq \lambda x.\, true$$

and we have an operator, $\sqsupseteq$, for comparing the relative strengths of two specifications, defined by

$$p \sqsupseteq q \triangleq \forall x.\, p\ x \Rightarrow q\ x$$

Other operators are specific to analogue circuits. Most of these operators are applicable to specifications for n-ported devices; here, for notational simplicity, we present the definitions as they relate to 2-ported devices defined using the *asymmetrical* convention.

Open and Closed Circuits We have the specifications for an open circuit and for its dual, a closed (or 'short') circuit:

$$open\ (v, i) \triangleq (i = 0) \qquad \text{and} \qquad closed\ (v, i) \triangleq (v = 0)$$

Serial and Parallel Composition We have an operator, $+\!\!+$, for forming the *serial composition* of two specifications:

$$(p +\!\!+ q)\ (v, i) \triangleq$$
$$\exists v_1, v_2 \colon voltage.$$
$$\quad p\ (v_1, i) \quad \wedge \quad q\ (v_2, i) \quad \wedge \quad (v_1 + v_2 = v)$$

and an operator, $\|$, for the dual notion, the *parallel composition* of two specifications:

$$(p \| q)\ (v, i) \triangleq$$
$$\exists i_1, i_2 \colon current.$$
$$\quad p\ (v, i_1) \quad \wedge \quad q\ (v, i_2) \quad \wedge \quad (i_1 + i_2 = i)$$

These operations are related by a number of laws. For example, parallel composition is commutative and associative and has *closed* as a zero and *open* as a unit:

$$p \| q = q \| p \qquad \qquad \text{and} \qquad p \| closed = closed$$
$$p \| (q \| r) = (p \| q) \| r \qquad \text{and} \qquad p \| open = p$$

Serial composition obeys a similar (dual) set of laws.

Example As an illustration, consider the parallel composition of:

- A specification, *ps*, (see Fig. 5(a)) for a 5.0*V*, 2.0*A* power supply having a tolerance of ±0.5*V* and a back-current limit of 1.0*A*.
- A specification, *res*, (see Fig. 5(b)) for a 3.0Ω resistor having a tolerance of ±1.0Ω. (Note that this specification is *not* a rectilinear one.)

The parallel composition, *ps* ‖ *res*, of these two specifications is as shown in Fig. 5(c). One can imagine is as being constructed from the graphs for *ps* and *res* by the pointwise addition of the currents.

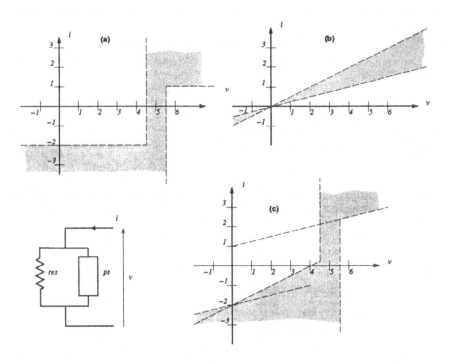

Fig. 5. Graphs of the specifications for: (a) a power supply, (b) a resistor, and (c) the parallel composition of these two specifications.

3.1 Relative Complement

A common task when undertaking goal-directed formal verification (as described later) is of the form:

> Given two specifications, *p* and *q*, determine the weakest specification, *r*, such that the parallel composition of *p* and *r* is stronger than *q*.

Expressed more formally, the task is: Given specifications p and q, determine the weakest specification r such that $(p \,\|\, r) \sqsupseteq q$.

We call the weakest relation r which satisfies $(p \,\|\, r) \sqsupseteq q$ the *relative complement for parallel composition of q with respect to p* and denote it by $q \,/_\| \, p$.

Definition of the relative complement We can infer a definition for the relative complement, $q \,/_\| \, p$, by working backwards. Given (see Fig. 6) relations p and q, we start off with the desired relation between them and their relative complement (which we shall call r). Thus, $(p \,\|\, r) \sqsupseteq q$. Expanding the two

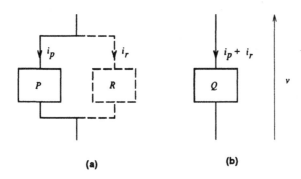

(a)   (b)

Fig. 6. The relative complement for parallel composition.

operators in this expression gives

$$\forall v; \ i_p, i_r.$$
$$p\,(v, i_p) \land r\,(v, i_r) \Rightarrow q\,(v, (i_p + i_r))$$

Rearranging (by currying the propositional form and moving one of the quantifiers inwards) gives

$$\forall v; \ i_r.$$
$$r(v, i_r) \Rightarrow (\forall i_p. \ p\,(v, i_p) \Rightarrow q\,(v, (i_p + i_r)))$$

Further rearrangement (by eta-conversion of the RHS) gives

$$r \sqsupseteq (\ \lambda v; \ i_r.$$
$$\forall i_p. \ p\,(v, i_p) \Rightarrow q\,(v, (i_p + i_r)))$$

Evidently, the weakest r that satisfies this proposition is obtained when the inequality is taken as an equality. Hence we arrive at the following definition:

$$q \,/_\| \, p \ \hat{=} (\ \lambda v; \ i_r.$$
$$\forall i_p. \ p\,(v, i_p) \Rightarrow q\,(v, (i_p + i_r)))$$

A similar (dual) definition can be formulated for $p \,/_{+\!\!+} \, q$, the *relative complement for serial composition*.

Example Here is a simple example illustrating the notion of the relative complement (for parallel composition). So as to allow the specifications to be shown graphically, the example is based on specifications of 2-ported devices and all specifications are rectilinear. The task is:-

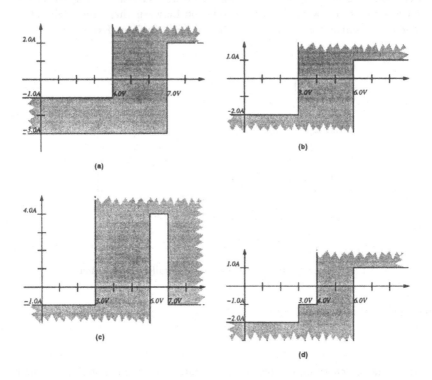

Fig. 7. Example on relative complements. (a) A specification p. (b) A specification q. (c) The specification r, defined as $p /_{\parallel} q$, the relative complement of q with respect to p. (d) $p \parallel r$, the parallel composition of p and r.

Assume that we have a power supply that satisfies a specification p and we wish to connect a component in parallel with it so as to realise a power supply satisfying a specification q. Determine the weakest specification, r, for that component.

- The specification p, shown in Fig. 7(a), is for a power supply with an output voltage in the range 4 to $7V$ (provided that a current of not more than $1A$ is drawn or a current of not more than $2A$ is supplied) and a guaranteed current limit of $3A$.

- The specification q, shown Fig. 7(b), is, on the other hand, for a power supply with an output voltage in the range 3 to $6V$ (provided that a current of not more than $2A$ is drawn or one of not more than $1A$ is supplied).

Evidently, the required specification, r, is the relative complement of q with respect to q, that is $r \triangleq q /_{\parallel} p$. It can be computed from the definition given above for the relative complement or, in a simple case like this, it can be inferred from first principles. The graph of r is shown in Fig. 7(c) and the graph for the resulting specification, $p \parallel r$, the parallel composition of p and r, is shown in Fig. 7(d). Notice that, as required, we have $(p \parallel r) \sqsupseteq q$.

Intuitively, the specification r, can be explained as follows:

- For voltages of less than $3V$, r contributes at least $1A$. This, in parallel with a similar contribution from specification p, meets the requirement (for a current of at least $2A$) imposed by specification q.
- For voltages of between $3V$ and $6V$ it is, like the specification q, indifferent to current flows.
- For voltages of between $6V$ and $7V$, it guarantees to consume a current of at least $4A$. Given that, in the same voltage range, p guarantees not to supply a current of more than $3A$ this means that the net overvoltage current will, as Q requires, be greater than $1A$.
- For voltages greater than $7V$, r can (since the specification q is *weaker* than the specification p in this region) even allow a current of up to $1A$ to be *supplied*.

4 Reasoning about Circuits

We now examine how one can reason about the properties of circuits. Circuits are, in general, hierarchically structured. Every circuit has a set of *ports* by which it can be connected to other circuits. A *primitive circuit* is termed a *component* and has no internal structure. A *compound circuit* (see Fig. 8(a)) has an internal structure consisting of a set of *nodes*, a set of *components* and a *wiring* (a function that describes how the components are connected to the nodes). Each node of the circuit has a port (of the overall circuit) associated with it.

4.1 Behavioural Extraction

Given a circuit, we can associate both a potential and a current with each of its ports and with each of its components' ports. We can assert that these potentials and currents are constrained by

- The behavioural predicates associated with each component; and
- The requirement that the potentials of interconnected ports be equal; and
- Kirchhoff's current law, which asserts that the algebraic sum of the currents incident on each node be zero.

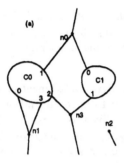 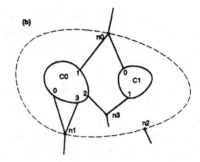

Fig. 8. (a) A circuit consisting of a set of nodes (n0, n1, n2 and n3), a set of components (C0 and C1) and a wiring.   (b) The same circuit but with node n3 hidden.

By writing down these assertions, existentially quantifying over the internal branch currents (since they are not externally visible) and abstracting on the voltages and currents at the ports of the circuit we can obtain a predicate, p, which characterises the external behaviour of the circuit. For example, for the circuit shown in Fig. 8, we have:

$$p((v_0, i_0), (v_1, i_1), (v_2, i_2), (v_3, i_3)) \triangleq$$
$$\exists i_{0,0}, i_{0,1}, i_{0,2}, i_{0,3}, i_{1,0}, i_{1,1}.$$
$$r_0((v_1, i_{0,0}), (v_0, i_{0,1}), (v_3, i_{0,2}), (v_1, i_{0,3})) \quad \wedge$$
$$r_1((v_0, i_{1,0}), (v_3, i_{1,1})) \quad \wedge$$
$$(i_0 = i_{0,1} + i_{1,0}) \wedge (i_1 = i_{0,0} + i_{0,3}) \wedge (i_2 = 0) \wedge (i_3 = i_{0,2} + i_{1,1})$$

Hiding The hiding (or internalising) of a node of a circuit involves two logically distinct operations which we term *disconnecting* the node and *disregarding* the node. The first involves asserting that there is no external flow of current through its associated port, the second involves hiding the potential of the port (by existential quantification). For example, the behavioural predicate q, for the circuit (see Fig. 8(b)) given by internalising node n3 is given by:

$$q((v_0, i_0), (v_1, i_1), (v_2, i_2)) \triangleq$$
$$\exists v_3.$$
$$\textbf{let } i_3 = 0 \textbf{ in}$$
$$p((v_0, i_0), (v_1, i_1), (v_2, i_2), (v_3, i_3))$$

4.2 Formal Verification

The principles of formal verification for non-ideal logic circuitry are virtually identical to those for ideal logic circuitry. Namely, if *circ* is a behavioural specification of a circuit (as determined above) and if *spec* is a given behavioural specification, then the circuit is said to be a *correct* implementation of the specification if the verification condition *circ* $\sqsupseteq$ *spec* holds. Equivalently stated in first-order form, this same condition is:

$$\forall \overline{(v, i)}. \ circ \ \overline{(v, i)} \Rightarrow spec \ \overline{(v, i)}$$

where, $\overline{(v, i)}$ is an informal notation for the tuple $((v_0, i_0), \ldots, (v_{n-1}, i_{n-1}))$. (In practice, it is often convenient to introduce a vector type for formulae of this kind rather than using n-tuples.)

4.3 Practical Considerations

An effective means of establishing a verification condition is to use a goal-directed approach. Assuming that the behavioural specification, *spec*, takes the form of an implication, $p \sqsupseteq q$, one attaches the predicate q to the appropriate ports of the circuit and then, using a *weakest precondition* approach, seeks to "push" this predicate back, incrementally, through the circuit. Success is achieved if, ultimately, it can be shown to follow from the predicate p. Using the weakest precondition technique involves both making use of the behavioural specifications of the components and of the *relative complement* operation described earlier.

As an illustration of this approach, we describe two typical manoeuvres that can be used to push assertions through a circuit.

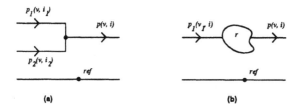

Fig. 9. Conditions and preconditions

Conductors incident on a node Consider (see Fig. 9(a)) a node at which three conductors are incident and assume that it is wished to establish an assertion $p(v, i)$ on the signals present on one of these conductors. Then it suffices to establish two further assertions, $p_1(v, i_1)$ and $p_2(v, i_2)$ on the signals present on the other two conductors, provided that

$$(p_1 \| p_2) \sqsupseteq p$$

In practice, one often uses intuition to define one of these preconditions (say p_1), in which case the weakest precondition for the other is given by the relative complement (for parallel composition) operation, viz. $p_2 \mathrel{\hat{=}} p /_{\|} p_1$.

A two-ported component Consider (see Fig. 9(b)) a two-ported component with a specification r and assume that it is wished to establish an assertion

$p(v, i)$ on the signals at one of its ports. Then it suffices to establish the condition $p_1(v_1, i)$ on the signals at its other port provided that

$$(r + p_1) \sqsupseteq p$$

Thus, the weakest precondition, p_1 is given by the relative complement (for serial composition) operation, viz. $p_1 \overset{\wedge}{=} p /_{+} r$.

4.4 Example

As an illustration, we briefly consider the specification and verification of the steady-state behaviour of a TTL NOT gate (Fig. 10(a)). The full analysis of this problem is lengthy, and even more so when the dynamic aspects of the same problem are addressed; here we present an outline only of the main steps for steady-state analysis.

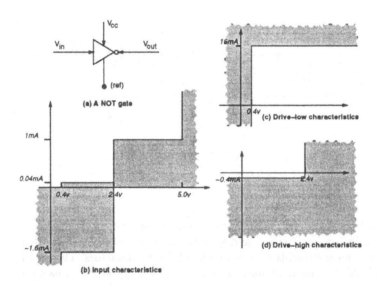

Fig. 10. A representation of a NOT gate along with its input and output specifications.

We begin by specifying the desired input and output characteristics of 'standard' TTL devices. This information (abstracted from manufacturers' data sheets) is expressed in the form of predicates over the voltages and currents at the ports of the device. Fig 10(b) shows the predicate $input\_load(v_{in}, i_{in})$ for an input port, and Figs 10(c and d) show the predicates *low_out* and *high_out* for an output port. We also specify predicates $low\_in \triangleq (v_{in} \le 0.8)$ and $high\_in \triangleq (v_{in} \ge 2.0)$ (that define the standard TTL thresholds) and a predicate *supply_ok* that is used to specify that the supply voltage v_{cc} is within bounds.

Using these predicates, we can then formulate a definition for the steady-state behaviour of a TTL NOT gate. It is

$$
\begin{aligned}
&notgate\_behav\ ((v_{cc}, i_{cc}), (v_{in}, i_{in}), (v_{out}, i_{out})) \triangleq \\
&\quad supply\_ok\ v_{cc} \Rightarrow \\
&\qquad input\_load\ (v_{in}, i_{in}) \wedge \\
&\qquad low\_in\ v_{in} \Rightarrow high\_out\ (v_{out}, i_{out}) \wedge \\
&\qquad high\_in\ v_{in} \Rightarrow low\_out\ (v_{out}, i_{out})
\end{aligned}
$$

If now one is given a proposed circuit for a TTL NOT gate (such as that shown in Fig 11) one can then set out to formally verify its correctness with respect to the above specification. This is done by attaching the consequents of the implications in the above definition to the appropriate points on the circuit and then, using weakest-precondition techniques, pushing these assertions back through the circuit and showing that they are implied by the antecedents.

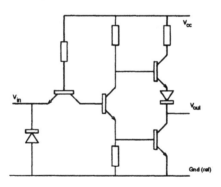

Fig. 11. A typical circuit for a NOT gate using TTL technology.

As a brief illustration, consider the first few steps involved in pushing one of the consequents, $low\_out\ (v_{out}, i_{out})$, in the above specification back through the circuit.

- Begin by labelling the output port of the circuit (see Fig 12(a)) with this predicate, called p for brevity, (see Fig 12(b)).
- Knowing that transistor T1 is meant to be "hard off" under these conditions, define a predicate q_1 (see Fig 12(b)) which specifies that this leakage current is less than 1 mA (a conservative bound).

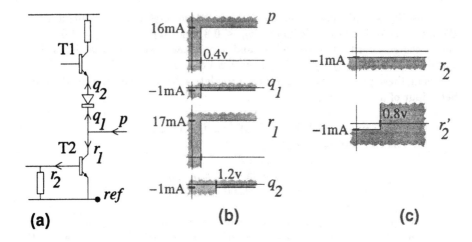

Fig. 12. Pushing preconditions through a circuit.

- Infer (using a relative complement operation) the weakest precondition, r_1, (see Fig 12(b)) that, together with q_1, will imply p.
- Infer from the specification for a junction diode (Fig 3(c), earlier) and the assertion q_1, the weakest precondition, q_2 (see Fig 12(b)).
- Infer, from the specification (given earlier in outline form only) of a junction transistor and from the assertion r_1, a sufficient precondition, r_2, (see Fig 12(c)). (Note: the specification for the transistor is assumed to specify a current gain, β, of at least 20.)
- Taking account of the diode characteristic of the base-emitter junction of transistor T2, the specification r_2 can be weakened to give r_2' (see Fig 12(c)).

By continuing this pattern of reasoning, one eventually establishes that the antecedents *supply_ok* v_{cc} and *high_in* v_{in} of the original specification imply the desired conclusion, *low_out* (v_{out}, i_{out})

5 Conclusions

This line of investigation is still at an early stage; as yet, no proofs have been subjected to the discipline of machine checking. We note, however, the prospect that much of the above kind of reasoning should be amenable to automation with tactics-directed theorem-proving techniques — especially if the tactics are given access to a representation of the circuit structure.

We expect the main impact of work in this field will be in allowing the design of faster, lower-powered digital circuitry by virtue of the enhanced degree of confidence the designer will have in being able to make much more use of pass-transistor circuitry and in being able to eliminate redundant stages of buffering.

252

The work may also contribute towards providing a sound semantic underpinning for the new "mixed-mode" hardware description languages (such as *Analog VHDL*[Vac93]) that are beginning to appear.

5.1 Open Problems

We conclude by mentioning two problems currently being addressed.

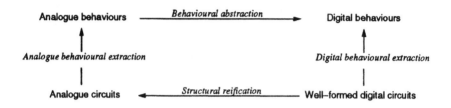

Fig. 13. Relating structure and behaviour at high and low levels of abstraction.

Dynamic behaviour The steady-state specifications (as described in this paper) are useful for only a limited range of purposes. Further, they are theoretically unsatisfactory since they depend on an unverified assumption that a steady-state condition actually exists. Extending the techniques described here to operate on dynamic behaviours (that is, on waveforms rather than just on signal levels) turns out to be quite realistically possible. It involves constructing assertions that hold over bounded intervals of time (rather than hold indefinitely in a steady state) in a manner similar to that described in [HD85].

Formalising structures Thus far, circuit diagrams have been handled informally — essentially as sketches on which behavioural annotations have been pencilled in. Techniques (see, for example [BHY92, Bou93, HD93]) exist, however, for treating circuits as values within a formal theory. This ability to represent and reason about both structural and behavioural information within the same formal theory opens up many interesting possibilities. A particularly fascinating one concerns the ability to be able to formalise the *design rules* (that is, those rules that define what it means for a digital circuit to be *well-formed*) for a particular technology and to be able to prove them correct. This task can be expressed as establishing the commutativity of a diagram (see Fig. 13) which relates analogue and digital behavioural extraction functions by means of structural reification and behavioural abstraction.

6 Acknowledgments

This work was partially sponsored by SERC grant GR/J78105.

References

[BHY92] B. C. Brock, W. A. Hunt, and W. D. Young. Introduction to a Formally Defined Hardware Description Language. In *Theorem Provers in Circuit Design*, volume A-10 of *IFIP Trans*, pages 3–35. North Holland, 1992.

[Bou93] Raymond Boute. Funmath illustrated; a declarative formalism and application examples. Technical report, Dept. of Informatics, Katholieke Universiteit Nijmegen, July 1993.

[Har92] John Harrison. Constructing the real numbers in hol. In Luc J. M. Claesen and Michael J. C. Gordon, editors, *Higher order logic theorem proving and its applications*, volume A-20 of *IFIP Transactions A: Computer Science and Technology*, pages 145–164. North Holland, September 1992.

[HD85] F. K. Hanna and N. Daeche. Specification and verification using higher-order logic. In Koomen and Moto-oka, editors, *Computer Hardware Description Languages and their Applications*, pages 418–433. North Holland, 1985.

[HD93] Keith Hanna and Neil Daeche. Strongly-Typed Theory of Structures and Behaviours. In *Correct Hardware Design and Verification Methods; IFIP Trans WG10.2*, pages 39–54. Springer-Verlag, 1993.

[HJ83] D. A. Hodges and H. G. Jackson. *Analysis and design of digital integrated circuits*. McGraw-Hill, 1983.

[Vac93] Alain Vachoux. IEEE VHDL subPAR 1076.1: Analog Extensions to VHDL; Design Objective Document. Technical Report 1.1, EPFL-DE-Lausanne, Switzerland, March 1993.

[WE85] N. H. E. Weste and K. Eshraghian. *Principles of CMOS VLSI Design*. Addison-Wesley, 1985.

Binary Decision Diagrams as a HOL Derived Rule

John Harrison

University of Cambridge Computer Laboratory
New Museums Site
Pembroke Street
Cambridge
CB2 3QG
England
jrh@cl.cam.ac.uk

Abstract. Exhaustive testing of boolean terms has long been held to be impractical for nontrivial problems, since the problem of tautology-checking is NP-complete. Nevertheless research on Ordered Binary Decision Diagrams, which was given a great impetus by Bryant's pioneering work, shows that for a wide variety of realistic circuit problems, exhaustive analysis is tractable. In this paper we seek to explore how these datastructures can be used in making an efficient HOL derived rule, and illustrate our work with some examples both from hardware verification and pure logic.

1 Binary Decision Diagrams

There are many ways to test whether a Boolean expression is a tautology (i.e. true for all truth assignments of the variables it involves), such as the use of truth tables or transformation to conjunctive normal form. The problem of tautology checking is NP-complete, and therefore all known algorithms have exponential worst-case performance; in fact the complementary operation of testing Boolean satisfiability was the original NP-complete problem given by Cook [6]. Nevertheless recent work has shown that in many practical situations, particularly in the verification of combinational logic circuits, binary decision diagrams do provide a practical means of tautology checking for expressions involving a large number of variables.

The basic idea of binary decision diagrams is to build up a 'decision tree' with the variables at the nodes and either 1 (true) or 0 (false) at the leaves. At each node the tree branches into two subtrees (often referred to as the 'then' and 'else' branches, regarding each node as a conditional) which represent the sub-functions formed by assuming that variable to be true or false, respectively. In that simple form, binary decision diagrams would merely be a way of organizing a truth table, and therefore have little interest. But the following two refinements, especially the latter, have led to an enormous increase in their efficiency.

- Rather than using trees, use directed acyclic graphs, sharing any common subexpressions which arise, as proposed by Lee [11] and Akers [2].

– Choose a canonical ordering of the variables at the outset, and arrange the graph such that the variables occur in that order down any path. This idea of Bryant's [5], in conjunction with maximal structure sharing (i.e. never duplicating nodes in the graph) gives a completely canonical representation of a boolean function, so boolean equivalence can be tested simply by comparing graphs. Such structures are often referred to as 'reduced ordered binary decision diagrams' (ROBDDs), but we'll just use the shorter title of BDDs.

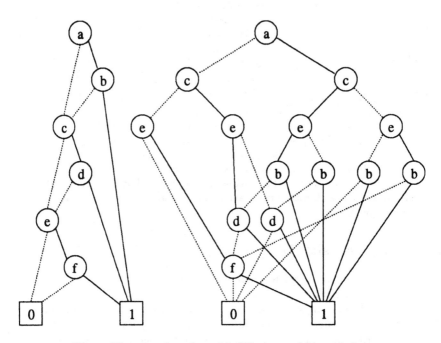

Fig. 1. The same function with different variable orderings

Efficiency and variable orderings

Figure 1 shows two BDDs for the same function, but with different variable orderings. One question which naturally arises when presented with a Boolean function is how to choose the canonical variable ordering required for BDDs. As the above picture illustrates, the choice of variable ordering can make a difference to the size of the BDDs produced. On large examples the difference can be between practicality and impracticality. Considerable work has been done on variable ordering heuristics [7, 12], and although nothing especially conclusive has emerged, there are a number of techniques which usually give satisfactory results in practice. For example, in verifying an adder it seems best to take the following order: first the carry-in, then the two low-order bits of the inputs, then the next input bits and so on up to the two high-order input bits. Other circuits can be handled with a similar ordering based on the network topology.

On the other hand there are circuits for which no variable ordering will give better than exponential performance. In his original paper [5], Bryant proved that the multiplier is one such circuit, and so BDD-based techniques are not easy to apply to large multipliers. Of course it is not surprising that an algorithm for solving an NP-complete problem will perform badly in some cases. Perhaps it is *more* surprising that such cases seem to occur so rarely in practice.

Complement Edges

Since the explosion of interest in BDDs, many extensions to the basic idea have been proposed. Though many of these are worthy of consideration, it was our intention to explore only the basic algorithm. Nevertheless the idea of *complement edges* as described by Brace et al. in [4] seemed worth incorporating in the implementation. The idea is simple: we allow each edge of the BDD graph to carry a tag, usually denoted by a small black circle in pictures. This denotes *the negation of* the subgraph it points to.

Complement edges have the great merit that complementing a BDD now takes constant time: one simply needs to flip the complement tag. In addition, greater sharing is achieved because a graph and its complement can be shared; only the edges pointing into it need differ. In particular we only need one terminal node, which we choose (arbitrarily) to be 1, with 0 represented by a complement edge into it. Notice that by a BDD we shall afterwards mean a pointer, complemented or otherwise, to a BDD node, which itself contains more such pointers.

Complement edges do create one small problem: without some extra constraints, canonicality is lost. This is illustrated in figure 2, where each of the four BDDs at the top is equivalent to the one below it. This ambiguity is resolved by ensuring that whenever we construct a BDD node, we transform between such equivalent pairs to ensure that the 'then' branch (i.e. that taken if the variable is 1) is uncomplemented. This choice is completely arbitrary.

Constructing BDDs

Assuming we are given a HOL boolean expression, the question arises of how to construct the corresponding BDD. Any top-down strategies based on case splits over variables would immediately have exponential performance and vitiate any advantages of the eventual BDD representation. Instead, a bottom-up strategy is used, where we first construct BDDs for the arguments of a given operator (e.g. conjunction) and then merge these together. Following Bryant's terminology, we will refer to each such step as an 'APPLY' operation.

Let us represent BDD nodes using HOL-like syntax for conditional expressions; i.e. $v \rightarrow t1|t2$ means 'if v then $t1$ else $t2$'. Suppose that we have two BDDs which we want to conjoin. If one node is 0 or 1 then we can easily reduce the node using a trivial propositional tautology, viz:

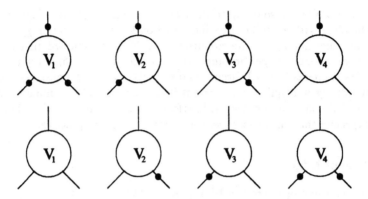

Fig. 2. Redundancy with complemented edges

$$node_1 \wedge 0 = 0$$
$$node_1 \wedge 1 = node_1$$
$$0 \wedge node_2 = 0$$
$$1 \wedge node_2 = node_2$$

If none of these special cases apply, then neither BDD is a terminal node. It may be that either or both is negated, but if so we can use the following transformation:

$$\neg(v \rightarrow t|e) = (v \rightarrow \neg t|\neg e)$$

Hence we may now assume that neither of the nodes is negated. The problem is now reduced to finding the BDD corresponding to:

$$(v_1 \rightarrow t_1|e_1) \wedge (v_2 \rightarrow t_2|e_2)$$

To do this, we just choose whichever variable comes first in the variable ordering. There are three possibilities. If v_1 and v_2 are identical, we apply the first rule below. If v_1 comes earlier in the canonical ordering we apply the second, and if v_2 comes earlier we apply the third.

$$(v_1 \rightarrow t_1|e_1) \wedge (v_2 \rightarrow t_2|e_2) = v_1 \rightarrow (t_1 \wedge t_2)|(e_1 \wedge e_2)$$
$$(v_1 \rightarrow t_1|e_1) \wedge (v_2 \rightarrow t_2|e_2) = v_1 \rightarrow (t_1 \wedge (v_2 \rightarrow t_2|e_2))|(e_1 \wedge (v_2 \rightarrow t_2|e_2))$$
$$(v_1 \rightarrow t_1|e_1) \wedge (v_2 \rightarrow t_2|e_2) = v_2 \rightarrow ((v_1 \rightarrow t_1|e_1) \wedge t_2)|(v_1 \rightarrow t_1|e_1) \wedge e_2)$$

This now gives us a simple recursive procedure to produce the BDD for any conjunction: we apply whichever of the above three rules applies till we reach a terminal node on one side or the other, when one of the four terminal cases given above terminates the recursion. This of course will not necessarily give us a completely reduced BDD. In Bryant's original treatment [5] the 'APPLY' operation was followed by a separate 'REDUCE' phase. This is complicated and inefficient, and in more recent implementations the two operations are combined. The above recursion is applied, but a global hash table of all nodes is maintained. Before a new node is created, the function looks in the table to see if it already exists. If so, a pointer (complemented as necessary) is used to the existing node. Maximal sharing is necessarily achieved because no node is ever duplicated.

There is one further optimization that is usually made. A second global hash table is maintained of all previous 'APPLY' operations together with their results. Before calling the 'APPLY' function recursively, we first check whether the desired result has already been computed, and if so, we use that result (which will itself be a BDD of course). Note further that when making this lookup we can, as described by Brace et al. [4] exploit special properties of the operator to be applied. In the present case of conjunction, we can exploit symmetry by looking up whichever of $b_1 \wedge b_2$ and $b_2 \wedge b_1$ satisfies some canonicality property (based on any fixed but arbitrary ordering among BDDs, such as pointer comparison). This cache of precomputed results has a major impact on efficiency, since it eliminates the need for a whole series of recursions down to a terminal node. Nevertheless at the frontiers of BDD research it sometimes proves necessary to garbage collect this cache, otherwise memory fills up completely.

2 Implementation

BDDs have hitherto mostly been implemented in low-level languages, mainly C. This allows efficient manipulation of pointers, and we can put in various low-level optimizations. For example, Brace et al. suggest using the bottom bit of a pointer to indicate complement status, rather than having a separate field. (This assumes that pointers to BDD nodes will be even: word-alignment can be relied on in most architectures.) Since it is our hope to implement the routine as a derived rule, we have to face two quite separate problems:

1. ML is inevitably less efficient than C, if only because tricky low-level optimizations like the above pointer hack are impossible to duplicate. Some dialects (e.g. Classic ML) lack the imperative features such as arrays which are necessary to implement fast lookup via hashing, though some structures such as AVL trees [1] might provide an acceptably efficient substitute.
2. We need to actually construct a HOL proof. This means that ad-hoc term manipulations need to be replaced by primitive inference.

We should notice first that producing a HOL derived rule does *not* necessarily mean that all the proof-finding has to be in ML. In fact there have been several experiments with linking HOL to external programs which produce proofs or checkable answers (for an unusual example see the papers by Laurent Théry and the present author [8, 9]). It would be quite possible to have an orthodox C implementation of BDDs produce a HOL proof which can be imported into ML.

We have not actively explored this possibility here, largely because of implementation complexity, but it does have some merit. In particular, it was mentioned above that choice of variable ordering can have an enormous impact of the size of the BDDs produced. An efficient external oracle might explore several ordering heuristics and only commit itself to producing a formal proof when a satisfactory ordering is found.

Three implementations

In order to clarify exactly where the efficiency problems arise, we have written three implementations:

1. An ANSI C implementation, using all optimizations which occurred to us, including the pointer trick mentioned above to represent complemented edges.
2. An implementation in Standard ML, not performing inference in the HOL sense.
3. An implementation in Standard ML producing a proof in `hol90`.

Considerable pains were taken to make the underlying algorithms in the three implementations as near identical as possible. This was mainly in order to allow meaningful comparisons, but as we shall explain later, it is important that the two Standard ML versions produce isomorphic BDD graphs.

More details of the algorithm

Here we fill in the details in the above sketch of the algorithm. The algorithm is based on two hash tables (implemented as arrays of lists):

1. A 'unique table' containing all the nodes in the BDD graph (represented as triples: the variable and the two branches). Before an 'APPLY' operation creates a node, it checks whether the resulting node would be trivial, i.e. has both branches identical. If so then the following transformation can be applied:

$$(v \rightarrow t|t) = t$$

If the branch is not trivial, it is looked up in the table, and inserted if it is not already there. However first, it is necessary to ensure canonicality in the presence of complemented edges, so as explained above, if the 'then' branch is negated, the BDD is transformed like this:

$$(v \rightarrow \neg t|e) = \neg(v \rightarrow t|\neg e)$$

The node (which is negated in the above expression) is then entered in the unique table.

2. A 'computed table' containing a record of previous computations. We have only implemented 'APPLY' for a single operation, namely conjunction. Remembering that we have a basic negation operation, which is almost cost-free, we can easily implement the other logical operations using that, viz:

$$x \vee y = \neg(\neg x \wedge \neg y)$$
$$x \implies y = \neg(x \wedge \neg y)$$
$$x \equiv y = (x \implies y) \wedge (y \implies x)$$
$$x \to y | z = (x \wedge y) \vee (x \wedge \neg z)$$

The last two are less efficient, but seem to arise less often in practice. It's more efficient to maintain a single 'computed' table for conjunctions, rather than effectively duplicate logically equivalent results for disjunction and implication. For a similar reason, the paper by Brace et al. takes the conditional (if-then-else) as basic, but using a ternary operator to implement binary ones seems to us rather wasteful. Equivalence and inequivalence can then be expressed directly via $x \to y | \neg y$, but the other connectives can be done equally efficiently in terms of conjunction and negation.

An 'APPLY' operation first checks for trivial cases, where one or other of the arguments is 0 or 1. These are dealt with as outlined above. Furthermore, cases where both arguments are identical, or one is the negation of the other, are also disposed of in an obvious way. Finally, based on an arbitrary ordering (pointer comparison in the C case, position in node table in the Standard ML case), the conjunction is oriented canonically, exploiting symmetry. Only when the arguments are placed in a canonical form does the 'APPLY' operation look to see if the result is in the computed table. If it is, then that result is used. Otherwise a recursion is performed as indicated above, and the eventual result is entered in the computed table.

Efficient implementation of complement edges

As noted above, it is easy to implement complement edges efficiently in C (albeit strictly nonportably). Standard ML has references (pointers), but these cannot be hacked with such abandon. Representing a BDD by a composite datastructure, such as a pointer-boolean pair, seemed likely to be highly inefficient, since it means creating more **cons** cells and performing multiple dereferences. Instead we exploit Standard ML's arrays to create our own bespoke pointers.

We wish in any case to have a global table of BDD nodes, indexed via hashing, and so we can store the nodes in a single array (in the present implementation this has a fixed size, which limits the total number of nodes, but this restriction could easily be lifted). Now the index in this array (simply an integer) can be used as a 'pointer' to the node, and to indicate negation we can simply numerically negate the number. (Of course we must avoid using element 0 of the array, as $-0 = 0$, but that is a small matter.) The terminal nodes are represented by numbers outside the array bounds. The slight extra cost of array indexing over pointer dereferencing seems well worthwhile, since we can manipulate BDDs in the form of integers very efficiently.

Performing primitive inference

The implementation which performs primitive inference is essentially a conversion which takes a HOL propositional formula E and produce a theorem:

$$\Gamma \vdash E = B$$

where B is a representation of a BDD. Obviously, a crucial question is how a BDD is to be represented inside the logic. We cannot simply represent the BDD in HOL as a tree of conditionals, or rather we *could* but we would immediately abandon sharing inside the HOL term. (Unshared BDDs are possible, but less attractive.) There are a number of ways of achieving such sharing. Perhaps the most obvious is to use a `let` binding, or what amounts to the same thing, a beta-redex. For example

```
let v = B1 in B2[v,..,v,..,v]
```

implements a BDD B2 which has three 'pointers' to a shared BDD B1. Experiments along these lines have been done, but the terms become rather complex because terms are shared among various 'levels' in the hierarchy of `let` bindings. Instead we adopted the following scheme. For all BDD nodes entered in the hash table, we invent stylized node names $n_1, n_2, n_3, \ldots$. Each such name will stand for a BDD node, and we have for each one a corresponding 'definition' term:

$$n_i = v \rightarrow b_j \mid b_k$$

where b_j and b_k are (possibly negated) other node variables or the special terminal node 1 which is of course represented by T in HOL.

The most obvious plan is then to carry all the relevant node definitions around as hypotheses in the theorem produced. Then, each 'APPLY' step simply needs to substitute the definition and do some very simple inference. However this proves unattractive because the hypotheses are stored as an unsorted list. It soon becomes hopelessly impractical to perform the implicit union of lists which occurs each time an 'APPLY' operation brings together subproofs which used different sets of hypotheses.

There is a simple way in which to avoid this inefficiency. Rather than having a list of assumptions, we can collect all the definitions into a single conjunctive term: all theorems carry just this one hypothesis. We can once and for all split up the conjuncts and put the 'definitional' theorems in an array for easy access; this only takes time $O(n)$ where n is the number of nodes in the BDD. Now all the algorithm can be implemented efficiently exactly as in the version without inference. Inference is necessary at every 'APPLY' operation, but only simple propositional reasoning. Hence the main algorithm can now be implemented with only a constant factor slowdown (albeit a significant constant factor) over an implementation which performs no inference.

How to get started

There is an obvious problem with the above scheme. We want to set up a big conjunction of all the necessary definitions at the beginning, and this entails knowing *in advance of running the algorithm* what all those definitions should be. Consequently a 2-pass implementation is required. First, the Standard ML version which does not do primitive inference is run; as it runs, it records all the nodes in its own node table. Then these are translated into theorems for the inference version, which can be run afterwards. Here it is crucial that the underlying steps of the two implementations are identical, or the nodes required might differ.

Although this 2-pass organization is an unwelcome increase in the complexity of the implementation, all the necessary code is at hand, and it proved quite easy to write the necessary interface. There is some advantage in that failures will tend to be trapped early by the first pass.

How to finish

A less obvious and more serious problem is how to get rid of all the BDD node definitions when the algorithm is finished. Of course, we are considering here cases where the given expression has been proved to be a tautology, i.e. the theorem produced by the BDD conversion is of the form:

$$\Gamma \vdash E = T$$

Logically, there is no difficulty in eliminating the assumptions. Recall that the BDD graph is acyclic. This means that among any set of nodes there will be a topmost one, which is not used in any other definition (and of course none of the node variables appear in the conclusion of the theorem). In fact we can always choose the latest node to be created, because of the bottom-up way in which the BDD was constructed. Hence by arranging the conjunction in order (we put the newest node as the leftmost conjunct in fact) we can always peel off the conjuncts in sequence.

Now consider how to eliminate such a topmost definition. It is a simple matter logically to discharge it and leave the remaining conjuncts as the new assumption:

$$\Delta \vdash (n_i = v \rightarrow b_i | b_j) \implies (E = T)$$

Since we have assumed that n_i is not free in Δ, we may instantiate n_i to $v \rightarrow b_i | b_j$ in this theorem. Since n_i is also not free in the initial expression E, this yields:

$$\Delta \vdash (v \rightarrow b_i | b_j = v \rightarrow b_i | b_j) \implies (E = T)$$

Now from reflexivity:

$$\vdash v \rightarrow b_i | b_j = v \rightarrow b_i | b_j$$

and Modus Ponens gives:

$$\Delta \vdash E = T$$

This process can be iterated, and so all the assumptions can be eliminated, leaving the bare theorem $\vdash E = T$ as required.

The above process obviously takes $O(n)$ primitive inferences, where n is the number of nodes. However a quadratic complexity in n lurks beneath the surface, since each instantiation must *check* that n_i is not free in the hypotheses. And the single hypothesis has size $O(n)$, all of which must be traversed! Consequently, this final innocuous cleanup has turned out to be the rate-determining step of the procedure. This is dramatically reflected in the results given below.

3 A logical application

Fast tautology checking is most useful for verification of circuits at a low level. In general theorem proving, tautology-checking applications are usually sufficiently simple that a naive case analysis algorithm is fast enough to prove them in an acceptable time. Nevertheless there are a few situations where fast tautology checking can be useful in a more abstract domain, and we address one such situation here by describing a decision procedure for a certain class of logical formulas.

We will consider first-order formulae involving no function symbols (but relation-symbols of arbitrary arities are allowed) and no equality symbol, with all universal quantifiers preceding existential ones, i.e.

$$\forall x_1 \ldots x_m. \, \exists y_1 \ldots y_n. \, P[x_1, \ldots, x_m, y_1, \ldots, y_n]$$

where $P[x_1, \ldots, x_m, y_1, \ldots, y_n]$ is quantifier-free. We will show how a decision procedure originally described by Bernays and Schönfinkel [3] may be implemented reasonably efficiently and applied to simple inequality reasoning. Our exposition will follow the nice discussion given by Hilbert and Ackermann [10].

Validity and satisfiability in finite domains

We will call a first-order formula ϕ *valid*, and write $\models \phi$, if it is true in all interpretations (in the usual sense). Now for any positive integer n, we shall call it *n-valid*, and write $\models_n \phi$, if it is valid in all interpretations with a domain size n (the domain being the set over which the quantifiers range). Now take a first-order formula ϕ conforming to the above restrictions:

$$\forall x_1 \ldots x_m. \, \exists y_1 \ldots y_n. \, P[x_1, \ldots, x_m, y_1, \ldots, y_n]$$

and consider the disjunction:

$$P[x_1, \ldots, x_m, x_{p_1}, \ldots, x_{p_n}] \vee \ldots \vee P[x_1, \ldots, x_m, x_{q_1}, \ldots, x_{q_n}]$$

where $x_{p_1}, \ldots, x_{p_n}$ run through all the possible choices of n elements from the set $x_1, \ldots, x_m$. Let us denote this disjunction by $Q[x_1, \ldots, x_m]$. We claim that the following four assertions are equivalent:

$$\models \quad \forall x_1 \ldots x_m. \, \exists y_1 \ldots y_n. \; P[x_1, \ldots, x_m, y_1, \ldots, y_n] \tag{1}$$

$$\models_n \quad \forall x_1 \ldots x_m. \, \exists y_1 \ldots y_n. \; P[x_1, \ldots, x_m, y_1, \ldots, y_n] \tag{2}$$

$$\models_n \quad \forall x_1 \ldots x_m. \, Q[x_1, \ldots, x_m] \tag{3}$$

$$\models \quad \forall x_1 \ldots x_m. \, Q[x_1, \ldots, x_m] \tag{4}$$

Obviously 1 implies 2 (it is a more general statement). Furthermore 4 implies 1, because one of the sets of terms in the disjunction provides a witness for the existential quantifier. It is a fact that 3 implies 4 for any quantifier-free statement $Q[x_1, \ldots, x_m]$ at all, not just of the special form we have arrived at. To see this, suppose 4 is false. Then we can find an interpretation and elements $a_1, \ldots, a_m$ such that

$$\models \neg Q[a_1, \ldots, a_m]$$

But then by restricting the interpretation to $\{a_1, \ldots, a_m\}$ it would follow that the original statement is not m-valid. (If the $a_1, \ldots, a_n$ are not all distinct, we can add new elements and identify them with a_1 as far as the action of the interpretation is concerned. Using this method we can always expand a domain while retaining a formula's validity — this depends of course on the fact that we do not have the special interpretation of the equality symbol.) Using this fact, it is similarly easy to show that 2 implies 3.

This means that to prove a theorem of the form 1, we need only prove one of the simpler form 4. Since we can choose the interpretation of predicates without restriction, we can assign all instances of the predicates independently. Consequently a statement of the form 4 is true iff its body is a substitution instance of a propositional tautology. Thus the whole problem is reduced to tautology checking; in general the tautology will be quite big, since there will be m^n disjuncts in our reduced form.

Apart from providing a ready source of tautologies to test, it is worth asking whether this sort of decision procedure is useful in everyday proofs. Certainly the form is quite restricted, but we believe it may be useful for simple forms of inequality reasoning, which arise quite often. For example, if we have an implication with universally quantified antecedent and consequent, it can be transformed into the above form. The basic properties of a partial or total order are all universal, and so this fits into the framework. Although we are not allowed a specially interpreted equality predicate, it is easy enough to simulate it by asserting some predicate's reflexivity, symmetry, transitivity and congruence properties, all universal sentences.

Note however that because of the lack of function symbols, it is not possible to decide algebraic theories such as group theory (in fact [13] no decision procedure exists for group theory). Certainly the axioms can all be rendered in relational form, i.e. $m(x, y, z)$ for $x.y = z$ etc., but the fact that the corresponding function is total:

$$\forall x \ y. \ \exists z. \ m(x, y, z)$$

is not a universal statement and so when negated (as it effectively will be, occurring as the antecedent of an implication) it does not fall within the compass of our procedure.

4 Results

In this section we present some results on tautology-checking examples. One of these examples was derived by applying the above logical decision procedure to a simple fact of inequality reasoning. This is the deduction of

$$\forall a \ b \ c. \ a \leq b \wedge c \nleq b \Longrightarrow a < c$$

from the set of 'axioms' for a total order (much larger examples could be done; the axioms are the dominant part of this term). The other four were taken from a public-domain suite of hardware verification examples produced by Diederik Verkest and others at IMEC in Belgium [14]. These circuits are as follows:

Name	Explanation	Number of inputs
add1.be	4-bit adder-subtractor	9
add2.be	4-bit ALU with 5 control signals	13
add3.be	8-bit ALU with 5 control signals	21
add4.be	12-bit ALU with 5 control signals	29

The fields of the table are as follows:

- Nodes: The number of nodes in the BDD generated (all algorithms give the same result, which is an independent check on their similarity).
- C: Time for the C implementation. The C version was compiled with the optimizer switched off.
- ML (initial): Time for the first ML pass which does no inference. Although it is only used to record the nodes for the inference pass, this pass does almost exactly the same as the C program, and can meaningfully be compared.
- ML (main): Time for the second ML pass doing inference to produce a tautology proof still decorated with assumptions.
- ML (cleanup): Time for the final cleanup phase, removing all the assumptions from the BDD theorem.

All timings are in CPU seconds on a Sparc 10. There were 1000 hash chains in the unique and computed tables; for the later examples this might profitably have been increased, but was kept fixed throughout.

Example	Nodes	C	ML (initial)	ML (main)	ML (cleanup)
add1.be	426	0.01	0.23	5.22	18.12
Inequality	1463	0.04	0.44	22.83	165.62
add2.be	1785	0.06	0.63	24.20	481.38
add3.be	5472	0.40	3.47	98.49	7425.85
add4.be	11252	0.80	8.71	354.84	(unable?)

Ideas for improving efficiency

The inference version was by no means implemented in the most efficient way possible. For example, all the lookup in the computed table was replicated, rather than recording results from the first pass. There is considerable potential for optimization.

However the most important object of attack is the final cleanup phase. As remarked above, there is a subtle quadratic dependency lurking here, and results manifest it in an unpleasant way.

There are plenty of ways in which this phase could be done more efficiently. The following idea has been started, but not finished, at time of writing. The difficulty is justifying the consistency of the set of node definitions, which in general might make incompatible recursive assignments. It is only because we have stuck to a variable ordering that we know the result to be admissible. Thus we have tried to perform a proof of this by generating a proforma theorem asserting that a list of definitions of a restricted form is admissible.

First, instead of using single variables for the node names, we use a function applied to a numeric argument; what we have written as n_i actually becomes $B(i)$ in HOL. Now we prove a proforma theorem by list recursion asserting that a list of such definitions may be satisfied subject to conditions. The list is a set of tuples which code the right hand sides of the definitions. The interpretation is that the ith element of the list defines $B(i)$, so automatically assuring us that no node is defined twice. Now the list is admissible provided that the tuple which is the ith element of the list has both its numeric fields less than i. This can be expressed as a simple list recursion, and to justify a given definition requires $O(n)$ steps.

Of course at each step we need to do some arithmetic reasoning on numerals, but that can be done in $O(log(n))$ time by representing the numbers as lists as in Tim Leonard's numeral library. The overall complexity is bounded by $O(n\ log(n))$, and since in practice n will certainly be bounded by the memory capacity of the machine, we effectively have an $O(n)$ bound over all feasible ranges. Consequently it should be possible to fix the problem of inefficiency in the cleanup phase.

5   Conclusion

We have shown that it is possible to implement BDDs as a HOL derived rule, which, except for the final phase, is tolerably fast. The slowness of this final cleanup phase is disappointing, but we believe that an approach like that outlined in the last section will alleviate it. In any case, the algorithms described here are much more efficient on large examples than naive tautology-checking algorithms based on repeated case splits. Almost certainly, add3.be is the largest tautology ever proved in HOL.

The derived rule can be used not only as a 'one-shot' tautology checker but in order to implement boolean expressions directly as BDDs and manipulate them efficiently in that form: this is what is done in mainstream uses of BDDs. It would be worth programming other operations on BDDs efficiently.

It is clear that a carefully optimized C implementation will always be much more efficient, and from a pragmatic point of view there is considerable merit in hardwiring BDDs in as a fundamental principle, as in Carl Seger's Voss system. Nevertheless we believe that implementing BDDs as a HOL derived rule is not a hopeless idea. For certain applications a constant factor of even several orders of magnitude in performance may be outweighed by the increased security of having a precise proof as a logician understands the word.

Acknowledgements

I would like to thank Mark Linderman, of Cornell University, who explained BDDs to me, as well as everyone else who made my stay there so enjoyable and stimulating. I am also very grateful to Catia Angelo for sending me the IMEC benchmarks, and to my supervisor Mike Gordon for his constant support and encouragement. Thanks are also due to the Natural Sciences and Engineering Research Council and the Isaac Newton Trust for financial support.

References

1. G. M. Adel'son-Vel'skii and E. M. Landis. An algorithm for the organization of information. *Soviet Mathematics Doklady*, 3:1259–1263, 1986.
2. S. B. Akers. Binary decision diagrams. *IEEE Transactions on Computers*, C-27:509–516, 1978.
3. P. Bernays and M. Schönfinkel. Zum entscheidungsproblem der mathematischen logik. *Mathematische Annalen*, 99:401–419, 1928.
4. Karl S. Brace, Richard L. Rudell, and Randall E. Bryant. Efficient implementation of a bdd package. *Proceedings of 27th ACM/IEEE Design Automation Conference*, pages 40–45, 1990.
5. Randall E. Bryant. Graph-based algorithms for boolean function manipulation. *IEEE Transactions on Computers*, C-35:677–691, 1986.
6. S. A. Cook. The complexity of theorem-proving procedures. In *Proceedings of the 3rd ACM Symposium on the Theory of Computing*, pages 151–158, 1971.
7. M. Fujita, H. Fujisawa, and N. Kawato. Evaluations and improvements of a boolean comparison program based on binary decision diagrams. In *Proceedings of the International Conference on Computer-Aided Design*, pages 2–5. IEEE, 1988.

8. John Harrison and Laurent Théry. Reasoning about the reals: the marriage of hol and maple. In *Logic programming and automated reasoning: proceedings of the 4th international conference, LPAR '93*, volume 698 of *Lecture Notes in Artificial Intelligence*, pages 351–353. Springer-Verlag, 1992.

9. John Harrison and Laurent Théry. Extending the hol theorem prover with a computer algebra system to reason about the reals. In *Proceedings of the 1993 International Workshop on the HOL theorem proving system and its applications*, volume 780 of *Lecture Notes in Computer Science*. Springer-Verlag, 1994.

10. D. Hilbert and W. Ackermann. *Principles of Mathematical Logic*. Chelsea, 1950.

11. C. Y. Lee. Representation of switching circuits by binary-decision programs. *Bell System Technical Journal*, 38:985–999, 1959.

12. S. Malik, A. Wang, R. K. Brayton, and A. Sangiovanni-Vincentelli. Logic verification using binary decision diagrams in a logic synthesis environment. In *Proceedings of the International Conference on Computer-Aided Design*, pages 6–9. IEEE, 1988.

13. A. Tarski. *Undecidable Theories*. Studies in Logic and the Foundations of Mathematics. North-Holland, 1953.

14. Diederik Verkest and Luc Claesen. Special benchmark session on tautology checking. In *Formal VLSI Correctness Verification*, pages 81–82. Elsevier Science Publishers, 1990.

Trustworthy Tools for Trustworthy Programs:
A Verified Verification Condition Generator

Peter V. Homeier and David F. Martin

Computer Science Department
University of California, Los Angeles
homeier@cs.ucla.edu and dmartin@cs.ucla.edu

Abstract. Verification Condition Generator (VCG) tools have been effective in simplifying the task of proving programs correct. However, in the past these VCG tools have in general not themselves been mechanically proven, so any proof using and depending on these VCGs might have contained errors. In our work, we define and rigorously prove correct a VCG tool within the HOL theorem proving system, for a standard while-loop language, with one new feature not usually treated: expressions with side effects. Starting from a structural operational semantics of this programming language, we prove as theorems the axioms and rules of inference of a Hoare-style axiomatic semantics, verifying their soundness. This axiomatic semantics is then used to define and prove correct a VCG tool for this language. Finally, this verified VCG is applied to an example program to verify its correctness.

1 Introduction

The most common technique used today to produce quality software without errors is testing. However, even repeated testing cannot reliably eliminate all errors, and hence is incomplete. To achieve a higher level of reliability and trust, programmers may construct proofs of correctness, verifying that the program satisfies a formal specification. This need be done only once, and eliminates whole classes of errors. However, these proofs are complex, full of details, and difficult to construct by hand, and thus may themselves contain errors, which reduces trust in the program so proved. Mechanical proofs are more secure, but even more detailed and difficult.

One solution to this difficulty is partially automating the construction of the proof by a tool called a *Verification Condition Generator* (VCG). This VCG tool writes the proof of the program, modulo a set of formulas called *verification conditions* which are left to the programmer to prove. These verification conditions do not contain any references to programming language phrases, but only deal with the logics of the underlying data types. This twice simplifies the programmer's burden, reducing the volume of proof and level of proof, and makes the process more effective. However, in the past these VCG tools have not in general themselves been proven, meaning that the trust of a program's proof rested on the trust of an unproven VCG tool.

In this work we define a VCG within the Higher Order Logic (HOL) theorem proving system [6], and prove that the truth of the verification conditions it returns suffice to verify the asserted program submitted to the VCG. This theorem stating the VCG's correctness then supports the use of the VCG in proving the correctness of individual programs with complete soundness assured. The VCG automates much of the work and detail involved, relieving the programmer of all but the essential task of proving the verification conditions. This enables proofs of programs which are both effective and trustworthy to a degree not previously seen together.

2 Previous Work

There has been very little work done on proving the correctness of expressions; an exception is Sokolowski's paper on a "term-wise" approach to partial correctness [10]. Even he does not treat expressions with side effects. Side effects appear commonly in "real" programming languages, such as in C, with the operators ++ and get_ch. In addition, several interesting functions are most naturally designed with a side effect; an example is the standard method for calculating random numbers, based on a seed which is updated each time the random number generator is run.

In this paper, we define a "verified" verification condition generator as one which has been proven to correctly produce, for any input program and specification, a set of verification conditions whose truth implies the consistency of the program with its specification. Preferably, this verification of the VCG will be mechanically checked for soundness, because of the many details and deep issues that arise. Many VCG's have been written but not verified; there is then no assurance that the verification conditions produced are properly related to the original program, and hence no security that after proving the verification conditions, the correctness of the program follows. Gordon's work below is an exception in that the security is maintained by the HOL system itself.

Igarashi, London, and Luckham in 1973 gave an axiomatic semantics for a subset of Pascal, and described a VCG they had written in MLISP2 [7]. The soundness of the axiomatic semantics was verified by hand proof, but the correctness of the VCG was not rigorously proven. The only mechanized part of this work was the VCG itself.

Larry Ragland, also in 1973, verified a verification condition generator written in Nucleus, a language Ragland invented to both express a VCG and be verifiable [9]. This was a remarkable piece of work, well ahead of its time. The VCG system consisted of 203 procedures, nearly all of which were less than one page long. These gave rise to approximately 4000 verification conditions. The proof of the generator used an unverified VCG written in Snobol4. The verification conditions it generated were proven by hand, not mechanically. This proof substantially increased the degree of trustworthiness of Ragland's VCG.

Michael Gordon in 1989 did the original work of constructing within HOL a framework for proving the correctness of programs [5]. He introduced new constants in the HOL logic to represent each program construct, defining them as functions directly denoting the construct's semantic meaning. This is known as a "shallow" embedding of the programming language in the HOL logic. The work included defining verification condition generators for both partial and total correctness as tactics. This approach yielded tools which could be used to soundly verify individual programs. However, the VCG tactic he defined was not itself proven. If it succeeded, the resulting subgoals were soundly related to the original correctness goal by the security of HOL itself. Fundamentally, there were certain limitations to the expressive power and proven conclusions of this approach, as recognized by Gordon himself:

> "$P[E/V]$ (substitution) is a meta notation and consequently the assignment axiom can only be stated as a meta theorem. This elementary point is nevertheless quite subtle. In order to prove the assignment axiom as a theorem within higher order logic it would be necessary to have types in the logic corresponding to formulae, variables and terms. One could then prove something like:
>
> $\vdash \ \forall P\,E\,V.$ **Spec** (**Truth**(**Subst**(P, E, V)), **Assign**$(V,$ **Value** $E)$, **Truth** P)
>
> It is clear that working out the details of this would be a lot of work." [5]

In 1991, Sten Agerholm [1] used a similar shallow embedding to define the weakest preconditions of a small **while**-loop language, including unbounded nondeterminism and blocks. The semantics was designed to avoid syntactic notions like substitution. Similar to Gordon's work, Agerholm defined a verification condition generator for total correctness specifications as an HOL tactic. This tactic needed additional information to handle sequences of commands and the **while** command, to be supplied by the user.

This paper explores the alternative approach described but not investigated by Gordon. It turns out to yield great expressiveness and control in stating and proving as theorems within HOL concepts which previously were only describable as meta-theorems outside HOL, as above. For example, we are able to prove the assignment axiom that Gordon cannot:

$$\vdash \ \forall q \, x \, e. \, \{ \, q \lhd [x := e] \, \} \ x := e \ \{q\}$$

where $q \lhd [x := e]$ is a substituted version of q, described later.

To achieve this expressiveness, it is necessary to create a deeper foundation than that used previously. Instead of using an extension of the HOL Object Language as the programming language, we create an entirely new set of datatypes within the Object Language to represent constructs of the programming language and the associated assertion language. This is known as a "deep" embedding, as opposed to the shallow embedding developed by Gordon. This allows a significant difference in the way that the semantics of the programming language is defined. Instead of defining a construct *as* its semantic meaning, we define the construct as simply a syntactic constructor of phrases in the programming language, and then separately define the semantics of each construct in a structural operational semantics [12]. This separation means that we can now decompose and analyze syntactic program phrases at the HOL Object Language level, and thus reason within HOL about the semantics of purely syntactic manipulations, such as substitution or verification condition generation, since they exist *within* the HOL logic.

This has definite advantages because syntactic manipulations, when semantically correct, are simpler and easier to calculate. They encapsulate a level of detailed semantic reasoning that then only needs to be proven once, instead of having to be repeatedly proven for every occurrence of that manipulation. This will be a recurring pattern in this paper, where repeatedly a syntactic manipulation is defined, and then its semantics is described and proven correct within HOL.

3 Higher Order Logic

Higher Order Logic (HOL) [6] is a version of predicate calculus that allows variables to range over functions and predicates. Thus denotable values may be functions of any higher order. Strong typing ensures the consistency and proper meaning of all expressions. The power of this logic is similar to set theory, and it is sufficient for expressing most mathematical theories.

HOL is also a mechanical proof development system. It is secure in that only true theorems can be proved. Rather than attempting to automatically prove theorems, HOL acts as a supportive assistant, mechanically checking the validity of each step attempted by the user.

The primary interface to HOL is the polymorphic functional programming language ML ("Meta Language") [4]; commands to HOL are expressions in ML. Within ML is a second language OL ("Object Language"), representing terms and theorems by ML abstract datatypes **term** and **thm**. A shallow embedding represents program constructs by new OL functions to combine the semantics of the constituents to produce the semantics of the combination. Our approach is to define a *third* level of language, contained within OL as concrete recursive datatypes, to represent the constructs of the programming language PL being studied and its associated assertion language AL. We begin with the definition of variables.

4 Variables and Variants

A variable is represented by a new concrete type `var`, with one constructor, `VAR:string->num->var`. We define two deconstructor functions, Base(VAR *str n*) = *str* and Index(VAR *str n*) = *n*. The number attribute eases the creation of variants of a variable, which are made by (possibly) increasing the number.

All possible variables are considered predeclared of type `num`. In future versions, we hope to treat other data types, by introducing a more complex state and a static semantics for the language which performs type-checking. Some languages distinguish between program variables and logical variables, which cannot be changed by program control. In this simple language, this is unnecessary. In our more recent work with procedure calls, we support logical variables; this is treated in our Category B paper being presented at this conference.

The *variant* function has type `var->(var)set->var`. *variant x s* returns a variable which is a variant of *x*, which is guaranteed not to be in the "exclusion" set *s*. If *x* is not in the set *s*, then it is its own variant. This is used in defining proper substitution on quantified expressions.

The definition of *variant* is somewhat deeper than might originally appear. To have a constructive function for making variants in particular instances, we wanted

$$variant\ x\ s = (x\ \text{IN}\ s => variant\ (mk\_variant\ x\ 1)\ s\ |\ x) \qquad (*)$$

where *mk_variant* (VAR *str n*) *k* = VAR *str* (*n+k*). For any finite set *s*, this definition of *variant* will terminate, but unfortunately, it is not primitive recursive on the set *s*, and so does not conform to the requirements of HOL's recursive function definition operator. As a substitute, we wanted to define the *variant* function using `new_specification` by specifying its properties, as

1) (*variant x s*) *is_variant x*, and
2) ~(*variant x s* IN *s*), and
3) $\forall z$. if (*z is_variant x*) $\wedge$ ~(*z* IN *s*), then Index(*variant x s*) $\leq$ Index(*z*),

where *y is_variant x* = (Base(*y*) = Base(*x*) $\wedge$ Index(*x*) $\leq$ Index(*y*)).

But even the above specification did not easily support the proof of the existence theorem, that such a variant existed for any *x* and *s*, because the set of values for *z* satisfying the third property's antecedent is infinite, and we were working strictly with finite sets. The solution was to introduce the function *variant_set*, where *variant_set x n* returns the set of the first *n* variants of *x*, all different from each other, so CARD (*variant_set x n*) = *n*. The definition of *variant_set* is

$$variant\_set\ x\ 0 = \text{EMPTY} \wedge$$
$$variant\_set\ x\ (\text{SUC}\ n) = \text{INSERT}\ (mk\_variant\ x\ n)\ (variant\_set\ x\ n).$$

Then by the pigeonhole principle, we are guaranteed that there must be at least one variable in *variant_set x* (SUC (CARD *s*)) which is not in the set *s*. This leads to the needed existence theorem. We then defined *variant* with the following properties:

1') (*variant x s*) IN *variant_set x* (SUC (CARD *s*)), and
2') ~(*variant x s* IN *s*), and
3') $\forall z$.if *z* IN *variant_set x* (SUC (CARD *s*)) $\wedge$ ~(*z* IN *s*),
       then Index(*variant x s*) $\leq$ Index(*z*).

From this definition, we then proved both the original set of properties (1)–(3), and also the constructive function definition given above (*), as theorems.

5 Programming and Assertion Languages

The syntax of the programming language PL is

exp:	e ::=	$n \mid x \mid ++x \mid e_1 + e_2 \mid e_1 - e_2$
bexp:	b ::=	$e_1 = e_2 \mid e_1 < e_2 \mid b_1 \wedge b_2 \mid b_1 \vee b_2 \mid \sim b$
cmd:	c ::=	**skip** | **abort** | $x := e$ | $c_1; c_2$ |
		if b **then** c_1 **else** c_2 | **assert** a **while** b **do** c

Table 1. Programming Language Syntax

Most of these constructs are standard. n is an unsigned integer; x is a variable; ++ is the increment operator; **abort** causes an immediate abnormal termination; the **while** loop requires an invariant assertion to be supplied. The notation used above is for ease of reading; each phrase is actually formed by a constructor function, e.g., ASSIGN:var->exp->cmd for assignment. We overload the same operator in different languages, asking the reader to disambiguate by context.

The syntax of the associated assertion language AL is

vexp:	v ::=	$n \mid x \mid v_1 + v_2 \mid v_1 - v_2 \mid v_1 * v_2$
aexp:	a ::=	**true** | **false** | $v_1 = v_2 \mid v_1 < v_2 \mid a_1 \wedge a_2 \mid a_1 \vee a_2 \mid \sim a \mid$
		$a_1 \Rightarrow a_2 \mid a_1 = a_2 \mid a_1 => a_2 \mid a_3 \mid$ **close** $a \mid \forall x. a \mid \exists x. a$

Table 2. Assertion Language Syntax

Again, most of these expressions are standard. $a_1 => a_2 \mid a_3$ is a conditional expression, yielding the value of a_2 or a_3 depending on the value of a_1. **close** a forms the universal closure of a, which is true when a is true for all possible assignments to its free variables. Again, the notation is for readability; e.g., the constructor AVAR:var->vexp creates a vexp from a variable.

6 Operational Semantics

The semantics of the programming language is expressed by the following three relations, where a state is a mapping from variables to num:

$E\ e\ s_1\ n\ s_2$: numeric expression e : exp evaluated in state s_1 yields numeric value n : num and state s_2.

$B\ b\ s_1\ t\ s_2$: boolean expression b : bexp evaluated in state s_1 yields truth value t : bool and state s_2.

$C\ c\ s_1\ s_2$: command c : cmd evaluated in state s_1 yields state s_2.

Here is the structural operational semantics [12] of the programming language PL, given as rules inductively defining the three relations E, B, and C. These relations are defined within HOL using Tom Melham's excellent rule induction package [2,8]. The notation $s[v/x]$ indicates the state s updated so that $(s[v/x])(x) = v$.

E	**Number:** $$\overline{E(n)\,s\,n\,s}$$ **Variable:** $$\overline{E(x)\,s\,s(x)\,s}$$	**Increment:** $$\frac{E\,x\,s_1\,n\,s_2}{E(++x)\,s_1\,(n+1)\,s_2[(n+1)/x]}$$
	Addition: $$\frac{E\,e_1\,s_1\,n_1\,s_2,\quad E\,e_2\,s_2\,n_2\,s_3}{E(e_1+e_2)\,s_1\,(n_1+n_2)\,s_3}$$	**Subtraction:** $$\frac{E\,e_1\,s_1\,n_1\,s_2,\quad E\,e_2\,s_2\,n_2\,s_3}{E(e_1-e_2)\,s_1\,(n_1-n_2)\,s_3}$$

B	**Equality:** $$\frac{E\,e_1\,s_1\,n_1\,s_2,\quad E\,e_2\,s_2\,n_2\,s_3}{B(e_1=e_2)\,s_1\,(n_1=n_2)\,s_3}$$	**Less Than:** $$\frac{E\,e_1\,s_1\,n_1\,s_2,\quad E\,e_2\,s_2\,n_2\,s_3}{B(e_1<e_2)\,s_1\,(n_1<n_2)\,s_3}$$	
	Conjunction: $$\frac{B\,b_1\,s_1\,t_1\,s_2,\quad B\,b_2\,s_2\,t_2\,s_3}{B(b_1\wedge b_2)\,s_1\,(t_1\wedge t_2)\,s_3}$$	**Disjunction:** $$\frac{B\,b_1\,s_1\,t_1\,s_2,\quad B\,b_2\,s_2\,t_2\,s_3}{B(b_1\vee b_2)\,s_1\,(t_1\vee t_2)\,s_3}$$	**Negation:** $$\frac{B\,b\,s_1\,t\,s_2}{B(\sim b)\,s_1\,(\sim t)\,s_2}$$

C	**Skip:** $$\overline{C\,\text{skip}\,s\,s}$$ **Abort:** (no rules) **Assignment:** $$\frac{E(e)\,s_1\,n\,s_2}{C(x:=e)\,s_1\,s_2[n/x]}$$ **Sequence:** $$\frac{C\,c_1\,s_1\,s_2,\quad C\,c_2\,s_2\,s_3}{C(c_1;c_2)\,s_1\,s_3}$$	**Conditional:** $$\frac{B\,b\,s_1\,T\,s_2,\quad C\,c_1\,s_2\,s_3}{C(\text{if } b \text{ then } c_1 \text{ else } c_2)\,s_1\,s_3}$$ $$\frac{B\,b\,s_1\,F\,s_2,\quad C\,c_2\,s_2\,s_3}{C(\text{if } b \text{ then } c_1 \text{ else } c_2)\,s_1\,s_3}$$ **Iteration:** $$\frac{B\,b\,s_1\,T\,s_2,\quad C\,c\,s_2\,s_3,\quad C(\text{assert } a \text{ while } b \text{ do } c)\,s_3\,s_4}{C(\text{assert } a \text{ while } b \text{ do } c)\,s_1\,s_4}$$ $$\frac{B\,b\,s_1\,F\,s_2}{C(\text{assert } a \text{ while } b \text{ do } c)\,s_1\,s_2}$$

Table 3. Programming Language Structural Operational Semantics

The semantics of the assertion language AL is given by recursive functions defined on the structure of the construct, in a directly denotational fashion:

$V\,v\,s$: numeric expression v : vexp evaluated in state s, yields a numeric value in num.
$A\,a\,s$: boolean expression a : aexp evaluated in state s, yields a truth value in bool.

V	$V\,n\,s = n$
	$V\,x\,s = s(x)$
	$V\,(v_1 + v_2)\,s = V\,v_1\,s + V\,v_2\,s$
	(–, * treated analogously)
A	A **true** $s = \mathrm{T}$
	A **false** $s = \mathrm{F}$
	$A\,(v_1 = v_2)\,s = (V\,v_1\,s = V\,v_2\,s)$ (< treated analogously)
	$A\,(a_1 \wedge a_2)\,s = (A\,a_1\,s \wedge A\,a_2\,s)$
	($\vee$, ~, $\Rightarrow$, $a_1{=}a_2$, $a_1{=}{>}a_2|a_3$ treated analogously)
	$A\,(\textbf{close}\ a)\,s = (\forall s_1.\,A\,a\,s_1)$
	$A\,(\forall x.\,a)\,s = (\forall n.\,A\,a\,s[n/x])$
	$A\,(\exists x.\,a)\,s = (\exists n.\,A\,a\,s[n/x])$

Table 4. Assertion Language Denotational Semantics

7 Substitution

We define proper substitution on assertion language expressions using the technique of *simultaneous substitutions*, following Stoughton [11]. The usual definition of proper substitution is a fully recursive function. Unfortunately, HOL only supports primitive recursive definitions. To overcome this, we use simultaneous substitutions, which are represented by functions of type `subst = var->aexp`. This describes a family of substitutions, all of which are considered to take place simultaneously. This family is in principle infinite, but in practice all but a finite number of the substitutions are the identity substitution ι. The virtue of this approach is that the application of a simultaneous substitution to an assertion language expression may be defined using only primitive recursion, not full recursion, and then the normal single substitution operation of [v/x] may be defined as a special case:

$$[v/x] = \lambda y.(y{=}x \Rightarrow v \mid \text{AVAR } y).$$

We apply a substitution by the infix operator $\triangleleft$. Thus, $a\triangleleft ss$ denotes the application of the simultaneous substitution ss to the expression a, where a can be either `vexp` or `aexp`. Therefore $a\triangleleft[v/x]$ denotes the single substitution of the expression v for the variable x wherever x appears free in a. Finally, there is a dual notion of applying a simultaneous substitution to a state, instead of to an expression; this is called *semantic substitution*, and is defined as $s\triangleleft ss = \lambda y.(V\,(ss\,y)\,s)$.

Most of the cases of the definition of the application of a substitution to an expression are simply the distribution of the substitution across the immediate subexpressions. The interesting cases of the definition of $a\triangleleft ss$ are where a is a quantified expression, e.g.:

$$(\forall x.\,a) \triangleleft ss = \textbf{let } free = \bigcup_{z \in (FV\,a) - \{x\}} FV\,(ss\,z) \quad \textbf{in}$$
$$\textbf{let } y = variant\ x\ free\ \textbf{in}$$
$$\forall\,y.\ a \triangleleft (ss[(\text{AVAR } y)\,/\,x])$$

Here FV is a function that returns the set of free variables in an expression, and *variant x free* is a function that yields a new variable as a variant of x, guaranteed not to be in the set *free*.

Once we have defined substitution as a syntactic manipulation, we can then prove the following two theorems about the semantics of substitution:

$$\vdash \forall v\, s\, ss.\ V(v \lhd ss)\, s\ =\ V v\, (s \lhd ss)$$

$$\vdash \forall a\, s\, ss.\ A(a \lhd ss)\, s\ =\ A a\, (s \lhd ss)$$

This is our statement of the Substitution Lemma of logic, and essentially says that syntactic substitution is equivalent to semantic substitution.

8 Translation

Expressions have typically not been treated in previous work on verification; there are some exceptions, notably Sokolowski [10]. Expressions with side effects have been particularly excluded. Since expressions did not have side effects, they were often considered to be a sublanguage, common to both the programming language and the assertion language. Thus one would see expressions such as $p \wedge b$, where p was an assertion and b was a boolean expression from the programming language.

One of the key realizations of this work was the need to carefully distinguish these two languages, and not confuse their expression sublanguages. This then requires us to *translate* programming language expressions into the assertion language before the two may be combined as above. In fact, since we allow expressions to have side effects, there are actually two results of translating a programming language expression e:

- an assertion language expression, representing the value of e in the state "before" evaluation, *and*
- a simultaneous substitution, representing the change in state from "before" evaluating e to "after" evaluating e.

For example, the translator for numeric expressions is defined using a helper function:

$VE1$: exp -> subst -> (aexp # subst):

$$
\begin{aligned}
VE1\,(n)\, ss &= n,\, ss &&\text{(where comma (,) makes a pair)}\\
VE1\,(x)\, ss &= ss\, x,\, ss\\
VE1\,(\texttt{++}x)\, ss &= (ss\, x) + 1,\, ss[((ss\, x) + 1)\, /\, x]\\
VE1\,(e_1 + e_2)\, ss &= (VE1\ e_1 \to \lambda v_1.\, (VE1\ e_2 \to \lambda v_2\, ss_2.\, (v_1 + v_2,\, ss_2)))\, ss\\
VE1\,(e_1 - e_2)\, ss &= (VE1\ e_1 \to \lambda v_1.\, (VE1\ e_2 \to \lambda v_2\, ss_2.\, (v_1 - v_2,\, ss_2)))\, ss
\end{aligned}
$$

where $\to$ is a "translator continuation" operator, defined as
$$(f \to k)\, ss\ =\ \textbf{let}\ (v,\, ss') = f\, ss\ \textbf{in}\ k\, v\, ss'$$

Then define
$$
\begin{aligned}
VE\ e &= \text{fst}\,(VE1\ e\ \iota) &&\text{(where } \iota \text{ is the identity substitution}\\
VE\_state\ e &= \text{snd}\,(VE1\ e\ \iota) &&\text{and fst and snd } select\ the\ members\ of\ a\ pair)
\end{aligned}
$$

We can then prove that these translation functions, as syntactic manipulations, are semantically correct, according to the following theorem:

$$\vdash \forall e\, s_1\, n\, s_2.\ (E\, e\, s_1\, n\, s_2) = (\ n = V(VE\, e)\, s_1\ \wedge\ s_2 = s_1 \lhd (VE\_state\ e)\,)$$

A similar set of functions are used to translate boolean expressions. We define the helper function *AB1* and the main translation functions *AB* and *AB_state*, and prove their correctness as

$$\vdash \ \forall \, b \, s_1 \, t \, s_2. \ (B \, b \, s_1 \, t \, s_2) = (\ t = A \, (AB \, b) \, s_1 \ \wedge \ s_2 = s_1 \triangleleft (AB\_state \, b))$$

These theorems mean that every evaluation of a programming language expression has its semantics completely captured by the two translation functions for its type. These are essentially small compiler correctness proofs.

As a product, we may now define the simultaneous substitution that corresponds to an assignment statement, overriding the expression's state change with the change of the assignment:

$$[x := e] = (VE\_state \, e)[(VE \, e) \, / \, x]$$

9 Axiomatic Semantics

We define the semantics of Floyd/Hoare partial correctness formulae as follows:

aexp:	$\{a\} = \textbf{close } a$	(the universal closure of a)
	$= \forall s. A \, a \, s$	(a is true in all states)
exp:	$\{p\}e\{q\} = \forall p \, q \, e \, n \, s_1 \, s_2. \ A \, p \, s_1 \wedge E \, e \, s_1 \, n \, s_2 \Rightarrow A \, q \, s_2$	
bexp:	$\{p\}b\{q\} = \forall p \, q \, b \, t \, s_1 \, s_2. \ A \, p \, s_1 \wedge B \, b \, s_1 \, t \, s_2 \Rightarrow A \, q \, s_2$	
cmd:	$\{p\}c\{q\} = \forall p \, q \, c \, s_1 \, s_2. \ A \, p \, s_1 \wedge C \, c \, s_1 \, s_2 \Rightarrow A \, q \, s_2$	

Table 5. Floyd/Hoare Partial Correctness Formulae Semantics

Given these formulae, we can now express the axiomatic semantics of the programming language, and *prove* each rule as a theorem from the previous structural operational semantics:

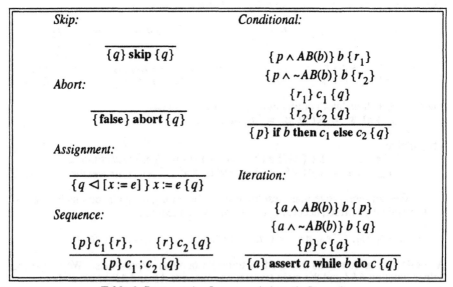

Table 6. Programming Language Axiomatic Semantics

The most interesting of these proofs was that of the **while**-loop rule. It was necessary to prove a subsidiary lemma first, by the strong version of rule induction for command semantics provided by Tom Melham's rule induction package. This lemma thus used versions of itself for "lower levels" in the relation built up by rule induction to prove each instance, and so needed strong induction to present as a usable assumption each hypothesized lower-level tuple in the relation. The subsidiary lemma was necessary because the **while**-loop rule as a theorem was not in the right syntactic form for the induction tactic. The lemma we proved is

$$\vdash \forall a\,b\,c\,p\,q.\ \{p\}c\{a\} \ \wedge \ \{a \wedge (AB\ b)\}b\{p\} \ \wedge \ \{a \wedge {\sim}(AB\ b)\}b\{q\} \ \Rightarrow$$
$$\forall w\,s_1\,s_4.\ C\,w\,s_1\,s_4 \Rightarrow$$
$$((w = \mathbf{assert}\ a\ \mathbf{while}\ b\ \mathbf{do}\ c\ ) \Rightarrow (A\,a\,s_1 \Rightarrow A\,q\,s_4))$$

Although we did prove analogous theorems as an axiomatic semantics for both the numeric and boolean expressions in the programming language, it turned out that there was a better way to handle them provided through the use of the translation functions. Using these translation functions, we may define functions to compute the appropriate precondition to an expression, given the postcondition, as follows.

vexp	ve_pre $e\ v$	$= v \vartriangleleft (VE\_state\ e)$
	vb_pre $b\ v$	$= v \vartriangleleft (AB\_state\ b)$
aexp	ae_pre $e\ a$	$= a \vartriangleleft (VE\_state\ e)$
	ab_pre $b\ a$	$= a \vartriangleleft (AB\_state\ b)$

Table 7. Expression Precondition Functions

We may now prove the following axiomatic semantics for expressions:

Numeric expression precondition:	*Boolean expression precondition:*
$\{\text{ae\_pre}\ e\ q\}\ e\ \{q\}$	$\{\text{ab\_pre}\ b\ q\}\ b\ \{q\}$

Table 8. Programming Language Expression Axiomatic Semantics

These precondition functions now allow us to revise the rules of inference for conditionals and loops, as follows.

Conditional:
$$\{r_1\}c_1\{q\}, \qquad \{r_2\}c_2\{q\}$$
$$\overline{\{AB\ b\ \Rightarrow\ \text{ab\_pre}\ b\ r_1\ |\ \text{ab\_pre}\ b\ r_2\}\ \mathbf{if}\ b\ \mathbf{then}\ c_1\ \mathbf{else}\ c_2\ \{q\}}$$

Iteration:
$$\{a \wedge AB\ b \Rightarrow \text{ab\_pre}\ b\ p\}$$
$$\{a \wedge {\sim}(AB\ b) \Rightarrow \text{ab\_pre}\ b\ q\}$$
$$\{p\}c\{a\}$$
$$\overline{\{a\}\ \mathbf{assert}\ a\ \mathbf{while}\ b\ \mathbf{do}\ c\ \{q\}}$$

Table 9. Programming Language Axiomatic Semantics (revisions)

10 Verification Condition Generator

We now define a verification condition generator for this programming language. To begin, we first define a helper function *vcg1*, of type `cmd->aexp-> (aexp # (aexp)list)`. This function takes a command and a postcondition, and returns a precondition and a list of verification conditions that must be proved in order to verify that command with respect to the precondition and postcondition. This function does most of the work of calculating verification conditions.

This is called by the main verification condition generator function, *vcg*, defined with type `aexp->cmd->aexp-> (aexp)list`. *vcg* takes a precondition, a command, and a postcondition, and returns a list of the verification conditions for that command.

In these definitions, comma (,) makes a pair of two items, square brackets ([]) delimit lists, semicolon (;) within a list separates elements, and ampersand (&) is an infix version of HOL's APPEND operator to join two lists.

vcg1	$vcg1$ (**skip**) q $= q,$ []
	
	 $vcg1$ (**abort**) q = **true**, []
	
	 $vcg1$ $(x := e)$ q $= q \triangleleft [x := e],$ []
	
	 $vcg1$ $(c_1 ; c_2)$ q = **let** $(r,h_2) = vcg1$ c_2 q **in**
	      let $(p,h_1) = vcg1$ c_1 r **in**
	        $p,$ $(h_1$ & $h_2)$
	
	 $vcg1$ (**if** b **then** c_1 **else** c_2) q =
	      let $(r_1,h_1) = vcg1$ c_1 q **in**
	      let $(r_2,h_2) = vcg1$ c_2 q **in**
	        $(AB$ b => ab_pre b r_1 | ab_pre b $r_2),$ $(h_1$ & $h_2)$
	
	 $vcg1$ (**assert** a **while** b **do** c) q =
	      let $(p,h) = vcg1$ c a **in**
	        $a,$ $[a \wedge$ AB $b \Rightarrow$ ab_pre b p ;
	          $a \wedge \sim(AB$ $b) \Rightarrow$ ab_pre b q] & h

vcg	vcg p c q = **let** $(r,h) = vcg1$ c q **in** $[p \Rightarrow r]$ & h

Table 10. Verification Condition Generator

The correctness of these VCG functions is established by proving the following theorems from the axioms and rules of inference of the axiomatic semantics:

VCG1_THM	$\vdash \forall c$ $q.$ **let** $(p,h) = vcg1$ c q **in** (**every close** h $\Rightarrow$ $\{p\}c\{q\}$)
VCG_THM	$\vdash \forall p$ c $q.$ **every close** $(vcg$ p c $q)$ $\Rightarrow$ $\{p\}c\{q\}$

Table 11. Verification of Verification Condition Generator

every P *lst* is defined in HOL as being true when for every element x in the list *lst*, the predicate P is true when applied to x. Accordingly, **every close** h means that the universal closure of every verification condition in h is true.

These theorems are proven from the axiomatic semantics by induction on the structure of the command involved. This verifies the VCG. It shows that the *vcg* function is *sound*, that the correctness of the verification conditions it produces suffice to establish the correctness of the annotated program. This does not show that the *vcg* function is *complete*, that if a program is correct, then the *vcg* function will produce a set of verification conditions sufficient to prove the program correct from the axiomatic semantics [3]. However, this soundness result is quite useful, in that we may directly apply these theorems in order to prove individual programs partially correct within HOL, as seen in the next section.

11 Example Programs

Given the *vcg* function defined in the last section and its associated correctness theorem, proofs of program correctness may now be partially automated with security. This has been implemented in an HOL tactic, called   VCG_TAC, which transforms a given program correctness goal to be proved into a set of subgoals which are the verification conditions returned by the *vcg* function. These subgoals are then proved within the HOL theorem proving system, using all the power and resources of that theorem prover, directed by the user's ingenuity.

As an example, we will take the quotient/remainder algorithm for integer division by repeated subtraction. The program to be verified, with the annotations of the loop invariant and pre- and postconditions, is

$$\{x0 = x \,\wedge\, y0 = y\}$$
$$r := x;$$
$$q := 0;$$
assert $x0 = q * y0 + r \,\wedge\, y0 = y$
while $\sim(r < y)$ **do**
$$\quad r := r - y;$$
$$\quad q := ++q$$
od
$$\{x0 = q * y0 + r \,\wedge\, r < y0\}$$

With the assumption that the program cannot change the values of $x0$ and $y0$, this specification means that if the program terminates, the final value of q must be the quotient of the division of $x0$ by $y0$, and r the remainder. (We could also prove that the final values of x and y are unchanged by this algorithm.)

Applying VCG_TAC to this goal produces the following three verification conditions:

VC1: $x0 = x \,\wedge\, y0 = y \,\Rightarrow\, x0 = 0 * y0 + x \,\wedge\, y0 = y$

VC2: $x0 = q * y0 + r \,\wedge\, y0 = y \,\wedge\, \sim(r < y)$
$\Rightarrow\, x0 = (q + 1) * y0 + (r - y) \,\wedge\, y0 = y$

VC3: $x0 = q * y0 + r \,\wedge\, y0 = y \,\wedge\, r < y$
$\Rightarrow\, x0 = q * y0 + r \,\wedge\, r < y0$

Here is a transcript of the application of VCG_TAC to this problem. We have written a parser for the subject language, using the parser library in HOL, invoked using the delimiters "[[" and "]]". The partial correctness goal is parsed and converted into the abstract syntax form used internally, which is printed as the current goal.   VCG_TAC then converts that goal into verification conditions in the Object Language of HOL.

```
#g [[ {x0 = x /\ y0 = y}
#       r := x;
#       q := 0;
#       assert x0 = q * y0 + r /\ y0 = y
#       while ~(r < y) do
#          r := r - y;
#          q := ++q
#       od
#    {x0 = q * y0 + r /\ r < y0}
# ]];;
"CSPEC
 (AAND
  (AEQ(AVAR(VAR `x0` 0))(AVAR(VAR `x` 0)))
  (AEQ(AVAR(VAR `y0` 0))(AVAR(VAR `y` 0))),
  SEQ
  (SEQ
   (ASSIGN(VAR `r` 0)(PVAR(VAR `x` 0)))
   (ASSIGN(VAR `q` 0)(NUM 0)))
  (WHILE
   (AAND
    (AEQ
     (AVAR(VAR `x0` 0))
     (APLUS
      (AMULT(AVAR(VAR `q` 0))(AVAR(VAR `y0` 0)))
      (AVAR(VAR `r` 0))))
    (AEQ(AVAR(VAR `y0` 0))(AVAR(VAR `y` 0))))
   (NOT(LESS(PVAR(VAR `r` 0))(PVAR(VAR `y` 0))))
   (SEQ
    (ASSIGN(VAR `r` 0)(MINUS(PVAR(VAR `r` 0))(PVAR(VAR `y` 0))))
    (ASSIGN(VAR `q` 0)(INC(VAR `q` 0))))),
  AAND
  (AEQ
   (AVAR(VAR `x0` 0))
   (APLUS
    (AMULT(AVAR(VAR `q` 0))(AVAR(VAR `y0` 0)))
    (AVAR(VAR `r` 0))))
  (ALESS(AVAR(VAR `r` 0))(AVAR(VAR `y0` 0))))"

() : void
Run time: 6.1s

#e(VCG_TAC);;
OK..
3 subgoals
"!x0 q y0 r y.
   ((x0 = (q * y0) + r) /\ (y0 = y)) /\ r < y ==>
   (x0 = (q * y0) + r) /\ r < y0"

"!x0 q y0 r y.
   ((x0 = (q * y0) + r) /\ (y0 = y)) /\ ~r < y ==>
   (x0 = ((q + 1) * y0) + (r - y)) /\ (y0 = y)"

"!x0 x y0 y.
   (x0 = x) /\ (y0 = y) ==> (x0 = (0 * y0) + x) /\ (y0 = y)"

() : void
Run time: 80.3s
Intermediate theorems generated: 5643
```

These verification conditions are each solved as a subgoal by normal HOL techniques.

The Object Language variables involved in these verification conditions are constructed to have names similar to the original program variable names; if there is a non-zero variant number, it is appended to the variable name. Thus, if one changed the name of program variable x to z in the example above, the verification conditions would be the same but with the OL variable z in place of x.

Here is the HOL definition of the VCG_TAC tactic:

```
    let VCG_TAC =
(a)       MATCH_MP_TAC vcg_THM
(b)       THEN REWRITE_TAC[vcg;vcg1]
          THEN CONV_TAC (DEPTH_CONV let_CONV)
(c)       THEN REWRITE_TAC[ab_pre;assign]
(d)       THEN REPEAT (CHANGED_TAC
                  (BETA_TAC THEN
                  REWRITE_TAC[VE1_DEF;VE_DEF;VE_state_DEF;
                              AB1_DEF;AB_DEF;AB_state_DEF;
                              IDENT_SS_var;trans_cont]))
          THEN REWRITE_TAC[a_subst_IDENT]
(e)       THEN CONV_TAC vcg_CONV
(f)       THEN REWRITE_TAC[APPEND_INFIX;APPEND;EVERY_DEF;CLOSE]
(g)       THEN CONV_TAC (TOP_DEPTH_CONV INTERPRET_aexp_CONV)
(h)       THEN REWRITE_TAC[V_DEF]
          THEN CONV_TAC (DEPTH_CONV var_BND_CONV)
          THEN REPEAT CONJ_TAC
          THEN ( GEN_TAC ORELSE ALL_TAC )
(i)       THEN INTERPRET_PROG_VARS_TAC;;
```

The VCG_TAC tactic first (a) applies the theorem VCG_THM, the last theorem of Table 11 of the previous section, to the current goal using the HOL tactic MATCH_MP_TAC to reason backwards from the program correctness statement to the invocation of the *vcg* function. By the theorem, the proof of these verification conditions will establish the proof of the original program correctness statement.

The next step of VCG_TAC is to "execute" the various syntactic manipulation functions mentioned in the current goal by symbolically rewriting the goal using the definitions of the functions. This applies (b) to the *vcg* function, (c) to the operators that create substitutions, (d) to the translation functions, (e) to the substitution functions, and others. Because the rewriting process is done symbolically, instead of actually executing a program, it is relatively slow, but complete soundness is assured. This "execution" converts the invocation of the *vcg* function on the annotated program into the actual set of verification conditions that the *vcg* function returns.

The tactic makes use at (e) of a set of conversions, culminating in vcg_CONV, to test the equality of variables (var_EQ_CONV), lookup a variable in a simultaneous substitution (var_BND_CONV), calculate a variant of a variable (variant_CONV), apply a substitution to an expression (subst_CONV), and reduce a term with nested "let" and substitution operators in an efficient order (vcg_CONV), among others.

After performing these conversions, the program correctness goal is left as a set of "constant" verification conditions in the assertion language. VCG_TAC then (f–i) uses the definitions of the semantics of the assertion language to rewrite these verification conditions into equivalent statements in the Object Language of HOL, beginning with (f) the definition of close, then proceeding with (g) the definitions of A and (h) V. In particular, (g) all quantification over assertion language variables, and (i) all references to assertion language variables within program states, are converted to references to similarly-named OL variables. These verification conditions are then presented to the user as the necessary subgoals that need to be solved in order to complete the proof of the program originally presented.

12 Future Work

In the future, we intend to extend this work to include several more language features, principally mutually recursive procedures and concurrency. In addition, we also intend to cover total correctness, beyond the partial correctness issues dealt with in this paper.

The work on mutually recursive procedures requires many new concepts and techniques to define the semantics and perform verification condition generator proofs. These include declarations of procedures, their collection into environments, their verification independent of actual use of the procedures, well-formedness conditions on programs, and the very delicate issue of parameter passing. We wish to find a method of proving the total correctness of systems of mutually recursive procedures which is efficient and suitable for processing by a VCG.

Concurrency raises a whole host of new issues, ranging from the level of structural operational semantics ("big-step" versus "small-step"), to dealing with assertions describing temporal sequences of states instead of single states, to issues of fairness. We believe that a proper treatment of concurrency will exhibit qualities of modularity and compositionality. *Modularity* means that a specification for a process should state both (a) the assumptions under which it should operate, and (b) the task (or commitment) which it should meet, given those assumptions. *Compositionality* means that the specification of a system of processes should be verifiable in terms of the specifications of the individual constituent processes.

13 Summary and Conclusions

The fundamental contribution of this work is the exhibition of a tool to ease the task of proving programs which is itself proven to be sound. This verification condition generator tool performs an automatic, syntactic transformation of the annotated program into a set of verification conditions. The verification conditions produced are themselves proven within HOL, establishing the correctness of the program within the same system wherein the VCG was verified.

This proof of the correctness of the VCG may be considered as an instance of a compiler correctness proof, with the VCG translating from annotated programs to lists of verification conditions. Each of these has its semantics defined, and the VCG correctness theorem closes the commutative diagram, showing that the truth of the verification conditions implies the truth of the annotated program.

The programming language and its associated assertion language are represented by new concrete recursive datatypes. This implies that they are completely independent of other data types and operations existing in the HOL system, without any hidden associations that might affect the validity of proof. This requires substantial work in defining their semantics and in proving the axioms and rules of inference of the axiomatic semantics from the operational semantics. However, this deeply embedded approach yields great expressiveness, ductility, and the ability to prove as theorems within HOL the correctness of various syntactic manipulations, which could only be stated as meta-theorems before. These theorems encapsulate a level of reasoning which now does not need to be repeated every time a program is verified, raising the level of proof from the semantic level to the syntactic. But the most important part of this work is the degree of trustworthiness of this syntactic reasoning. Verification condition generators are not new, but we are not aware of any other proofs of their correctness to this level of rigor. This enables program proofs which are both trustworthy and effective to a degree not previously seen together.

References

1. Sten Agerholm, "Mechanizing Program Verification in HOL", in *Proceedings of the 1991 International Workshop on the HOL Theorem Proving System and its Applications, Davis, August 1991*, edited by M. Archer, J. J. Joyce, K. N. Levitt, and P. J. Windley (IEEE Computer Society Press, 1992), pp. 208–222.

2. J. Camilleri and T. Melham, "Reasoning with Inductively Defined Relations in the HOL Theorem Prover", Technical Report No. 265, University of Cambridge Computer Laboratory, August 1992.

3. Stephen A. Cook, "Soundness and Completeness of an Axiom System for Program Verification", in *SIAM Journal on Computing*, Vol. 7, No. 1, February 1978, pp. 70–90.

4. G. Cousineau, M. Gordon, G. Huet, R. Milner, L. Paulson, and C. Wadsworth, *The ML Handbook* (INRIA, 1986).

5. Michael J. C. Gordon, "Mechanizing Programming Logics in Higher Order Logic", in *Current Trends in Hardware Verification and Automated Theorem Proving*, ed. P.A. Subrahmanyam and Graham Birtwistle, Springer-Verlag, New York, 1989, pp. 387–439.

6. Michael J. C. Gordon, and T. F. Melham, *Introduction to HOL, A theorem proving environment for higher order logic*, Cambridge University Press, Cambridge, 1993.

7. S. Igarashi, R. L. London, and D. C. Luckham, "Automatic Program Verification I: A Logical Basis and its Implementation", *ACTA Informatica* 4, 1975, pp. 145–182.

8. Tom Melham, "A Package for Inductive Relation Definitions in HOL", in *Proceedings of the 1991 International Workshop on the HOL Theorem Proving System and its Applications, Davis, August 1991*, edited by M. Archer, J. J. Joyce, K. N. Levitt, and P. J. Windley (IEEE Computer Society Press, 1992), pp. 350–357.

9. L. C. Ragland, "A Verified Program Verifier", Technical Report No. 18, Department of Computer Sciences, University of Texas at Austin, May 1973.

10. Stefan Sokolowski, "Partial Correctness: The Term-Wise Approach", *Science of Computer Programming*, Vol. 4, 1984, pp. 141–157.

11. Allen Stoughton, "Substitution Revisited", *Theoretical Computer Science*, Vol. 59, 1988, pp. 317–325.

12. Glynn Winskel, *The Formal Semantics of Programming Languages, An Introduction*, The MIT Press, Cambridge, Massachusetts, 1993.

S : A Machine Readable Specification Notation based on Higher Order Logic

J. Joyce, N. Day and M. Donat

Department of Computer Science
University of British Columbia
Vancouver, British Columbia, Canada

Abstract. This paper introduces a new notation called S which is based on higher order logic. It has been developed specifically to support the practical application of formal methods in industrial scale projects. The development of S has occurred in the context of an investigation into the possibility of using formal specification techniques in the development of a \$400 million air traffic control system. We were motivated to develop this notation after reaching the conclusion that existing notations such as Z are not suitable for use in this particular project. In addition to providing an introduction to S, this paper describes a public domain software tool called "Fuss" which has been implemented to support the use of S as a specification language.

1 Introduction

S[1] is a machine readable notation developed specifically to serve as a specification language for the practical application of formal methods in industrial scale projects. The notation is based upon the formalism of higher order logic and, in this respect, shares much in common with the machine readable "object language" of the HOL system [5]. The development of S has also been influenced by the Z notation [10]. Like Z, and unlike the HOL object language, S includes constructs for the declarations and definitions of types and constants. S also includes a specification construct called a "template" which serves as a packaging mechanism in the same way that "schemas" are often used in Z specifications. Although the development of S has been influenced by our desire to use S to specify software, hardware and mixed software-hardware systems, S is a "pure" notation in the sense that it does not include any built-in concepts of hardware or software.

To support the use of S as a specification notation, we have developed a prototype implementation of a tool called "Fuss" that checks an S specification in a manner analogous to the checking of a Z specification by a commercial tool called "Fuzz" [11]. In particular, our tool checks for conformance to rules of syntax and typing. Additionally, we have prototyped an extension to Fuss that translates an S specification into an ML script (for HOL88). When used as input to the HOL system, the ML script generates a HOL theory consisting of

---

[1] Its name comes from thinking of it as a kinder, gentler Z.

declarations and definitions from the S specification. Thus, the S notation and Fuss could be used as a front-end for a formal verification methodology which involves the use of the HOL system. Similarly, we are interested in the possibility of developing extensions to Fuss that translate S specifications into the notations of other higher-order logic theorem-proving tools and perhaps even other kinds of verification tools such as model checkers.

Section 2 elaborates on the context, motivation and objectives of the development of S. Section 3 provides a brief description of S [2]. In Section 4, we illustrate the use of S in terms of the translation of a (hypothetical) software requirement for an air traffic control system into S. Section 5 briefly describes Fuss, our prototype implementation of a tool for checking specifications written in S. This is followed by a description of a tool, implemented as an extension to Fuss, that translates S specifications into an ML script. Related work is discussed in Section 7. Section 8 briefly mentions some of our plans for the future development of Fuss and related tools. Finally, some advantages of the use of S and Fuss are considered in the conclusion.

2 Context, Motivation and Objectives

The development of S has occurred in the context of a project in which we are investigating the possible use of formal specification techniques in a $400 million air traffic control system. This system, called CAATS (Canadian Automated Air Traffic System), is reputed to be the largest software project in Canada. It has been under development by Hughes Aircraft of Canada since 1989 and is scheduled for delivery in late 1996. The software requirements for this system are expressed in approximately 3,100 pages of highly structured natural language. The CAATS Software Requirements Specifications (SRSs) is based on the concept of a thread, namely, "a path through a system that connects an external event or stimulus to an output event or response" [4]. This threads-based approach differs significantly from other traditional approaches to software requirements specification such as Structured Analysis. Given that much of the software requirements specification effort for this project was nearing completion when our investigation began, we have focused our efforts on the possibility of retrofitting formal specification techniques with the existing natural language based specification documents. In this context, we have realized the importance of using a notation that supports the development of formal specifications that closely match the level of abstraction, organization, and notational conventions of a pre-existing natural language specification.

Initially we experimented with Z for capturing CAATS software requirements in a mathematical notation. In a preliminary study, we translated a very small fragment of the software requirements specification for CAATS into approximately 60 pages of liberally documented Z. Although we found that there was a very good match in *style* between Z specification and the threads-based

---

[2] We are currently preparing a complete user manual for the notation and the Fuss tool [8].

approach used in the CAATS SRSs, we soon became disenchanted with Z for several reasons. Given that Z involves many non-ASCII characters and a graphical presentation format (where components of the specification are presented as combinations of text and various kinds of box-like shapes), it is not possible to directly create and edit a Z specification using a standard text editor. Instead, one must typically create and edit a Z specification indirectly in terms of a sequence of text formatting commands which, upon processing by a suitable text formatter, will produce the desired Z specification. We found this particularly cumbersome – and, in our opinion, an approach based on Z would be difficult to retrofit into an existing software process. We also found that the semantics of Z can be difficult to explain to others who do not have prior experience with formal methods – and sometimes, we even found it difficult within our team to reach agreement on the meaning of certain Z fragments.

We also considered some alternatives to Z. For instance, we briefly considered the use of the HOL object language (a formulation of higher order logic) and the HOL system. Although we felt that a HOL approach, given its higher-order logic foundation, offers the advantage of semantics that is easier to understand and explain to others, we also felt that it would be unreasonable to propose a practical specification methodology for use in an industrial context that would depend on the use of the HOL system as a tool for checking specifications. Since the HOL object language does not include constructs for the declaration and definitions of types and constants, one must also become familiar with the HOL meta-language – namely, ML – to create formal specifications that can be checked by the HOL system. Not only does this place the additional burden of understanding of ML onto the shoulders of the specifier, but moreover, the distinction between the meta and object languages of HOL are frequently a source of confusion for new users of the HOL system.

Given that we were unable to find a notation and supporting tools that would suit our investigation into possible application of formal specification techniques in CAATS, we were motivated to develop our own notation and tools. Our development of S and Fuss was based upon a number of objectives including:

- the notation must be based entirely on ASCII characters so that it can be edited directly using standard text editors and incorporated into other documents without processing;
- the semantics must be relatively simple so that specifications can be understood by users who do not necessarily have expertise in formal methods;
- the notation must be supported by an easy-to-use software tool that automatically checks a specification with respect to rules of syntax and typing.

In addition to these main objectives, we have incorporated a number of features into S that we found useful in our efforts to formally specify selected aspects of CAATS. Some of these features are described in the next section of this paper.

3   A Brief Description of S

S can be described as a superset of the HOL object language in the sense that
expressions of the HOL object language are also expressions of S. There are a
few minor syntactic differences. For example, the reserved words "**function**",
"**forall**", "**exists**" and "**select**" are used in S instead of "\\", "!", "?" and "**@**"
for lambda expressions, universal quantification, existential quantification and
Hilbert's choice operator respectively. The notation "**if A then B else C** is
used in S instead of "**A => B | C**" for conditional expressions. The type operator
for Cartesian product in S is "*****" instead of "**#**".

The S notation also includes constructs for the declaration and definition of
types and constants. In this respect, S is similar to Z which also includes constructs for the declaration and definition of types and constants. However, the
inclusion of these constructs in S represents a significant difference between S
and the HOL object language which does not provide constructs for the declaration of types and constants. In the case of HOL, constructs for the declaration
and definition of types and constants are part of the HOL meta-language – in
particular, they are functions defined in the ML programming language which
is used to implement the HOL system. We have included constructs for the
declaration and definition of types and constants in S because it is essential to
have such constructs to write formal specifications — and we desire the notation
to be "stand-alone"', that is, independent of any meta-language or supporting
software tool.

3.1   Declaration and Definition of Types

A *type declaration* in S may be used to introduce one or more new types. For
example, the paragraph,

    : flight_plan, aerodrome, customs_office;

is an example of a type declaration paragraph. This example introduces three
new types, "**flight_plan**", "**aerodrome**" and "**customs_office**". Types introduced by means of a type declaration paragraph are "uninterpreted" in the sense
that nothing is said about their composition. A type declaration paragraph may
be used in an S specification in the same way that the declaration of one or
more "basic types" may be used in a Z specification to achieve a desired level
of abstraction. A type declaration paragraph may also be used to introduced
parameterized types. For example, the type declaration,

    : (tyvar) set;

introduces a parameterized type, "**set**", which is parameterized by a single type
variable, "**tyvar**".

A *type definition* may also be used to introduce a new type. Unlike a type
declaration, the definition of a type describes its composition in terms of a set

of constructors. These constructors are implicitly declared as S constants. The constructors of a type definition are functions that map zero or more objects to objects of the new type. For example,

```
: ty := c1 | c2 :num | c3 :bool :bool*num | c4 :ty;
```

defines a new type "ty" in terms of four constructors, "c1", "c2", "c3" and "c4".

The type definition may be recursive – that is, the new type may appear on the right-hand side of the type definition provided that at least one of the branches does not contain a reference to the new type. The limitations on recursive type definitions are the same as those found in HOL.

A type definition may also be parameterized by one or more more type variables. For example,

```
: (tyvar) list := NULL | CONS :tyvar :(tyvar)list;
```

is a recursive type definition parameterized by a type variable, "tyvar".

In addition to type declarations and type definitions, S provides a construct for the introduction of *type abbreviations*. In this case, a new type is not declared or defined – instead, a new name is introduced which can be used in place of potentially more complicated type expression. For example,

```
: ty == num -> bool;
```

results in the introduction of "ty" as an abbreviation for "num -> bool".

3.2   Declaration and Definition of Constants

A *constant declaration* in S, for example,

```
c : num;
```

may be used to introduce one or more new constants. In this case, a new constant, "c", has been declared as a constant with type "num". Constants may also be declared to be infix constants by surrounding the name of the new constant with underscores as illustrated, for instance, by the following declaration:

```
(_ is_greater_than _) : num -> num -> bool;
```

Another way to introduce constants is by means of a *constant definition*. A constant definition is equivalent to the declaration of a new constant and the introduction of a definitional axiom for the new constant. For instance, the following constant definition,

```
plustwo n := n + 2;
```

is an example of a simple, non-recursive constant definition. They may be recursively defined in terms of a recursively defined type. Constant definitions, along with constant declarations, may also be parameterized by type variables. This is illustrated by the following definition,

```
(:tyvar) length NULL := 0 |
         length (CONS (x:tyvar) s) := 1 + length s;
```

which defines a new constant, "length", in terms of a recursively defined type, "list", whose two constructors are "NULL" and "CONS". The recursive definition of "length" is parameterized by a type variable, "tyvar". This definition also illustrates the use of pattern matching in constant definitions.

We consider the explicit declaration of type variables in declarations and definitions to be useful as a mechanism for discouraging a specifier from blindly parameterizing a declaration or definition by an excessive number of type parameters.

3.3 Templates

The S notation also includes a construct called a *template* which we have found to be particularly useful as a packaging mechanism when specifying software systems. The notion of a template was originally inspired by the idea of a schema in Z. But aside from the possibility of using S to write Z-like specifications, the template construct is a solution to a problem often seen in formal specifications of large systems where logical expressions grow to unreadable proportions because of long lists of parameters associated with various operators (e.g., functions and predicates).

A template can be described as a predicate with "implicit parameters". These parameters are "implicit" in the sense that they do not appear as actual parameters in an expression when the template is referenced. For instance, a template "A" with two implicit parameters, "x" and "y", is referenced as just "A" instead of "A(x,y)". The names of the implicit parameters of a template are given in the definition of the template together with zero or more Boolean expressions that specify constraints on these implicit parameters and/or global constants. For example,

```
A := {x,y:num; x < y};
```

is a definition of template "A" with the constraint that "x" is less than "y".

A template may be referenced in any expression that follows the definition of the template. A template reference may appear anywhere that a Boolean expression may appear. For example, the expression,

```
forall y. ~(y = 0) ==> exists x. A
```

expresses the arithmetic fact that every natural number except zero is greater than some other natural number. In the above expression, "x" and "y" refer to the implicit parameters of template "A".

Another example is the following template definition which includes a reference to template "**A**":

```
B := {z:bool; A ==> z; x > 9};
```

This template definition consists of a single variable declaration and two constraint expressions. The first of these two constraint expressions contains a reference to template "**A**".

By means of a simple syntactic transformation called "template expansion", any S specification can be converted into an equivalent S specification where all of the template definitions have been replaced by constant definitions, and references to the defined templates replaced by references to these constants. For example, the above template definitions would be transformed into the following two constant definitions,

```
A(x,y) := x < y;
B(x,y,z) := ((x < y) ==> z) /\ (x > 9);
```

and expressions such as,

```
forall y. ~(y = 0) ==> exists x. A
forall x y z. A \/ B
```

would be transformed by means of template expansion into the following expressions:

```
forall y. ~(y = 0) ==> exists x. x < y
forall x y z. (x < y) \/ (((x < y) ==> z) /\ (x > 9))
```

Thus, the template construct of S can be seen as mechanism for defining constants where the parameters are not given explicitly. Although not very interesting in theoretical terms, this construct has considerable practical value when used in the formal specification of large systems. For instance, the template construct allows a top level specification of a requirement, as illustrated by our example in Section 4, to be expressed in a manner that is not cluttered with parameters.

Another Z-like feature of S is that templates can be "decorated" with one or more decoration symbols – namely, "**'**", "**!**" and "**?**". (The use of "**!**" and "**?**" in S is entirely different than the use of these symbols in the HOL object language where they are used as quantifiers. In S, as in Z, the decoration of a name with one of these symbols is merely a naming convention.) The decoration of a template reference has the effect, upon template expansion, of decorating each of the implicit parameters of the template and each of the template references that appear in the definition of the decorated template reference. For example, an expression such as,

```
forall x y. A \/ B'
```

where the reference to "B" has been decorated with a single "'" would result, upon template expansion, in the following expression:

```
forall x y. (x < y) \/ (((x' < y') ==> z') /\ (x' > 9))
```

where x', y' and z' are free.

3.4 Names of Types, Constants and Templates

Another useful feature of S is the ability to use phrases such as "this is a very long name" as the names of types, constants and templates. With the exception of a small set of standard mathematical and logical symbols such as "+", "=" and "<", a name containing characters that are neither upper/lower case letters, digits nor the underscore character must be enclosed inside a pair of matching double quotes. For example:

```
"the successor of three" := 4;
```

There are several reasons why we allow names in S to be arbitrary strings of characters. In part, this is a reflection of the fact that arbitrary strings are often used as names in the natural language specification of a large software system. For example, names such as,

```
<operator enter filed flight plan>
[filed flight plan]
\Validate IFR/CVFR Flight Plan Route\
```

are all examples of names that appear in the CAATS software requirements specification. In addition to the separation of words with whitespace characters, the above examples include characters such as "<", ">", "[", "]" and "\" which have special significance with respect to naming conventions for CAATS documentation. For example, a string enclosed inside a matching pair of angle brackets, "<" and ">", indicates a reference to an entry in the project data dictionary. Given this use of arbitrary strings in natural language specifications, we have allowed arbitrary strings to be used as names in S specifications so that a greater correspondence between the natural language specification and its formalization in S can be achieved.

3.5 Post-Fix Function Application

A software specification frequently involves the representation of data objects with multiple attributes. For instance, the example in the next section of this paper involves the representation of flight plans in a hypothetical air traffic control system where a flight plan has a number of attributes including the names of the departure and destination aerodromes. As illustrated in the next section, these attributes can be represented formally in S as functions. For instance, we can introduce a function,

```
"<departure aerodrome>" : flight_plan -> aerodrome;
```

that maps a given flight plan to an aerodrome. Although the application of such functions to objects can be expressed in terms of simple juxtaposition,

```
"<departure aerodrome>" new_flight_plan
```

we have found that a specification can be made more readable by using a post-fix form of function application where the operator appears after the operand and is separated from the operand by a dot:

```
new_flight_plan."<departure aerodrome>"
```

This post-fix form of function application suggests the idea of a record type with a field called ""<departure aerodrome>"" – or in the case of an object-oriented approach, the informal idea of an "flight_plan" object with ""<departure aerodrome>"" as one of its attributes. Mathematically, this syntax is nothing more than post-fix application of a function to a value, but this syntax is very helpful when attempting to draw upon the intuition of specification readers familiar with these more established kinds of specification and programming notations.

4 Example

In this section, we illustrate the use of S as a notation for the formalization of a software requirement for a hypothetical air traffic control system. Our example is motivated by our familiarity with the CAATS software requirements specification – however, this fragment is *not* intended to be a representation of any requirement in the CAATS software requirements specification.

Consider the following natural language specification of a software requirement for part of the system responsible for processing flight plans as they are entered into the system.

Requirement 57:
"If a new flight plan has been entered and the <departure aerodrome> of the new flight plan is outside Canada and the <destination aerodrome> is inside Canada, then send the new flight plan to the Canada Customs office serving the <destination airport>".

We begin our formalization of this requirement by focusing on the high level logical structure of the above natural language sentence – in particular, by identifying the main logical connectives of the requirement, breaking up the natural language specification into phrases, and re-assembling these phrases into an S expression based on the high level logical structure of the sentence. In this case, the natural language specification can be broken up into four separate phrases,

"a new flight plan has been entered"
"the <departure aerodrome> of the new flight plan is outside Canada"
"the <destination aerodrome> of the new flight plan is inside Canada"
"send the new flight plan to the Canada Customs office serving the
<destination aerodrome>"

and then assembled into an expression using three logical connectives that we
can define in S – namely, NOT for logical negation, AND for logical conjunction
and a "split" connective, if ... then ..., for logical implication. This yields the
following constant definition which represents the top level formalization of the
above natural language requirement:

```
Requirement 57 :=
  if (
    "a new flight plan has been entered"
    AND "the <departure aerodrome> of the new flight plan is
      outside Canada"
    AND "the <destination aerodrome> of the new flight plan is
      inside Canada")
  then
    "send the new flight plan to the Canada Customs office serving
    the <destination aerodrome>";
```

In the above constant definition, each of the four phrases is formalized as a
template reference. The second step in the formalization of this requirement is
the definition of the four phrases as templates:

```
"a new flight plan has been entered" :=
  {
    source : sender;
    new_flight_plan : flight_plan;
    source has_sent (new_flight_plan_message(new_flight_plan))
  };

"the <departure aerodrome> of the new flight
      plan is outside Canada" :=
  {
    new_flight_plan : flight_plan;
    NOT("is inside Canada"
            (new_flight_plan."<departure aerodrome>"))
  };

"the <destination aerodrome> of the new flight
      plan is inside Canada" :=
  {
    new_flight_plan : flight_plan;
    "is inside Canada"(new_flight_plan."<destination aerodrome>")
```

```
};
```

```
"send the new flight plan to the Canada Customs office serving the
<destination aerodrome>" :=
  {
    new_flight_plan : flight_plan;
    new_flight_plan_message(new_flight_plan) is_sent_to
      (new_flight_plan."<destination aerodrome>".customs_office)
  };
```

The third step in the formalization of this hypothetical requirement is the declaration of constants and types used in the above template definitions. In particular, this involves the declaration of the following constants,

```
(_ has_sent _) : sender -> message -> bool;
(_ is_sent_to _) : message -> receiver -> bool;
new_flight_plan_message : flight_plan -> message;
"<departure aerodrome>" : flight_plan -> aerodrome;
"<destination aerodrome>" : flight_plan -> aerodrome;
"is inside Canada" : aerodrome -> bool;
customs_office : aerodrome -> receiver;
```

which, in turn, are based on the declaration of the following types:

```
: flight_plan, message, sender,
    receiver, aerodrome;
```

Since these types and constants are likely to be used to specify other related requirements, the declaration of these types and constants would be part of the supporting infrastructure shared by various components of the formal specification.

5 Fuss – A Tool for Specification in S

To support the use of S as a notation for formal specification, we have developed a prototype implementation of a tool called "Fuss" that checks an S specification for conformance to rules of syntax and typing. The tool may be used in batch mode (in the same way that Fuzz may be used to check a Z specification). Fuss may also be used in an interactive manner in combination with a text editor to incrementally develop an S specification. A user may interactively query Fuss about the current state of the S specification under development – for example, a user can inquire about the type of a particular constant as determined by a previous type definition, constant declaration or constant definition.

Fuss does not currently check that certain rules of definition are satisfied by type definitions and constant definitions. However, we expect to extend Fuss to make these additional kinds of checks.

In addition to its function as a specification checker, Fuss builds an internal representation of an S specification that can be accessed through a C programming language interface. Thus, Fuss can be used as the front-end for user-defined applications that accept S specifications as input and apply some specific processing or transformation to the S specifications. An example of a user-defined application would be a tool that transforms an S specification into the specification notation of a verification tool. The 'S to HOL' translator described in the next section is implemented in this manner.

6 S and HOL

We have developed an extension to Fuss that translates an S specification into an ML file — the meta-language of the HOL system – which can then be used as input to the HOL system. It generates a theory consisting of declarations and definitions of types and constants corresponding to the declarations and definitions of types, constants and templates of the S specification. The first step of this transformation involves the replacement of template definitions and template references with constant definitions and constant references as described earlier in Section 3.3. The rest of the transformation is straightforward given the fact that S and the HOL object language share a common logical foundation, i.e., higher-order logic and moreover, a very similar syntax. When an S specification involves names that would be rejected by the HOL system — for example, the use of "<destination aerodrome>" as the name of a constant – these are transformed algorithmically into a name that will be accepted by the HOL system.

Currently, this extension serves as a "one-way" interface from S to HOL. We plan to eventually implement a "HOL to S" extension to the HOL system so that the declarations and definitions of a HOL theory and its ancestors could be translated into S and dumped into an input file for Fuss. This would allow a S specification to be created on top of some existing HOL infrastructure such as a HOL library.

The transformation of an S specification into an ML file for use as input to the HOL system allows the HOL system to be used as a verification tool for an S based approach to formal specification and verification of a software system.

In CAATS, we are experimenting with the use of the HOL system as a mechanism for generating test cases for a major part of the CAATS development program called System Integration Testing (SIT). In this approach, test cases are formally derived as logical consequences of one or more software requirements. The use of HOL for this purpose, as a means of increasing the level of automation in this aspect of developing a large software system, may offer very significant benefits in terms of cost reduction over the traditional approach to test case generation which is manually intensive and prone to human error. Our initial experiments with this approach have been very encouraging and we are now examining how this technique could be developed into a production process.

7 Related Work

Although we regard S as an independent notation, its development is clearly related to efforts by others which may be described as efforts to combine Z with the HOL system. ICL [7] has developed an "industrial strength" version of the HOL system, called ProofPower, which includes a "deep" semantic embedding of Z in the underlying logical framework of the HOL system, namely, higher-order logic. A simpler approach taken by Bowen and Gordon [1] is based upon a "shallow" embedding of a subset of Z for the HOL system. Both of these approaches demonstrate the possibility of a semantic link between Z and the HOL system which allows a Z specification to be mechanically translated into a representation within the logical framework of the HOL system. One possible problem with this kind of approach is that a complete and "safe" Z-to-HOL translation process may depend on a substantial amount of logical infrastructure to bridge the considerable gap between the mathematical foundation of Z and the mathematical foundation of the HOL system. While it may be possible to create this infrastructure, it is unclear to us how successfully this infrastructure can be hidden to allow the HOL system to be used in a practical way to reason about a Z specification. This kind of approach may be contrasted with our scheme with translating S into input for the HOL system which only involves a tiny amount of infrastructure since both S and the HOL system share a common mathematical foundation, namely, higher-order logic.

VDM and Larch are two other formal specification languages that can be used for similar purposes as S. As with Z, VDM uses some built-in non-ASCII symbols as operators. In Larch, the user defines their own set of symbols possibly drawing from a library [12]. This means that like S, no special, non-ASCII symbols have to be used.

In both VDM and Larch the specification is usually formatted in terms of pre- and post-conditions explicitly unlike Z or S. A Larch specification is further partitioned into an interface specification, giving pre- and post-conditions for each operation, and a "trait", where the function symbols are defined algebraically [12]. With S, there is more flexibility to adapt the format of the specifications to existing styles when retrofitting formal specification techniques into an on-going project.

8 Future Work

We are now considering extensions to S and Fuss which would allow descriptive (extra-logical) information to be associated with fragments of an S specification. For instance, this may include the ability to annotate fragments of S specifications with information that records links to "higher level" requirements and other project information.

In addition to some of the planned refinements of Fuss already mentioned in this paper, we are interested in the possibility of providing support for the execution of an "operational subset" of S.

We also plan to investigate the integration of S with a state machine formalism which is better suited to modeling concurrent components. SpecCharts [9] is a graphical, hierarchical formalism similar to Statecharts [6] except that the actions take place within a state (known as a behaviour in SpecCharts). These actions are described in VHDL. For higher levels of abstraction or software systems, S would be a better language for describing the actions than VHDL. By using an executable subset of S, it would be possible to model check requirements of these specifications similar to the manner in which a model checker was created for Statecharts [3].

Another possibility is to develop extensions to S that allow decision tables in a graphical format which could be compiled into S for analysis purposes.

9 Conclusion

S, together with support provided by Fuss, offers several advantages over the development of formal specifications directly in the HOL object language using the HOL system as a specification checker. For instance, the Fuss tool can be used effectively to check a specification without any expertise beyond an understanding of the S notation. This contrasts with the use of the HOL system as a specification checker which requires some understanding of the HOL system meta-language in addition to an understanding of the HOL object language. Another very important practical advantage of S is the fact that the Fuss system is a relatively small C program which we expect will be easily ported to a wide variety of platforms including those often found in engineering environments such as PCs.

Similarly, S and Fuss offer several advantages over the use of Z and Z-based specification checking tools such as Fuzz. For instance, one very important advantage – particularly, when one considers the possibility of using formal specifications in a large project involving non-academic document preparation tools – is that S is based entirely on printable ASCII characters. An S specification can be created directly using an ordinary text editor and incorporated into other documents without any special text formatting tools. This contrasts with Z where one must typically create a machine-readable specification in terms of a sequence of text formatting commands which, after processing by a suitable text formatting tool, results in a graphical representation of the Z specification. This is a benefit for either retrofitting formal specification techniques into existing requirements documents or when starting a new project since the supporting natural language description is very important for interpreting the meaning of the formal specification.

Another advantage of S is the existence of a verification tool with a substantial user community, namely the HOL system, which can be used to reason almost directly about an S specification using the same proof mechanisms that would likely be used to reason about the same specification expressed directly in the HOL object language.

Therefore, we believe S satisfies our goal of creating a specification language for the practical application of formal methods in industrial scale projects while leaving the door open to apply current verification techniques to specifications written in this language. So far, S, and the benefits of formal methods in general, have been welcomed with interest at Hughes. We have given fragments of specifications written in S to Hughes employees with no knowledge of formal methods and they have been able to understand them. Our next step is to see how Hughes employees react to working with the notation themselves.

10 Acknowledgements

In addition to the authors of this paper, work was completed on this project by S. Kahan, M. Wong Cheng In, and Z. Zhu. This work has been supported by funds from the B.C. Advanced Systems Institute, Hughes Aircraft of Canada and the Natural Science and Engineering Research Council of Canada.

References

1. Jonathan Bowen and Mike Gordon. Z and HOL. Draft copy.
2. D. Craigen, S. Gerhart and T. Ralston. An International Survey of Industrial Applications of Formal Methods (2 Volumes). Technical Report #NRL/FR/5546-93-9581, Naval Research Laboratory, Washington, D.C. .
3. Nancy Day. A Model Checker for Statecharts. Technical Report 93-35, Department of Computer Science, University of British Columbia, October, 1993.
4. Michael S. Deutsch and Ronald R. Willis. *Software Quality Engineering - A Total Technical and Management Approach.* Prentice Hall Series in Software Engineering, Englewood Cliffs, New Jersey, 1988.
5. M. J. C. Gordon and T. F. Melham (eds.,). *Introduction to HOL: a theorem proving environment for higher order logic.* Cambridge University Press, 1993.
6. David Harel. Statecharts: A visual formalism for complex systems. *Science of Computing*, 8:231-274, 1987.
7. R.B. Jones. ICL ProofPower. BCS FACS FACTS, 1(1): 10 13, 1992. Series III.
8. J. Joyce and N. Day. S: A General Purpose Specification Notation. In preparation.
9. Sanjiv Narayan, Frank Vahid, and Daniel D. Gajski. System Specification with the SpecCharts Language. *IEEE Design and Test of Computers*, pages 6-13, December, 1992.
10. J.M. Spivey. *The Z Notation: A Reference Manual.* 2nd edition, Prentice-Hall, 1992.
11. J.M. Spivey. *The fuzz Manual.* 2nd edition, Computer Science Consultancy.
12. Jeannette M. Wing. A Specifier's Introduction to Formal Methods. *Computer*, 23(9):8-22, September, 1990.

An Engineering Approach to
Formal Digital System Design

Mats Larsson

Department of Computer and Information Science,
Linköping University, S-581 83 Linköping, Sweden
Email: mla@ida.liu.se

Abstract. This paper describes a first attempt at building design tools
that amalgamate theorem proving and engineering methods. To gain
acceptance such a tool must focus on the engineering task and proof
steps must be hidden. From these ideas a prototype system based on the
HOL proof assistant has been designed. The key features of this system
are threefold. First, we use window reasoning for modelling the design
process; Second, we have defined a set of application specific derived
inference rules that implement common design tasks; Third, we have
extended the design representation in logic with annotations to support
efficient algorithmic reasoning.

1 Introduction

As production technology has continued to improve during the last decades,
the complexity of digital systems has increased accordingly. This ever increasing complexity has led to digital systems having shorter life cycles and to the
design task growing more difficult. At the same time, the use of digital systems
in safety critical applications — such as pacemakers, aeroplane stability control,
and nuclear plant control — has led to a need to be able to ensure correctness.
Correctness is of course an important criteria for all designs but these applications have served to emphasise the problem. Thus, we need to reduce the design
time while at the same time design more complex systems, and ensure their
correctness.

These requirements have initiated a growing interest in applying formal methods to digital system design. There are two main approaches — *post-design
verification* and designing by *correctness-preserving transformations* — where
post-design verification was the first but has failed to make any significant impact on industry practices. One reason for this is the intrinsic complexity of
formal verification, amplified by the often large semantic gap between the two
models to be verified and the lack of information sharing with the design process. By using correctness-preserving transformations we avoid these problems
since the proof steps are hidden in the transformations, information sharing is
automatic, and the semantic gap is bridged by taking small steps guided by the
designers knowledge. The advantage of post-design verification is that the verification problem is separate from the design process thus making it possible to

focus on one aspect at a time. A formal design method[1] on the other hand, must support **both** the design and the verification process.

The formal design method presented here uses correctness-preserving transformations and is targeted towards the reification of a, possibly partial, formal design specification in logic into a formal specification of a complete design that implements the initial design specification. Furthermore, it supports the implementation of efficient algorithmic design functions.

Previous work in this area has focused on methods where transformations are either proved correct using logic [5, 8] or stated correct according to a design algebra [6]. The issue of efficient algorithmic reasoning has however not been emphasised.

The main contribution of this paper is demonstrating the applicability of modelling the digital system design process using window inference and making the method accessible to designers by supplying a set of application specific design transformations. A second contribution is a method to extend the design representation with information necessary to support the design task without cluttering the logical representation used for verification and make efficient algorithmic reasoning possible.

2 Our Approach

We regard design as an iterative process that progresses stepwise by transforming the design specification until a satisfactory design has been reached. A schematic view of this process is depicted in Fig. 1. This is a monotone process and corrections can only be introduced at the state where the error was introduced. To support this we can traverse the window stack backwards undoing design steps in the process, and we can reuse design transformations just as we can with proofs.

To be able to use computer-assisted formal reasoning we must model the process of design in a formal system. We do this by specifying a satisfaction relation that must hold between consecutive design specifications. This way we do not formalize the design process as such but rather a property of it.

Our model of the design process gives no guidance in choosing which design transformation to apply at a certain design state. It only rules out some choices for not satisfying the satisfaction relation or the design checks. Therefore it is important that the designer has access to a set of design analysis functions to assist him in making a good choice. This kind of algorithmic reasoning necessary in any realistic design system should be possible to perform in an efficient way. We identify the following criteria to be important for a formal design system:

- enforce consistency;
- computationally efficient;
- support designer.

---

[1] A method based on correctness-preserving transformations.

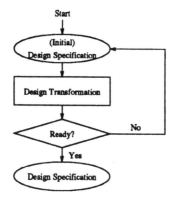

Fig. 1. The design process

When designing we must capture other aspects of the design than when we verify, e.g. physical properties and constraints. This leads to formalisms aimed at design in general being more complex than formalisms for verification [9]. This together with the efficiency criteria implies that only information that will be used for verification should be represented in logic. Our solution is a two-level design representation where we represent additional properties of the design objects outside the logic. We use annotations of the design objects (*design annotations*) to represent such properties. In the design annotations we record information that can be used to check and analyse designs. For example, assume the designer makes a connection between two ports. In the logic interconnection is modelled as two devices sharing the name of a port. By annotating the port as representing a line we make this information accessible for design analysis, e.g. list all unconnected ports, and design checks, e.g. that two lines with the same name are not introduced.

To support the design task we define transformation procedures for common design tasks, thus providing the designer with a set of application specific procedures for transforming the design or, in other words, a design algebra. These transformations are behaviourally correct by proof and incorporates design checks that checks, possibly application specific, design constraints and make use of information from design analysis.

3 A Prototype System

To try out the ideas presented in Sect. 2 we have implemented a prototype system [7], a schematic view of which is depicted in Fig. 2. It is based on the HOL proof assistant and uses the window inference package [4] extended with support for design annotations. We use the HOL proof assistant because it is a mature and well supported proof system with properties such as safe symbolic reasoning about design objects, safe extensions to the basic logic via definitions, and a programmable interface to the logic [2].

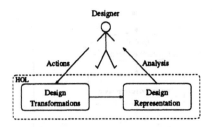

Fig. 2. The prototype system

3.1 Design Representation

We use predicates in higher-order logic to model behaviour in the style presented
by Gordon [3]. This choice is natural since we want our satisfaction relation to
ensure functional correctness, i.e. what most approaches to post-design verification do. An example is the definition of the adder shown below[2]

```
|- !del in1 in2 out.
    ADD del(in1,in2)out = (!t. out(t + del) = (in1 t) + (in2 t))
```

where !t. out(t + del) = (in1 t) + (in2 t) is the term specifying the behaviour
of the adder, ADD del(in1,in2)out is the predicate abbreviating the specification,
and del in1 in2 out is the list of parameters. In this approach a specification
treats a device as a black box, i.e. the behaviour is defined only in terms of the
values that can be observed externally. A *design specification* is a conjunction
of predicates specifying behaviour, $P_1 \wedge \ldots \wedge P_n$. A *partial specification* does
not specify the behaviour of a device in full detail, e.g. not for all possible values
on the external variables.

Design Annotations. When reifying abstract specifications into concrete specifications we have to take into consideration physical attributes — such as limited
resources, direction of ports, and the existence of physical carriers to transport
signals on — of the concrete design objects. The tasks that are to be performed
when reifying a design specification are scheduling, allocation, binding, and interconnection. Scheduling is the task of assigning computations to time steps.
By allocation we mean the introduction of concrete resources — such as devices and lines — into our design specification. Binding is the mapping of the
unlimited set of abstract operations onto the limited set of allocated concrete
resources. Interconnection is the task of allocating and binding physical carriers
to transport data between allocated devices.

To support these tasks in the prototype system we have defined a basic
set of design annotations to keep track of resources, such as *devices* and *lines*,
and properties, such as *direction of ports*. We also define a set of functions to

---

[2] ! is the ASCII representation of universal quantification in HOL.

manipulate the design annotations. We make a distinction between updating functions, *design annotation updates*, such as:

```
new_device, new_line, del_line and new_time
```

and querying functions, *design annotation queries*, such as:

```
get_device, list_device, list_devices, list_inputs, list_outputs,
get_line, list_lines, list_sources and list_drains
```

With the help of these functions we can write functions to perform design checks and analysis. We can, for example, write a boolean function to test if a variable is an input port. Such a function could look like this in ML[3]:

```
let is_input name anno = mem name (list_inputs anno);;
```

3.2 Window Inference

A key component of our approach is the use of the *window inference* package in HOL to model the transformational design process. We only sketch window reasoning here. A more detailed account can be found elsewhere [4, 7].

In window inference the state of a design is recorded in a window. Among the information recorded is a *design theorem* that relates two design specifications by a relation, R, and a set of *assumptions* under which the design theorem holds:

$$\Gamma \vdash specification_n \ R \ specification_0$$

Where Γ is the assumptions, $specification_0$ is the initial, and $specification_n$ the current, design specification. The only requirement on R is that it is a preorder[4]. A design derivation consists of a stack of windows where the top window denotes the current state of design and the bottom window denotes the initial design state. This stack can be traversed using window commands. To modify the contents of a window we can apply a window transformation command, wtc:

$$\text{wtc: } window \times theorem \rightarrow window$$

Where *theorem* is the logical justification of the modification and the resulting window is pushed onto the window stack. To transform the previous window we could apply a *justification theorem* of the form:

$$\Gamma \vdash specification_{n+1} \ R \ specification_n$$

This results, thanks to the transitivity of R, in a new window holding the following design theorem (*see* also Fig. 3):

$$\Gamma \vdash specification_{n+1} \ R \ specification_0$$

---

[3] let is a declaration of variables and mem $x \ l$ returns true if some element of l is equal to x, otherwise it returns false.

[4] A preorder is a reflexive, transitive relation.

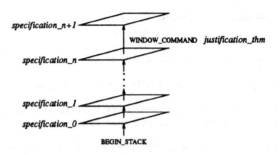

Fig. 3. A window stack

Derivation of Subdesigns. An important property of window inference is that it supports partitioning the design tree into manageable subdesigns while still having access to information from the subdesigns environment. Consider that we would like to transform a subexpression, s_0, in the design theorem below:

$$\Gamma \vdash specification_n \, [s_0] \, R \, specification_0$$

We then open a subwindow on s_0 and get a new (but related) window stack with an initial window holding the design theorem:

$$\Gamma, \gamma \vdash s_0 \, r \, s_0$$

Where γ is the additional set of assumptions that hold in, and r the relation preserved by, the subwindow. We transform this theorem as above in m steps giving us the following design theorem:

$$\Gamma, \gamma \vdash s_m \, r \, s_0$$

This process is depicted in Fig. 4. When we are satisfied with the form of s_m we close the subwindow, thus substituting s_0 for s_m in $specification_n$:

$$\Gamma \vdash specification_n \, [s_m] \, R \, specification_0$$

Note that when we close a sub-derivation window, the entire sub-derivation collapses into a single design transformation at the next level of stacks. This is depicted in Fig. 5. When we are satisfied with the current design specification we close the window stack. The result of the design derivation is the final window's design theorem.

Extended Window Inference. Since the window inference package does not support the use of design annotations we must extend the definition of a window in the package. We do this by implementing the design annotations as an abstract data type in ML and add this to the definition of a window together with interface functions for accessing and updating annotations in a window. Note that it is essential that we keep design annotations on the window stack in order to keep the design state consistent when traversing the window stack.

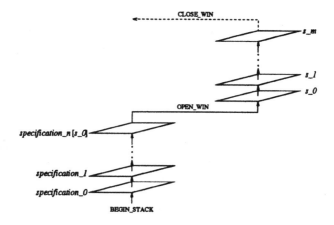

Fig. 4. A hierarchy of window stacks

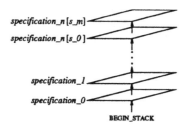

Fig. 5. A collapsed window stack hierarchy

To begin reasoning with the extended window inference system the user creates a window stack as before. Creating a window stack to transform our specification under the assumptions Γ, while preserving R, now results in a single window containing:

$$\Gamma \vdash specification_0 \ R \ specification_0 \quad \text{and} \quad annotation_0$$

Modifying the design must now result in a transformation of the current design specification while preserving R, **and** in an update of the design annotation.

$$\Gamma \vdash specification_1 \ R \ specification_0 \quad \text{and} \quad annotation_1$$

After the transformation the window theorem records the proof that achieves the initial specification, and the window annotation records the current design annotation.

Satisfaction Relation. Since we want to do design reification we must use a weaker satisfaction relation than equality. Unfortunately, this means that we do not get the total correctness property inherent in equational reasoning.

In the prototype system we have chosen to use logical implication as satisfaction relation. The motivation for this is that implication is a simple, well understood relation that intuitively fits the role of satisfaction relation for design and there already exists much support for proving theorems about implication in HOL.

3.3 Window Transforms

To facilitate formal design we have defined a set of derived inference rules corresponding to basic design steps, and we use the name *window transforms* to denote them. Every window transform applies a built-in or derived justification theorem to the current design theorem, and an update to the current design annotation to create a new design state. If either the justification or the update fails, the transform fails. This way the design state is kept consistent.

We can build libraries of window transforms to support different needs— such as applications, technologies, design styles, and company rules. The fact that the small set implemented here is enough for the designs we have experimented with indicates that a library of basic transforms need not have an intimidating size.

Apart from these application specific window commands we also make use of built-in window commands as well as many theorems and inference rules in HOL. The basic transforms are presented next.

Allocation and Binding of Devices. In the prototype system there are three kinds of devices — computational devices, multiplexors, and delays — each of which have their own set of transforms for allocation and binding.

Allocation is modelled by introducing a predicate defining a physical resource and annotating this as a device with a unique name. The justification is created by specializing the theorem $\forall t_1\, t_2.\, t_1 \wedge t_2 \supset t_2$. Allocation only fails if the given name is already used in the design. The generic window transform for allocation is ALLOC_TRANS.

Binding is modelled by associating an operator with a device in the design specification. The association has the form of a set of constraints on the ports of the device that results from matching the body of the device with the current design specification. If the matching succeeds the current design specification is replaced by the resulting set of constraints. The justification is a simple proof of equivalence. Every time we bind a computational device to an operator a new (unique) variable is introduced to denote the time at which the device will perform the operation. This variable is later used for scheduling. The window transforms for binding are BIND_TRANS, BIND_MUX_TRANS, and BIND_DELAY_TRANS.

Often we want to perform allocation and binding in one step, i.e. we introduce a device and bind it to the current design specification. The window transforms for allocation and binding are ALLOC_BIND_TRANS, MUX_TRANS, and DELAY_TRANS.

To further automate the design task we have defined a transform that traverses a design specification and allocates and binds a delay wherever possible. This window transform is named MAP_DELAY_TRANS.

Scheduling. We model scheduling the computations by placing them on a common time frame. To do this we have written a procedure to analyse the data-flow of the current design specification using the design annotations and come up with a schedule according to the asap (as soon as possible) strategy. This procedure returns a mapping from the unique variables denoting time that was introduced by the binding to a common time variable. If this succeeds then the variables in the design specification is renamed according to this mapping. This is justified by the rule INST in HOL. This rule instantiates free variables in a theorem. The window transform for scheduling is SCHEDULE_TRANS.

Interconnection. Interconnection is modelled as two devices sharing the name of a port variable which is annotated as being a line. This is implemented by replacing the names of the original ports with the line name in the current design specification. This is justified by the rule INST in HOL. The annotation update fails when we try to create a line with a name that is already used in the design or if we try to extend a line that does not exist in the design. It will also fail if we try to connect two output ports. The window transforms for interconnection are CONNECT_TRANS that creates a new line from a set of ports and ATTACH_TRANS that adds a port to an existing line.

To further automate the design task we have defined a transform that traverses a design specification and interconnects ports wherever possible. We name this MAP_CONNECT_TRANS.

4 An Example

To illustrate formal design with the prototype system, we sketch the derivation of an infinite input response (IIR) filter. A detailed presentation of this example can be found elsewhere [7].

We start by entering the initial design specification at the top-level of a window stack with the BEGIN_STACK window command. The resulting output from the system is:

```
==> * (output m = 0) /\
       (output(SUC(t + m)) = (a * (input t)) + (b * (output(t + m)))))
```

The notation used in the example is as follows. The current design specification is denoted by $\star$, and the relation that the current window preserves is denoted by ==> which is the ASCII representation of implication in HOL.

In the specification m is the latency of the system which will be generated during the design and a and b are constants we instantiate the specification with. The latency tells us how many time units we have to wait for the first output to appear whereas the delay tells us with which frequency we will generate new results after the initial latency. In this example, the delay of the specification is fixed to 1 time unit by the use of SUC in the specification of the filter. The number of time units it takes for an input to have an effect on the output is thus determined by the sum of the latency and the delay.

Since the delay of the filter specification is 1 time unit and the devices we will use to implement the filter have delays of 2 and 4 time units (arbitrary choice) we realise that the (macro) time unit used in the specification and the (micro) time unit used in the implementation can not be the same. Thus we have used a more abstract notion of time when we defined the filter than what we did when we specified the behaviour of the adder and the multiplier. The use of abstract notions of time is a convenient method to specify the temporal ordering of events in situations where the time between events is not important. In the case of the filter we have specified that the current output of the filter will depend on the previous input and the previous output with out saying anything about how long ago previous was. To be able to design by refinement from such abstract specifications we need a method to relate two notions of time. We do this with abstraction predicates:

$$\vdash \text{REFINE } n \text{ } abs \text{ } ref = (abs = \lambda t. \, ref(n * t))$$

This predicate relates two functions of time with a variable n. By introducing this predicate with one argument bound to a port variable and the other argument bound to a new variable of the same type, we can use the definition of the predicate to rewrite the macro time specification to be expressed in terms of micro time. We introduce a REFINE predicate and instantiate it so that it relates *output* to a new variable, *out*. Note that REFINE does not represent a device but a timing constraint.

```
==> * Refine n output out /\
    (output m = 0) /\
    (output(SUC(t + m)) = (a * (input t)) + (b * (output(t + m)))))
```

We introduce an abstraction predicate for the input too and rewrite our original specification with the bodies of the abstraction predicates:

```
==> * Refine n input in /\
    Refine n output out /\
    ((\t. out(n * t))m = 0) /\
    ((\t. out(n * t))(SUC(t + m)) =
    (a * ((\t. in(n * t))t)) + (b * ((\t. out(n * t))(t + m)))))
```

After applying beta conversion to all subterms and expanding the successor function we have the following design:

```
==> * Refine n input in /\
    Refine n output out /\
    (out(n * m) = 0) /\
    (out(n * ((t + m) + 1)) =
    (a * (in(n * t))) + (b * (out(n * (t + m))))))
```

When we look at the design specification we see that the output is defined by two conjuncts that together form a conditional statement on the value of the time variable. This is easier to see if we state it in the following way:

$$out\,(n*(t+m)) = ((t = 0) \rightarrow 0 \mid (a*in(n*(t-1))) + (b*out(n*((t-1)+m))))$$

We have defined one possible implementation of this behaviour, using a MUX device, as a design transform, INIT_MUX_TRANS. We apply this to the design specification:

```
==> * Refine n input in /\
      Refine n output mux1_out /\
      MUX(mux1_sel,mux1_in1,mux1_in2)mux1_out /\
      (mux1_sel(n * m) = T) /\
      (mux1_sel(n * ((t + m) + 1)) = F) /\
      (mux1_in1(n * m) = 0) /\
      (mux1_in2(n * ((t + m) + 1)) =
      (a * (in(n * t))) + (b * (mux1_out(n * (t + m))))))
```

We have now completed the initialization phase of the design. A graphical representation of the current design can be seen in Fig. 6 where constraints on ports are denoted by dotted lines.

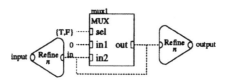

Fig. 6. The design after the initialization phase

The next design phase will be to synthesise the data path specified by the last conjunct in the current design specification. We start by opening a subwindow on that conjunct, and proceed to allocate computational resources. Since we can identify the outermost operator in the conjunct to be an addition we need a device that can perform addition. We use the adder defined in Sect. 3.1. In order to use this device we must allocate a device of this kind and bind it to this instance of addition. We do this in one step with the window transform ALLOC_BIND_TRANS 'add1' "ADD 2" where 'add1' is the string annotating this device instance and "ADD 2" identifies and partially instantiates the parametrized device definition. The introduced device is scheduled to perform the operation at some time point denoted by a new variable. The new design specification is:

```
    ! Refine n input in
    ! Refine n output mux1_out
    ! MUX(mux1_sel,mux1_in1,mux1_in2)mux1_out
    ! mux1_sel(n * m) = T
    ! mux1_sel(n * ((t + m) + 1)) = F
    ! mux1_in1(n * m) = 0
    | input = (\t. in(n * t))
    | !t. mux1_out t = (mux1_sel t => mux1_in1 t | mux1_in2 t)
==> * ADD 2(add1_in1,add1_in2)add1_out /\
    (mux1_in2(n * ((t + m) + 1)) = add1_out(t_2 + 2)) /\
    (a * (in(n * t)) = add1_in1 t_2) /\
    (b * (mux1_out(n * (t + m))) = add1_in2 t_2)
```

The window inference interface precedes the assumptions of a subdesign by ! and lemmas derived from the assumptions by |. In this case the new variable denoting time is t_2. The resulting design specification can not as a whole be bound to a device even though it still holds unbound operators. To proceed we focus on subexpressions with unbound operators and allocate and bind devices to them just as we did above with the add operation. In this case the unbound operators are multiplications in conjuncts three and four. We allocate one multiplier for each of these and terminate all subdesigns which results in the following design specification:

```
==> * Refine n input in /\
    Refine n output mux1_out /\
    MUX(mux1_sel,mux1_in1,mux1_in2)mux1_out /\
    (mux1_sel(n * m) = T) /\
    (mux1_sel(n * ((t + m) + 1)) = F) /\
    (mux1_in1(n * m) = 0) /\
    ADD 2(add1_in1,add1_in2)add1_out /\
    (mux1_in2(n * ((t + m) + 1)) = add1_out(t_2 + 2)) /\
    MUL 4(mul1_in1,mul1_in2)mul1_out /\
    (add1_in1 t_2 = mul1_out(t_3 + 4)) /\
    (a = mul1_in1 t_3) /\
    (in(n * t) = mul1_in2 t_3) /\
    MUL 4(mul2_in1,mul2_in2)mul2_out /\
    (add1_in2 t_2 = mul2_out(t_4 + 4)) /\
    (b = mul2_in1 t_4) /\
    (mux1_out(n * (t + m)) = mul2_in2 t_4)
```

An alternative implementation would be to let the two multiplications share one multiplier. This would save one multiplier at the cost of additional multiplexors and delays. Such an example can be found elsewhere [7].

We have now allocated a computational resource for every operator in the specification, and the data-path design phase is thereby concluded. The result can be seen in Fig. 7.

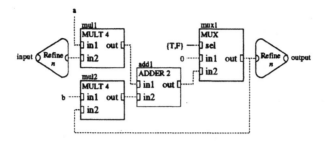

Fig. 7. The completed design after data-path design

We now turn our attention to scheduling the operations. SCHEDULE_TRANS with a parameter $(n * (t + m))$ to indicate the common time frame, calculates and introduces an asap schedule automatically.

```
==> * Refine n input in /\
      Refine n output mux1_out /\
      MUX(mux1_sel,mux1_in1,mux1_in2)mux1_out /\
      (mux1_sel(n * m) = T) /\
      (mux1_sel(n * ((t + m) + 1)) = F) /\
      (mux1_in1(n * m) = 0) /\
      ADD 2(add1_in1,add1_in2)add1_out /\
      (mux1_in2(n * ((t + m) + 1)) = add1_out((n * (m + t)) + 6)) /\
      MUL 4(mul1_in1,mul1_in2)mul1_out /\
      (add1_in1((n * (t + m)) + 4) = mul1_out((n * (t + m)) + 4)) /\
      (a = mul1_in1(n * (t + m))) /\
      (in(n * t) = mul1_in2(n * (t + m))) /\
      MUL 4(mul2_in1,mul2_in2)mul2_out /\
      (add1_in2((n * (t + m)) + 4) = mul2_out((n * (t + m)) + 4)) /\
      (b = mul2_in1(n * (t + m))) /\
      (mux1_out(n * (t + m)) = mul2_in2(n * (t + m)))
```

Having scheduled all computations, the retiming variable n and the latency variable m can be fixed. From the current design specification we extract the following constraints on these variables:

$$n * ((t + m) + 1) = (n * (t + m)) + 6$$
$$n * t = n * (t + m)$$

Solving the first constraint gives the value 6 to n and solving the second give the value 0 to m. We first instantiate n to 6 but due to the fact that the latency variable is used in the original specification we cannot instantiate it. Instead we introduce a predicate constraining its value.

```
==> * m = 0
    Refine 6 input in /\
    Refine 6 output mux1_out /\
    MUX(mux1_sel,mux1_in1,mux1_in2)mux1_out /\
    (mux1_sel 0 = T) /\
    (mux1_sel(6 * (t + 1)) = F) /\
    (mux1_in1 0 = 0) /\
    ADD 2(add1_in1,add1_in2)add1_out /\
    (mux1_in2((6 * t) + 6) = add1_out((6 * t) + 6)) /\
    MUL 4(mul1_in1,mul1_in2)mul1_out /\
    (add1_in1((6 * t) + 4) = mul1_out((6 * t) + 4)) /\
    (a = mul1_in1(6 * t)) /\
    (in(6 * t) = mul1_in2(6 * t)) /\
    MUL 4(mul2_in1,mul2_in2)mul2_out /\
    (add1_in2((6 * t) + 4) = mul2_out((6 * t) + 4)) /\
    (b = mul2_in1(6 * t)) /\
    (mux1_out(6 * t) = mul2_in2(6 * t))
```

All computations are now scheduled and all constraints on ports are in a form
making interconnection possible. We connect ports together by repeated applications of CONNECT_TRANS that annotate some port variables as lines and rewrite
judiciously. At this stage we also allocate and bind devices to represent constants
with the CONST_TRANS transform:

```
==> * CONST 0 const3_vf /\
    CONST b const2_vf /\
    CONST a const1_vf /\
    (m = 0) /\
    Refine 6 input in /\
    Refine 6 output mux1_out /\
    MUX(mux1_sel,const3_vf,add1_out)mux1_out /\
    (mux1_sel 0 = T) /\
    (mux1_sel(6 * (t + 1)) = F) /\
    ADD 2(mul1_out,mul2_out)add1_out /\
    MUL 4(const1_vf,mul1_in2)mul1_out /\
    (in(6 * t) = mul1_in2(6 * t)) /\
    MUL 4(const2_vf,mux1_out)mul2_out
```

The design is now completed and can be seen in Fig. 8. Lines are denoted as
filled-in lines between ports. The resulting design theorem held by the window,
rewritten on the form:

$$constraints \vdash implementation \supset specification$$

is then:

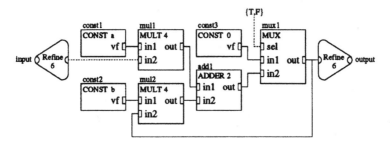

Fig. 8. The completed design

```
(m = 0), (mux1_sel 0 = T), (mux1_sel(6 * (t + 1)) = F),
Refine 6 input in, Refine 6 output mux1_out
|- (CONST 0 const3_vf /\ CONST b const2_vf /\ CONST a const1_vf /\
    MUX(mux1_sel,const3_vf,add1_out)mux1_out /\
    ADD 2(mul1_out,mul2_out)add1_out /\
    MUL 4(const1_vf,input_v)mul1_out /\
    MUL 4(const2_vf,mux1_out)mul2_out)
    ==> ((output m = 0) /\
        (output(SUC(t + m)) = (a * (input t)) + (b * (output(t + m)))))
```

The fact that we get as a result of the data-path design a specification of the requirements necessary for the derived device implementation to satisfy the specification is an example of how well this method applies to digital system design.

5 Conclusions

In this paper we have demonstrated the applicability of window inference reasoning to digital system design and we have proposed an architecture for such a system based on the concepts of design transformations and annotations.

We conjecture that window inference reasoning has some important properties for design applications:

- Supports partitioning of the design task.
- Allows the definition of application specific design transformations.
- Supports the preservation of arbitrary reflexive and transitive relations when transforming expressions.

We have also presented a two-level design representation to support both formal reasoning and the efficient implementation of design algorithms. Although the annotations used in this example were basic they still proved to be valuable both to avoid making trivial errors in using the window transforms as well as in writing the algorithms underlying the more complex window transforms. We argue that in order for formal design techniques to become acceptable for real-world designers, an appropriate balance between formal proof and less strict but computationally efficient reasoning must be established and both types of reasoning supported. The list of possible future work includes:

- Implement more analysis functions, e.g. a critical path analyser [9] and a pipeline extraction tool [1].
- Use design representations that better capture algorithmic synthesis tasks [9].
- Use design annotations for storing redundant instead of additional information, i.e. information deduced from the logical representation.

Acknowledgements. This work was supported by the Swedish Board of Technical Development (NUTEK). I also gratefully acknowledge the support of the Prince Bertil Scholarship. Part of this work was carried out during a visit to Cambridge University and I am grateful for the help I received there. I thank Andy Gordon and the referees of this paper for their valuable comments.

References

1. B. Fjällborg. *Pipeline Extraction for Pipeline Synthesis*. PhD thesis, Department of Computer and Information Science, Linköping University, S-581 83 Linköping, Sweden, May 1992.
2. M. Gordon and T. Melham, editors. *Introduction to HOL*. Cambridge University Press, Cambridge, England, Mar. 1993.
3. M. J. C. Gordon. Why Higher-Order Logic is a Good Formalism for Specifying and Verifying Hardware. In G. J. Milne and P. A. Subrahmanyam, editors, *Formal Aspects of VLSI Design: Proceedings of the Edinburgh Workshop on VLSI*, pages 153–177, Edinburgh, Scotland, 1985. North Holland.
4. J. Grundy. Window Inference in the HOL System. In P. J. Windley, M. Archer, K. N. Levitt, and J. J. Joyce, editors, *The Proceedings of the International Tutorial and Workshop on the HOL Theorem Proving System and its Applications*, pages 177–189, University of California at Davis, United States, Aug. 1991. ACM/IEEE, IEEE Computer Society Press.
5. F. K. Hanna, M. Longley, and N. Daeche. Formal Synthesis of Digital Systems. In L. J. M. Claesen, editor, *Formal VLSI Specification and Synthesis, VLSI Design Methods-I*, volume 1, pages 153–169. IFIP, North Holland, 1990.
6. S. D. Johnson and B. Bose. DDD — A System for Mechanized Digital System Design Derivation. In *Proceedings of the 1991 International Workshop on Formal Methods in VLSI Design*, Miami, United States, Jan. 1991. ACM.
7. M. J. P. Larsson. A Transformational Approach to Formal Digital System Design. Licentiate Thesis 378, Department of Computer and Information Science, Linköping University, May 1993.
8. E. M. Mayger and M. P. Fourman. Integration of Formal Methods with System Design. In A. Halaas and P. Denyer, editors, *VLSI91*, Edinburgh, Scotland, 1991. Elsevier.
9. Z. Peng and K. Kuchcinski. Automated Transformation of Algorithms into Register-Transfer Level Implementations. *IEEE Transactions on Computer-Aided Design of Integrated Circuits and Systems*, 13(2):150–166, Feb. 1994.

Generating Designs Using an Algorithmic Register Transfer Language with Formal Semantics

Juin-Yeu Lu [*] and Shiu-Kai Chin

Dept. of Electrical & Computer Engineering, Syracuse University, Syracuse NY

Abstract. ARTL (Algorithmic Register Transfer Language) is a language used to describe and specify synchronous hardware at the algorithmic and register-transfer levels. Its syntax and natural semantics are formalized in higher-order logic using HOL. An ARTL simulation engine (abstract machine) and compiler are described and verified within HOL. The machine and compiler for ARTL is fully implemented. Also, we present the principles of ARTL synthesis using to standard cells and field programmable gate arrays (FPGAs).

1 Introduction

Our objective is to develop tools and techniques which link register-transfer level descriptions with more abstract algorithmic descriptions. In ARTL we attempt to use a single language to describe algorithmic state machine behavior and register-transfer behavior. The use of formal semantics and verification supports the design of correct hardware. Formal semantics allows rigorous statements about hardware descriptions to be made. The abstract machine and formally verified compiler supports simulation.

Previous work focused on the register-transfer level, [1, 2, 3]. ARTL supports an algorithmic state machine style of description. Designers describe behavior in a completely symbolic fashion. Lower-level descriptions are related to algorithmic ones by using abstract data types and the semantics of ARTL.

The natural semantics of ARTL is based on a finite state machine model and the operational behavior of RTL constructs. ARTL, its simulation engine and compilation rules are deeply embedded in HOL. Our approach is similar to [4]. In [4], ELLA and SILAGE are shallowly embedded in HOL, and post hoc verification on the VHDL simulation cycle is presented.

An ARTL program is a text representation of an algorithmic state machine flow chart. In ARTL, the control-flow and the datapath logic are separate. This makes existing tools, e.g. [9, 10], easier to use in the synthesis process. ARTL does not create a state encoding. Rather, it uses the results of other tools, e.g. [5, 6, 7, 8]. This means the same ARTL description can be mapped to different encodings depending on the technology being used. For example, a "one-hot"

---

[*] Supported by NY State Center for Advanced Technology at Syracuse University.

state encoding might be used for field programmerable gate arrays (FPGAs) whereas a binary state encoding might be used for a standard-cell VLSI circuit.

ARTL descriptions are synthesizable to CMOS standard cells and FPGAs. When synthesizing an algorithmic and register transfer specification, the synthesizer needs to provide a means to perform state encoding for the control path and resource allocation for its data path using the design library. Our previous work [17] provides a means to transform HOL verified circuit descriptions into CMOS parameterized cells, such as arithmetic and logic circuits, so that the enriched library is used to implement high level operations (e.g. addition, multiplication, etc.)

The organization of the rest of this paper is as follows. In Section 2, we give the abstract syntax and natural semantics of ARTL. Section 3 gives the simulation model and correctness theorems. Section 4 discusses the synthesis of ARTL using standard cells and FPGAs.

2 Algorithmic Register Transfer Language

ARTL is a cycle-driven[2] based hardware specification language. It allows an engineer to describe the control using finite state machines and register primitives to describe data paths. ARTL is defined by an 8-tuple:

$ARTL = (S,I,R,O,s,\delta,\lambda,\theta)$, where

- S is the nonempty set of state symbols,
- I is the set of input data which range over signal words and natural numbers,
- R is the set of data registers,
- O is the set of output signals,
- s is the initial state symbol,
- δ is the state-transition function, $S(t+1) = \delta(S(t),I(t),R(t))$,
- λ is the output function, $O(t) = \lambda(S(t),I(t),R(t))$ and
- θ is the register transfer function, $R(t+1) = \theta(S(t),I(t),R(t))$.

An ARTL program is a text representation of an algorithmic state machine (ASM) flow chart [18], as shown in Figure 1. An algorithmic state machine is a collection of ASM blocks. An ASM block has a state box which contains a state symbol (mnemonic) and unconditional signal transfer expressions; and a network of decision boxes and conditional signal transfer expressions. Each path from one state to the next is called a *link path*. A boolean expression is associated with each link path.

A valid ARTL program meets four requirements. 1) All states are unique. 2) Control is modeled as a synchronous and deterministic finite-state machine. 3) Signals are assigned to output ports and registers. 4) Multiple-signal assignments to an output port or register are not allowed. These criteria are formally specified by the natural semantics.

---

[2] A cycle corresponds to one clock tick.

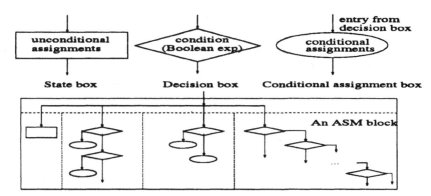

Fig. 1. Algorithmic state machine flow chart

2.1 Data Representation and Timing Model

Every object in ARTL has an associated type. This section presents all the object types we use. We first introduce the ones inherited from the HOL [12, 13] predefined types, then the concrete types follow. Throughout this work we use the following syntax to represent types. The left hand side of the "=" is the type being defined and the left hand side of ":" is a type abbreviation for structures made from existing types. In this work, five HOL predefined types are used: *num, bool, string, list* and *word*. ARTL introduces a data type *bsig* that ranges over all the single-bit signal values. *X* stands for "don't care". By incorporating the word library [14], we extend signals to multiple-bit words.

```
bsig = Hi | Lo | X
```

Signal values are assigned to output ports and registers. A port or a register has a constructor and a name. There are two kinds of I/O ports, *inport* and *outport*.

There are three signal carriers: *in_signal, out_signal* and *reg_signal* representing input, output and register signals respectively. They are defined as functions as shown below, where *ty1 − > ty2* denotes a function from the domain *ty1* to the co-domain *ty2* and $x \sharp y$ represents a 2-tuple. A signal can be multi-bit. It is represented by the built-in type *word* and its size is specified by *psize*.

```
inport = Ip string
outport = Op string
regport = Rp string
state_symbol = Sv string
in_signal:inport -> time -> ((bsig)word # psize)
out_signal:outport -> ((time # (bsig)word)list # psize # sig_flag)
reg_signal:regport -> ((time # (bsig)word)list # psize # sig_flag)
```

If a signal value is assigned to an output port, then the output port changes its value at the current cycle. The output port will be reset to *Lo* at the next cycle unless a value is assigned to it. If a signal is assigned to a register, then the register changes its value at the next cycle. The register will hold its value until a value is assigned to it. The signal transfer behavior is specified by the semantic relations *a_sem* and *asgn_sem* in Section 2.4.

2.2 Syntax

ARTL consists of signal expressions, boolean expressions, state-transition expressions and ASM expressions. The abstract syntax is specified using new types. The type *sig* defines the syntax for signal expressions, *bexp* for boolean expressions, *a_exp* and *asgn_exp* for signal assignment expressions, *str* for state-transition expressions and *artl* for ASM expressions.

```
sig=Bs (bsig)word|In inport|Not sig|And sig sig|Or sig sig|Reg regport
    |Add sig sig|Sub sig sig|Shr bsig sig|Shl sig bsig|N num psize
bexp=Bcons bool|Bnot bexp|Band bexp bexp|Bor bexp bexp|Beq_sig sig sig
a_exp=Asgn_o outport sig |Asgn_r regport sig|Asgn_null
    |Asgn_list a_exp a_exp
asgn_exp=Cond_asgn bexp asgn_exp asgn_exp|Asgn a_exp
    |Asgn_comp asgn_exp asgn_exp|Asgn_exp_null
str=If bexp state_symbol str|Else stvar
artl=Asm state_symbol asgn_exp str |Asms state_symbol asgn_exp str artl
```

2.3 Example

Figure 2 shows an ASM diagram and its corresponding ARTL description follows. Each dashed box corresponds to an ASM block and within an ASM block the state box is represented by a rectangular box labelled by *fy*, *hg*, *hy*, or *gf*. Dashed lines represent concurrency.

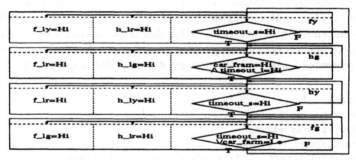

Fig. 2. Traffic Light Controller

```
⊢ tlc =
Asms (Sv 'fy')
 (Asgn (Asgn_list (Asgn_o(Op 'f_ly')(Bs(bit Hi)))
                  (Asgn_o(Op 'h_lr')(Bs(bit Hi)))))
 (If(Beq_sig(In(Ip 'timeout_s'))(Bs(bit Hi))) (Sv 'hg') (Else(Sv 'fy')))
(Asms (Sv 'hg')
 (Asgn (Asgn_list (Asgn_o(Op 'f_lr')(Bs(bit Hi)))
                  (Asgn_o(Op 'h_lg')(Bs(bit Hi)))))
 (If(Band (Beq_sig(In(Ip 'car_farm'))(Bs(bit Hi)))
    (Beq_sig(In(Ip 'timeout_l'))(Bs(bit Hi)))) (Sv 'hy') (Else(Sv 'hg')))
(Asms(Sv 'hy')
 (Asgn(Asgn_list (Asgn_o(Op 'f_lr')(Bs(bit Hi)))
                 (Asgn_o(Op 'h_ly')(Bs(bit Hi)))))
 (If(Beq_sig(In(Ip 'timeout_s'))(Bs(bit Hi))) (Sv 'fg') (Else(Sv 'hy')))
(Asm (Sv 'fg')
 (Asgn(Asgn_list (Asgn_o(Op 'f_lg')(Bs(bit Hi)))
                 (Asgn_o(Op 'h_lr')(Bs(bit Hi)))))
 (If(Band (Beq_sig(In(Ip 'car_farm'))(Bs(bit Lo)))
  (Beq_sig(In(Ip 'timeout_s'))(Bs(bit Hi)))) (Sv 'fy') (Else(Sv 'fg')))))))
```

2.4 Formal Semantics

We define the meaning of each ARTL construct in terms of a set of configuration transition rules. A configuration (labelled with _config) of an ARTL expression is defined as a tuple and its natural semantics is specified as a set of relations (predicates) with the type of _config − > _config − > bool. A transition rule is an implication relation consisting of a conclusion (below the horizontal line) and premisses (above the horizontal line), where "*initial configuration* $\implies$ *final configuration*" denotes an initial-final configuration relation. The following sections give the semantics for each construct.

Signal Expressions. The *sig_sem* rules assign meaning to signal expressions including logical and arithmetic operations. The configuration of a signal expression is a 4-tuple:

```
sig_config:in_signal # reg_signal # sig # time
sig_sem:sig_config -> sig_config -> bool
```

For example, consider the configuration transition rule of a logical *Or*. If signal expressions *se1* and *se2* evaluate to signal constants *bs1* and *bs2*, *Or se1 se2* evaluates to a constant whose value at time *t* is the operation *Or* of *bs1* and *bs2*. The operation *Or* is defined in terms of the function *wsor*, the *Or* function mapped over the bits of a word.

$$\frac{(ins, regs, se1, t) \overset{sig\_sem}{\implies} (ins, regs, Bs\ bs1, t);\quad (ins, regs, se2, t) \overset{sig\_sem}{\implies} (ins, regs, Bs\ bs2, t)}{(ins, regs, Or\ se1\ se2, t) \overset{sig\_sem}{\implies} (ins, regs, Bs(wsor\ bs1\ bs2), t)}$$

Boolean Expressions. The natural semantics for boolean expressions is similar to that for signal expressions.

```
bexp_config:in_signal # reg_signal # bexp # time
bexp_sem:bexp_config -> bexp_config -> bool
```

For example, consider negation. If a boolean expression evaluates to a boolean constant, then the negation over the expression is performed on the resulting constant.

$$\frac{(ins,regs, be,t) \overset{bexp\_sem}{\Longrightarrow} (ins,regs,Bcons\ bc,t)}{(ins,regs,Bnot\ be,t) \overset{bexp\_sem}{\Longrightarrow} (ins,regs,(Bcons\ \neg bc),t)}$$

Assignment Expressions. The semantics of signal assignment expression is defined by two rules *a_sem* and *asgn_sem*. The rule *a_sem* defines the semantics for the unconditional assignment expressions, while the rule *asgn_sem* defines the semantics for the compound assignment expressions including the conditional and unconditional expressions. For signal transfers to output ports (or registers), two functions *o_bound* and *r_bound*[3] bind signal values to output ports (or registers). Within the functions *o_bound* and *r_bound*, if the signal word does not fit with its destination, the function *word_alignment* will truncate the most significant bits or extend with *Lo* values to fit the word size of its destination. The configuration of a signal assignment expression is a 5-tuple.

```
a_config:in_signal # reg_signal # a_exp # time # out_signal
a_sem:a_config -> a_config -> bool
asgn_config:in_signal # reg_signal # asgn_exp # time # out_signal
asgn_sem:asgn_config -> asgn_config -> bool
⊢ ∀bs opt os t b.o_bound bs opt os t b =
   (λopt'.(((opt' = opt) ∧ (b = T)) →
   (CONS(t,word_alignment bs(SND(os opt')))(FST(os opt')),SND(os opt'))
   | os opt'))
```

For signal assignments, if the signal expression evaluates to a signal constant, then the constant is assigned to an output port (or a register).

$$\frac{(ins,regs,se,t) \overset{sig\_sem}{\Longrightarrow} (ins,regs,Bs\ bs,t)}{(ins,regs,Asgn\_o\ opt\ se,t,os) \overset{a\_sem}{\Longrightarrow} (ins,regs,Asgn\_null,t,o\_bound\ bs\ opt\ os\ t\ T)}$$

$$\frac{(ins,regs,se,t) \overset{sig\_sem}{\Longrightarrow} (ins,regs,Bs\ bs,t)}{(ins,regs,Asgn\_r\ rpt\ se,t,os) \overset{a\_sem}{\Longrightarrow} (ins,r\_bound\ bs\ rpt\ regs\ t\ T,Asgn\_null,t,os)}$$

---

[3] The function *o_bound* schedules to change the signal value of an output port signal at the current clock cycle, while the function *r_bound* schedules to change the signal value of a register at the next clock cycle.

Two compound assignment expressions are simultaneously evaluated based on the same initial input and register status. The individual results of each evaluation are combined as a whole (by the application to the functions *union_os* and *union_rs*). Note that an output port (or register) is not assigned multiple signals. That is enforced by the predicate *mu_disjoint1* within the semantic rules which follow. The function *asgn_meta_o (asgn_meta_r)* extracts output (register signal transfer) expressions from a signal assignment expression and creates an intermediate form which is a list of 2-tuples, $< bexp\|output\_port\_name >$ ($<$ $bexp\|register\_port\_name >$), consisting of an output port (register) name and the associated guard expression. Two 2-tuples are disjoint if their port (register) names are different or the associated guard expressions do not both evaluate to *T*. That all tuples are mutually disjoint means that the corresponding assignment expression assigns at most one signal to an output port (or register).

$$
\frac{
\begin{array}{l}
mu\_disjoint1\ ins\ regs\ t\ asgn\_disjoint\ (asgn\_meta\_o\ (Asgn\_comp\ ae1\ ae2)) \\
mu\_disjoint1\ ins\ regs\ t\ asgn\_disjoint\ (asgn\_meta\_r\ (Asgn\_comp\ ae1\ ae2)) \\
(ins, regs, ae1, t, os) \overset{asgn\_sem}{\Longrightarrow} (ins, regs', Asgn\_exp\_null, t, os') \\
(ins, regs'', ae2, t, os) \overset{asgn\_sem}{\Longrightarrow} (ins, regs'', Asgn\_exp\_null, t, os'')
\end{array}
}{
\begin{array}{l}
(ins, regs, Asgn\_comp\ ae1\ ae2, t, os) \overset{asgn\_sem}{\Longrightarrow} \\
(ins, union\_rs\ regs\ regs'\ regs'', Asgn\_exp\_null, t, union\_os\ os\ os'\ os'')
\end{array}
}
$$

Given a conditional signal assignment expression *Cond_asgn be ae1 ae2*, at time *t*, if the *boolean* expression *be* evaluates to *T* and the assignment expression *ae1* reduces to a null expression, then output ports and registers change their values based on the evaluation of the expression *ae1*; otherwise, if the *boolean* expression *be* evaluates to *F*, output ports and registers change their values based on the evaluation of the assignment expression *ae2*.

$$
\frac{
\begin{array}{l}
mu\_disjoint1\ ins\ regs\ t\ asgn\_disjoint\ (asgn\_meta\_o\ (Cond\_asgn\ ae1\ ae2)) \\
mu\_disjoint1\ ins\ regs\ t\ asgn\_disjoint\ (asgn\_meta\_r\ (Cond\_asgn\ ae1\ ae2)) \\
(ins, regs, be, t) \overset{bexp\_sem}{\Longrightarrow} (ins, regs, Bcons\ T, t) \\
(ins, regs, ae1, os) \overset{asgn\_sem}{\Longrightarrow} (ins, regs', Asgn\_exp\_null, t, os')
\end{array}
}{
(ins, regs, Cond\_asgn\ be\ ae1\ ae2, t, os) \overset{asgn\_sem}{\Longrightarrow} (ins, regs', Asgn\_exp\_null, t, os')
}
$$

$$
\frac{
\begin{array}{l}
mu\_disjoint1\ ins\ regs\ t\ asgn\_disjoint\ (asgn\_meta\_o\ (Cond\_asgn\ ae1\ ae2)) \\
mu\_disjoint1\ ins\ regs\ t\ asgn\_disjoint\ (asgn\_meta\_r\ (Cond\_asgn\ ae1\ ae2)) \\
(ins, regs, be, t) \overset{bexp\_sem}{\Longrightarrow} (ins, regs, Bcons\ F, t) \\
(ins, regs, ae2, t, os) \overset{asgn\_sem}{\Longrightarrow} (ins, regs', Asgn\_exp\_null, t, os')
\end{array}
}{
(ins, regs, Cond\_asgn\ be\ ae1\ ae2, t, os) \overset{asgn\_sem}{\Longrightarrow} (ins, regs', Asgn\_exp\_null, t, os')
}
$$

State-Transition Expressions. State-transition expressions determine the next state symbols for each ASM block. Its structure is like a tail-nested *If_else* statement. As previously mentioned, only one next state may exist at a given time. This is checked by the predicates *str_disjoint* and *mu_disjoint1*.

```
str_config:in_signal # reg_signal # str # time
str_sem:str_config -> str_config -> bool
```

$$mu\_disjoint1\ ins\ regs\ t\ str\_disjoint(str\_meta(If\ be\ sv\ str1)(Bcons\ T))$$
$$(ins, regs, be, t) \overset{bexp\_sem}{\Longrightarrow} (ins, regs, Bcons\ T, t)$$

$$\overline{(ins, regs, If\ be\ sv\ str1, t) \overset{str\_sem}{\Longrightarrow} (ins, regs, Else\ sv, t)}$$

$$mu\_disjoint1\ ins\ regs\ t\ str\_disjoint(str\_meta(If\ be\ sv\ str1)(Bcons\ T))$$
$$(ins, regs, be, t) \overset{bexp\_sem}{\Longrightarrow} (ins, regs, Bcons\ F, t)$$
$$(ins, regs, str1, t) \overset{str\_sem}{\Longrightarrow} (ins, regs, Else\ sv', t)$$

$$\overline{(ins, regs, If\ be\ sv\ str1, t) \overset{str\_sem}{\Longrightarrow} (ins, regs, Else\ sv', t)}$$

If a state-transition expression reduces to the form *Else state_symbol*, then the next-state symbol is determined.

$$\overline{}$$
$$(ins, regs, Else\ sv, t) \overset{str\_sem}{\Longrightarrow} (ins, regs, Else\ sv, t)$$

ARTL Programs. An ARTL program is a collection of ASM blocks. Its meaning is the transition relation from one ASM block to the other. An ASM block corresponds to a state within a finite state machine. Each ASM block has its own signal assignment and state-transition expressions. The rule *artl_sem* rule includes the predicate *semantic_chk* that ensures the validity of an ARTL program in such a way that it transforms an ARTL program into an intermidate form and checks if all states appearing in the state-transition expressions are defined and if all defined states are unique.

An ARTL program configuration is defined as an 8-tuple consisting of input signals, registers, output signals, an ARTL program, assignment expressions for the current ASM block, state-transition expressions for the current ASM block, the current cycle time and the state symbol for the current ASM block. Note that in this structure, the current ASM block is extracted from the ARTL program, where the corresponding signal assignments and state-transition expressions are placed at the 5th and 6th positions within the tuple.

```
artl_config:in_signal # reg_signal # out_signal # artl # asgn_exp # str
    # time # state_symbol
artl_sem:artl_config -> artl_config -> bool
```

For an ASM block, if the next-state symbol is determined and the assignment expressions are exhausted, then the ARTL program will change its configuration from one ASM block to another. Within this transition, five actions are done during this change: 1) update output signals, 2) update registers, 3) change the current state symbol, 4) advance the current time, and 5) initiate the next ASM block. The functions *update_os* and *update_rs* update the output ports and registers in a way that if an output port has not been assigned a value, it is given a "Lo" for the current cycle. If a register is not scheduled to be assigned a value for the next cycle, it will inherit its current value.

semantics_chk artll ins regs t
init_exists artll cs asgn1 strl
$(ins, regs, asgn1, t, os) \overset{asgn\_sem}{\Longrightarrow} (ins, regs', Asgn\_exp\_null, t, os')$
$(ins, regs, strl, t) \overset{str\_sem}{\Longrightarrow} (ins, regs, Else\ sv, t)$

───────────────────────────────

$(ins, regs, os, artll, asgn1, strl, t, cs) \overset{artl\_sem}{\Longrightarrow}$
$(ins, update\_rs\ regs'\ t, update\_os\ os'\ t, artll, init\_asgn\ artll\ sv, init\_str\ artll\ sv, t+1, sv)$

Correctness Theorems. The deterministic behavior of ARTL programs requires that assignment expressions are deterministic and state-transition expressions give unique next-state symbols. The evaluation of the an assignment expression is deterministic if the associated signal expressions and the boolean expressions are deterministic. The evaluation of a state-transition expression is deterministic if the associated boolean expression is deterministic. All these properties are proved. The relationships between them are shown in Figure 3.

```
artl_deterministic
```
$\vdash \forall (a1{:}artl\_config)\ a2.artl\_sem\ a1\ a2 \Rightarrow (\forall a3.artl\_sem\ a1\ a3 \Rightarrow (a2{=}a3))$
```
asgn_deterministic
```
$\vdash \forall (a1{:}asgn\_config)\ a2.asgn\_sem\ a1\ a2 \Rightarrow (\forall a3.asgn\_sem\ a1\ a3 \Rightarrow (a2{=}a3))$
```
str_deterministic
```
$\vdash \forall (s1{:}str\_config)\ s2.str\_sem\ s1\ s2 \Rightarrow (\forall s3.str\_sem\ s1\ s3 \Rightarrow (s2{=}s3))$
```
a_exp_deterministic
```
$\vdash \forall (a1{:}a\_config)\ a2.a\_sem\ a1\ a2 \Rightarrow (\forall a3.a\_sem\ a1\ a3 \Rightarrow (a2{=}a3))$
```
bexp_deterministic
```
$\vdash \forall (b1{:}bexp\_config)\ b2.bexp\_sem\ b1\ b2 \Rightarrow (\forall b3.bexp\_sem\ b1\ b3 \Rightarrow (b2{=}b3))$
```
sig_deterministic
```
$\vdash \forall (s1{:}sig\_config)\ s2.sig\_sem\ s1\ s2 \Rightarrow (\forall s3.sig\_sem\ s1\ s3 \Rightarrow (s2{=}s3))$

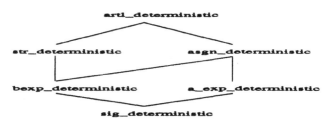

Fig. 3. Determinacy Theorems

3 Simulation Model

The MAC machine is an abstract machine used to support a fine grain computational model for ARTL. The machine has its own instruction set which is defined as a new type *code*. A machine state is defined as an 8-tuple:

```
mac_config: (I,R,O,Acs,Ac,time,s,V)
code = PUSH_SIG (bsig)word | ... | SYNS state_symbol | ... | JOINT
I:in_signal
R:reg_signal
O:out_signal
Acs:asms_codes
Ac:asm_codes
asm_codes:(code)list
asms_codes:(state_symbol # asm_codes)list
s:state_symbol
V:val_stack
val_stack = BOOL  bool | SIG (bsig)word | OUT_REG out_signal reg_signal
```

Acs stores the code for the ARTL program and *Ac* stores the code for the current ASM block. Intermediate results are recorded on the stack *V*. The operational semantics of the abstract machine is described by set of machine configuration-transition rules. A configuration transition has the form:

$$(I,R,O,Acs,Ac,time,s,V) \longmapsto (I,R',O',Acs',Ac',time',s',V')$$

"$\longmapsto$" denotes a one-step transition. Note that the natural semantic relations described in Section 2.4 only show the relation between the initial and final configuration of a computation. To execute an ARTL program on the MAC machine, the source language is transformed to the machine code. The function *Trans_artl* performs this task. It takes as input an ARTL program and returns a list of 2-tuples where each tuple consists of a state symbol and the code for the associated ASM block.

```
Trans_artl:artl -> (state_symbol # (code)list)list
```

Compiler Correctness. The compilation function is proved correct with respect to the mapping between the natural semantics for ARTL and the behavior of the abstract machine. The key notion is:

For all valid programs
if 1.$(Env,artl,asm,time,state\_symbol) \stackrel{artl\_sem}{\Longrightarrow} (Env',artl,asm',time',state\_symbol')$,
    2. Trans_artl artl = codes,
    3. get_initial_asm codes asm = init_asm, and
    4. $(Env,codes,init\_asm,time,state\_symbol) \stackrel{\star}{\longmapsto}$
        $(Env'',codes,init\_asm',time'',state\_symbol'')$
then 1. Env' = Env'',
     2. time' = time'', and
     3. state_symbol' = state_symbol'', where
Env includes I,O and R, and $\stackrel{\star}{\longmapsto}$ is the reflexive-transitive closure of a relation.

The following theorem reflects the correctness relationship as shown above. It can be read as "For every valid initial ARTL program configuration *a1*, if it evolves to the final configuration based on its natural semantic rule *artl_sem*, then the compiled code running on the abstract machine will have the corresponding initial and final state".

```
⊢ ∀(a1:artl_config) (a2:artl_config).artl_sem a1 a2 ⇒
  (∀(vsk:(val_stack)list)odes). let ins=get_artl_config_ins a1 in
  let regs=get_artl_config_regs a1 in  let os=get_artl_config_os a1 in
  let artl1=get_artl_config_artl a1 in
  let asgn1=get_artl_config_asgn a1 in
  let str1=get_artl_config_str a1 in let t=get_artl_config_time a1 in
  let cs=get_artl_config_sv a1 in let regs'=get_artl_config_regs a2 in
  let os'=get_artl_config_os a2 in let t'=get_artl_config_time a2 in
  let cs'=get_artl_config_sv a2 in
  let ac=SND(HD(Trans_artl(Asm cs asgn1 str1))) in
  let acs=Trans_artl artl1 in let ac'=sel_code acs cs' in
  Rtc mac (ins,regs,os,acs,ac,t,cs,vsk) (ins,regs',os',acs,ac',t',cs',[]))
```

From this result, we see that the method of equivalence checking [11] for two sequential machines is extended to the problem of correctness transformation between a source language and its execution model.

Correctness of Two Equivalent Evaluations on MAC. We have proved that the compiled code for two equivalent ARTL descriptions running on the MAC machine will behave the same way in terms of their output signals. It means that two designs (ARTL descriptions) may be judged for equivalence by the running their compiled code on the MAC machine and checking the corresponding outputs at the end of each clock cycle.

The following theorem is read as "If two descriptions have the same meaning, then running the two compiled codes on the MAC machine will yield the same output signals for all corresponding ASM blocks".

```
⊢∀(a1:artl_config) (a2:artl_config).
  artl_sem a1 a2 ⇒  (∀ a3 a3'. artl_sem a3 a3' ⇒
  (get_artl_config_os a1=get_artl_config_os a3) ∧
  (get_artl_config_os a2=get_artl_config_os a3') ∧
  (get_artl_config_ins a1=get_artl_config_ins a3) ∧
  (get_artl_config_time a1=get_artl_config_time a3) ∧
  (get_artl_config_time a2=get_artl_config_time a3') ⇒
  (∀ vsk vsk3. let ins=get_artl_config_ins a1 in
  let regs=get_artl_config_regs a1 in
  let os=get_artl_config_os a1 in
  let artl1=get_artl_config_artl a1 in
  let asgn1=get_artl_config_asgn a1 in
  let str1=get_artl_config_str a1 in
  let t=get_artl_config_time a1 in
  let cs = get_artl_config_sv a1 in
```

```
let regs'=get_artl_config_regs a2 in
let os'=get_artl_config_os a2 in
let cs'=get_artl_config_sv a2 in
let ac=SND(HD(Trans_artl(Asm cs asgn1 str1))) in
let acs=Trans_artl artl1 in
let regs3=get_artl_config_regs a3 in
let artl3=get_artl_config_artl a3 in
let asgn3=get_artl_config_asgn a3 in
let str3=get_artl_config_str a3 in
let cs3=get_artl_config_sv a3 in
let regs3'=get_artl_config_regs a3' in
let cs3'=get_artl_config_sv a3' in
let ac3=SND(HD(Trans_artl(Asm cs asgn3 str3))) in
let acs3=Trans_artl artl3 in
  (Rtc mac (ins,regs3,os,acs3,ac3,t,cs3,vsk3)
           (ins,regs3',os',acs3,sel_code acs3 cs3',t + 1,cs3',[]) ∧
   Rtc mac (ins,regs,os,acs,ac,t,cs,vsk)
           (ins,regs',os',acs,sel_code acs cs',t + 1,cs',[]))))
```

4   Synthesis and Technology Mapping

Control path and data path logic are separated in ARTL. Control expressions
are of type *boolean*, while datapath expressions are of type *sig*. Data paths are
guarded by *boolean* expressions and state symbols. Assuming that the state symbols are encoded using binary state or "one-hot" encoding method, the synthesis
of a data path is "ANDing" the datapath logic with the *boolean* logic and the
encoded state.

For output expressions, given a state *S1*, if a signal expression *se1* is assigned
the output port *port1* according to the conditions *be1* and *be2*:

Cond_asgn be1 (Cond_asgn be2 (Asgn (Asgn_o (Op 'port1') se12)) ...

the implementation will have the structure as shown in Figure 4.

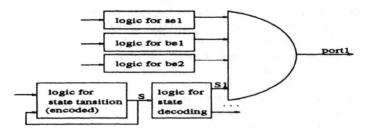

Fig. 4. Synthesis for Output Expressions

For register transfer expressions, given a state *S2*, if a signal expression *sel_not2* is assigned the register *r1* according to the conditions *be1* and *be2*:

Cond_asgn be1 (Cond_asgn be2 (Asgn (Asgn_ ...)) (Asgn(Asgn_r (Rp 'r1') sel_not2))) (...

the implementation will have the structure as shown in Figure 5.

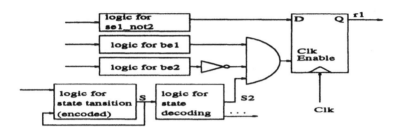

Fig. 5. Synthesis for Register Transfer Expressions

As mentioned in Section 2, each *link path* is associated with a *boolean* expression describing a condition that may cause a FSM change from one state to the next. To synthesize the state-transition logic, an encoding must be chosen. Existing techniques to obtain an optimal state assignment include [5, 6, 7, 8]. When a state encoding is chosen, the state transition logic is easily obtained. If we use D flip-flops, the input of a state register is the sum of the product terms in which a product term is the product of the *boolean* expression and the current state code of the *link path* that contributes to a condition of setting the state register for the next state. For clarity, we represent a *link path* as a 3-tuple:

$$L:(boolean, current\_state\_symbol, next\_state\_symbol)$$

For example, consider two *link paths*, L_1 and L_4:

$L_1:(B_1,S_2,S_4)$
$L_4:(B_4,S_4,S_6)$

Suppose we use three registers, $r_1r_2r_3$, to encode the states as follows:

S_2 010
S_4 110
S_6 101

then we find inputs, $D_1D_2D_3$, of the state registers:

$D_1 = B_1 \neg r_1 r_2 \neg r_3 \vee B_4 r_1 r_2 \neg r_3$
$D_2 = B_1 \neg r_1 r_2 \neg r_3$
$D_3 = B_4 r_1 r_2 \neg r_3$

4.1 Standard Cells

Standard cell generator libraries are common within a VLSI CAD system, e.g. GDT [19], where logic cells are parameterized such that a specific cell is built by giving the necessary information to the generator. Signal attributes of ARTL, such as data type and port size, provide guidance for datapath allocation. However, cell generators that support high level operations like arithmetic operations may not be available in the library initially. Thus, the library needs to be enriched so as to cover all the functions realizable from the given information. Currently, we can easily add cell generators for high level operations by translating the existing verified HOL descriptions to corresponding parameterized cell generators [17].

4.2 FPGAs

When using register-rich FPGAs with narrow fan-in gates (such as Xilinx) to implement finite state machines, one-hot encoding is recommended [8, 7]. One-hot encoding produces simpler next-state logic functions. We refer the interested reader to [8] for details of one-hot-encoding optimization techniques.

It is straightforward to map the control part of an ARTL program to an FGPA architecture as mentioned above in such a way that each state symbol is assigned to a state register. The state encoding and decoding logic in Figure 4 and Figure 5 will disappear. For example, let's consider the traffic light controller, where there are eight *link paths*:

L_1 :$(timeout\_s, S_{fy}, S_{hg})$
L_2 :$(\neg timeout\_s, S_{fy}, S_{fy})$
L_3 :$(car\_farm \wedge timeout\_l, S_{hg}, S_{hy})$
L_4 :$(\neg(car\_farm \wedge timeout\_l), S_{hg}, S_{hg})$
L_5 :$(timeout\_s, S_{hy}, S_{fg})$
L_6 :$(\neg timeout\_s, S_{hy}, S_{hy})$
L_7 :$(\neg car\_farm \wedge timeout\_l, S_{fg}, S_{fy})$
L_8 :$(\neg(\neg car\_farm \wedge timeout\_l), S_{fg}, S_{fg})$

Based on the method described in Section 4 and one-hot encoding scheme, we use four registers ,$r_{hg} r_{fy} r_{hy} r_{fg}$,to represent four state symbols. The inputs to the state registers will be:

$D_{hg} = (timeout\_s \wedge r_{fy}) \vee (\neg(car\_farm \wedge timeout\_l) \wedge r_{hg})$
$D_{fy} = (\neg timeout\_s \wedge r_{fy}) \vee ((\neg car\_farm \wedge timeout\_l) \wedge r_{fg})$
$D_{hy} = (car\_farm \wedge timeout\_l \wedge r_{hg}) \vee (\neg timeout\_s \wedge r_{hy})$
$D_{fg} = (timeout\_s \wedge r_{hy}) \vee (\neg(\neg car\_farm \wedge timeout\_l) \wedge r_{fg})$

5 Conclusions

We have described an algorithmic and register transfer level language ARTL with formal semantics and abstract machine to support its execution. The operational semantics of the machine and the natural semantics of ARTL are formally related by the compilation functions and proved correct. All the definitions and theorems are embedded in HOL. The definitions and compilation rules provide complete information on the language and its implementation.

Existing state encoding schemes easily map ARTL descriptions to standard cell VLSI circuits and FPGAs due to the separation of control and datapath expressions. Because of ARTL's simplicity, ARTL programs are easily synthesized.

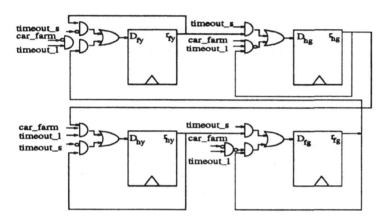

Fig. 6. The Control Part of TLC with One-Hot Encoding

References

1. Sanjiv Naryan, Frank Vlhid and Daniel D. Gajske, "System Specification with the SpecChart Language," IEEE Design & Test of Computer, December 1992.
2. Mandayam Srivas, Mark Bickford and Ian Sutherland, Spectool: A Computer-Aided Verification Tool for Hardware Designs, ORA Corp. Tech. Report RL-TR-91-339, Vol. 1, December 1991.
3. Michael Monachino, "Design Verification System for Large-Scale LSI Designs," IBM Journal of Research and Development, Vol. 26, January 1982.
4. R. Boulton, A. Gordon, M. Gordon, J. Harrison, J. Herbert and J. V. Tassel, "Experience with enbedding hardware description languages in HOL," Tech. Report, University of Cambridge Computer Lab., 1992.
5. Giovanni De Micheli, Robert K. Brayton, Alberto Sangiovanni-Vincentelli, "Optimal State Assignment for Finite State Machines," IEEE Trans. on Computer-Aided Design Vol. CAD-4, No. 3, July 1985.
6. Srinvas Devadas, Richard Newton, "Exact Algorithm for Output Encoding, State Assignment, and Four-Level Boolean Minimization," IEEE Trans. on Computer-Aided Design Vol. 10, No. 1, January 1985.
7. Martine Schlag, Pak K. Chan, and Jackson Kong, "Empirical Evaluation of Multi-level Logic Minimization Tools for an FPGA Technology," FPGAs, edited by Will Moore and Wayne Luk, Abingdon EE&CS Books, England, 1991.

8. Dave Allen, "Automatic One-hot Re-encoding for FPGAs," Field Programmable Gate Array Conference, 1992.

9. R. Brayton, G. D. Hachtel, C. McMullen, and A. L. Sangiovann-Vincentelli, Logic Minimization Algorithms for VLSI Synthesis, Hingham, MA: Kluwer Academic, 1984.

10. R. K. Brayton, R. Ruldell, A. Sangiovanni-Vincentelli, and A. Wang, "MIS: A multiple level logic optimization system," IEEE Trans. Computer-Aided Design, vol. CAD-6, Nov. 1987.

11. Srinvas Devadas, Hi-Keung Ma and A. Richard Newton, "On the verification of sequential machines at different levels of abstraction," IEEE Trans. Computer-Aided Design, vol. 7, June 1988.

12. Mike Gordon, "A proof Generating System for Higher-Order Logic," in VLSI Specification, Verification and Synthesis, edited by Graham Birtwistle and P.A. Subrahmanyam, Kluwer, 1987.

13. Thomas F. Melham, Automating Recursive Type Definitions in Higher Order Logic, Tech. Report No. 146, University of Cambridge Computer Lab., Jan. 1989.

14. Wai Wang, Modelling Bit Vectors in HOL: the word Libray, Higher Order Logic Theorem Proving and Its Applicationsedited, HUG'93, Proceedings, Jeffrey J. Joyce and Carl-Johan H. Seger (eds), Lecture Notes in Computer Science 780, Springer-Verlag, 1994.

15. Hanne Riis Neilson and Flemming Nielson, Semantics with Applications: A Formal Introduction to Computer Science, Weily, 1991.

16. Juin-Yeu Lu and Shiu-Kai Chin, Using HOL and Operational Semantics to Describe and Verify a Programming Language and its Implementation, CASE center Tech. Report No. 9212, Syracuse University, December 1992.

17. Juin-Yeu Lu and Shiu-Kai Chin, "Linking HOL to a VLSI CAD System," Higher Order Logic Theorem Proving and Its Applicationsedited, HUG'93, Proceedings, Jeffrey J. Joyce and Carl-Johan H. Seger (eds), Lecture Notes in Computer Science 780, Springer-Verlag, 1994.

18. Christopher R. Clare, Design Logic System Using State Machines, McGraw-Hill, 1973.

19. Mentor Graphics Corporation, GDT Manuals V.5, CA, 1990.

20. Xilinx, Inc., The Programmable Gate Array Data Book, CA, 1989.

A HOL Formalisation of the Temporal Logic of Actions

Thomas Långbacka

Åbo Akademi University, Department of Computer Science
Lemminkäinengatan 14–18, SF–20520 Åbo, Finland.
Email: langback@aton.abo.fi

Abstract. We describe an attempt to formalise the semantics of the Temporal Logic of Actions (TLA) by means of the HOL theorem prover. Special concern has been devoted to trying to formalise the rules that govern refinement mappings and data hiding in the TLA framework.

1 Introduction

The HOL system [7] has recently become more and more popular as a basis for mechanising logics and programming semantics. Examples include, but are not restricted to, Refinement calculus [3, 10], CSP [4] and UNITY [2]. This paper describes an attempt at formalising the temporal logic of actions (TLA) [8] using the HOL system.

TLA is a logic designed mainly for reasoning about concurrent algorithms. TLA is based on temporal logic, but allows refinement of specifications by ensuring (via it's syntax) that all TLA formulas are invariant under stuttering. In TLA algorithms are expressed directly as formulas in the TLA logic, rather than as programs in a separate programming language. However, reasoning about algorithms on the level of the logic involves a lot of tedious proof details. Hence, much would be gained if the reasoning could be (partly) automated using a mechanical proof assistant.

In this paper, we base ourselves on previous work reported in [9, 12]. Contrary to the work mentioned above we deal with the full TLA logic, i.e. we have made an attempt at formalising the rules that deal with refinement mappings [1] and data hiding in the TLA framework. Mechanising TLA reasoning without these provides a good base for verifying program properties but does not adequately support proofs of correctness of program development steps.

Our formalisation is done by defining the semantics of TLA directly in HOL (a so called "shallow" embedding). The proof rules of TLA are then proved as HOL theorems. After this, these rules can be used for reasoning about programs. This means that we can show that a program has a certain property, or that one program implements another one, by proving a corresponding HOL theorem.

There has been work done on automating TLA by means of the Larch Prover [6]. This paper is not very detailed, making comparisons between that work and our work difficult. The paper does, however, present a translator from high-level TLA formulas to the syntax of the theorem prover. Such a tool would be very convenient for our HOL formalisation as well. Chou [5] formalises parts of TLA in HOL, but like in the previous work in [9, 12], refinement mappings and data hiding are not dealt with.

We assume that the reader is familiar with the HOL system. When referring to HOL-terms and interaction with the HOL system we use the syntax of HOL. To make formulas more readable, we omit type information which can be inferred from the context or is given elsewhere in the text.

2 The Temporal Logic of Actions

The Temporal Logic of Actions (TLA) is a logic for reasoning about concurrent algorithms. One of the main ideas of TLA is that algorithms are expressed directly in the logic, and not as programs in a separate programming language.

TLA is based on a simple temporal logic, with □ ("always") as the only primitive temporal operator. TLA allows refinement of algorithms by guaranteeing (through its syntax) that all well-formed TLA formulas are invariant under stuttering.

2.1 Predicates, Actions and Temporal formulas

The TLA logic has three different levels. One can express *predicates, actions* and *temporal formulas*. Predicates are interpreted as in ordinary predicate logic. The semantics of temporal formulas is defined by means of infinite state sequences, *behaviours*. Given some behaviour, the validity of a temporal formula is evaluated over all the states in the behaviour. Thus the meaning of a temporal formula is the set of behaviours that satisfy that formula.

An *action* is a boolean expression that relates two states (the *initial* and *final* state of that action) to each other. As an example,

$$(x' = x + 1) \wedge (y' = y)$$

is an action that increments x by 1 and leaves y unchanged (thus primed variables always stand for final states).

Predicates can always be interpreted as actions. In this case they put no restriction on the final state (or the initial state in case they appear primed as in the rule INV1 in Figure 1). Similarly, both predicates and actions can be interpreted as temporal formulas. In this case a predicate asserts something about the first state of a behaviour (or the second if primed) while an action relates the first two states of a behaviour.

If $\mathcal{A}$ is an action and f is a state function, then $[\mathcal{A}]_f$ is the action

$$[\mathcal{A}]_f \triangleq \mathcal{A} \vee (f = f')$$

where f' is the same as f but with all occurrences of program variables primed (The symbol $\triangleq$ is to be read "is defined to be equal to"). For example, if $\mathcal{A}$ is the action $(x' = x + 1) \wedge (y' = y)$ then the action $[\mathcal{A}]_{(x,y)}$ permits arbitrary changes of all other variables than x and y. In this way stuttering can be modeled. By allowing actions to appear only in the form above invariance under stuttering is achieved[1].

2.2 Expressing algorithms in TLA

An algorithm is described by a formula

$$\Pi = Init \wedge \Box[\mathcal{N}]_f \wedge F \tag{1}$$

where $Init$ is a predicate that characterises the permitted initial states, $\mathcal{N}$ is the disjunction of all the actions of the algorithm and F is a formula that describes a fairness condition (thus safety and liveness are treated within a single framework). Note that the predicate $Init$ is used as such in the TLA-formula (1). This shows how TLA permits predicates, actions and temporal formulas to be combined.

The safety part of (1) is $\Box[\mathcal{N}]_f$ which states that every step of the program must be permitted by $\mathcal{N}$ or it must leave f unchanged. Thus $\Box[\mathcal{N}]_f$ states that every step is an $\mathcal{N}$-step or a stuttering step.

The liveness part F is typically a conjunction of fairness conditions on actions of the algorithm. Fairness requirements are easily expressed in TLA. Weak and strong fairness with respect to an action $\mathcal{A}$ is expressed as

$$WF_f(\mathcal{A}) \triangleq (\Box\Diamond\langle\mathcal{A}\rangle_f) \vee (\Box\Diamond\neg Enabled\langle\mathcal{A}\rangle_f)$$
$$SF_f(\mathcal{A}) \triangleq (\Box\Diamond\langle\mathcal{A}\rangle_f) \vee (\Diamond\Box\neg Enabled\langle\mathcal{A}\rangle_f)$$

where $\langle\mathcal{A}\rangle_f \triangleq \neg[\neg\mathcal{A}]_f$ and $Enabled(\mathcal{A})$ is true of those states in which it is possible to perform $\mathcal{A}$. The meaning of the symbol $\Diamond$ ("eventually") is defined as $\Diamond\mathcal{A} \triangleq \neg\Box\neg\mathcal{A}$.

As a proof system, TLA has only a small set of *basic* proof rules (see [8] for all details). However, a set of *additional* rules are added for convenience. In addition to the set of pre-defined rules, temporal reasoning as well as ordinary mathematical reasoning is used. In Figure 1 we list some frequently used rules of TLA. The meaning of the symbol $\leadsto$ ("leads-to"), that appears in the LATTICE rule, is defined as $\mathcal{A} \leadsto B \triangleq \Box(\mathcal{A} \Rightarrow \Diamond B)$.

In TLA, proving that program Π has property $\Box\Phi$ means proving that the TLA formula $\Pi \Rightarrow \Box\Phi$ is valid. Similarly, proving that Π is implemented by another program Π' means proving that the formula $\Pi' \Rightarrow \Pi$ is valid.

2.3 Variables and types

TLA assumes that there is an infinite supply of program variables, though a given algorithm always mentions only a finite number of them. The values that

---

[1] The state function f in a formula of the form $[\mathcal{A}]_f$ has to be a tuple containing all program variables in $\mathcal{A}$ in order to guarantee invariance under stuttering.

<u>LATTICE.</u>

$\succ$ well-founded partial order on nonempty set S

$$\frac{F \wedge (c \in S) \;\Rightarrow\; (H_c \rightsquigarrow (G \vee \exists d \in S : (c \succ d) \wedge H_d))}{F \;\Rightarrow\; ((\exists c \in S : H_c) \rightsquigarrow G)}$$

<u>STL4.</u>

$$\frac{F \Rightarrow G}{\Box F \Rightarrow \Box G}$$

<u>INV1.</u>

$$\frac{I \wedge [\mathcal{N}]_f \Rightarrow I'}{I \wedge \Box[\mathcal{N}]_f \Rightarrow \Box I}$$

<u>TLA2.</u>

$$\frac{P \wedge [\mathcal{A}]_f \;\Rightarrow\; Q \wedge [\mathcal{B}]_g}{\Box P \wedge \Box[\mathcal{A}]_f \;\Rightarrow\; \Box Q \wedge \Box[\mathcal{B}]_g}$$

where

F, G, H_c, H_d are TLA formulas $\qquad$ P, Q, I are predicates

$\mathcal{A}$, $\mathcal{B}$, $\mathcal{N}$ are actions $\qquad$ f, g are state functions

Fig. 1. Some frequently used proof rules of TLA.

---

variables can take are not organised in types. Instead, TLA assumes that there exists only one single set of values, and a variable can take any value from this set. Often, however, it is assumed that variables only hold values of a certain predefined subset of the set of values (e.g. the set of integers etc.). In this case the type correctness of variables has to be proven, i.e. that a formula of the form $\Pi \Rightarrow \Box T$ holds (assuming Π is a specification of an algorithm and T is a predicate asserting that the program variables belong to the sets of values that they are supposed to).

2.4 An example algorithm

As an example we consider a simple algorithm, which increments variables x and y indefinitely. The initial state is characterised by the predicate

$$Init_1 \;\triangleq\; (x = 0) \wedge (y = 0)$$

The incrementation is defined by the following two actions:

$$\mathcal{M}_1 \;\triangleq\; x' = x + 1 \wedge Unch\ y \tag{2a}$$

$$\mathcal{M}_2 \;\triangleq\; y' = y + 1 \wedge Unch\ x \tag{2b}$$

A predicate of the form *Unch x* means that $x' = x$. In TLA the fact that the value of program variables aren't changed has to be expressed explicitly, otherwise their value in the next state is not known.

The fairness requirement of the algorithm in 2 is weak fairness with respect to both actions, so the algorithm is described by the following formula:

$$\Pi_1 = Init_1 \wedge \Box[\mathcal{M}_1 \vee \mathcal{M}_2]_{(x,y)} \wedge \mathrm{WF}_{(x,y)}(\mathcal{M}_1) \wedge \mathrm{WF}_{(x,y)}(\mathcal{M}_2) \qquad (3)$$

2.5 Refinement mappings and data hiding

In TLA program variables can be hidden using existential quantification as in: $\exists x : \Pi$ (here Π is assumed to be a specification of an algorithm in the normal sense) Informally this means that we do not constrain the value of the variable x at all, it is sufficient that there is some sequence of values such that Π holds.

Figure 2 describes the formal semantics of the existential quantifier. Given this rather complicated definition one can prove that for any TLA formula F, any variable x and any behaviours s_1 and s_2 the following holds

$$\natural s_1 = \natural s_2 \Rightarrow s_1[\exists x : F] = s_2[\exists x : F] \qquad (4)$$

stating that all existentially quantified TLA formulas are invariant under stuttering. The validity of the formula is actually independent of whether F is invariant under stuttering and the properties of the function $\natural^2$.

$$\langle\langle s_0, s_1, \ldots \rangle\rangle =_x \langle\langle t_0, t_1, \ldots \rangle\rangle \quad \triangleq \quad \forall n \in \mathrm{Nat} : \forall \text{'}v\text{'} \neq \text{'}x\text{'} : s_n[\![v]\!] = t_n[\![v]\!]$$

$$\natural\langle\langle s_0, s_1, s_2, \ldots \rangle\rangle \quad \triangleq \quad \text{if } \forall n \in \mathrm{Nat} : s_n = s_0$$
$$\text{then } \langle\langle s_0, s_0, s_0, \ldots \rangle\rangle$$
$$\text{else if } s_1 = s_0 \text{ then } \natural\langle\langle s_1, s_2, s_3, \ldots \rangle\rangle$$
$$\text{else } \langle\langle s_0 \rangle\rangle \circ \natural\langle\langle s_1, s_2, \ldots \rangle\rangle$$

$$\sigma[\exists x : F] \quad \triangleq \quad \exists \rho, \tau \in \mathrm{St}^\infty : (\natural\sigma = \natural\rho) \wedge (\rho =_x \tau) \wedge \tau[F]$$

where
o denotes concatenation of sequences σ is a behavior
St^∞ denotes the set of all behaviours $s, s_0, t_0, s_1, t_1, \ldots$ are states

Fig. 2. Semantics of quantification in TLA.

Although the existential quantifier $\exists$ clearly differs from the quantifier of predicate logic, it still obeys the laws from predicate logic (the laws can be derived using the basic rules governing quantification that are given in Figure 3). Typically, proving that an algorithm implements another means proving a

---

[2] Of course the question as to whether (4) actually correctly captures the notion of invariance under stuttering or not is dependant on the properties of the function $\natural$.

formula of the form $\exists y : \Pi' \Rightarrow \exists x : \Pi$. Assuming that y does not occur free in Π this formula is proven by giving a refinement mapping [1] (that we can call $\overline{x}$) that expresses x by means of the variables in Π' and then proving the formula $\Pi' \Rightarrow \overline{\Pi}$ where $\overline{\Pi}$ stands for Π with every occurrence of x replaced by $\overline{x}$. After this we simply apply rule E1 to deduce $\Pi' \Rightarrow \exists x : \Pi$ and then by applying rule E2 we are done.

<u>E1.</u>
$\vdash F(g/x) \Rightarrow \exists x : F$

<u>E2.</u>
$F \Rightarrow G$
x does not occur free in G
$$\overline{(\exists x : F) \Rightarrow G}$$

where
x is a variable F, G are TLA formulas
g is a state function

Fig. 3. Rules for quantification in TLA.

---

3 TLA in HOL

3.1 States, Predicates and Actions in HOL

In our formalisation, we assume that every variable has a well-defined type. We formalise states as tuples where every component corresponds to one variable of the state (the type of the component indicates what type its values must have). The state is made potentially infinite by adding a final component with polymorphic type. Thus, this final component ("the rest of the universe") can be instantiated to any tuple. This is needed to allow refined versions of the algorithm to add new variables to the state if needed. Our choice of representation of states was inspired by the experiences gained by the developers of a HOL formalisation of the Refinement Calculus in HOL (see [3] vs. [10]).

As an example, we consider the state space containing the variables x and y which range over natural numbers. The corresponding state space in HOL has type `:num#num#'s` (`'s` is a type variable).

Predicates and actions are formalised as boolean expressions with lambda-bound program variables. For example, in the above mentioned state space, the action $\mathcal{M}_1$ of (3) is formalised as

$$\lambda(x,y,z)(x',y',z').\ (x'=x+1) \wedge (y'=y)$$

and the state predicate $x > 0$ is formalised as

$$\lambda(x,y,z).\ x>0$$

Note that the variables are anonymous, in the sense that the previous action is equivalently expressed by

$$\lambda(\mathbf{a,z,x})(\mathbf{c,t',z'}). \quad (\mathbf{c=a+1}) \wedge (\mathbf{t'=z})$$

However, we avoid confusion if we use the TLA rules for priming as a convention when writing actions in HOL.

Constants (called *rigid variables* in TLA) are formalised as variables that are not bound by λ-abstraction. This representation directly mirrors the meaning of rigid variables in TLA, which are not constants in the normal meaning of the word, but rather variables whose value is fixed but unknown.

The boolean connectives are lifted to predicates and actions in a straightforward way. The connectives for predicates are called **pnot, pand, por, pimp** etc. Similarly, the connectives for actions are called **anot, aand**, etc. As examples, we show the theorems that define the lifted conjunctions:

$\vdash_{def}$ p pand q $= \lambda$s. p s $\wedge$ q s
$\vdash_{def}$ a aand b $= \lambda$s s'. a s s' $\wedge$ b s s'

where **s** and **s'** are states. We also define functions that represent validity:

$\vdash_{def}$ pvalid p $= \forall$s. p s
$\vdash_{def}$ avalid a $= \forall$s s'. a s s'

3.2 Temporal Logic in HOL

We have formalised the TLA logic by semantically embedding in HOL in a straightforward way. Behaviours have type :**num**$\rightarrow$**state** where :**state** is the type of the state (in the generic case, :**state** is just a polymorphic type). Temporal formulas are represented as *temporal properties*, i.e., as boolean functions on behaviours. The following theorem defines the $\square$-operator:

$\vdash_{def}$ box f t $= \forall$i. f(λn. t(i+n))

for temporal formula **f** and behaviour **t**.

We also lift the boolean connectives to temporal formulas, giving them the names **tnot, tand**, etc. For example, the following theorems define lifted conjunction and lifted validity:

$\vdash_{def}$ f tand g $= \lambda$t. f t $\wedge$ g t
$\vdash_{def}$ tvalid f $= \forall$t. f t

where **t** is a behaviour.

3.3 TLA Formulas in HOL

We have formalised TLA in HOL by directly defining the semantics of TLA. HOL also permits another approach, where one defines a new type corresponding to TLA formulas and then define the semantics separately. However, the syntax of TLA is such that it would be quite involved to define TLA formulas as a new type.

One consequence of our approach is that we do not formalise the notion of "well-formed TLA formula" at all. This is justified by a pragmatic viewpoint: our aim is to mechanise TLA-reasoning about algorithms, not to reason about the properties of the TLA logic itself.

In TLA, every predicate can also be interpreted as an action or as a temporal formula. In our formalisation this is not true. Instead, predicates, actions and temporal formulas have distinct types. We define functions that do the "lifting" from predicates to the action and temporal levels (see Fig. 4).

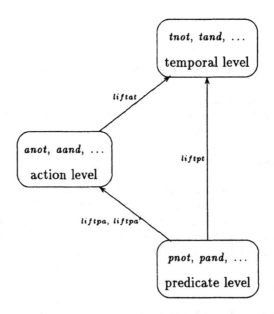

Fig. 4. Levels of the TLA logic

The functions **liftpa** and **liftpa'** that lift predicates to actions are defined as follows:

$\vdash_{def}$ liftpa p $= \lambda$s s'. p s
$\vdash_{def}$ liftpa' p $= \lambda$s s'. p s'

Similarly, the functions **liftpt** and **liftat** that lift predicates and actions to the temporal level are defined as follows:

$\vdash_{def}$ liftpt p t = p (t 0)
$\vdash_{def}$ liftpa a t = a (t 0) (t 1)

The square function $[\mathcal{A}]_f$ and the *Enabled* predicate are formalised by the following defining theorems:

$\vdash_{def}$ square a f = a aor (λs s'. f s = f s')
$\vdash_{def}$ enabled a = λs. $\exists$s'. a s s'

where f is a function from states to an arbitrary type.

Finally, we consider how an algorithm is represented. We assume that we are given a predicate (the initialisation init), an action (action) and a temporal formula (the liveness part live). The algorithm is then described by the formula

(liftpt init) tand (box (liftat (square action w))) tand live

where w represents the function $\lambda(x_1, ..., x_n, z).(x_1, ..., x_n)$ which takes the global state as argument and returns the part of the state that the algorithm works on (i.e., the state with the last component removed).

3.4 An Example Algorithm.

As an example we consider the algorithm described in (3). The following defining theorems show how this algorithm is formalised:

$\vdash_{def}$ Init1 = λ(x,y,z).(x=0) $\wedge$ (y=0)
$\vdash_{def}$ M1 = λ(x,y,z)(x',y',z').(x'=x+1) $\wedge$ (y'=y)
$\vdash_{def}$ M2 = λ(x,y,z)(x',y',z').(y'=y+1) $\wedge$ (x'=x)
$\vdash_{def}$ w = λ(x,y,z).(x,y)
$\vdash_{def}$ F = (WF M1 w) tand (WF M2 w)
$\vdash_{def}$ Prog1 = (liftpt Init1) tand
          (box (liftat (square (M1 aor M2) w))) tand F

Here WF stands for the HOL equivalent of weak fairness in TLA.

4 The Simple TLA Logic in HOL

The proof rules of the simple TLA logic (TLA without refinement mappings and data hiding) are represented as theorems in HOL which are proved using the definition of the semantics. We will not show how all of the rules are formulated in HOL. Instead, we describe the rules from Figure 1. Below we list the HOL theorems representing the rules INV1, STL4 and TLA2. This rule is frequently used when proving invariance properties of programs. It falls into the category of additional rules, i.e. rules that can be proven from the basic ones.

$\vdash$ avalid (((liftpa I) aand (square N f)) aimp (liftpa' I)) $\Rightarrow$
  tvalid (((liftpt I) tand (box (square N f))) timp (box (liftpt I)))

⊢ p pimplies q ⇒ box (liftpt p) timplies box (liftpt q)

⊢ (liftpa p aand square a f) aimplies (liftpa q aand square b g) ⇒
  (box (liftpt p) tand box (liftat (square a f))) timplies
  (box (liftpt q) tand box (liftat (square b g))))

The **pimplies** function above is shorthand for **pvalid(p pimp q)** and likewise for the **timplies** function. Note how the HOL implication corresponds to the meta-implication in the rule. The proofs (in HOL) of the above rules are quite straightforward, using rewriting and other basic HOL proof methods.

An important (basic) rule in liveness proofs is the LATTICE rule (see Figure 1), which encodes the principle of well-founded ranking. This rule presupposes the existence of a well-founded order on some set involved in the reasoning. To allow the formulation of this rule in HOL reasoning, we have used a separate theory of well-founded sets[3] where the notion of well-foundedness is defined and basic properties of well-founded sets are proved. In HOL the theorem representing the LATTICE rule is as follows:

⊢ wellf po
  ⇒ (∀c.f timplies ((h c) leads_to (g tor (λt.∃d.po d c ∧ h d t))))
  ⇒ f timplies ((λt.∃c. h c t) leads_to g)

At first sight the rules in HOL may look ugly, since we cannot mix predicates, actions and temporal formulas as neatly as TLA does; we have to use the lifting functions. However, this is sometimes an advantage, since we avoid the confusion which can arise when the same term is sometimes a predicate and sometimes a temporal formula.

In the revised version of TLA [8] some of the basic as well as additional rules have changed. We have not yet proved all of the changed additional rules.

5   Refinement mappings and data hiding in HOL

We have formalised the notion of data hiding in a way resembling the method used in [11] for defining data refinement in the refinement calculus. Thus, the state space is extended to consist of two parts: a *hidden* and a *visible* part. A behaviour over such an extended state space is a function of type **num->('h#'s)** in the generic case.

The way the quantifier ∃ is defined in [8] it actually obeys the same laws as the ordinary existential quantifier of predicate logic. The way we define quantification below may not be as general, but we feel that it (together with our formalisation of the rules E1 and E2) is sufficient (with suitable type instantiations in the rules) for proving refinement steps involving refinement mappings.

---

[3] Due to Jockum von Wright

5.1 Basic definitions

Using the above representation we define the existential quantifier ∃ as a function **ex** according to the following:

⊢ ∀f s. ex f s = (∃t x. (flat s = flat t) ∧ f(λn. x n,t n))

where **f** is a formula over an extended state space and **s** is a behaviour over a non-extended state space. In this representation multiple quantification can be expressed by instantiating the hidden component as a tuple. The function **flat** (representing the function ♮ from Figure 2) is defined as follows:

⊢ ∀b. flat b =
    (λn.(∃m. m − stut b m = n) → (b (εm. m − stut b m = n)) | (b n))

Here **b** is a behaviour over a simple state space and **stut** is a recursively defined function that given a behaviour **b** and a variable **m** of type **num** returns the number of stuttering steps that the behaviour has performed during the time period 0..m.

5.2 Proof rules

We proceed by formulating the HOL equivalents of the proof rules E1 and E2 from Figure 3 and prove them to be HOL theorems. Given the definitions above the proof rule E1 becomes:

⊢ ∀f p g. f (λn. g (p n),SND (p n)) ⇒ ex f (λn. SND (p n))

where **f** is a formula over an extended state space, **p** is a behaviour over an extended state space and **g** is a function of type ('hh#'a)->'h (a refinement mapping). The rule E2 has the following format:

⊢ ∀f g. ((∀s. f s ⇒ g (λn. SND (s n))) ∧
        (∀p p'. (flat p = flat p') ⇒ (g p = g p'))) ⇒
        ∀t. (ex f t ⇒ g t)

Here we have to add the second conjunct (on the left hand side of the meta-implication of the rule) in order to get a sound rule. In the rule **f** is a formula over an extended state space, **g** is a formula over a simple state space, **s** is a behaviour over an extended state space and **t, p, p'** are behaviours over a simple state space. The second conjunct asserts the the formula **g** is invariant under stuttering. This requirement follows from the fact that our definition of the semantics of TLA allows defining formulas that are not invariant under stuttering (however, there is an implicit assumption that only formulas allowed by the syntax of TLA are defined). However, if the formula **g** in the rule is of the form **ex h** (as it usually is) then invariance under stuttering is trivially true (as shown in section 2 in equation (4)). If the formula **g** is not existentially quantified we need a function "lifting" it to the type of extended behaviours. We also actually need to prove invariance under stuttering.

The way these rules are used is, in general, as follows. Suppose we want to prove that a specification **ex g** is correctly implemented by **ex f**. Then we need to prove the following HOL theorem:

$\vdash \forall t.\ (\text{ex f t} \Rightarrow \text{ex g t})$

Here **f** and **g** are formulas over extended state spaces (they may be the same but are usually different). In order to prove this we actually prove the following theorem:

$\vdash \text{f timplies g}'$

where **g'** stands for the original formula **g** with the hidden component expressed by means of an refinement mapping of the type required by rule E1. The main work in proving a refinement step is done in proving the above theorem. It is of course essential that the theorem that is to be proven is actually on the level our temporal logic (as it is above) so that the proof rules can be used (in the same way as they would be used in a paper and pen proof).

By rewriting the above theorem using the semantical definitions it is shown that the following is a theorem:

$\vdash \forall s.\ \text{f s} \Rightarrow \text{ex g } (\lambda n.\ \text{SND (s n)}))$

after which we are done since both premises of rule E2 have been established.

5.3 Applying the rules to a simple example

To show how to use the rules above in practice we consider a simple example. We define two algorithms Φ and Ψ as follows:

$$Init_1 \triangleq (x, i) = (0, 0)$$
$$Init_2 \triangleq (x, y, i, j) = (0, 0, 0, 0)$$
$$A1 \triangleq (x', i') = (x + 1, i + 1)$$
$$A2 \triangleq (x', y', i', j') = (x + 1, y + 2, i + 1, j + 2)$$
$$\Phi \triangleq Init_1 \wedge \Box[A1]_{(x,i)}$$
$$\Psi \triangleq Init_2 \wedge \Box[A2]_{(x,y,i,j)}$$

What we want to prove is the validity of the formula $\exists x : \exists y : \Psi \Rightarrow \exists x : \Phi$ (i.e. that $\exists x : \exists y : \Psi$ implements $\exists x : \Phi$). The algorithms are expressed in HOL as follows:

$\vdash_{def}$ Init1 $= \lambda(x,(i,z)).(x=0) \wedge (i=0)$
$\vdash_{def}$ A1 $= \lambda(x,(i,z))(x',(i',z')).(x'=x+1) \wedge (i'=i+1)$
$\vdash_{def}$ v $= \lambda(x,(i,z)).(x,i)$
$\vdash_{def}$ Prog1 $=$ (liftpt Init1) tand (box (liftat (square A1 v)))
$\vdash_{def}$ Init2 $= \lambda((x,y),(i,z)).(x=0) \wedge (i=0)$
$\vdash_{def}$ A2 $= \lambda((x,y),(i,j,z))((x',y'),(i',j',z')).$
$\qquad\qquad (x'=x+1) \wedge (y'=y+2) \wedge (i'=i+1) \wedge (j'=j+1)$
$\vdash_{def}$ w $= \lambda((x,y),(i,j,z)).((x,y),(i,j))$
$\vdash_{def}$ Prog2 $=$ (liftpt Init2) tand (box (liftat (square A2 w)))

and the theorem we want to prove in order to establish that **Prog2** correctly implements **Prog1** is

$\vdash \forall s.$ ex Prog2 s $\Rightarrow$ ex Prog1 s

Using the same reasoning as above this theorem is proven by proving the validity of the following formula:

$\vdash$ Prog2 timplies $\lambda s.$Prog1 $(\lambda n.(\text{FST}(\text{FST}(s\ n)),\text{SND}(s\ n)))$

A proof of a theorem as the one above typically follows the same proof outline as pen and paper proofs in TLA. However, a lot of effort goes into transforming goals so that they match the rules. Typically this means proving lots of simple tautologies on the predicate, action as well as temporal levels.

6 Conclusions

We have described an attempt at formalising TLA in HOL with the purpose of allowing mechanised reasoning of (concurrent) algorithms. Especially, we have formalised the rules that govern data hiding and refinement mappings which was not considered in the earlier work described in [9, 12].

There are however many ways in which the work presented here could be improved. Focusing specifically on the part concerning the formalisation of data hiding and refinement mappings, it needs to be tested with more realistic examples. Also there is a need for a proof procedure for proving invariance under stuttering (for formulas that follow the TLA syntax), to simplify the use of rule E2.

More generally, all the additional rules that have changed in the revised version of TLA [8], and haven't yet been proved, should be proved. These rules, that deal with fairness (which is the most difficult part in TLA reasoning), are essential in simplifying proofs of refinement.

An interface of the type described in [6], translating some higher level representation of the TLA syntax to the somewhat complicated one of HOL (as well as handling the grouping of program variables into tuples etc.), would be a great help. Also any amount of automation that could be added to the general level of TLA reasoning (in HOL) would certainly be of use.

Acknowledgements

Thanks are due to Jockum von Wright for all his help and advice. I also want to thank anonymous referees for their comments and helpful suggestions. Finally I want to thank Michael Butler for his comments on the final version of this paper.

References

1. M. Abadi and L. Lamport. The existence of refinement mappings. *Theoretical Computer Science*, 82(2):253–284, 1991.
2. F. Andersen. *A Theorem Prover for UNITY in Higher Order Logic*. PhD thesis, Technical University of Denmark, Lyngby, 1992.
3. R. J. R. Back and J. von Wright. Refinement concepts formalized in higher order logic. *Formal Aspects of Computing*, 2:247–272, 1990.
4. A. Camillieri. Mechanizing CSP trace theory in Higher Order Logic. *IEEE Transactions on Software Engineering*, 16(9):993–1004, 1990.
5. C-T Chou. Mechanical verification of distributed algorithms in higher-order logic. In these proceedings.
6. U. Engberg, P. Groenning, and L. Lamport. Mechanical verification of concurrent systems with TLA. In G. v. Bochmann and D. K. Probst, editors, *Computer Aided Verification – Fourth International Workshop. CAV '92. Montreal. Canada. June 29 – July 1. 1992*, volume 663 of *Lecture Notes in Computer Science*. Springer Verlag, 1993.
7. M.J.C. Gordon and T.F. Melham, editors. *Introduction to HOL*. Cambridge University Press, 1993.
8. L. Lamport. The temporal logic of actions. Research Report 79, DEC, Systems Research Center, December 1991. A revised version of the paper will appear in *ACM Transactions on Programming Languages and Systems*.
9. J. von Wright. Mechanising the temporal logic of actions in HOL. In *Proceedings of the 1991 HOL Tutorial and Workshop*, August 1991.
10. J. von Wright. Program refinement by theorem prover. In *BCS FACS Sixth Refinement Workshop – Theory and Practise of Formal Software Development. 5th – 7th January, City University, London, UK.*, 1994.
11. J. von Wright, J. Hekanaho, P. Luostarinen, and T. Långbacka. Mechanising some advanced refinement concepts. *Formal Methods in System Design*, 3:49–81, 1993.
12. J. von Wright and T. Långbacka. Using a theorem prover for reasoning about concurrent algortihms. In G. v. Bochmann and D. K. Probst, editors, *Computer Aided Verification – Fourth International Workshop. CAV '92. Montreal. Canada. June 29 – July 1. 1992*, volume 663 of *Lecture Notes in Computer Science*. Springer Verlag, 1993.

Studying the ML Module System in HOL

Savi Maharaj[*1] and Elsa Gunter[2]

[1] University of Edinburgh, Department of Computer Science
JCMB, King's Buildings, Edinburgh EH9 3JZ.
phone: 44 31 650 5183 email: svm@dcs.ed.ac.uk
[2] AT&T Bell Laboratories, Rm.#2A-432
600 Mountain Ave., Murray Hill, N.J. 07974, USA
phone: 1 908 582 5613 email: elsa@research.att.com

Abstract. In an earlier project [5] the dynamic semantics of the Core of Standard ML (SML) was encoded in the HOL theorem-prover. We extend this by adding the dynamic Module system. We then develop a possible dynamic semantics for a Module system with higher-order functors and encode this as well. Next we relate these two semantics via embeddings and projections and discuss how we can use these to state and to prove that evaluation in the proposed system is a conservative extension, in an appropriate sense, of evaluation in the SML Module system.

1 Introduction

The project described in this paper is part of an ongoing program of work on encoding the semantics of the SML programming language within HOL in order to support formal reasoning about the semantics and about SML programs. The encoding is done via definitions and derived properties, as opposed to using an axiomatization. This approach necessitates the use of a higher-order language, but was chosen because it provides us with induction principles, powerful tools for proving meta-theoretic properties of the semantics and facts about evaluation that follow from such properties. Examples of these are negative facts such as that certain expressions do not evaluate.

In earlier work [5] Myra VanInwegen and Elsa Gunter encoded the dynamic semantics of Core SML and then formally verified the fact that dynamic evaluation is deterministic. In our project we build upon this foundation by adding the dynamic semantics of the SML Module system (Modules). We then propose a speculative dynamic semantics for a Module system with higher-order functors for an extension of SML. To the best of our knowledge this is the first time that such a semantics has been written down. By presenting this semantics within a theorem prover we gain assistance in proving theorems that give confidence in the proposed semantics. Such theorems include that the extended Module system is, in a sense which we shall define, a conservative extension of the original system.

---

* Partially supported by AT&T

In Section 2 we first describe the fragments of SML syntax and semantics concerned with dynamic evaluation of Modules, and then discuss the way in which we represent these in HOL. In Section 3 we describe our proposed extended Module system with higher-order functors and discuss various issues that arise in deciding upon a semantics for this system. We also describe the HOL encoding of the extended system. In Section 4 we discuss in detail how to state and prove that the proposed system is a conservative extension of Modules. Finally we give our concluding remarks in Section 5.

2 Mod-ML

The semantics of SML is presented in *The Definition of Standard ML* [4], which we shall refer to as the *Definition*. It is divided into four parts, giving the static and then the dynamic semantics of first the Core and then the Modules. Roughly speaking, static semantics concerns well-typedness and the satisfaction of sharing requirements, while dynamic semantics concerns the evaluation of terms under the assumption that static properties have already been checked.

Syntax

The dynamic semantics makes use of a reduced syntax for SML in which information that is not relevant to dynamic evaluation has been removed. The syntax classes for Modules consist of Core identifiers and declarations, extended by signatures, structures and functors. The abstract syntax for Modules is shown in Fig. 1 along with a small part of the Core abstract syntax. Signatures (*valdesc*, $\cdots$, *sigdec*) are allowed to specify values, exceptions and structures and can be formed in several ways, including extracting the signatures from a list of structures (OPENspec). Structures (*strexp*, $\cdots$, *strbind*) can contain Core declarations and structure declarations and can be built up from other structures in various ways. Functors (*funbind, fundec*) are defined by binding a functor identifier to a structure identifier and signature, specifying the input to the functor, an optional output signature, and a structure expression which makes up the functor body.

Semantic objects   The *Definition* gives the meanings of programs of the Module system in terms of *semantic objects* which we present in Fig. 2. We use the notation Fin(X) and $X \xrightarrow{\text{fin}} Y$ to represent finite subsets and finite functions, respectively. Module declarations are evaluated with respect to a basis, which contains information about previously declared modules. Signature bodies (Spec) are evaluated to interfaces (Int) which contain information needed to later thin structures to fit the signature. This includes the names of variables and exceptions and the internal interface environment. Structure declarations evaluate to give environments (Env) which contain information about the variables, exceptions and structures contained in the declared structure. Coincidentally, these environments are the same as those used in Core evaluation (the structure information is needed to evaluate long identifiers), so the Core semantic object

$longx ::= x \mid strid_1.\cdots.strid_n.x$ $(n \geq 1, x$ in any class marked "long")   (Core)

var $::= $ VAR $string$ (long) variable (Core)

$excon ::= $ EXCON $string$ (long) exception constructor (Core)

dec $::= $ (omitted)   declaration (Core)

$strid$ $::= $ STRID $string$ (long) structure identifier (Core)

$sigid$ $::= $ SIGID $string$ signature identifier

$funid$ $::= $ FUNID $string$ functor identifier

$valdesc ::= $ VARvaldesc var $\langle valdesc \rangle$ value description

$exdesc$ $::= $ EXCONexdesc $excon$ $\langle exdesc \rangle$ exception description

$sigexp$ $::= $ SIGsigexp $spec$ $\mid$ SIGIDsigexp it $sigid$ signature expression

$spec$ $::= $ VALspec $valdesc$ $\mid$ EXCEPTIONspec $exdesc$ $\mid$ specification
            STRUCTUREspec $strdesc$ $\mid$
            LOCALspec $spec$ $spec$ $\mid$
            OPENspec $longstrid_1 \cdots longstrid_n$ $\mid$ $(n \geq 1)$
            INCLUDEspec $sigid_1 \cdots sigid_n$ $\mid$ $(n \geq 1)$
            EMPTYspec $\mid$ SEQspec $spec$ $spec$

$strdesc ::= $ STRIDstrdesc $strid$ $sigexp$ $\langle strdesc \rangle$ structure description

$sigbind ::= $ BINDsigbind $sigid$ $sigexp$ $\langle sigbind \rangle$ signature binding

$sigdec ::= $ SIGNATUREsigdec $sigbind$ $\mid$ EMPTYsigdec $\mid$ signature declaration
          SEQsigdec $sigdec$ $sigdec$

$strexp$ $::= $ STRUCTstrexp $strdec$ $\mid$ structure expression
          LONGSTRIDstrexp $longstrid$ $\mid$
          APPstrexp $funid$ $strexp$ $\mid$
          LETstrexp $strdec$ $strexp$

$strdec$ $::= $ DECstrdec dec $\mid$ STRUCTUREstrdec $strbind$ $\mid$ structure declaration
          LOCALstrdec $strdec$ $strdec$ $\mid$ EMPTYstrdec $\mid$
          SEQstrdec $strdec$ $strdec$

$strbind ::= $ BINDstrbind $strid$ $\langle sigexp \rangle$ $strexp$ $\langle strbind \rangle$ structure binding

$funbind ::= $ BINDfunbind $funid$ $strid$ $sigexp$ $\langle sigexp \rangle$ strexp   functor binding
                  $\langle funbind \rangle$

$fundec$ $::= $ FUNCTORfundec $funbind$ $\mid$ EMPTYfundec $\mid$ functor declaration
          SEQfundec $fundec$ $fundec$

$topdec$ $::= $ STRDEC $strdec$ $\mid$ SIGDEC $sigdec$ $\mid$ FUNDEC $fundec$ top declaration

Fig. 1. Modules abstract syntax

can be re-used. Functor declarations are evaluated to give a FunEnv mapping the functor identifier to a functor closure consisting of an identifier and interface for the input structure, an interface for the output, if supplied, the body of the functor, and the basis in which applications of the functor are to be evaluated.

Functions on the semantic objects The semantics makes use of various auxiliary functions which operate upon the semantic objects. These include looking up identifiers in the various environments and bases, denoted by e.g. *G (sigid)*,

$$E \in \text{Env} = \text{StrEnv} \times \text{VarEnv} \times \text{ExConEnv} \qquad \text{environment (Core)}$$

$$SE \in \text{StrEnv} = strid \overset{\text{fin}}{\to} \text{Env} \qquad \text{structure env. (Core)}$$

$$VE \in \text{VarEnv} = var \overset{\text{fin}}{\to} \text{Val} \qquad \text{value environment (Core)}$$

$$EE \in \text{ExConEnv} = excon \overset{\text{fin}}{\to} \text{ExName} \qquad \text{excon environment (Core)}$$

$$\text{Val} = \text{(omitted)} \qquad \text{value (Core)}$$

$$\text{ExName} = \text{(omitted)} \qquad \text{exception name (Core)}$$

$$I \in \text{Int} = \text{IntEnv} \times \text{Fin}(var) \times \text{Fin}(excon) \qquad \text{interface}$$

$$IE \in \text{IntEnv} = strid \overset{\text{fin}}{\to} \text{Int} \qquad \text{interface environment}$$

$$G \in \text{SigEnv} = sigid \overset{\text{fin}}{\to} \text{Int} \qquad \text{signature environment}$$

$$IB \in \text{IntBasis} = \text{SigEnv} \times \text{Intenv} \qquad \text{interface basis}$$

$$\text{FunctorClosure} = (strid \times \text{Int}) \times (\text{StrExp} (\times \text{Int})) \times \text{Basis} \qquad \text{functor closure}$$

$$F \in \text{FunEnv} = funid \overset{\text{fin}}{\to} \text{FunctorClosure} \qquad \text{functor environment}$$

$$B \in \text{Basis} = \text{FunEnv} \times \text{SigEnv} \times \text{Env} \qquad \text{basis}$$

Fig. 2. Modules semantic objects

which means look up *sigid* in the signature environment G; projecting the components of semantic objects, denoted by e.g. E of B; injecting objects as components of larger objects, written as e.g. E in Basis, which denotes the basis $(\{\}, \{\}, E)$; updating environments and bases, written as e.g. $G + \{sigid \mapsto I\}$. Functions are lifted in the obvious way to operate on larger objects which contain the domain of the function as a component. For example, one can look up signature identifiers in bases (by looking them up in the SigEnv component).

The semantics of OPENspec, which allows a signature to be formed by extracting the signatures from a list of structures, demands a special function to perform this extraction. The structures are evaluated to produce environments, and then from each environment we must extract an interface. This is done by the function Inter : Env → Int defined as:

$$\text{Inter } (SE, VE, EE) = (IE, \text{domain } VE, \text{domain } EE)$$

where

$$IE = \{strid \to \text{Inter } E; SE(strid) = E\}.$$

That is, for structure environments, the interface we extract is the set of mappings from the *strids* they contain to interfaces for the *envs* they associate with the *strids*. For variable and exception environments, the extracted interfaces are just the sets of identifiers they contain.

Another important function, $\downarrow$: Env × Int → Env, thins an environment to fit a given interface. This is needed whenever a signature is explicitly given to a structure. It is defined by:

$$(SE, VE, EE) \downarrow (IE, vars, excons) = (SE', VE', EE')$$

where

$$SE' = \{strid \to E \downarrow I; SE(strid) = E \text{ and } IE(strid) = I \}$$

and *VE'* and *EE'* are obtained by restricting the domains of *VE* and *EE* to *vars* and *excons* respectively.

The evaluation relations   The main part of the semantics consists of inference rules giving the evaluation relations by which the phrase classes are related to the semantic objects. We shall not give these in detail here.

Encoding the Semantics in HOL

We encode the semantics of the Module system in HOL and call the resulting package of definitions Mod-ML. We shall write the names of Mod-ML types and terms in teletype font. Mod-ML is an extension of the system HOL-ML described in [5], which is an encoding of the dynamic semantics of a large subset of the Core language of SML. All of the definitions that make up HOL-ML must be loaded into HOL in order for Mod-ML to be loaded. In both encodings, SML syntax and semantic objects are represented in HOL either as new inductive types or by means of objects (such as sets and lists) already present in the HOL libraries. Evaluation relations, relating the terms of each syntactic class to the corresponding semantic objects, are defined inductively as the smallest relations satisfying various evaluation-relation predicates representing the rules of the dynamic semantics.

Syntax   Each of the phrase classes in Fig. 1 is represented by a type in HOL. Rather than introducing all of these types at once as one huge mutually recursive type definition, we have chosen to separate them out into mutually dependent groups and define each of these separately. We believe that by avoiding unnecessary mutual recursion among the types the presentation of definitions and theorems will be made clearer and inductive proofs will be simpler. To make these definitions we used a package developed by Elsa Gunter and Healfdene Goguen which supports simultaneously mutually recursive and nested recursive definitions of types. This package both extends and depends upon a previous package for mutual recursion developed by Elsa Gunter and Myra VanInwegen and first used in the development of HOL-ML.

We re-use various techniques developed in [5]: optional arguments to constructors in the grammar are encoded by defining a type 'a option with constructors NONE and SOME 'a; the lists of identifiers used by OPENspec and INCLUDEspec are represented by defining a type 'a nonemptylist. For long identifiers we define a new type 'a long.

Semantic objects and functions   To encode the semantic objects we must make choices about the representation of finite sets and finite function spaces. For the former we use the sets supplied by the HOL library since finiteness can be added later as an axiom to theorems which require it. For the latter we use lists of pairs of the appropriate identifier and value types. When we encode functions that are intended to operate on finite function spaces we are always careful to ensure that the lists are maintained in lexicographical ordering by identifiers.

Our aim here is to make the list structure transparent so that we adequately represent finite functions.

Evaluation To complete the encoding we define the evaluation relations which say how syntactic terms evaluate to the appropriate semantic object. Again we re-use the techniques of [5]. For those phrase classes that form individual recursive types we use the HOL command `new_inductive_definition` to define the evaluation relation. This command defines a relation from a family of rules giving an inductive description of the relation. Unfortunately it is capable of defining only a single relation, not a mutually recursive family of such relations. Therefore for phrase classes that form a nested or mutually inductive group, and thus have evaluation relations defined by mutual recursion, we must do the corresponding job by hand. For example, the phrase classes *strexp*, *strdec* and *strbind* form a mutually inductive group. To define their evaluation relations, we first define an evaluation-relation predicate for the group. This predicate, `ModML_eval_structures_pred` is defined over possible evaluation relations for the three phrase classes, and is true if the possible evaluation relations satisfy all the rules for evaluating the three phrase classes. Then we define each evaluation relation as the logical intersection of all relations that satisfy the evaluation-relation predicate. For example, the evaluation relation for *strexp* is defined as (simplified):

```
eval_strexp strexp:strexp B:basis E:env =
  ∀ poss_eval_strexp poss_eval_strdec poss_eval_strbind.
    ModML_eval_structures_pred
              poss_eval_strexp poss_eval_strdec poss_eval_strbind ⇒
         poss_eval_strexp strexp B E
```

Defining the evaluation relations in this manner has the advantage that it readily gives us an induction principle for proving facts about them. The last thing we do is to prove that each evaluation relation satisfies the appropriate evaluation-relation predicate. This and the proof of the induction principles were done by two tactics which were a concise composition of built-in tactics parameterized by the definitions.

3 Higher order functors

It has been proposed (Section 8.5 of [3], [1]) to extend SML by allowing functors to take functors as arguments and to be declared within structures (and therefore to be specified in signatures). No definitive semantics has yet been proposed for these "higher-order functors" though a possible static semantics is outlined in [1]. Here we use HOL to work out what the dynamic semantics of this extension should be, and then to explore the relationship between the extended system and the original system. For readability we present the new semantics in the informal notation used for the SML semantics, but we should like to stress that this semantics was developed *within* HOL.

Syntax A possible syntax for higher-order functors is given in [1]. However this differs from SML syntax in various idiosyncratic ways ([1] is a speculative, draft paper) which obfuscate the relationship between terms in the two languages. Therefore we decided to develop our own grammar by starting with the grammar of SML and making changes for the new constructs. The grammar for the Core language remains unchanged. Changes to the grammar of the Module system are listed in Fig. 3 and explained here.

$spec$ ::= (as in Fig. 1) | FUNCTORspec *funid strid sigexp sigexp*
$strexp$ ::= (as in Fig. 1) | APPstrexp *longfunid strexp* | (as in Fig. 1)
$moddec$ ::= DECmoddec *dec* | STRUCTUREmoddec *strbind* |
                LOCALmoddec *moddec moddec* |
                OPENmoddec $longstrid_1 \cdots longstrid_n$ | $(n \geq 1)$
                EMPTYmoddec | SEQmoddec *moddec moddec* |
                FUNCTORmoddec *funbind*
$funbind$ ::= (as in Fig. 1) | REBINDfunbind *funid longfunid*
$topdec$ ::= MODDEC *moddec* | SIGDEC *sigdec*

Fig. 3. Abstract syntax for higher-order functors (additions and changes)

Specifications (*spec*) Functors can now be specified in signatures. This is done by giving a *funid*, a *strid* and signature for the input structure, and a signature for the output structure.
Structure expressions (*strexp*) We now have long *funids*, referring to functors declared within structures, and we can apply these to form new structures.
Module declarations (*moddec*) Functor declarations are now to be treated as a special kind of structure declaration, so to support this we amalgamate the two phrase classes into a new class of module declarations.
Core SML provides a feature (**open**) by which the declarations within a structure can be exposed to the top-level. When we extend to higher-order functors we choose to keep this syntax with its original semantics — which means that it does not expose the functor bindings within a structure. To allow functor bindings to be exposed we add a version of **open** (OPENmoddec) to the language of module declarations.
Functor bindings (*funbind*) It appears to be an omission in the SML grammar that no syntax is supplied for rebinding a functor to another functor identifier. We remedy this since this language feature is important to us as we can use it to give top-level names to functors declared within structures and to rebind functors passed through a functor's parameter.
Top declarations (*topdec*) The change here reflects the fact that structure declarations and functor declarations have been combined.

Semantic objects The main difficulty in deciding upon semantic objects is figuring out what environments should be. As we have seen, environments play a double role in the dynamic semantics of the SML Module system. They tell us the values associated with long identifiers during the evaluation of a Core expression, and they are the values returned by the evaluation of structure expressions. Since SML structures contain only structure declarations and Core language declarations, the environments they generate contain precisely the information needed by Core evaluation. However once we allow functor bindings within structure bodies, the situation changes. The environments needed for Core evaluation require only enough information to allow long identifiers to be looked up; they need no information concerning functors and functor bindings. However the environments returned by structures now do need to contain information concerning the functors bound in the body of the structure. Therefore, we are faced with two alternatives: either use environments in the Core dynamic semantics which have excess information, or define two kinds of environment, one for Core evaluation being the one we already have, and one for structure values. In the *Definition* there are already two kinds of "environments": environments for Core evaluation and bases for Module evaluation. We therefore decided that the second option was most in keeping with the spirit of the *Definition*, and this is the choice we have pursued here. In future work, we intend to encode both approaches and prove that the two are essentially the same.

The choice to have two different kinds of environments has ramifications elsewhere. One of these is the need to have two different kinds of **open** declarations: one which throws away functor information (for the Core language) and one which exposes it (for the Module system.) Another ramification is that we are obliged to define how to cut a Module-level environment down to a Core-level environment to enable the passing of evaluation between the Module system and the Core. We shall use the notation E of ME for this function. Its definition is straightforward.

The semantic objects for higher-order functors are those defined in Fig. 4, plus the classes SigEnv, FunctorClosure and FunEnv which remain unchanged from Fig. 2.

$$I \in \text{Int} = \text{FunIntEnv} \times \text{StrIntEnv} \times \text{Fin}(var) \times \text{Fin}(excon)$$
$$SIE \in \text{StrIntEnv} = strid \xrightarrow{\text{fin}} \text{Int}$$
$$FIE \in \text{FunIntEnv} = funid \xrightarrow{\text{fin}} \text{Int}$$
$$IB \in \text{IntBasis} = \text{FunIntEnv} \times \text{SigEnv} \times \text{StrIntEnv}$$
$$ME \in \text{ModEnv} = \text{FunEnv} \times \text{ModStrEnv} \times \text{VarEnv} \times \text{ExConEnv}$$
$$MSE \in \text{ModStrEnv} = strid \xrightarrow{\text{fin}} \text{ModEnv}$$
$$B \in \text{Basis} = \text{SigEnv} \times \text{ModEnv}$$

Fig. 4. Semantic objects for higher-order functors (additions and changes)

Interfaces (Int) Interfaces are prescriptions for how to thin the view of a structure. Since structures may now contain functors, interfaces must now prescribe how to thin the view of a functor. Therefore they contain a new component: a functor interface environment.

Structure Interface Environments (StrIntEnv) These are the equivalent of interface environments in SML. We have renamed them to more accurately reflect their function in the semantics of higher-order functors.

Functor Interface Environments (FunIntEnv) These contain interface information to be used in thinning functors. The nature of this information is discussed at length later.

Interface Bases (IntBasis) These now have a new component: a functor interface environment.

Module-level environments (ModEnv) These are the environments obtained as the result of evaluating structures, which now can contain functors. To reflect this, these environments contain a functor environment (FunEnv) component. In the rest of this paper we will refer to these objects as "environments" unless there is a possibility of confusion with Core-level environments.

Module-level structure environments (ModStrEnv) These are the Module-level counterparts of the Core-level structure environments (StrEnv).

Bases (Basis) Bases no longer need to contain a separate functor environment since this has been moved into the ModEnv component.

Functions on semantic objects Most of the projection, injection, and modification functions on the new semantic objects can be defined by straightforward changes to the corresponding SML functions. Here we describe those functions that are significantly different:

Extracting interfaces Interfaces and environments now contain information about functors, so we must change the definition of Inter which extracts an interface from an environment. The new definition is as follows:

$$\text{Inter } (FE, \, MSE, \, VE, \, EE) = (FIE, \, SIE, \, \text{domain } VE, \, \text{domain } EE)$$

where

$$FIE = \{funid \mapsto \text{Inter\_funclos } (funclos) \; ; \; FE \, (funid) = funclos\}$$

and, as before,

$$SIE = \{strid \mapsto \text{Inter } ME \, ; \, MSE \, (strid) = ME\}$$

This is fine, except that we haven't defined Inter_funclos yet. Inter_funclos is a significant complication that arises in the dynamic semantics of the higher-order Module system that is not present in the original system, since interfaces there did not need to make mention of functors. The interface information we have decided to keep for functors is the interface of the output structure. We discuss why this is the right choice (as opposed to using the interface for the input structure, or both, or neither) in the next subsection. If the functor closure is constrained (by an interface arising from an original constraining signature), then

Inter_funclos extracts the interface constraining the output structure. However, if the functor closure is unconstrained, then we must calculate the interface from the structure expression describing the output structure of the functor. That is, we must be able to extract an interface from syntax. Now we are on a slippery slope, because structure expressions can and do contain every other category in the grammar, except top declarations. Therefore, we have to define the contribution of each grammatical category (except top declarations) to interfaces. Making these definitions is long and rather tedious, and we omit any further discussion of how it is done here. It is worth commenting that using automated assistance to type check the terms in our definitions and to warn us of any cases we had missed did speed the process of making the definitions and increased our confidence that we have made them correctly.

Let us reflect for a moment on which feature of the language necessitates the function Inter for extracting interfaces from environments, and all the other interface-extraction functions it requires. An interface is the semantic equivalent of a signature expression. An environment is the semantic equivalent of a structure expression. So when do do we syntactically express the act of turning a structure into a signature? This occurs when we open a structure within a signature (*viz.* OPENspec). This is intended to add the signature of the structure to the signature containing the **open**. One might reasonably ask if this is a desirable language feature. However, this language feature is clearly present in the *Definition*, and we felt we would not be carrying out the task of extending the specification if we simply chose to omit it.

Thinning environments   Interfaces become more complicated in the setting of higher-order Modules because they must contain information concerning how to thin the view of a functor. In Fig. 3 we defined interfaces but did not explain how we decided what their functor components should be. Here we explain how we arrived at our choice.

Functor closures are thinned in a manner prescribed by functor specifications.[3] These provide us with two interfaces (signatures) : one describing the input taken by a functor and another describing the structures produced by a functor. There are three possibilities for using these interfaces in thinning a functor closure.

The first possibility is to replace the first interface of the functor closure by the (larger) first interface of the functor specification. This has the effect of guaranteeing that the functor body will receive a larger environment with more bindings from its input structure. This means that when the functor body is evaluated, more values will be looked up in the input environment. There is no change to the bindings available in the resulting output environment. The second possibility is to replace the second interface of the functor closure by the (smaller) second interface of the functor specification. Functors thinned in this manner will take exactly the same inputs as they did before thinning. However when applied, the resulting environments will have fewer components than those produced by applying the unthinned versions. The third possibility is to combine

---

[3] These are *specs* of the form (FUNCTORspec *funid strid sigexp sigexp*).

both the first and the second definitions of thinning.

We believe that the second definition of thinning is the correct one. The first method of thinning (and consequently also the third) can result in the wrong environment being used for final computations. Consider the example:

```
val x = 5;
functor F(I:sig end) = struct open I val z = x end
signature SIG = sig functor F(I:sig val x : int end)
                               : sig val z:int end end
structure A = struct functor F = F end
structure B:SIG = A
structure I = struct val x = 6 end
structure A1 = A.F(I)
structure A2 = B.F(I)
val test = A1.z = A2.z
```

If we use the first method of thinning, we find that $A1.z = 5$ and $A2.z = 6$, when it should be the case that $A1.z = 5 = A2.z$. For computing z, F requires the x in the top-level environment be used, and this should remain the case if we subsequently thin F. Thinning should change only the visibility of identifiers, not the underlying computations, and hence not the environments used for identifier lookup.

We therefore chose to record in the functor interface only the second (i.e. output) interface provided by a functor specification, and to thin functor closures by replacing only their output interfaces. Here is our new definition of thinning an environment by an interface:

$$(FE, MSE, VE, EE) \downarrow (FIE, SIE, vars, excons) = (FE', MSE', VE', EE')$$

where

$$FE' = \{ funid \mapsto (strid, int, strexp, int_2, B) \;;$$
$$\quad\quad FE\,(funid) = (strid, int, strexp, int_1, B) \text{ and } FIE\,(funid) = int_2 \}$$

and, as in SML,

$$MSE' = \{ strid \mapsto ME{\downarrow}I \;;\; MSE\,(strid) = ME \text{ and } SIE\,(strid) = I \}$$

and VE' and EE' are obtained by restricting the domains of VE and EE to $vars$ and $excons$, respectively.

Evaluation Generally we obtain the evaluation rules for the new language by modifying the evaluation rules of SML to work with the new semantic objects and functions in the obvious way. We must also add new rules to deal with the syntax we have added, and make significant changes to some other rules. We give these rules in Fig. 5, along with pointers to relevant rules in the SML semantics for readers with access to a copy of the *Definition*. Here we describe the rules in Fig. 5:

1. This rule defines how to evaluate the application of a *longfunid* to a *strexp*. As in SML, evaluating a *strexp* can result in an exception packet p. These cause no complications so we do not discuss them. [Replaces rule 162.]
2. This rule shows how to evaluate a Core declaration in a basis B: we must first extract a Core environment from B, evaluate the declaration in this environment, and then lift any resulting Core environment to a Module environment to get the final result. [Replaces rule 164.]
3. This give the semantics of the Module-level **open** declaration. [Follows rule 166.]
4. This rule shows how to evaluate a functor declaration, now that functor declarations are specific instances of module declarations. [Follows rule 168.]
5. This describes how to evaluate a functor specification. [Follows rule 183].
6. This rule gives the semantics of rebinding a functor to a new identifier. [Follows rule 187.]

$$\boxed{B \vdash strexp \Rightarrow ME/p}$$

$$\frac{B\,(longfunid) = (strid : I, strexp'\langle : I'\rangle, B') \quad}{B \vdash strexp \Rightarrow ME_1 \quad B' + \{strid \mapsto ME_1 \! \downarrow \! I\} \vdash strexp' \Rightarrow ME_2}{B \vdash longfunid(strexp) \Rightarrow ME_2\langle \! \downarrow \! I'\rangle} \tag{1}$$

$$\boxed{B \vdash moddec \Rightarrow ME/p}$$

$$\frac{E \text{ of } (ME \text{ of } B) \vdash dec \Rightarrow E'}{B \vdash dec \Rightarrow E' \text{ in Modenv}} \tag{2}$$

$$\boxed{B \vdash moddec \Rightarrow ME/p}$$

$$\frac{B(longstrid_1) = ME_1 \; \cdots \; B(longstrid_n) = ME_n}{B \vdash \text{open } longstrid_1 \; \cdots \; longstrid_n \Rightarrow ME_1 + \cdots + ME_n} \tag{3}$$

$$\boxed{B \vdash moddec \Rightarrow ME/p}$$

$$\frac{B \vdash funbind \Rightarrow F}{B \vdash \text{functor } funbind \Rightarrow F \text{ in Modenv}} \tag{4}$$

$$\boxed{IB \vdash spec \Rightarrow I}$$

$$\frac{IB \vdash sigexp \Rightarrow I_1 \quad IB + \{strid \mapsto I_1\} \vdash sigexp' \Rightarrow I_2}{IB \vdash \text{functor } funid(strid : sigexp) : sigexp' \Rightarrow \{funid \mapsto I_2\} \text{ in Int}} \tag{5}$$

$$\boxed{B \vdash funbind \Rightarrow F}$$

$$\frac{B\,(longfunid) = funclos}{B \vdash funid \text{ = } longfunid \Rightarrow \{funid \mapsto funclos\}} \tag{6}$$

Fig. 5. Evaluation rules for higher-order functors

Encoding higher-order functors We encode the new semantics into HOL by the same techniques used to encode the semantics of SML Modules. The encoding is called HOF-ML. Some of the new phrase classes and semantic objects are identical to the old ones, so for these we simply re-use the types used to encode these in Mod-ML. For other phrase classes and semantic objects we distinguish the names of the HOF-ML constructors and types from those of Mod-ML by appending _h to them. Thus, for example, the type representing HOF-ML structure descriptions is **strdesc_h** with constructor **STRIDstrdesc_h**. We do this because we will eventually want to have both Mod-ML and HOF-ML present in HOL together so that we can prove theorems about the relationship between them.

4 Relating the two semantics

So far we have described the encoding of two possible Module systems to extend the Core language of SML. It is our claim that the system specified by HOF-ML is a conservative extension of the system specified by Mod-ML (and the *Definition*). At present, we have not finished proving this fact. Before we discuss how we prove such a result, we need to discuss what result we are trying to prove. Most concisely, we aim to prove that there is a function **embed_topdec** mapping top-level declarations of Mod-ML into top-level declarations of HOF-ML, a function **embed_basis** mapping bases from Mod-ML into bases of HOF-ML, and a function **proj_basis_h** mapping bases (or exception packets) of HOF-ML back to bases (or exception packets) from Mod-ML such that

- For each basis B_1, state s_1 and top-level declaration **top_dec** of Mod-ML, if there exists a state s_2 and a basis (or packet) B_2 of Mod-ML such that

$$\text{eval\_topdec top\_dec } s_1 \; B_1 \; s_2 \; B_2$$

holds, then there exists a state s_2' and a basis (or packet) B_2' of HOF-ML such that

$$\text{eval\_topdec\_h (embed\_topdec top\_dec) } s_1 \; (\text{embed\_basis } B_1) \; s_2' \; B_2'$$

also holds, and
- For each basis B_1, states s_1 and s_2, top-level declaration **top_dec** of Mod-ML, and basis (or packet) B_2 of HOF-ML, if

$$\text{eval\_topdec\_h (embed\_topdec top\_dec) } s_1 \; (\text{embed\_basis } B_1) \; s_2 \; B_2$$

holds, then

$$\text{eval\_topdec top\_dec } s_1 \; B_1 \; s_2 \; (\text{proj\_basis\_h } B_2)$$

also holds.

Informally, this states that if you wish to evaluate a top-level declaration of Mod-ML, it suffices to translate into HOF-ML, evaluate there and translate the result back. This statement of conservative extension focuses on top-level declarations and bases. However just to define the functions **embed_topdec**, **embed_basis**, and **proj_basis_h**, we need to define the corresponding functions for all categories of syntax and semantics in Mod-ML and HOF-ML.

Embedding Mod-ML in HOF-ML

Embedding the syntax of Mod-ML into that of HOF-ML is generally straightforward. Some phrase classes, such as identifiers are embedded by the identity function since they are represented by the same HOL types in both Mod-ML and HOF-ML. We give the flavor of the embedding by showing three of the mutually recursive clauses for structure expressions, declarations and bindings:

```
embed_strexp (STRUCTstrexp strdec) = STRUCTstrexp_h (embed_strdec strdec))
embed_strdec (STRUCTUREstrdec strbind) =
        STRUCTUREmoddec_h (embed_strbind strbind))
embed_strbind (BINDstrbind strid strexp) =
        BINDstrbind_h strid (embed_strexp strexp))
```

The only clause whose embedding is not trivial is **APPstrexp**. There the functor identifier that is applied must be lifted to a long functor identifier. Both functor and structure declarations are mapped to the appropriate kinds of HOF-ML Module declarations. Similarly, top-level functor declarations must be mapped to top-level Module declarations. The embeddings are trivial for all other cases. Defining an embedding of the semantic objects of Mod-ML into those of HOF-ML is also easy.

Projecting HOF-ML back to Mod-ML

It might appear that we only need to project the semantic objects of HOF-ML into Mod-ML, and can forget about the syntax, since the conservativity result only uses the projection of semantic objects (bases, to be precise). Unfortunately, this is not so. To project bases we need to project functor environments, and hence functor closures. To project functor closures we need to project structure expressions — syntax. With the exception of this dependency, the definition of the projection functions for the semantic objects is straightforward. The only complication is that when we project a basis we must first pull the environment it contains into its constituent parts to access the functor environment and the structure environment and project them to acquire the corresponding components of a basis in Mod-ML.

Projecting HOF-ML syntax back into Mod-ML is somewhat more complicated. This is because we have merged two classes, **strdec** and **fundec**, in Mod-ML into one class, **moddec** in HOF-ML, and now we are going to have to tease them apart again. In fact, this problem prevents the collection of functions briefly described above from actually being true embeddings: both **EMPTYstrdec** and

EMPTYfundec get mapped to **EMPTYmoddec**. In attempting to "project" HOF-ML back to Mod-ML, we cannot necessarily determine which **EMPTYmoddec**s came from **EMPTYstrdec** and which from **EMPTYfundec**. Sequences of Module declarations are another source of ambiguity. What do we do with a sequence of Module declarations that contains both structure declarations and functor declarations? In an arbitrary manner we choose to map the sequence to a sequence of the same kind as the leftmost declaration in the sequence, mapping declarations of a different kind to an empty declaration. While this complicates the definition of the projection function for declarations, and causes a loss of information, there is no harm in it since no such mixed sequence could be the result of an embedded sequence from Mod-ML.

Throughout the definitions of the embedding and projection functions, just as with the functions for extracting interfaces, we relied heavily on the package for nested mutually recursive types, and its support for generating definitions for functions from primitive mutually recursive specifications over those types.

Proving Conservativity

Although the result relating the evaluation of top-level declarations in the two Module systems mentioned above is our main statement of conservative extension, in order to prove such a result we need to prove corresponding results for all layers of the evaluation relations. Therefore, to simplify the process we begin with showing the corresponding results for signature expressions, descriptions, and specifications, and work our way up through the syntax classes. To further simplify the process, we show each of the two parts of the main conservativity theorem separately. The next step in proving the results is to coerce each of the implications into a form that we can use with our induction principles. For example, the second clause for **topdec** becomes:

$\forall$ topdec_h s_1 B s_2 bp. eval_topdec_h topdec_h s_1 B s_2 bp $\Rightarrow$
$\forall$ topdec B'. ((topdec_h = embed_topdec topdec) $\wedge$ (B = embed_basis B')) $\Rightarrow$
eval_topdec topdec s_1 B' s_2 (proj_basis_pack bp)

Once we perform this transformation, we can apply our induction principle to reduce our problem to showing that the conclusion of the resulting implication holds for all the evaluation rules. To show these results, we have various tools at our disposal, including structural induction over both the syntax and the semantics, rewriting with theorems that state the distinctness of all the constructors, and rewriting with the equations stating the mutually recursive "definitions" of the embedding and projection functions, the functions for extracting interfaces, *etc.* Moreover, by proving the results in a bottom-up fashion, starting with the earliest syntax classes, we have these results also at our disposal when proving the later results.

While there is a great deal of regularity involved in carrying out these proofs, it is not apparent at present that we could write a general purpose tactic that would automatically prove all of these theorems. Each case seems to have just enough that is distinct about it to benefit from interactive guidance.

5 Conclusion

We have outlined how we used the interactive theorem prover HOL to give the dynamic specification of a higher-order Module system for SML, and then to relate it to the SML Module system specification. It is our belief that this task is too large to be done by hand, the *Definition* notwithstanding. Using the expressiveness of HOL, the packages built into it, and packages we added to it, we were able to formulate the specification as fast, maybe faster, with the theorem prover, as we could formulate it by hand. Moreover, we have received some assurances that our specification makes sense from the type-checking of the terms, the checks that no clauses were omitted from our function definitions, and other checks that were performed automatically by HOL. Most importantly, by encoding the specification in a theorem prover, we are now able to formally prove facts about the specification and about programs written in complying implementations. Not only did we receive benefits from the theorem prover, but so did the theorem prover receive benefits from us. The specification task has motivated us to improve HOL's handling of mutually recursive types, and to write a general purpose tactic suitable for defining mutually recursive families of relations and deriving the appropriate induction principles.

Both the benefits to the specification task and to the theorem prover were made possible by a combination in HOL of an expressive language capable of developing much general mathematics, with an open yet secure system which allows users to develop theorem-proving methodologies to suit their needs.

References

1. David B. MacQueen, Mads Tofte. *A Semantics for Higher-order Functors*, unpublished draft. Contact Dave MacQueen at AT&T Bell Labs, Room 2a-431, 600 Mountain Ave, Murray Hill, NJ 07974, USA.
2. Thomas F. Melham. *A Package for Inductive Relation Definitions in HOL*, Proceedings of the 1991 International Workshop on the HOL Theorem Proving System and its Applications. IEEE Computer Society Press, 1992.
3. Robin Milner, Mads Tofte. *Commentary on Standard ML*, The MIT Press, Cambridge, Mass, 1991.
4. Robin Milner, Mads Tofte, Robert Harper. *The Definition of Standard ML*, The MIT Press, Cambridge, Mass, 1990.
5. Myra VanInwegen, Elsa Gunter, HOL-ML. Proceedings of HUG '93. Lecture Notes in Computer Science 780, Springer-Verlag, 1994.

Towards a Mechanically Supported and Compositional Calculus to Design Distributed Algorithms

I.S.W.B. Prasetya

Rijksuniversiteit Utrecht, Vakgroep Informatica
Postbus 80.089, 3508 TB Utrecht, Nederland
Email: wishnu@cs.ruu.nl

Abstract. This paper presents a compositional extension of the programming calculus UNITY, which is used to design distributed programs. As the extension is compositional, we can use it to derive a program 'on the fly'. That is, we can shape a program at the same time as we manipulate and decompose its given specification, and each time we apply a compositionality theorem we basically add a detail to the shape. Safety properties are known to be compositional in UNITY, but progress in general are not. So, we define a class of progress properties which are compositional. In addition, for programs that are constructed from components that do not write each other's write variables, the compositionality of this new class of progress can be expressed elegantly.

We also have formalized and verified the resulting calculus using the theorem prover HOL. Together with the available tools in HOL this provides a mechanical support in designing distributed programs.

1 Introduction

In a calculus for programming, we usually have a number of program composition operators —for example sequential composition, if, and while constructs— to construct a more complicated program from simpler ones. Compositionality is a property of a calculus that enables us to derive the properties of a program —for example, safety and progress properties in the case of distributed programming, or pre-post specification and termination in the case of sequential programming— from the properties of its constituents. For example, we have in sequential programming the well known law of sequential composition:

$$\frac{\{P\}\, S\, \{Q\}}{\{P\}\, S;T\, \{R\}}$$

which states how a pre-post specification (in terms of a Hoare triple) of the composite program $S;T$ can be expressed in terms of the pre-post specifications of its constituents S and T.

A compositional calculus is preferable because it allows us to construct a program in a large number of smaller steps, each of which preserves the meaning of the formula. So, in principle no verification after construction is needed.

In contrast to sequential programming, constructing a distributed program is often made difficult by the lack of laws for program composition. The greatest obstacle is that progress is not a compositional property with respect to parallel composition.

In this paper we extend UNITY, a simple calculus for distributed programming invented by Chandy and Misra [CM88], with a number of program compositions, and define a new class of compositional progress properties. Property refinement is well supported in standard UNITY, but not program construction, basically because the notion of progress in UNITY is too liberal to be compositional[1]. Designing a distributed program in UNITY is usually done by first refining an initial specification down to a primitive level; in the second step one has to guess what the program should be and post-verify it. Suppose we end up with n primitive level specifications. Guessing a UNITY program that will satisfy all of them is like solving n equations, which may not be easy for a sufficiently large n. However, the designer may already have a good candidate in mind, which he silently constructs during the design. Compositionality can be seen as a way to formally keep track of the 'shape' of a program which gradually grows more detailed as we proceed in the design process. By exploiting compositionality we can lessen the burden of post-verification. A simple calculation at the end of Section 4 shows this.

There are already a number of compositional calculi for distributed programming, for example as in [PJ91, dBvH94]. They are intended to reason about programs that communicate through channels. Compositionality is achieved by recording the history of the channels. Our results assume no such explicit underlying mechanism and therefore should be more generally applicable.

We have verified the correctness of our calculus with respect to UNITY in a theorem prover called HOL and the correctness of the UNITY rules itself has also been verified in HOL [And92, Pra93b]. HOL is a general purpose and interactive theorem prover developed by Mike Gordon. HOL supports reasoning in *higher order logic* (hence the name HOL), that is, a version of predicate calculus in which variables can range over functions. The logic is also typed. Actually, the name *theorem prover* is somewhat misleading —perhaps, *proof-assistant* is a better word— as it suggests that the computer is capable of proving a theorem all by itself. Only for a certain class of formulas a fully automatic proof is possible, otherwise the human operator has to manually guide the computer. Still, a mechanically verified proof increases our confidence, which is very much sought-after especially in distributed programming where informal reasoning can be very error prone. To help us writing a mechanical proof, HOL provides a wide range of tools and a meta language called ML which can be used to add one's own proof tools. By verifying our calculus in HOL we also make it available as an extension of the HOL logic. Together with available HOL tools it provides a mechanical support when designing distributed programs.

The contribution described here is twofold: (1) we propose a variant of UNITY which is compositional, and (2) we have verified the logic with HOL

---

[1] To be more precise: progress by ensures is compositional but not progress by leads-to.

and made it available as an extension of the HOL logic[2].

The rest of this paper is organized as follows. Section 2 explains some notational conventions. Section 3 gives a brief introduction of UNITY. Section 4 explains our extension to UNITY. Finally, in Section 5 we give some conclusions.

For presentational reasons we will not write our formulas in HOL notation. We may also be less formal and hide some details in presenting our formulas. The formal presentation and proof (in HOL formulas and codes) of the results mentioned in this paper is available at request.

To read this paper familiarity with UNITY or some other temporal programming logics will be helpful. Familiarity with the HOL system is not required.

2 Notation

The application of a function f to x is denoted as $f.x$. The projection of a function f with respect to a subset A of the domain of f is denoted by $f \restriction A$. The identity function is denoted with id.

For a commutative and associative binary operator $\oplus$ with unit e, $(\oplus i : P.i : f.i)$ denotes $f.i \oplus f.j \oplus f.k \oplus \cdots$ for all $i, j, k, \ldots$ satisfying P. If P is empty then $(\oplus i : P.i : f.i) = e$. If P is a singleton $\{x\}$ then $(\oplus i : P.i : f.i) = x$.

A predicate over a set A is an object of type $A \rightarrow \mathbf{bool}$. People often overload boolean operators to define their lift to the predicate level. To emphasize the distinction we add a *dot* above a boolean symbol to denote its lifted version. The definition of the lifted operators is given below. For a predicate p over A, $[p]$ denotes that p holds at any s in A. That is, $[p] = (\forall s : s \in A : p.s)$.

Definition 2.1 : Predicate Operators

tt	$= (\lambda s.\ \mathsf{true})$		ff	$= (\lambda s.\ \mathsf{false})$
$\dot{\neg} p$	$= (\lambda s.\ \neg p.s)$		$p \mathbin{\dot{\wedge}} q$	$= (\lambda s.\ p.s \wedge q.s)$
$p \mathbin{\dot{\vee}} q$	$= (\lambda s.\ p.s \vee q.s)$		$p \mathbin{\dot{\Rightarrow}} q$	$= (\lambda s.\ p.s \Rightarrow q.s)$

Set notation is used as standard. Set complement is denoted with superscript c like A^c. Membership notation $a, b \in A$ abbreviates $a \in A \wedge b \in A$. For a set V, $\mathcal{P}(V)$ denotes the power set of V, that is, the set of all subsets of V.

3 A Brief Review of UNITY

The UNITY view of a program is simply a collection of atomic and terminating actions. An execution of a UNITY program is an infinite execution where at each execution step an action is non-deterministically selected and executed. The selection process is required to be fair in the sense that every action should

---

[2] Ours is a different formalization of UNITY in HOL than that of Andersen [And92]. The distinction will be emphasized later.

be executed infinitely often. Using a UNITY program, we can describe a parallel and fair system. The simplicity of these notions of program and execution is the obvious advantage of UNITY.

Instead of giving a precise syntax of a UNITY program, below we give an example of one.

Prog   Mdist
Var d
Init $(\forall i : i \in V : d.i.i = 0)$
Assign
       $([\![i,j : i,j \in V \wedge i \neq j : d.i.j := (\min k : E.i.k : d.k.j) + 1)$

A UNITY program P is described by a set of variables that P can read and write (the **Var** section) —later we will however make an explicit distinction between read and write variables—, a predicate describing the allowed initial states (the **Init** section), and a list of actions separated by the $[\![$ symbol (the **Assign** section).

The program is executed as explained above. The absence of an explicit order in the execution of the actions may make a UNITY program confusing in the first place. But this is only UNITY's way to *model* parallelism. A UNITY program can be implemented in full parallel, or fully sequential, or anything in between, as long as the atomicity and fairness conditions are met. This enables us to separate some of the implementation concerns from the design.

A program can be specified by its safety and progress properties. For example, in above program $d.i.j$ will eventually contain the minimal distance between vertices i and j in the network described by the set of vertices V and edges E, provided the network is connected, after which the system configuration will not change (stable).

Before we give the formal definition of a proper UNITY program, first let us carefully elaborate on how we are going to represent some more primitive programming notions such as program states, state predicates, and actions.

3.1 States, State Predicates, Actions, and Variables

We will *assume* a space of all available program variables and a space of values. Let us call them, respectively, Var and Val. That is, the variables of any program in our universe are members of Var and their values range over Val.

A *program state* is simply an object of type Var $\to$ Val. We denote the space of *all* program states with State. A state predicate is a predicate over State. For example $(\lambda s. \ 0 < s.x)$ is a state predicate which is true in all state s in which $s.x$, the value of x, is greater than 0. We denote the space of all state predicates with Pred.

In a similar way as we have lifted the boolean operators we can also lift constants, variables, functions, and relations to the state level. For $X \in A$,

$v \in \mathsf{Var}$, $f \in A \to B$, and $R \in A \to A \to \mathsf{bool}$ we define:

$$X^{\cdot} = (\lambda s.\ X) \qquad\qquad v^{\cdot} = (\lambda s.\ s.v)$$
$$f^{\cdot} = (\lambda g, s.\ f.(g.s)) \qquad R^{\cdot} = (\lambda f, g, s.\ R.(f.s).(g.s))$$

Using above notation $(\lambda s.\ 0 < s.x)$ can also be written $0^{\cdot} <^{\cdot} x^{\cdot}$.

An *action* is a relation on State, that is, it is an object of type State $\to$ State $\to$ bool. For any action a, $a.s$ describes the set of possible states after executing a at state s^3. So, $a.s.t$ means that executing a at s may bring the system to state t. For example, assignment $x := x + y$ can be modelled by a relation a satisfying $a.s.t = (t.x = s.x + s.y) \land (\forall v : v \neq x : t.v = s.v)$.

The space of all actions is denoted with Action. An action a is called always enabled, denoted by $\Box_{\mathsf{En}} a$, if each state s is related to some next state by a. Always enabledness is required in a UNITY program to explicitly exclude non-terminating and aborting actions (else the logic may falsely conclude that any progress is possible).

A Hoare triple specification can be formulated as follows: for any $a \in$ Action and $p, q \in$ Pred:

$$\{p\}\ a\ \{q\} = (\forall s, t :: p.s \land a.s.t \Rightarrow q.t)$$

An action a *ignores* a set of variables V, denoted by $a \not\rightarrow V$, if no execution of a can alter the values of variables in V. That is, if a can bring the system from state s to state t, the value of any variable $x \in V$ in s remains unchanged in t. Using projection, this can be defined as follows.

Definition 3.1 : Ignores
For any $V \subseteq$ Var and $a \in$ Action:

$$a \not\rightarrow V = (\forall s, t :: a.s.t \Rightarrow (s \restriction V = t \restriction V))$$

A set of variables V is *invisible* to an action a, denoted by $a \leftarrow V$, if changing the values of variables in V does not alter the effect of a on variables outside V.

Definition 3.2 : Invisible Variables
For any $V \subseteq$ Var and $a \in$ Action, $a \leftarrow V$ holds iff for any $s, t, s', t' \in$ State:

$$(s \restriction V^c = s' \restriction V^c) \land (t \restriction V^c = t' \restriction V^c) \land (s' \restriction V = t' \restriction V) \land a.s.t \Rightarrow a.s'.t'$$

For example, $\{y, z\}$ is ignored by $x := x + y$ but $\{x\}$ is not; $\{z\}$ is invisible to $x := x + y$ but $\{x, y\}$ is not. One may ask what would happen if an action a happens to be the empty relation. In this case, any set of variables is ignored and invisible to a. However, in a UNITY program all actions are required to be

---

^{3} So, an action can in principle be non-deterministic, whereas in [CM88] this is not the case. We discovered that determinism at the action level is not necessary to derive the basic theorems of UNITY mentioned in [CM88].

always enabled, so they cannot be empty (unless of course in the trivial case where we have an empty state space).

We use the notion 'ignored' and 'invisible' variables to define the meaning of write and read variables. A variable is *not* a (significant) write variable of a program P if it is ignored by every action in P. It is *not* a (significant) read variable of P if it is invisible to every action in P.

3.2 UNITY Program

Now we can define formally what we mean by a UNITY program. We will represent a UNITY program with a quadruple consisting of a set describing the actions, a predicate describing the initial state, a set describing the read variables, and a set describing the write variables of the program. A UNITY program is required to be non-empty and to consist exclusively of always-enabled actions. Most descriptions of UNITY, including the original [CM88], although admitting the notion of write and read variables, remain informal in their treatment. As pointed out in [Pra93a] and [UHK94], variable accessibility plays an important role in supporting compositionality. So, our first extension to UNITY is to explicitly require a program to respect their variables declarations. That is, a program should only write to its declared write variables (*write constraint*), and read from its declared read variables (*read constraint*). In addition we also assume that any write variable is also readable. Below we give the definition of a predicate Unity that characterizes the space of all UNITY programs, that is, P is a UNITY program iff Unity.P holds.

Definition 3.3 : UNITY Program
For any $A \subseteq$ Action, $J \in$ Pred, $R \subseteq$ Var, and $W \subseteq$ Var, Unity.(A, J, R, W) holds iff:

$$(A \neq \emptyset) \wedge (\forall a : a \in A : (\Box_{\mathsf{En}} a) \wedge (a \nrightarrow W^c) \wedge (a \nleftarrow R^c)) \wedge (W \subseteq R)$$

To access each component of a UNITY program we introduce the following *destructors*. For any UNITY program P:

$$P = (\mathsf{a}P, \mathsf{ini}P, \mathsf{r}P, \mathsf{w}P)$$

In the sequel, let P, Q, R range over UNITY programs, a, b, c range over Action, and p, q, r, s, J range over Pred.

In the next subsection we give the meaning of the primitive operators of the UNITY logic, by which we can express the behavior of a distributed program.

3.3 Safety, Progress, and Parallel Composition

One can divide the properties of a distributed program into two classes: safety properties, which tell us what the program should do and not do, and progress properties, which tell us what the program will do. An example of a safety property is "J is invariant in program P"; an example of a progress property is "program P will eventually compute the correct output".

In UNITY safety is expressed by **unless** operator and progress by **ensures** and $\mapsto$ (read: 'leads-to'). Intuitively, p **unless** q holds in program P means that once p holds during the execution of P, then it will remain to hold until q holds. p **ensures** q encompasses p **unless** q, but in addition it also states the existence of an action that can establish q. By fairness this action will eventually be executed, and thus establish q. **ensures** expresses a progress property that can be established by the effect of a single action, but not if it has to be achieved by a cooperation of several actions. For the latter we have to use $\mapsto$, which is defined as the the least transitive and left-disjunctive closure of **ensures**. Using these three operators, one can express various kinds of interesting properties of a distributed program. The formal definition of **unless** and **ensures** is given below. The definition of $\mapsto$ follows later.

Definition 3.4 : Unless

p **unless** q **in** P $=$ $(\forall a : a \in aP : \{p \dot\wedge(\neg q)\} \, a \, \{p \dot\vee q\} \,)$

Definition 3.5 : Ensures

p **ensures** q **in** P $=$ $(p$ **unless** q **in** $P) \wedge (\exists a : a \in aP : \{p \dot\wedge(\neg q)\} \, a \, \{q\} \,)$

Here are some examples of program properties that can be expressed using the above two operators: (1) $(\forall X :: p$ **unless** $(x \neq X\dot{}) $ in $P)$ states that P can only disrupt p by writing to variable x; and (2) tt **ensures** p in P implies that eventually P will reach p.

Let Trans.R mean that R is a transitive relation and Ldisj.R mean that R is a left disjunctive relation. The formal definition of Trans is as usual, the definition of Ldisj is given below. $\mapsto$ is defined as the least transitive and left-disjunctive closure of **ensures**.

Definition 3.6 : Left Disjunctive Relation
For any $U \in A \to A \to$ bool:

Ldisj.U $=$ $(\forall W, y : W \neq \emptyset : (\forall x : x \in W : U.x.y) \Rightarrow U.(\dot\vee x : x \in W : x).y)$

Definition 3.7 : Leads-to

$p \mapsto q$ $=$ $(\forall R : (\lambda r, s. \, r$ **ensures** s in $P) \subseteq R \wedge$ Trans.$R \wedge$ Ldisj.$R : R.p.q)$

For example —let dist.$i.j$ be the minimum distance between vertices i and j in network (V, E)— in the program Mdist at beginning of this section, tt $\mapsto$ $((d.i.j)\dot{} = (\text{dist}.i.j)\dot{})$ is a progress property of Mdist, stating that eventually $d.i.j$ will be equal to dist.$i.j$. However, tt **ensures** $((d.i.j)\dot{} = (\text{dist}.i.j)\dot{})$ is not a valid property of Mdist because this situation will in general not be reachable in one step.

Because of UNITY's simple notion of program and program execution, parallel composition can be modelled by merging the actions and variables of the component programs (see below). unless and **ensures** have a very nice compositionality property with respect to ‖ but, as is pointed in [CM88] and [Pra93a], the definition of $\mapsto$ is too liberal to be compositional. In the next section we will define a more restricted class of progress which is compositional. This is also where our work differs from that of Andersen. Andersen's formalization of UNITY in HOL [And92] only has **ensures** and $\mapsto$ to describe progress and hence suffers the above mentioned limitation.

Definition 3.8 : Parallel Composition

$$P \| Q \;=\; (\mathrm{a}P \cup \mathrm{a}Q, \mathrm{ini}P \wedge \mathrm{ini}Q, \mathrm{r}P \cup \mathrm{r}Q, \mathrm{w}P \cup \mathrm{w}Q)$$

Theorem 3.1 : unless Compositionality

$$\frac{(p \text{ unless } q \text{ in } P) \;\wedge\; (p \text{ unless } q \text{ in } Q)}{p \text{ unless } q \text{ in } (P \| Q)}$$

Theorem 3.2 : ensures Compositionality

$$\frac{(p \text{ ensures } q \text{ in } P) \;\wedge\; (p \text{ unless } q \text{ in } Q)}{p \text{ ensures } q \text{ in } (P \| Q)}$$

---

4 Extending UNITY

In this section we will introduce a new progress operator called *reach* which is compositional. Compositionality is achieved by restricting it to describe only the progress of the writable part of a program. The motivation is that if a program can make some progress at all, this can be observed by observing its write variables. We also define a stronger progress operator called *convergence*. Convergence can be used to implement eventual stability ($\Diamond\Box$ in temporal logic) and it has many nice calculational properties. Compositionality of reach with respect to ‖ is formulated by the so-called Singh Law due to A.K. Singh [Sin89][4]. Even nicer compositionality results can be achieved by restricting ‖ to work on write-disjoint programs, that is, programs that do not write to each other's write variables.

4.1 Reach Operator

Before we can give the definition of our new progress operator we first need some definitions.

---

[4] The original formulation uses $\mapsto$, and thus in-correct.

A state predicate p is confined by a set of variables V, denoted by $p \in \mathsf{Pred}.V$, if p does not restrict the value of any variable outside V. For example $0^\cdot <^\cdot x^\cdot$ is confined by $\{x, y\}$, but $(0^\cdot <^\cdot x^\cdot) \wedge (0^\cdot <^\cdot z^\cdot)$ is not because it restricts the value of z.

Definition 4.1 : Confinement

For any $V \subseteq \mathsf{Var}$:

$$p \in \mathsf{Pred}.V = (\forall s, t.(s \restriction V = t \restriction V) \Rightarrow (p.s = p.t))$$

Because in HOL sub-typing is not standardly supported, $p \in \mathsf{Pred}.V$ is just our way to encode the intended type of p, which is $(V \to \mathsf{Val}) \to \mathsf{bool}$.

tt and ff are confined by any set, and confinement is also preserved by predicate operators. So for example if $p, q \in \mathsf{Pred}.V$, then so are $\neg p$ and $p \wedge q$.

A predicate p is stable in a program P if it is preserved by any action in P, that is, if p unless ff holds in P. Note that the conjunction and disjunction of two stable predicates is again stable.

Definition 4.2 : Stable Predicate

$$\mathsf{stable}.P.p = p \text{ unless ff in } P$$

We define a variant of ensures operator —we will call it Ens— in which the described progress property only concerns the write variables of a program, since these are the only part in the program that will be affected by its execution. The progress can be based on some knowledge of the states of other variables and since the program cannot alter the value of these variables, this knowledge is captured by a stable predicate.

Definition 4.3 : Ens

$$\mathsf{Ens}.P.J.p.q = ((p \wedge J) \text{ ensures } q \text{ in } P) \wedge \mathsf{stable}.P.J \wedge p, q \in \mathsf{Pred}.(\mathbf{w}P)$$

So, for example, let $\mathbf{w}P = \{x, y\}$. Then $\mathsf{Ens}.P.J.\mathsf{tt}.(0^\cdot <^\cdot x^\cdot)$ is a permissible Ens expression whereas $\mathsf{Ens}.P.J.\mathsf{tt}.((0^\cdot <^\cdot x^\cdot) \wedge (0^\cdot <^\cdot z^\cdot))$ is not because the predicate in last argument is not confined by $\mathbf{w}P$.

The new operator is called *reach*, denoted by $\rightarrowtail$. $p \rightarrowtail q$ in P under J is read "q is reachable from q in P given J holds and is stable". It is defined as the least transitive and left-disjunctive closure of Ens. As a consequence, progress by $\rightarrowtail$ restricts itself to the writable part of a program.

Definition 4.4 : Reach

$$p \rightarrowtail q \text{ in } P \text{ under } J = (\forall R : \mathsf{Ens}.P.J \subseteq R \wedge \mathsf{Trans}.R \wedge \mathsf{Ldisj}.R : R.p.q)$$

It is not too difficult to show that $\rightarrowtail$ itself is left-disjunctive, transitive, and includes **Ens**. In addition, being the least closure it also induces an induction principle: any relation R that is left-disjunctive, transitive, and includes **Ens** also includes $\rightarrowtail$. This follows directly from the definition of $\rightarrowtail$.

Using this induction principle, one can show that $p \rightarrowtail q \in P$ under J implies $(p \wedge J) \rightarrowtail q$ in P. The converse does not always hold. For example if $x \notin \mathbf{w}P$ then $(0^{\cdot} <^{\cdot} x^{\cdot}) \rightarrowtail (0^{\cdot} <^{\cdot} x^{\cdot})$ in P under tt is not a valid progress expression whereas $(0^{\cdot} <^{\cdot} x^{\cdot}) \mapsto (0^{\cdot} <^{\cdot} x^{\cdot})$ in P is. Using the induction, one can also show that $p \rightarrowtail q$ in P under J implies J is stable and $p, q \in$ Pred.$(\mathbf{w}P)$. Below is a list of other, interesting properties of $\rightarrowtail$. They can be proven in a similar line, albeit more complicated, as the proofs of the corresponding properties for $\mapsto$ in [CM88].

Theorem 4.1 : $\Rightarrow$ Lifting

$$\frac{p, q \in \text{Pred.}(\mathbf{w}P) \ \wedge \ [(J \wedge p) \Rightarrow q]}{p \rightarrowtail q \text{ in } P \text{ under } J}$$

Theorem 4.2 : $\rightarrowtail$ Bounded Progress

Let $\prec$ be a *well founded* relation over *non-empty* A and m be some metric function that maps State to A. Let $q \in$ Pred.$(\mathbf{w}P)$. Then we have:

$$\frac{(\forall M : M \in A : p \wedge (m^{\cdot}.\text{id} =^{\cdot} M^{\cdot}) \ \rightarrowtail \ (p \wedge (m^{\cdot}.\text{id} \prec^{\cdot} M^{\cdot})) \vee q \text{ in } P \text{ under } J)}{p \rightarrowtail q \text{ in } P \text{ under } J}$$

Theorem 4.3 : $\rightarrowtail$ Progress Safety Progress (PSP)

$$\frac{(p \rightarrowtail q \text{ in } P \text{ under } J) \ \wedge \ r \in \text{Pred.}(\mathbf{w}P) \ \wedge \ ((r \wedge K) \text{ unless } s \text{ in } P)}{(p \wedge r) \rightarrowtail ((q \wedge r) \vee s) \text{ in } P \text{ under } (J \wedge K)}$$

Theorem 4.4 : $\rightarrowtail$ Stable Shift

$$\frac{(p \rightarrowtail q \text{ in } P \text{ under } (J \wedge r)) \ \wedge \ \text{stable.}P.J \ \wedge \ r \in \text{Pred.}(\mathbf{w}P)}{(p \wedge r) \rightarrowtail q \text{ in } P \text{ under } J}$$

Theorem 4.5 : $\rightarrowtail$ Completion Law

For any finite and non-empty set of indices W and $r \in$ Pred.$(\mathbf{w}P)$:

$$\frac{(\forall i : i \in W : f.i \rightarrowtail g.i \text{ in } P \text{ under } J) \ \wedge \ (\forall i : i \in W : (g.i \wedge J) \text{ unless } r \text{ in } P)}{(\wedge i : i \in W : f.i) \rightarrowtail ((\wedge i : i \in W : g.i) \vee r) \text{ in } P \text{ under } J}$$

As we have remarked above, $p \rightarrowtail q$ in P under J can only be valid to describe progress within the writable part of program P, which should suffice as this is the only part which will ever be affected by any progress within P. The stable predicate J can be used to describe the non-writable part of P, on which the progress $p \rightarrowtail q$ may depend. The reader may note that $\rightarrowtail$ appears very much like the subscripted progress operator introduced by Sanders [San91]. However, since Sanders' operator is based on $\mapsto$ it also suffers the same problem with compositionality.

As corollaries of the transitivity of $\rightarrowtail$ and $\Rightarrow$ Lifting law, $. \rightarrowtail .$ in P under J is monotonic with respect to $\Rightarrow$ in its second argument, anti-monotonic in its first argument, and also reflexive, provided the stability of J and the confinement constraint are respected.

In general, $\rightarrowtail$ satisfies the following compositionality law for parallel composition. The law, due to A.K. Singh [Sin89], states that if under condition r program Q will announce any update it makes to P's read variables by setting s true, and if we have a progress $p \rightarrowtail q$ in P then starting from $p \wedge r$ the composite program $P\|Q$ will either reach q or Q makes an update on the way, in which case s will hold. It should be noted that this law cannot be derived if we are not explicit in what we mean with read and write variables.

Theorem 4.6 : Singh Law for $\rightarrowtail$

For any $r, s \in \mathsf{Pred}.(\mathbf{w}(P\|Q))$:

$$\frac{(p \rightarrowtail q \text{ in } P \text{ under } J) \;\wedge\; \mathsf{stable}.Q.J \qquad (\forall C :: (r \;\wedge\; (\wedge\, x : x \in \mathbf{r}P \cap \in \mathbf{w}Q : x = C_x')) \text{ unless } s \text{ in } Q\,)}{(p \wedge r) \rightarrowtail (q \vee \neg r \vee s) \in (P\|Q) \text{ under } J}$$

The Singh Law describes how two arbitrary parallel programs can influence each other's progress. However, in many cases we know more about how the component programs interact with one another through their shared variables. That knowledge can be exploited to derive nicer compositionality results. We will return to this later.

4.2 Convergence

One of standard methods to prove progress is by exploiting a well-founded relation. This is done by defining a bound function over the program states. If one can prove that the execution of a program either establishes q or decreases the value of the bound function with respect to a well-founded relation, then since the well-foundedness implies that the decrease cannot go on forever, q eventually must hold. This is in fact what Theorem 4.2 states.

Another, closely related, method is to divide the execution in rounds which are ordered by a well-founded relation. The program tries to establish a stable predicate $p.i$ in the current round i, then proceeds to the next rounds in the ordering. The well-foundedness guarantees the existence of some initial rounds to start the computation and since each $p.i$ is stable, eventually $(\wedge\, i :: p.i)$ will be established. Problems that can be solved this way are called *round-solvable problems* due to Lentfert and Swierstra [Len93]. An example of a round-solvable problem is the problem of computing the minimal path between vertices in a network.

To formalize this principle we need to model progress towards stability in our logic. For this purpose we introduce *convergence* operator, denoted with $\leadsto$, which can be considered as an implementation of $\Diamond\Box$ in UNITY. Convergence has been identified as a useful notion for specifying self-stabilizing programs [Her91, AG92, Len93]. It also satisfies many useful calculational properties.

Definition 4.5 : Convergence
For any $q \in \mathsf{Pred}.(\mathbf{w}P)$:

$$p \leadsto q \text{ in } P \text{ under } J$$
$$=$$
$$(\exists q' :: (p \longmapsto (q \dot\wedge q') \text{ in } P \text{ under } J) \wedge \mathsf{stable}.P.(J \dot\wedge q \dot\wedge q'))$$

Intuitively, $p \leadsto q$ in P under J means that P can progress from $p \dot\wedge J$ to q, which will remain to hold afterward. With other words, $p \dot\wedge J$ converges to q. Notice that tt $\leadsto q$ in P under tt states that P will reach q and remain there regardless its initial state. In other words, P is self-stabilizing.

Clearly, $\leadsto$ implies $\longmapsto$, and thus it also implies $\mapsto$. Below are some basic properties of $\leadsto$. Due to space limitation we do not present their proofs but they can be proven using the afore mentioned properties of **stable** and $\longmapsto$.

Theorem 4.7 : $\Rightarrow$ Promotion

$$\frac{\mathsf{stable}.P.(p \dot\wedge J) \ \wedge \ p,q \in \mathsf{Pred}.(\mathbf{w}P) \ \wedge \ [(J \dot\wedge p) \Rightarrow q]}{p \leadsto q \text{ in } P \text{ under } J}$$

Theorem 4.8 : Accumulation

$$\frac{(p \leadsto q \text{ in } P \text{ under } J) \ \wedge \ (q \leadsto r \text{ in } P \text{ under } J)}{p \leadsto (q \dot\wedge r) \text{ in } P \text{ under } J}$$

Theorem 4.9 : Stable Shift

$$\frac{\mathsf{stable}.P.J \ \wedge \ r \in \mathsf{Pred}.(\mathbf{w}P) \ \wedge \ p \leadsto q \text{ in } P \text{ under } J \wedge r}{(p \dot\wedge r) \leadsto q \text{ in } P \text{ under } J}$$

Theorem 4.10 : Disjunction

$$\mathsf{Ldisj}.(\lambda p, q.\ p \leadsto q \text{ in } P \text{ under } J)$$

Theorem 4.11 : $\leadsto$ Conjunction
For any non-empty and finite set of indices W, and $f, g \in W \to \mathsf{Pred}$:

$$\frac{(\forall i : i \in W : f.i \leadsto g.i \text{ in } P \text{ under } J)}{(\wedge i : i \in W : f.i) \leadsto (\wedge i : i \in W : g.i) \text{ in } P \text{ under } J}$$

The most distinguishing property of $\leadsto$ is that it is both $\vee$ and $\wedge$-junctive

(Disjunction and Conjunction Law) whereas $\rightarrowtail$ is only $\vee$-junctive. This makes $\rightsquigarrow$ calculationally attractive.

In addition, $\rightsquigarrow$ is also *transitive*. As corollaries of $\Rightarrow$ Promotion and $\rightsquigarrow$ transitivity, . $\rightsquigarrow$. in P **under** J is monotonic with respect to $\Rightarrow$ in its second argument and anti-monotonic in its first, provided the stability of J and the confinement constraint is respected.

The principle of dividing an execution into rounds is formalized in the theorem below. However, instead of ordering the rounds with a well-founded relation we require them to be organized as a dag (directed acyclic graph) which seems to be a more familiar structure in distributed programming. There is no loss of generality since a well founded relation also defines a dag. A dag over set A can be represented by a relation $\lhd \in A \rightarrow A \rightarrow \mathbf{bool}$ such that the transitive closure $\lhd$, which we will denote by $\lhd+$, is acyclic. Note that if A is finite and $\lhd$ is a dag then $\lhd+$ is well-founded.

Theorem 4.12 : Round Decomposition
Let $\lhd \in A \times A$ define a dag over a finite and non-empty set A. For any $f \in A \rightarrow$ Pred we have:

$$\frac{(\forall y : y \in A : (\wedge\, x : x \lhd+ y : f.x) \rightsquigarrow f.y)}{\mathrm{tt} \rightsquigarrow (\wedge\, y : y \in A : f.y)}$$

The compositionality of $\rightsquigarrow$ with respect to parallel composition can be derived from Singh Law. But as we have remarked earlier, nicer compositionality results can be obtained if we know more about how the component programs interact. The next sub-section will present compositionality results of so-called *write-disjunct* networks of programs.

4.3 Write-Disjunct Composition

Two programs, P and Q, are called write-disjunct, denoted by $P \div Q$, if their set of write variables are disjoint. Consequently, P only writes to its own variables and to Q's *input* variables, that is, those variables read but not written by Q. The same holds for Q. As the reader recalls, $p \rightarrowtail q$ in P under J describes some progress of the writable part in P. The state of the non-writable part, on which the progress may depend, is described by the stable predicate J. Because P and Q are write-disjunct, and hence Q does not write to the writable part of P, so, provided the stability of J is respected, Q cannot destroy any progress by $\rightarrowtail$ in P. This transparency of progress is a very nice property, which is also why we have defined $\rightarrowtail$ as it is. We call parallel composition of write-disjunct programs *write-disjunct composition*. The definition of $\div$ is given below.

Definition 4.6 : Write-Disjunct Programs

$$P \div Q = ((\mathbf{w}P \cap \mathbf{w}Q) = \emptyset)$$

It is easy to see that $\div$ is commutative. The basic compositionality property of $\div$ is given by the *Transparency* principle which we have motivated some paragraphs earlier.

Theorem 4.13 : Transparency

$$\frac{(P \div Q) \ \wedge \ \mathsf{stable}.Q.J \ \wedge \ (p \rightarrowtail q \text{ in } P \text{ under } J)}{p \rightarrowtail q \text{ in } (P[\![Q) \text{ under } J}$$

As a corollary, a similar property also applies to $\rightsquigarrow$. Below are some useful basic laws that can be derived from the Transparency principle.

Theorem 4.14 : Spiral Law

$$\frac{(P \div Q) \ \wedge \ \mathsf{stable}.P.(J \wedge q) \ \wedge \ \mathsf{stable}.Q.J}{(p \rightarrowtail q \text{ in } P \text{ under } J) \ \wedge \ (\mathsf{tt} \rightarrowtail r \text{ in } Q \text{ under } (J \wedge q))}{p \rightarrowtail (q \wedge r) \text{ in } (P[\![Q) \text{ under } J}$$

Theorem 4.15 : Conjunction

$$\frac{(P \div Q) \ \wedge \ (p \rightsquigarrow q \text{ in } P \text{ under } J) \ \wedge \ (r \rightsquigarrow s \text{ in } Q \text{ under } J)}{(p \wedge r) \rightsquigarrow (q \wedge s) \text{ in } (P[\![Q) \text{ under } J}$$

In the Spiral law, progress to r is implemented by splitting a program in two where the first one takes care of progress to some q and the second completes the task by progressing from q to r. In Conjunction law, progress to $q \wedge s$ is realized by two parallel components, one converging to q and the other to s. Indeed, the Spiral law suggests a sequential division of tasks, whereas Conjunction law suggests a parallel division.

Write-disjunct composition is useful in practice. The parallel composition of two programs is said to form a *fork* if their set of input variables —that is, those variable read but not written by a program— is equal and they do not write to each other. It is said to form a *free parallel* if they have no variable in common. Clearly these are two instances of write-disjoint composition. Forking and free parallel are the typical constructions used when applying the Conjunction law above.

Another useful instance of write-disjunct composition is *layering*. It is a well-known technique in constructing self-stabilizing programs [AG90, Sch93]. In layering we have two programs called *lower layer* and *upper layer*. The lower layer is independent from the upper layer, so it does not read from the latter. The upper layer, on the contrary, bases its computation on the results of the lower layer. If we also insist that the lower layer only writes to the upper layer's input variables then the layers are write-disjoint. Layering suggests a sequential division of tasks and it is typically used when applying the Spiral law above.

4.4 How Much Do You Save?

Compositionality may lessen our proof obligation, which is especially true in verifying a progress property. To verify whether a UNITY program satisfies a progress specification first we have to decompose the specification into a set of **ensures** properties and then verify each one of them. Suppose we come up with m **ensures** properties and we have a program with k actions. Without using compositionality (of $\rightarrowtail$) we have to verify something like km Hoare triples. However, if somehow we can use, for example, the Transparency Law by recognizing that the progress is actually realized by only a part of the program consisting k_1 actions $(k_1 < k)$, then we only have to verify mk_1 Hoare triples and save ourselves $m(k - k_1)$ verifications. This is a rough estimation, but the point is that by identifying part of the program that does not contribute to a progress property we can save ourselves from some superfluous verifications. If the difference between k and k_1 is too small the estimation is too gross and we may find that the application of a compositionality law increases the amount of the proof obligation instead. This suggests the use of compositionality in the design to break a program into components of a sufficiently small size, but not too small, and then post-verify the specifications of each component.

5 Conclusion

The goal of our research is to be able to mechanically verify distributed algorithms for the simple reason that reasoning about distributed systems is inherently complicated and error prone. As in many cases a specification of a distributed program is expressed in higher order formulas and their verification will be hard to automate. Still, a mechanical proof is almost essential to gain confidence. In our view, a theorem prover is mainly another sort of medium on which we are doing our proofs; much like another sort of pencil and paper but with an additional advantage that a machine will take over much of the burden of the proof administration from us. For being able to do so, we must first have a good calculus to reason about distributed systems. Our choice was UNITY, due to its simplicity. Still, UNITY lacks certain things to facilitate a convenient calculation, which is why we worked on its extension which is presented in this paper.

To summarize what we have done: We have extended UNITY with two notions of progress which are compositional and present various basic laws to use in calculation. We have also defined a special case of parallel composition called write-disjunct program composition. The newly introduced progress operators have been defined in such a way that their compositionality with respect to write-disjunct composition can be nicely expressed.

We have verified the results we mentioned here in HOL theorem prover, and they are available as an extension of the HOL logic and thus provides one of the bases for a mechanically supported verification of distributed algorithms.

6 Acknowledgement

I wish to thank Doaitse Swierstra for his ideas and effort to improve my notation and my English.

References

[AG90]   A. Arora and M.G. Gouda. Distributed reset. In *Proceedings of the 10th Conference on Foundation of Software Technology and Theoretical Computer Science*, 1990. Also in *Lecture Notes on Computer Science* vol. 472.

[AG92]   A. Arora and M.G. Gouda. Closure and convergence: A foundation for fault-tolerant computing. In *Proceedings of the 22nd International Conference on Fault-Tolerant Computing Systems*, 1992.

[And92]   Flemming Andersen. *A Theorem Prover for UNITY in Higher Order Logic*. PhD thesis, Technical University of Denmark, 1992.

[CM88]   K.M. Chandy and J. Misra. *Parallel Program Design – A Foundation*. Addison-Wesley Publishing Company, Inc., 1988.

[dBvH94] F.S. de Boer and M. van Hulst. A proof system for asynchronously communicating deterministic processes, 1994. Submitted to a conference.

[Her91]   Ted Herman. *Adaptivity through Distributed Convergence*. PhD thesis, University of Texas at Austin, 1991.

[Len93]   P.J.A. Lentfert. *Distributed Hierarchical Algorithms*. PhD thesis, Utrecht University, April 1993.

[PJ91]   P.K. Pandya and Mathai Joseph. P-a logic –a compositional proof system for distributed programs. *Distributed Computing*, (5):37–54, 1991.

[Pra93a] I.S.W.B. Prasetya. Formalization of variables access constraints to support compositionality of liveness properties. In *Proceeding HUG 93, HOL User's Group Workshop*, pages 326–339. University of British Columbia, 1993.

[Pra93b] I.S.W.B Prasetya. *UU_UNITY: a Mechanical Proving Environment for UNITY Logic*. University of Utrecht, 1993. Draft. Available at request.

[San91]   B.A. Sanders. Eliminating the substitution axiom from UNITY logic. *Formal Aspects of Computing*, 3(2):189–205, 1991.

[Sch93]   Marco Schneider. Self-stabilization. *ACM Computing Surveys*, 25(1), March 1993.

[Sin89]   A.K. Singh. Leads-to and program union. *Notes on UNITY*, 06-89, 1989.

[UHK94]   R. Udink, T. Herman, and J. Kok. Compositional local progress in unity. to appear in the proceeding of IFIP Working Conference on Programming Concepts, Methods and Calculi, 1994., 1994.

Simplifying Deep Embedding:
A Formalised Code Generator*

Ralf Reetz and Thomas Kropf

Institut fuer Rechnerentwurf und Fehlertoleranz (Prof. D. Schmid)
Universitaet Karlsruhe
Zirkel 2
Postfach 6980
76128 Karlsruhe
Germany
e–mail:{reetz,kropf}@ira.uka.de

Abstract. A tool is described, which simplifies the formalisation method "deep embedding" for a system with syntax and semantics in the following way: a formalisation of the syntax described by a context–free grammar is automatically generated and a formalisation of the complete semantics for the given syntax is simplified by automatically generating a formal code generator out of a formalised context–restricted semantics.

1 Introduction

Formal systems Φ usually consist of a syntax, semantics for all syntactical sentences and a set of rules to produce new sentences of the system out of given ones. "Deep embedding" is a well-known method for mechanising such formal systems in HOL. The syntax and semantics are mechanised *within* the logic so that reasoning over both using HOL is possible:

- A set of formulas Σ is defined as the (usually abstract) syntactic domain of Φ by using fitting type definitions.
- A set of formulas Γ is defined as the semantic domain of Φ by using fitting type definitions.
- Semantics of syntactic objects of Φ is usually defined by a semantic function $\Psi : \Sigma \to \Gamma$.

Numerous works have used this formalisation method, e.g. for a process term language [1], π–calculus [6], microSR [10], to name only some of them. Up to now, most approaches using deep embedding have used small syntactic domains and therefore "manageable" semantic functions, so they were able to realise deep embedding manually.

As more complex systems come into the scope of deep embedding, (e.g. our experience [7] in formalising the hardware description language VHDL with more

---

* This work was financed by the German research society (DFG) under contract SFB 358.

than 100 syntactic constructs and an informal semantics description of more than 130 pages [9]), realising deep embedding, i.e. defining Σ, Γ and Ψ completely manually is not feasible with justifiable effort.

Two ideas to simplify deep embedding are presented here:

1. The up to now manually used method for defining syntactic domains is generalised to context–free grammars and automated by using the mutually recursive type definition package from Gunter [3].
2. Defining semantic functions Ψ is simplified by formalising a code generator from Ripken [8].

1.1 Notation

The reader is assumed to be familiar with context–free grammars (see, e.g. [4]) and the HOL library `mutrec` for mutually recursive type definitions [3].

A context-free grammar $G = (V, T, P, S)$ consists of a finite set of variables V, a finite set of terminals T, a finite set of productions P and the startsymbol $S \in V$. A symbol is a variable or a terminal. For variables we use $A, B, \ldots$, for terminals $a, b, \ldots$, for symbols $\alpha_i, \alpha_j; \ldots$, for words $\alpha \in (V \cup T)^*$ of length n we use $\alpha_1 \alpha_2 \ldots \alpha_n$, for the empty word ϵ and $L(G)$ for the context–free language of G. Here, a production has the form $p : A \rightarrow \alpha_1 \alpha_2 \ldots \alpha_n$, where p is an arbitrary name for the production, A the left side and $\alpha_1 \alpha_2 \ldots \alpha_n$ the right side of the production.

For mutually recursive type definitions, we use pseudo–code from the implementation of the library `mutrec` in HOL90.

2 Algorithm

2.1 Defining syntax using mutually recursive types for derivation trees

Defining an abstract syntax does not mean formalising the words of a context–free language $L(G)$, but formalising its syntactic structure. This is represented by derivation trees, which are defined as follows:

1. Every internal vertex has a label $A \in V$.
2. The label of the root is S.
3. Every leaf has a label $a \in T$ or ϵ.
4. If an internal vertex v has label A and vertices $v_1, \ldots, v_n$ are children of vertex v, in order from the left, with labels $\alpha_1, \ldots, \alpha_n$, respectively, then $p : A \rightarrow \alpha_1 \ldots \alpha_n$ is a production of P.
5. If vertex v has label ϵ, v is a leaf and is the only child of its father.

Formulas for trees can be build as follows: the formula for a leaf is an atomic formula (i.e. a constructor without arguments) and the formula for an internal vertex is a formula consisting of a function (i.e. a constructor) with the formulas for the children as its arguments. So, an abstract syntactic domain Σ is a set of formulas for derivation trees build as described before.

As HOL–types ca not be empty, for simpliticity we restrict ourselves to a context–free grammar G, where the empty string ϵ does not belong to the language of the grammar $\epsilon \notin L(G)$. Furthermore, G should be a grammar without useless[2] variables and without ϵ–productions[3]. For every context–free grammar G with $\epsilon \notin L(G)$, an equivalent grammar without useless variables and without ϵ–productions can be computed [4].

In the following, attributed derivation trees are considered: every node of a derivation tree gets an additional, arbitrary label $\mathfrak{a}$ of type $\mathfrak{A}$. So, an attributed abstract syntactic domain $\Sigma(\mathfrak{A})$ is considered.

Because the number of variables, terminals and productions is finite, the types can be defined automatically, using the library for defining mutually recursive types:

1. Divide the productions of G into sets of productions in the following way:

$$P_A = \{(p_A : A \to \alpha) \in P\}$$

$$P = \bigcup_{A \in V} P_A$$

2. Define mutually recursive types as follows:

   (a) For every terminal $a \in T$, define a type **a** for a with one type constructor a with an arbitrary attribute represented by a HOL type variable $'\mathfrak{A}$ as its only argument:

   ```
   {type_name    = "a",
    constructors = [{name     = "a",
                     arg_info = [existing (=='':'𝔄'==)]
                   ]}}
   ```

   (b) For every variable $A \in V$, for all k productions define a type **A** with a type constructor $p_A^i : A \to \alpha_{i_1} \ldots \alpha_{i_{n_i}} \in P_A$ with $|P_A| = k$, where the types of the arguments of the type constructor are a HOL type variable $'\mathfrak{A}$ for an arbitrary attribute together with the types for every symbol α_i of the right side α of the production p_A^i:

   ```
   {type_name    = "A",
    constructors = [{name      = "p_A^1",
                     arg_info = [existing (=='':'𝔄'==),
                                 being_defined α_{1_1},
                                 ...
                                 being_defined α_{1_{n_1}}]},
   ```

   $\ldots$

---

[2] A is useful if there is a derivation $S \overset{*}{\Rightarrow} \alpha A \beta \overset{*}{\Rightarrow} w$ with $w \in T^*, \alpha, \beta \in (V \cup T)^*$. Otherwise A is useless.

[3] productions of the form $p : A \to \epsilon$

```
{name      = "p_A^k",
 arg_info = [existing (==':' '==),
             being_defined α_{k_1},
             ...
             being_defined α_{k_{n_k}}]}
]}}
```

As a result, we get a type constructor for every different kind of vertex of the derivation tree: every constructor stands for a certain label of a vertex and for the associated production of an internal vertex. Because useless variables were assumed to be eleminated before, the type operator S with its type argument '$\mathfrak{A}$ for the startsymbol S formalises the set of all arbitrary attributed derivation trees of G and therefore the attributed abstract syntactic domain $\Sigma(\mathfrak{A})$.

As a "toy" example used throughout the paper, consider the set of arithmetic expressions with variables, addition and multiplication, where multiplication has a higher priority than addition. Grammar $G = (V, T, P, S)$ defines the syntax Σ with:

$$V = \{E, T, F\} \qquad\qquad P = \begin{cases} p_E^0 : E \rightarrow E \pm T, \\ p_E^1 : E \rightarrow T, \\ p_T^0 : T \rightarrow T * F, \\ p_T^1 : T \rightarrow F, \\ p_F^0 : F \rightarrow (\ E\ ), \\ p_F^1 : F \rightarrow \underline{I} \end{cases}$$

$$T = \{\underline{\pm}, \underline{*}, \underline{I}, \underline{(}, \underline{)}\}$$

$$S = E$$

Let strings be the attributes, i.e. $\mathfrak{A}$ is **string**. Let the terminal $\underline{I}$ be the only symbol with a non–empty string as its attribute. E.g. the HOL representation of an attributed derivation tree for $\underline{I} * \underline{I} \in L(G)$ with attributes "X", "Z", respectively, is (see fig. 1):

$$p_E^1\ "" \ (p_T^0\ "" \ (p_T^1\ "" \ (p_F^1\ "" \ (\underline{I}\ "X"))) \ (\underline{*}\ "") \ (p_F^1\ "" \ (\underline{I}\ "Z")))$$

2.2 Defining semantics by a formal code generator

The key idea presented here is to define a semantic function Ψ as a formal compiler and to compute a definition of a formal compiler by using well–known methods for compiler–building. In a compiler, one distinguishes between several phases (although they might be interleaved): first, the lexical and the syntactical phases are used to build a derivation tree and a symbol table. Second, the derivation tree will be attributed, using the symbol table. Finally, the code of the target language is generated.

In our case, we do not need the lexical and syntactical phases, since our syntactic domain already consists of the derivation trees. Furthermore, we do not consider attributation of the derivation trees here. So we assume that our syntactic domain consists of all attributed derivation trees.

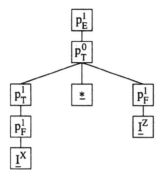

Fig. 1. attributed derivation tree for $\underline{I} * \underline{I}$

The key approach to get a fast and easy–to–define code generation phase is to narrow the scope of a translation step of attributes into code to a very local scope, in fact to the attributes of one node and its children in the derivation tree (i.e. one production). In that local scope, one distinguishes between distribution of an attribute by sending it down to the children and synthesizing code from the children back to the node.

This motivates the code generator from Ripken [8]. Here the code generation phase is logically divided into two steps. First, a traversing scheme is defined, which traverses the derivation tree in a combined top–down/bottom–up order, reflecting distribution and synthesis. The traversing scheme defines a list of visited nodes $l = v_1 \dots v_n$ with the following properties:

1. $v_1 = v_n$ is the root.
2. Every node v appears twice in l.
3. If the second appearence of a node v in l is deleted, a list of the nodes in preorder is obtained.
4. Between the first and the second appearence of a node v, all nodes of its subtrees appears twice.

An example is shown in fig. 2. The following list is obtained:

$$p_E^1 \; p_T^0 \; p_T^1 \; p_F^1 \; \underline{I}^X \; \underline{I}^X \; p_F^1 \; p_T^1 \; * \; * \; p_F^1 \; \underline{I}^Z \; \underline{I}^Z \; p_F^1 \; p_T^0 \; p_E^1$$

Now, for the second step, a local translation function $f(v_i, v_j)$ is defined for every possible pair (v_i, v_j) within such traversing lists l, which is then applied successivly through the list l (in fact, f is applied to the attributes of the nodes):

1. Start with $i = 1$
2. Compute $f(v_i, v_{i+1})$, replace v_{i+1} with the computed result, increment i and repeat (2.) until the end of l is reached.

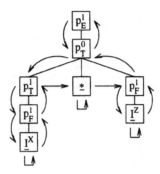

Fig. 2. the traversing scheme for $\underline{I} * \underline{I}$

For the example above, the result of the translation is described by:

$$f(f(f(f(f(f(f(f(f(f(f(f(f(f(f(p_E^1,p_T^0),p_T^1),\\
p_F^1),\underline{I}^X),\underline{I}^X),p_F^1),p_T^1),*),*),p_F^1),\underline{I}^Z),\underline{I}^Z),p_F^1),p_T^0),p_E^1)$$

A first approach to formalise this code generator in HOL might be to define mutually recursive functions over the types representing derivation trees, which computes the list of attributes as described above and then to define the translation scheme above over list of attributes. But then, the formal validation of that code generator would also be divided into two steps, which makes proving difficult. So, an approach where these two steps are "merged" into one is used instead.

A set of functions f_x for local translations is defined by the user. Then, a set of mutually recursive functions g_x over the types representing derivation trees is automatically defined, which realises the traversing scheme and calls the fitting functions f_x.

Let $\mathfrak{A}$ be the type of the attributes and $\mathfrak{T}$ be the type of the target language, i.e. the type for the semantic domain Γ choosen by the user. There are four cases on which the user defines local translations f_x:

1. For every terminal a:

$$f_{aa} : \mathfrak{T} \to \mathfrak{A} \to \mathfrak{T}$$

2. For every production $p_A : A \to \alpha_1 \dots \alpha_n$:

$$f_{p_A \alpha_1} : \mathfrak{T} \to \mathfrak{A} \to \mathfrak{T}$$

3. For every two neighboured symbols α_i, α_{i+1} of the right side of all production rules $p_A : A \to \alpha_1 \dots \alpha_i \alpha_{i+1} \dots \alpha_n$:

$$f_{p_A \alpha_i \alpha_{i+1}} : \mathfrak{A} \to \mathfrak{T} \to \mathfrak{A} \to \mathfrak{T}$$

4. For every production $p_A : A \to \alpha_1 \ldots \alpha_n$:

$$f_{\alpha_n p_A} : \mathfrak{T} \to \mathfrak{A} \to \mathfrak{T}$$

Now, the mutually recursive traversing functions g_x are automatically defined:

1. For every termial a, a traversing function g_a is defined as:

$$g_a \ (a \ \mathfrak{a}) = \\ \lambda f. f_{a\mathfrak{a}} \ (f \ \mathfrak{a}) \ \mathfrak{a}$$

That means if the traversation of a leaf a by g_a is called, it gets a translation $f : \mathfrak{A} \to \mathfrak{T}$ from its caller, applies f to the attribute $\mathfrak{a}$ of the terminal a and then applies the local translation $f_{a\mathfrak{a}}$ to the previous result in $\mathfrak{T}$ and the attribute $\mathfrak{a}$, thus computing a new result in $\mathfrak{T}$.

2. For every production $p_A : A \to \alpha_1 \ldots \alpha_n$, a traversing function g_A is partially specified as:

$$g_A \ (p_A \ \mathfrak{a} \ \alpha_1 \ldots \alpha_n) \ = \\ \lambda f. \\ f_{\alpha_n p_A} \\ (g_{\alpha_n} \ \alpha_n \\ \vdots \\ (f_{p_A \alpha_2 \alpha_3} \ \mathfrak{a} \\ (g_{\alpha_2} \ \alpha_2 \\ (f_{p_A \alpha_1 \alpha_2} \ \mathfrak{a} \\ (g_{\alpha_1} \ \alpha_1 \\ (f_{p_A \alpha_1}(f \ \mathfrak{a})))))) \ldots)$$

That means if the traversation of a subtree with root p_A by g_A is called, it gets a translation $f : \mathfrak{A} \to \mathfrak{T}$ from its caller, applies f to the attribute of the root $\mathfrak{a}$, applies the local translation for going down from the root to its leftmost child $f_{p_A \alpha_1}$, calls the traversation of the leftmost child g_{α_1}, then applies the local translation for going from the leftmost child to its right neighbour $f_{p_A \alpha_1 \alpha_2}$, calls the traversation of this right neighbour g_{α_2} and so on. Finally, the local translation for going up from the rightmost child back to the root $f_{\alpha_n p_A}$ is applied. For an example, see fig. 3.
Note that an extention is made here: when going from a child to its right neighbour, the local translation gets the attribute of the root, too. This can be useful as practical experiences have shown.

Finally, the translation for an attributed derivation tree $d : \mathfrak{A} \ S$ with attribute type $\mathfrak{A}$, startsymbol S, and a user-choosen initial target value $\mathfrak{t} : \mathfrak{T}$ is defined as:

$$g \ d \ = \\ g_S \ d \ (\lambda \mathfrak{a}. \mathfrak{t})$$

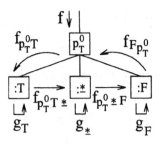

Fig. 3. partial specification of g_T for p_T^0

The reason why the translation is called to be restricted to the scope of one production becomes clear by considering the partial specification of a traversing function g_A above: the current production p_A as the root of the current derivation subtree for variable A is known and this information is coded into different local translation functions. Furthermore, the symbols of the right side of p_A are known, i.e. the types of the children of the current root are known and coded into the different local translation functions, too. But the value of the children, i.e. the productions for the children are unkown for g_A.

We return to our example. A postfix notation is taken as the semantics Γ: a list of tokens formalised as strings in postfix notation of the syntactical sentence. So, the type of the target language $\mathfrak{T}$ is **string list**. In this simple example, only three local translations f_x are needed. The other ones simply passes their input to the output and are not listed here. Furthermore, the empty list $[]$:**string list** is used as the initial target value.

$$f_{Tp_E^0}\ (t:\mathfrak{T})\ (a:\mathfrak{A})\ =$$
APPEND $t\ ["+"]$

$$f_{Fp_T^0}\ (t:\mathfrak{T})\ (a:\mathfrak{A})\ =$$
APPEND $t\ ["*"]$

$$f_{II}\ (t:\mathfrak{T})\ (a:\mathfrak{A})\ =$$
APPEND $t\ [a]$

$$g\ d\ =$$
$$g_S\ d\ (\lambda a.[\ ])$$

E.g. the semantics of $(\ \underline{I}^X\ \underline{+}\ \underline{I}^Y\ )\ \underline{*}\ \underline{I}^Z$ is the result $[X;Y;+;Z;*]$ of a translation as shown in fig. 4.

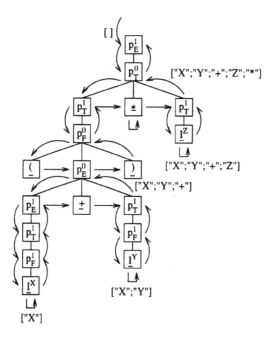

Fig. 4. Translation of $(\ \underline{I}^X \pm \underline{I}^Y\ ) \ast \underline{I}^Z$

3 Formal Validation of Semantics

One of the most important advantages of deep embedding is the possiblity to formally validate the defined semantics, since both syntax and semantics are presented within the logic. Validating semantics definitions is done by proving semantic properties of syntactical objects.

In the approach presented here, validating the defined semantics means to verify the formal code generator. Only the verification of safety properties of the code generator is discussed here. Verifying a safety property P here means to show that $P : \mathfrak{T} \to$ **bool** holds for the result of the translation of all attributed derivation trees d:

$$\forall d.P(g\ d)$$

As [5], [2] have already shown, structural induction is an appropriate proof method for proving properties of syntactical structures, i.e. derivation trees. To show that P is a safety property of the code generator by structural induction, the user has to specify fitting safety properties P_{α_i} for every symbol α_i, i.e. a safety property has to be specified for every sub–derivation tree with a different symbol as its root. Note that $P = P_S$ for startsymbol S.

E.g. to show that the postfix notation of the arithmetic expressions is never the empty list, the following properties can be used:

$$P_(\; t = T$$
$$P_) \; t = T$$
$$P_* \; t = T$$
$$P_+ \; t = T$$
$$P_I \; t = \neg(t = [\;])$$
$$P_E \; t = \neg(t = [\;])$$
$$P_F \; t = \neg(t = [\;])$$
$$P_T \; t = \neg(t = [\;])$$

Using structural induction and some additional simple proof steps, the following is obtained:

$$\vdash \forall d. \neg((g \; d) = [\;])$$

This holds, because every arithmetic expression contains a variable which by definition of f_{II} causes the code generator to compute a list with at least one element containing the name of that variable.

4 Implementation

The described method was implemented in a library called **generator**, which realises a functor **new_generator_definition**. The user supplies the grammar as a structure, which is then mapped by **new_generator_definition** to a new structure containing the functions below for the supplied grammar. Furthermore, the definitions of the types representing the derivation trees for the supplied grammar are stored into the current theory.

```
- val output_local_translations :
    {info_filename             : string,
     attribute_type            : hol_type,
     result_type               : hol_type,
     prefix_local_translation : string} -> unit;
```

This is intended to help the user with defining local translations f_x. For a supplied name **prefix_local_translation** for f, all needed local translations with its parameters are written into a text file.

```
— val new_translation_definition :
     {attribute_type              : hol_type,
      result_type                 : hol_type,
      init_result_value           : term,
      prefix_local_translation    : string,
      prefix_translation          : string,
      local_translations          : thm list} -> translation_info;
```

For a supplied name **prefix_translation** for g, a supplied name
prefix_local_translation for f, a supplied attribute type
$\mathfrak{A} =$ **attribute_type**, a supplied target type $\mathfrak{T} =$ **result_type** and an
initial target value $t_0 =$ **init_result_value**, the functions g_x are defined
and a data structure describing this semantics definition needed by other
functions below is returned.

Note that different semantics definitions for the same syntax are possible.

```
— val mk_translation :
     translation_info -> term -> term
```

mk_translation takes an attributed derivation tree and returns a term describing the application of the code generator to the attributed derivation
tree.

```
— val translation_CONV : translation_info -> conv -> conv
```

translation_CONV derives the semantics of a derivation tree. It traverses the
derivation tree bottom–up and derives the semantics by specialising the fitting definitions and putting these definitions together in the fitting order. In
Order to keep the needed memory small during translation, the supplied conversion is applied after every translation step, which is supposed to simplify
a currently computed target code after an application of a local translation,
which might extend the previously computed target code, thus comsuming
more memory for larger theorems. This can be useful as practical experiences
have shown.

```
— val mk_property_translation : translation_info -> term -> term
```

mk_property_translation sets up a safety goal for validating a semantics
definition.

```
— val property_translation_INDUCT_THEN :
     translation_info -> {symbol:string,property:term} list
     -> tactic -> tactic -> tactic
```

For a supplied list of safety properties for every symbol,
property_translation_INDUCT_THEN applies structural induction to a safety goal set up by **mk_property_translation** as described above. The first

supplied tactic is then applied to the resulting subgoals for the terminals and the second supplied tactic is then applied to the resulting subgoals for the variables.

5 Conclusion and future works

We have shown how validated deep embedding can be simplified by formalising methods from compiler building. We assumed that the syntax of a formal system consists of completely attributed derivation trees. To increase the scope of deep embedding, future work could be the formalisation of attribution systems from compiler building to obtain derivation trees without attributes as the formal syntax. Furthermore, to obtain sentences of the language as the formal syntax, the lexical and the syntactical phases might be automatically formalised or, as a faster, but less secure alternative, the term pretty printer of the HOL–system might be extended automatically.

6 Acknowledgements

Elsa Gunter's library for mutually recursive type definitions made this work possible. She especially helped us by setting up an implementation of the library "mutrec" for HOL90.

References

1. E. de Barros Lucena. Reasoning about petri nets in HOL. In M. Archer, J.J. Joyce, K.N. Levitt, and P.J. Windley, editors, *HOL Theorem Proving System and its Applications*, pages 384–394, Davis, California, August 1991. IEEE Computer Society, ACM SIGDA, IEEE Computer Society Press.
2. P. Curzon. A verified compiler for a structured assembly language. In M. Archer, J.J. Joyce, K.N. Levitt, and P.J. Windley, editors, *HOL Theorem Proving System and its Applications*, pages 253–262, Davis, California, August 1991. IEEE Computer Society, ACM SIGDA, IEEE Computer Society Press.
3. E. Gunter. A broader class of trees for recursive type definitions for HOL. In *HUG93, HOL User's Group Workshop*, Vancouver, Canada, August 1993. Springer Verlag.
4. J.E. Hopcroft and J.D. Ullman. *Introduction to automata theory, languages, and computation*. Addison Wesley, 1979.
5. D.F. Martin and R.J. Toal. Case studies in compiler correctness using HOL. In M. Archer, J.J. Joyce, K.N. Levitt, and P.J. Windley, editors, *HOL Theorem Proving System and its Applications*, pages 242–252, Davis, California, August 1991. IEEE Computer Society, ACM SIGDA, IEEE Computer Society Press.
6. T.F. Melham. A mechanized theory of the π-calculus in HOL. Technical Report 244, University of Cambridge, Computer Laboratory, Cambridge, England, January 1992.

7. R. Reetz and Th. Kropf. Formalisierung eines Flussgraphmodells in Logik hoeherer Ordnung und dessen Anwendung in der Hardware–Verifikation. In D. Monjau, editor, *Rechnergestuetzter Entwurf und Architektur mikroelektronischer Systeme*, pages 193–202, Oberwiesenthal, Germany, May 1994. Gesellschaft fuer Informatik e.V. (GI).

8. R. Ripken. Generating an intermediate–code generator in a compiler–writing system. In E. Gelenbe and D. Potier, editors, *International Computing Symposium*, pages 121–127. North Holland, 1975.

9. IEEE standard VHDL language reference manual. Std 1076, IEEE, 1987.

10. C. Zhang, R. Shaw, R. Olsson, K. Levitt, M. Archer, M. Heckman, and G. Benson. Mechanizing a programming logic for the concurrent programming language microSR in HOL. In *HUG93, HOL User's Group Workshop*, Vancouver, Canada, August 1993. Springer Verlag.

Automating Verification by Functional Abstraction at the System Level*

Klaus Schneider[1], Ramayya Kumar[2] and Thomas Kropf[1]

[1] Universität Karlsruhe, Institut für Rechnerentwurf und Fehlertoleranz,
(Prof. D. Schmid), P.O. Box 6980, 76128 Karlsruhe, Germany,
e-mail:{schneide|kropf}@ira.uka.de
[2] Forschungszentrum Informatik, Haid-und-Neustraße 10-14,
76131 Karlsruhe, Germany, e-mail:kumar@fzi.de

Abstract. The verification of digital circuits at higher levels of abstraction still suffers from complex and unstructured proofs. In this paper, we present a class of circuits that can be used for the implementation of arbitrary processes without shared memory. These processes communicate with each other according to a handshake protocol. We have proven general theorems to automatically derive correctness theorems for composed handshake circuits. The contribution of this paper is therefore a new design style based on handshake circuits and a highly automated approach to verification at the system level based on functional abstraction.

1 Introduction

Proof-based approaches to hardware verification can be roughly classified into two categories, namely formal synthesis and post verification. Using *post verification*, an arbitrary design is given with its specification, and the task is to prove the design to be correct relative to the given specification. As there is in general no restriction on the design, this approach to hardware verification is certainly the most general one. However, no common structure of correctness proofs has been found at the system level up to now, and therefore the automation of the verification at this level is very restricted. *Formal synthesis*, on the other hand, offers a completely different approach for achieving verified designs. Given a specification, a set of transformation rules is used to refine the specification until an implementable design is obtained. The advantage of this approach is that only the transformation rules have to be proven correct to assure the correctness of all resulting designs. However, in all known approaches, the transformation rules have to be applied manually, and therefore the entire design has to be done manually, too.

In this paper, we present a new approach to hardware verification based on a class of circuits that communicate with each other according to a handshake protocol. Although these handshake circuits are synchronous circuits, they

---

* This work has been partly financed by a german national grant, project Automated System Design, SFB No.358.

cover some of the ideas of the synthesis of delay-insensitive circuits developed by Martin [BuMa87, Mart90a]. Martin adapted Hoare's 'Communicating Sequential Processes' [Hoar78] for synthesizing quasi-delay-insensitive circuits from a formal specification. The synthesis of handshake circuits, i.e. the transformation of a given specification into a register-transfer structure can be done similarly as outlined there. Hence, the presented verification method – together with an automatic synthesis procedure – is a new approach to formal synthesis of register transfer structures. In this paper, we will however not describe the synthesis of handshake circuits, instead we focus on the verification of given structures. From the viewpoint of post-verification, we can therefore state that we have developed a highly automated approach to the verification of given designs using these handshake circuits.

Handshake processes can be composed in various ways: given two handshake processes P_1 and P_2, the process that is obtained by synchronizing P_1 and P_2 is again a handshake process. If the processes P_1 and P_2 are executed one after the other, then this sequence is again a handshake process. Thus, the class of handshake processes is closed with respect to *synchronization* and *sequentialization*. Furthermore, each loop (data dependent or not) can be viewed as a handshake process, as the output data is only available after the termination of the loop. Loops are therefore considered basic handshake processes, while we call the handshake processes resulting from a synchronisation or a sequentialisation a composed handshake process.

The key to the automation of the verification of handshake designs is the use of preproven correctness theorems. In case of a synchronisation or a sequentialisation, the correctness of the handshake process can be automatically reduced to the correctness of the used components similar to composition principle of Abadi and Lamport [AbLa90]. In [AbLa90], the composition of specifications depends on a suitable environment which is in our case provided by the handshake protocol, which is retained after each module composition step. Additionally, this restriction leads to a proof methodology for verifying the function and the handshake behaviour of basic components. However, this verification task cannot be completely automated. In the domain of software verification, data-dependent loops are usually verified by invariant rules of so-called Floyd-Hoare calculi [Hoar69]. We use a similar rule for the verification of basic handshake circuits [ScKK94a]. As a result, a method for the verification of basic handshake circuits is obtained that can be automated almost up to the construction of suitable invariants.

As the correctness theorems of composed handshake structures are independent of the function of the circuit and depend only on the kind of composition (sequence, synchronization or pipeline) the underlying principle of this approach is called *functional abstraction*. Besides structural abstraction, data abstraction, and time abstraction [Melh88a], functional abstraction is therefore an additional abstraction mechanism which is important for the automation of verification at the system level.

The outline of the paper is as follows: In the next section, we define the class of handshake circuits. We also present the basic composition structures and the corresponding abstract correctness theorems for their verification. In section 3, we consider some case studies. The paper concludes with a short summary and a discussion of future work.

2 Handshake Circuits and Functional Abstraction

Handshake circuits could be used for the implementation of arbitrary processes, but we add the constraint that they do not communicate via shared memory. This constraint is the basis for a hierarchical verification of handshake circuits, where the circuits are verified separately one after the other. If shared memory would be allowed, the verification of a component using this shared memory had to consider all other components which share the memory as well. Therefore, the use of shared memory disallows local verification of single components, and therefore it prevents hierarchical verification.

A general handshake circuit is shown in figure 1. The circuit is divided into a control and a data path, where both the data path and the control path has one input and one output. The ports of the data path have polymorphic types such that they model a general input-output behavior. As we consider synchronous circuits and model the discrete time by natural numbers, the types of the ports are of the form $N \to \alpha$, where α is a type variable and N is the type of the natural numbers. If, for example a real circuit with two boolean valued inputs has to be modeled, then the input port will be of type $N \to (B \times B)$, where B is the boolean type.

Handshake circuits have two global states: a 'wait'-state, where old results are stored and new inputs can be read, and a 'computation-state', where new inputs are ignored and a computation of previously read inputs is performed. The output f of the handshake circuit signals that the circuit is available for new computations. The input req of the controller indicates that a new task is requested. New inputs are read only if the circuit has finished a previous computation ($f^{(t)} = T$) and a new task is requested ($req^{(t)} = T$). A computation

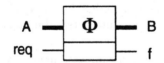

Fig. 1. A general handshake circuit.

according to the handshake-protocol is as follows (figure 2): if the circuit is ready ($f^{(t_0)} = T$) at time t_0 and a new task is given ($req^{(t_0)} = T$), then the inputs $A^{(t_0)}$ are read and the circuit switches for some time to the computation state,

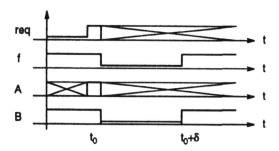

Fig. 2. The handshake protocol.

where new inputs are ignored. If the task is done at time $t_0 + \delta$, then the output f becomes high and the result of the computation can be read from the output B. The circuit remains in the wait-state and stores the result until a new task is requested.

In contrast to usual handshake protocols, a handshake circuit does not wait until another component has read its computed data, i.e. it does not wait for an acknowledge. Alternatively, handshake circuits could also be provided with an additional acknowledge input indicating that the data has been read by another component. However, handshake circuits would then require an additional state, where the output has been computed, but not been read so far. As this 'acknowledge'-problem only occurs in sequences of handshake circuits, we decided to construct sequences of handshake circuits in such a manner, that this problem is solved by the use of the sequence controller (cf. subsection 2.1).

The specification of handshake circuits requires reasoning about temporal relations of the inputs and outputs of the handshake circuit. These temporal relations can be described by higher order temporal operators [ScKK94a] which can be obtained from common propositional temporal operators [Kroe87]. For example, the WHEN operator and the UNTIL operator are defined as follows:

$$[x \text{ WHEN } b]^{(t_0)} := \Big(\forall \delta . (\forall t. t < \delta \rightarrow \neg b^{(t+t_0)}) \wedge b^{(\delta + t_0)} \rightarrow x(\delta + t_0)\Big)$$

$$[x \text{ UNTIL } b]^{(t_0)} = \left(\begin{array}{l} \big((\forall t. \neg b^{(t+t_0)}) \rightarrow (\forall t. x^{(t+t_0)})\big) \wedge \\ \big(\forall \delta . (\forall t. t < \delta \rightarrow \neg b^{(t+t_0)}) \wedge b^{(\delta + t_0)} \\ \qquad\qquad \rightarrow (\forall t. t < \delta \rightarrow x^{(t+t_0)})\big) \end{array} \right)$$

According to the above definitions, $[x \text{ WHEN } b]^{(t_0)}$ holds, iff x holds when b becomes true for the first time after t_0. $[x \text{ UNTIL } b]^{(t_0)}$ holds, iff x holds until b holds, or b never becomes true. It has to be noted, that the above definitions specify so-called weak operators, i.e. the operators do not require that the event will actually happen. Alternatively, we could also define strong operators, but as weak and strong operators can be defined by each other, it makes no difference, what operators are used.

The specification of handshake circuits using these temporal operators is as follows:

Definition 2.1 (Specification of Handshake-Circuits)
The specification of a handshake circuit with request input req, ready output f, data input A and data output B, that performs the task Φ is as follows:

$$\text{PROTOCOL}\,(req, A, B, f, \Phi) := f^{(0)} \wedge$$
$$\left[\forall t_0. \begin{array}{l} \left(f^{(t_0)} \rightarrow \left[\left(\lambda t.\, (B^{(t)} = B^{(t_0)}) \wedge (f^{(t)} = f^{(t_0)}) \right) \text{ UNTIL } req \right]^{(t_0)} \right) \wedge \\ \left(f^{(t_0)} \rightarrow \left[\left(\lambda t.\, (B^{(t)} = B^{(t_0)}) \wedge (f^{(t)} = f^{(t_0)}) \right) \text{ WHEN } req \right]^{(t_0)} \right) \wedge \\ \left(f^{(t_0)} \rightarrow \left[\left(\lambda t_1.[(\lambda t_2.\Phi(A, B, t_1, t_2)) \text{ WHEN } f]^{(t_1+1)} \right) \text{ WHEN } req \right]^{(t_0)} \right) \end{array} \right]$$

The specification describes that

1. the circuit is ready at time 0,
2. the outputs are stored until the next task is requested,
3. the outputs are stored at the time of the new request,
4. and a new task is performed correctly when f becomes high for the first time after the new request.

We describe the task Φ by a predicate whose arguments are the input stream, the output stream, the point of time where the new request occurred and the point of time where the result is available. For example, the following formula is the specification of a handshake circuit, whose task is to apply the operator $\otimes$ to the inputs A and B which are read at time t_0. It is furthermore required that the result is available exactly after $n + 1$ cycles, where n is the bitwidth of the inputs.

$$\text{PROTOCOL}\left(req, \lambda t.(A^{(t)}, B^{(t)}), C, f, \lambda t_0\, t_1. \left(\begin{array}{l} (C^{(t_1)} = A^{(t_0)} \otimes B^{(t_0)}) \\ \wedge\, (t_1 - t_0 = n + 1) \end{array} \right) \right)$$

In the rest of this section, we describe how handshake circuits can be implemented by standard controllers. We will also state abstract theorems about these standard controllers in order to verify the resulting handshake circuit. Correctness theorems of handshake circuits are of the following form (Circuit is the circuit description):

$$\text{Circuit}(req, A, B, f) \rightarrow \text{PROTOCOL}\,(req, A, B, f, \Phi)$$

The theorems about the standard controllers can be used by a tactic similar to MATCH_MP_TAC to reduce the goal to the correctness of the used components. As the theorems about the standard controllers are independent of the algorithms performed by used components, these theorems are correctness theorems at the system level.

2.1 Composition of Handshake Circuits

As already mentioned, handshake processes are closed with respect to sequentialization and synchronisation. Given two handshake circuits, we implement a sequence of two handshake circuits by a standard controller, which controls the request and finish outputs of the used handshake circuits. Similarly we implement the synchronization of two handshake circuits by a another standard contoler.

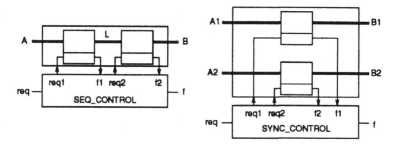

Fig. 3. The standard controllers for sequences and synchronisations.

The SEQUENCE_CONTROL controller is used for the implementation of the sequentialization operation of two handshake-processes. This controller has three internal states WW, CW, WC as shown in figure 4. Initially, the circuit is in state WW where both handshake circuits are in the wait-state. If a request occurs in this state, the state CW is reached, where the first handshake circuit is in its computation state and the second one is in the wait-state. If the computation of the first handshake circuit terminates, then the state WC is reached where the first circuit is in the wait-state and the second one performs its computation. If the computation of the second circuit also terminates, then the entire circuit reaches again the state WW.

The sequentialization controller is defined by the following equations, where s_1 and s_2 are used to encode the states. req_1 and req_2 are the request inputs of the used handshake circuits and f is the finish output of the resulting handshake circuit.

$$- s_1^{(t+1)} := req_1^{(t)} \vee \left(s_1^{(t)} \wedge \neg f_1^{(t)} \right)$$

$$- s_2^{(t+1)} := req_2^{(t)} \vee \left(s_2^{(t)} \wedge \neg f_2^{(t)} \right)$$

$$- f^{(t)} := \left(\neg s_1^{(t)} \wedge \neg s_2^{(t)} \right) \vee \left(s_2^{(t)} \wedge f_2^{(t)} \right)$$

$$- req_1^{(t)} := f^{(t)} \wedge req^{(t)}$$

$$- req_2^{(t)} := s_1^{(t)} \wedge f_1^{(t)}$$

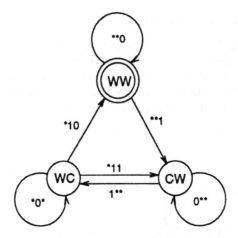

Fig. 4. Transition diagram of the sequence controller.

The correctness of a sequence of handshake circuit can be easily reduced to the correctness of the used handshake components by the theorem below.

$$
\begin{aligned}
&\forall \text{Circuit } \Phi_1\ \Phi_2\ \text{Circuit}_1\ \text{Circuit}_1\ req\ A\ L\ B\ f. \\
&\left(\begin{array}{l}
\text{Circuit}(req, A, B, f) = \\
\quad \exists req_1\ f_1\ req_2\ f_2\ L. \\
\qquad \text{Circuit}_1(req_1, A, L, f_1)\wedge \\
\qquad \text{Circuit}_2(req_2, L, B, f_2)\wedge \\
\qquad \text{SEQUENCE\_CONTROL}(req, req_1, f_1, req_2, f_2, f)
\end{array}\right) \wedge \\
&(\forall req_1\ A\ L\ f_1.\text{Circuit}_1(req_1, A, L, f_1) \rightarrow \text{PROTOCOL}\,(req_1, A, L, f_1, \Phi_1)) \wedge \\
&(\forall req_2\ L\ B\ f_2.\text{Circuit}_2(req_2, L, B, f_2) \rightarrow \text{PROTOCOL}\,(req_2, L, B, f_2, \Phi_2)) \\
&\qquad \left(\begin{array}{l}
\text{Circuit}(req, A, B, f) \rightarrow \\
\rightarrow \quad \text{PROTOCOL}\left(req, A, B, f, \lambda A\ B\ t_1\ t_2.\left(\begin{array}{l}\exists L\ \delta. \\ \Phi_1(A, L, t_1, \delta)\wedge \\ \Phi_2(L, B, \delta, t_2)\end{array}\right)\right)
\end{array}\right)
\end{aligned}
$$

The above theorem states that if given a circuit Circuit is implemented by the circuits SEQUENCE_CONTROL, Circuit$_1$ and Circuit$_2$, and Circuit$_1$ performs the task Φ_1, and Circuit$_2$ performs the task Φ_2, then the entire circuit is correct. The task of the composed handshake circuit is given as the conjunction of the tasks of the used components. δ is the point of time, where the first component finishes its computation and where the result L is given to the second component. We have given the specification in a relational manner, hiding the intermediate result by existential quantification. As a special case, that occurs very frequently, the tasks are given in a functional manner, i.e. $\Phi_1(A, L, t_1, \delta)$ is of the form $(L^{(\delta)} = \varphi_1(A^{(t_1)}))$ and $\Phi_2(L, B, \delta, t_2)$ is of the form $(B^{(t_2)} = \varphi_2(L^{(\delta)}))$. In this special case, the entire task is equivalent to $B^{(t_2)} = \varphi_2(\varphi_1(A^{(t_1)}))$, i.e. the existential quantification can be eliminated.

The synchronization of two given handshake circuits is implemented by the SYNC_CONTROL controller. This controller has no internal states and consists simply of two AND gates:

- $f^{(t)} := f_1^{(t)} \wedge f_2^{(t)}$
- $req_1^{(t)} := req^{(t)} \wedge f^{(t)}$
- $req_2^{(t)} := req^{(t)} \wedge f^{(t)}$

The entire circuit is available for new computations iff all used handshake components are available. A global request is passed to the used handshake components only in this case. The correctness of a handshake circuit which results of a synchronisation of other handshake circuits can be derived from the correctness of the used components due to the following theorem:

$$
\begin{aligned}
&\forall \text{Circuit } \text{Circuit}_1 \text{ Circuit}_2 \ \Phi_1 \ \Phi_2 \ req \ A_1 \ B_1 \ A_2 \ B_2 \ f. \\
&\left(\begin{array}{l}
\text{Circuit}(req, A_1, B_1, A_2, B_2, f) = \\
\quad \exists req_1 \ f_1 \ req_2 \ f_2. \\
\quad\quad \text{Circuit}_1(req_1, A_1, B_1, f_1) \wedge \\
\quad\quad \text{Circuit}_2(req_2, A_2, B_2, f_2) \wedge \\
\quad\quad \text{SYNC\_CONTROL}(req, req_1, f_1, req_2, f_2, f)
\end{array}\right) \wedge \\
&\left(\begin{array}{l}
\forall req_1 \ A_1 \ B_1 \ f_1. \\
\quad \text{Circuit}_1(req_1, A_1, B_1, f_1) \rightarrow \text{PROTOCOL}(req_1, A_1, B_1, f_1, \Phi_1)
\end{array}\right) \wedge \\
&\left(\begin{array}{l}
\forall req_2 \ A_2 \ B_2 \ f_2. \\
\quad \text{Circuit}_2(req_2, A_2, B_2, f_2) \rightarrow \text{PROTOCOL}(req_2, A_2, B_2, f_1, \Phi_2)
\end{array}\right) \wedge \\
&(\forall A \ B \ t_0 \ t.[B^{(t+1)} = B^{(t)}] \rightarrow [\Phi_1(A, B, t_0, t+1) = \Phi_1(A, B, t_0, t)]) \wedge \\
&(\forall A \ B \ t_0 \ t.[B^{(t+1)} = B^{(t)}] \rightarrow [\Phi_2(A, B, t_0, t+1) = \Phi_2(A, B, t_0, t)]) \\
&\rightarrow \left(\begin{array}{l}
\text{Circuit}(req, A_1, B_1, A_2, B_2, f) \rightarrow \\
\quad \text{PROTOCOL}\left(req, \lambda t.(A_1^{(t)}, A_2^{(t)}), \lambda t.(B_1^{(t)}, B_2^{(t)}), f,\right. \\
\quad\quad \left. \lambda A \ B \ t_1 \ t_2. \left(\begin{array}{l} \Phi_1\left(\lambda t.\text{FST}(A^{(t)}), \lambda t.\text{FST}(B^{(t)})), t_1, t_2\right) \wedge \\ \Phi_2\left(\lambda t.\text{SND}(A^{(t)}), \lambda t.\text{SND}(B^{(t)})), t_1, t_2\right) \end{array}\right)\right)
\end{array}\right)
\end{aligned}
$$

Similar to the correctness theorem of the sequentialisation, we use relational descriptions of the tasks. In the special case of functional descriptions, the above theorem can be simplified, as the assumptions 4 and 5 become trivial in this case.

2.2 Basic Handshake Circuits

Basic handshake circuits can be considered as hardware implementations of loops. We distinguish between iterations of other handshake circuits and iterations without any handshake circuit. The ONE_LOOP_CONTROL controller (figure 5) is used for the creation of a basic handshake circuit of a given operation unit OP_UNIT without other handshake circuits. We assume, that the operation unit has the schematic form as given on the right hand side of figure

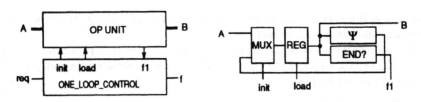

Fig. 5. The basic standard controller for iterations without handshake circuits.

5. At the initialisation of the loop, the controller sets the *init* and the *load* signal and therefore the inputs are read into registers. After that, a combinatorical circuit checks, if iterations are required at all. If this is the case, the output of another combinatorical circuit Ψ is loaded into the register until the termination condition is fulfilled. The termination of the iteration is signaled by f_1 to the controller. After that, the controller sets $load = \text{F}$ such that the circuit is storing the result. The **ONE_LOOP_CONTROL** controller is defined by the following equations, where the wait-state and the computation state of the circuit are encoded by the state variable q:

- $f^{(t)} := \neg q^{(t)} \vee f_1^{(t)}$,
- $q^{(t+1)} := req^{(t)} \vee \neg f^{(t)}$,
- $init^{(t)} := req^{(t)} \wedge f^{(t)}$,
- $load^{(t)} := q^{(t+1)} = req^{(t)} \vee \neg f^{(t)}$.

As already mentioned, the verification of basic handshake circuits cannot be completely automated. However, the following theorem allows to reduce the correctness goal to simpler subgoals:

$$
\forall \Phi \text{ OP\_UNIT Circuit } req\,A\,B\,f.
$$
$$
\left(\begin{array}{l} \text{Circuit}(req, A, B, f) = \\ \quad \exists init\ load\ f_1. \\ \quad\quad \text{OP\_UNIT}(A, init, load, f_1, B)\wedge \\ \quad\quad \text{ONE\_LOOP\_CONTROL}(req, init, load, f_1, f) \end{array}\right) \wedge
$$
$$
\left(\begin{array}{l} \text{Circuit}(req, A, B, f) \rightarrow \\ \quad \forall t_0.f^{(t_0)} \wedge req^{(t_0)} \rightarrow [\lambda t_1.\Phi(A, B, t_0, t_1) \text{ WHEN } f]^{(t_0+1)} \end{array}\right) \wedge
$$
$$
\left(\begin{array}{l} \forall load\ init\ f_1. \\ \quad \text{OP\_UNIT}(A, init, load, f_1, B) \rightarrow \\ \quad\quad\quad \left[\forall t.\neg init^{(t)} \wedge \neg load^{(t)} \rightarrow \left(B^{(t+1)} = B^{(t)}\right)\right] \end{array}\right)
$$
$$
\rightarrow (\text{Circuit}(req, A, B, f) \rightarrow \text{PROTOCOL}(req, A, B, f, \Phi))
$$

The above theorem reduce the correctness goal of a basic handshake circuit to three subgoals corresponding to the assumptions of the theorem. The first subgoal can be easily proven, as it only requires that the corresponding standard controller is used. The third subgoal can also easily proven as it only requires to prove that the registers of the operation store their data if *load* and *init* are false. The second subgoal asserts that if the circuit is ready $(f^{(t_0)})$ at time t_0 and a new task is requested $(req^{(t_0)})$, then the task Φ is performed correctly. In order to prove these subgoals, we use an invariant tactic [ScKK94a] similar to Floyd-Hoare calculi for software verification [Hoar69]. This tactic is based on the following characterization of the WHEN operator:

$$
[x \text{ WHEN } b]^{(t_0)} = \begin{pmatrix} \exists J. \\ J^{(t_0)} \wedge \\ (\forall t. \neg b^{(t+t_0)} \wedge J^{(t+t_0)} \rightarrow J^{(t+t_0+1)}) \wedge \\ (\forall d. b^{(d+t_0)} \wedge J^{(d+t_0)} \rightarrow x^{(d+t_0)}) \end{pmatrix}
$$

The invariant tactic reduces a goal of the form $\Gamma \vdash [x \text{ WHEN } b]^{(t_0)}$ to three subgoals, where the first one corresponds to the initialization of the loop, the second one corresponds to the iteration phase and the third one corresponds to the termination of the loop. In general, further lemmata about used data types are required for the proof, such that further manual interaction is required. In many cases, however, these subgoals are first order, i.e. they can be proven automatically by first order provers such as FAUST [ScKK92a, ScKK93a].

Finally, the NESTED_LOOP_CONTROL controller is used for iterations using handshake circuits. As a basic handshake circuit can be viewed as a hardware-implementation of a loop, this controller can be viewed as the implementation of nested loops. Similar to the ONE_LOOP_CONTROL, this controller has signals for the initialization of both the inner and the outer loops (*init*), a signal for demanding further iterations (req_1), and two inputs indicating the termination of the inner (f_1) and outer loops (f_2). The controller is defined by the following equations similar to the above ones:

- $f^{(t)} := \neg q^{(t)} \vee \left(f_1^{(t)} \wedge f_2^{(t)} \right)$
- $q^{(t+1)} := req^{(t)} \vee \neg f^{(t)}$
- $init^{(t)} := req^{(t)} \wedge f^{(t)}$
- $req_1^{(t)} := init^{(t)} \vee \left(q^{(t)} \wedge f_1^{(t)} \wedge \neg f_2^{(t)} \right)$

Basic handshake circuits using the above controller can be verified by the help of the following theorem:

$$
\begin{aligned}
&\forall \Phi \text{ OP\_UNIT Circuit } req \; A \; B \; f. \\
&\begin{pmatrix} Circuit(req, A, B, f) = \\ \exists init \; req_1 \; f_1 \; f_2. \\ \quad OP\_UNIT(A, init, req_1, f_1, f_2, B) \wedge \\ \quad NESTED\_LOOP\_CONTROL(req, init, req_1, f_1, f_2, f) \end{pmatrix} \wedge \\
&\begin{pmatrix} Circuit(req, A, B, f) \rightarrow \\ \quad \forall t_0. f^{(t_0)} \wedge req^{(t_0)} \rightarrow [\lambda t_1. \Phi(A, B, t_0, t_1) \text{ WHEN } f]^{(t_0+1)} \end{pmatrix} \wedge \\
&\begin{pmatrix} \forall init \; req_1 \; f_1 \; f_2. \\ \quad OP\_UNIT(A, init, req_1, f_1, f_2, B) \rightarrow [\forall t. \neg req_1^{(t)} \rightarrow \left(B^{(t+1)} = B^{(t)} \right)] \end{pmatrix} \\
&\rightarrow (Circuit(req, A, B, f) \rightarrow PROTOCOL(req, A, B, f, \Phi))
\end{aligned}
$$

Similar to the correctness theorem of the ONE_LOOP_CONTROL controller, the above theorem allows to reduce correctness goals to three subgoals, where only the second one is not trivial to prove. This subgoal can also be proven by the invariant approach as outlined above.

3 Case Studies

In this section, we show some examples we have successfully verified by our approach. The first example, a russian multiplier circuit exemplifies the use of invariants for proving the correctness of basic handshake circuits. The second example, a circuit for computing the approximate square root for natural numbers uses nested iterations and the third example uses both circuits for the computation of the euclidian metric. We use the word library implemented by W. Wong [Wong93] for the specification of the circuits, the data abstraction function BN-VAL of this library is however omitted in the following. Instead, we indicate the application of a BNVAL operation by a special font ($\mathcal{ABC}$...), i.e.

$$\mathsf{BNVAL}\ \underbrace{\left([b_n{}^{(t)},\ldots,b_0{}^{(t)}]\right)}_{=:\ B^{(t)}} := \underbrace{\sum_{i=0}^{n}\mathsf{BV}\left(b_i{}^{(t)}\right) \times 2^i}_{=:\mathcal{B}^{(t)}}$$

3.1 Russian Multiplier

The implementation RUSS_MULT_IMP(n, req, A, B, OUT, f) of a russian multiplier circuit that computes the product of two natural numbers by successively halving the one and doubling the other one is given in figure 6. It has to be noted that the required time for one multiplication depends on the given data, but is never longer than the specified bitwidth n. The specification of the multiplier handshake circuit is as follows:

RUSS_MULT_SPEC(n, req, A, B, OUT, f) :=
  PROTOCOL $\left(req, \lambda t.(A^{(t)}, B^{(t)}), OUT, f,\right.$
  $$\left.\lambda t_0\ t_1.\left(\begin{pmatrix}\mathcal{OUT}^{(t_1)} = \mathcal{A}^{(t_0)} \cdot \mathcal{B}^{(t_0)}\end{pmatrix} \atop \wedge\,(t_1 - t_0 \le n + 1)\right)\right)$$

The application of the tactic for the basic standard controllers reduces the correctness goal to the following subgoal:

RUSS_MULT_IMP(n, req, A, B, OUT, f) $\vdash$
$$\forall t_0. req^{(t_0)} \wedge f^{(t_0)} \rightarrow \left[\lambda t.\left(\begin{pmatrix}\mathcal{OUT}^{(t)} = \mathcal{A}^{(t_0)} \cdot \mathcal{B}^{(t_0)}\end{pmatrix} \atop \wedge\,(t - t_0 \le n + 1)\right)\ \mathsf{WHEN}\ f\right]^{(t_0+1)}$$

The proof of the above subgoal can be established by the invariant tactic using the following invariant:

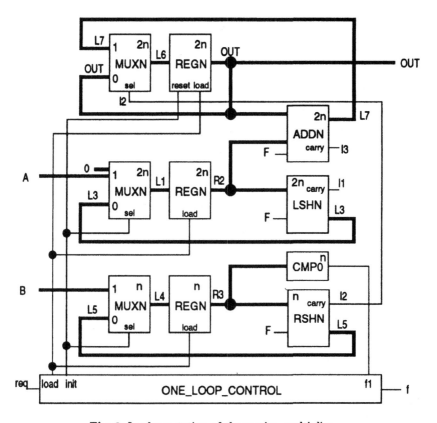

Fig. 6. Implementation of the russian multiplier.

$$\lambda t.\left(\mathbf{OUT}^{(t)} + \mathcal{R}_2^{(t)} \cdot \mathcal{R}_3^{(t)} = \mathcal{A}^{(t_0)} \cdot \mathcal{B}^{(t_0)}\right)$$
$$\wedge\left(\mathcal{R}_2^{(t)} = 2^{t-(t_0+1)} \cdot \mathcal{A}^{(t_0)}\right)$$
$$\wedge\left(\mathcal{R}_3^{(t)} = \left(\mathcal{B}^{(t_0)} \text{ DIV } 2^{t-(t_0+1)}\right)\right)$$
$$\wedge\left(t \leq n + t_0 + 1\right)$$

The invariant above states that the product of the registers R_2 and R_3, added to the value of register OUT is equal to the product of the input numbers during the computation. Moreover, R_2 is obtained by successively doubling the input $A^{(t_0)}$, and R_3 is obtained by successively halving the input $B^{(t_0)}$. The resulting subgoals are easy to prove, where we need the following lemmata which are the basis of the algorithm:

- $(y \text{ MOD } 2) = 0 \vdash x \cdot y = (2 \cdot x) \cdot ((y \text{ DIV } 2))$
- $(y \text{ MOD } 2) = 1 \vdash x \cdot y = x + (2 \cdot x) \cdot ((y \text{ DIV } 2))$

Similar to the russian multiplier, a handshake divider circuit can be constructed and verified. The circuit has the following specification:

$$
\begin{aligned}
&\mathsf{DIVIDER\_SPEC}(n, req, A, B, OUT, f) := \\
&\quad \mathsf{PROTOCOL}\left(req, \lambda t.(A^{(t)}, B^{(t)}), OUT, f, \right. \\
&\qquad\qquad \left. \lambda t_0\, t_1.(0 < \mathcal{B}^{(\delta)}) \rightarrow \left(\begin{matrix} \left[\mathcal{D}^{(t_1)} = \left(\mathcal{A}^{(t_0)}\ \mathsf{DIV}\ \mathcal{B}^{(t_0)}\right)\right] \wedge \\ \left[\mathcal{M}^{(t_1)} = \left(\mathcal{A}^{(t_0)}\ \mathsf{MOD}\ \mathcal{B}^{(t_0)}\right)\right] \wedge \\ (t_1 - t_0 = n + 1) \end{matrix}\right)\right)
\end{aligned}
$$

3.2 Computation of the Square Root

The square root for natural numbers is defined as follows using Hilbert's choice operator:

$$
\sqrt{x} := \varepsilon y.y \cdot y \le x \wedge x < (y+1) \cdot (y+1)
$$

If real numbers are considered, the square root of a number x is a solution of the equation $x^2 - a = 0$, i.e. we can use Newton's iteration to obtain the following iteration:[1]:

$$
x_{n+1} := \frac{x_n + \frac{a}{x_n}}{2}.
$$

One can prove that this iteration for natural numbers either converges or alternates between two values y and $y + 1$ after some iterations. Therefore, the computation can be stopped after a finite number of iterations, and the the approximate square root of the natural number is been obtained. A hardware implementation of the algorithm is given in figure 7. The specification of a handshake circuit performing the computation of the square root is as follows:

$$
\begin{aligned}
&\mathsf{HERON\_SPEC}(n, req, A, B, OUT, f) := \\
&\quad \mathsf{PROTOCOL}\left(req, A, B, f, \right. \\
&\qquad\qquad \left. \lambda t_0\, t_1.\left(0 < \mathcal{A}^{(t_0)}\right) \rightarrow \left(\mathcal{B}^{(t_1)} = \sqrt{\mathcal{A}^{(t_0)}}\right)\right)
\end{aligned}
$$

The circuit can be verified using the tactic for the basic standard controllers and the invariant tactic with the following invariant:

---

[1] This is the following iteration for computing solutions of $f(x) = 0$:

$$
x_{n+1} := x_n - \frac{f(x_n)}{f'(x_n)}.
$$

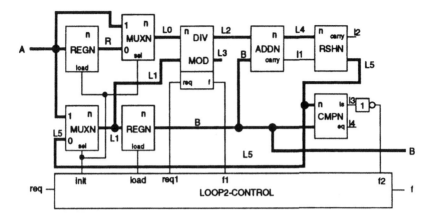

Fig. 7. Newton iteration for computing the square root.

$$
\lambda t.
$$
$$
q^{(t)} \wedge
$$
$$
\left[
\begin{pmatrix}
\lambda \delta. \\
q^{(\delta)} \wedge \\
\left(0 < \mathcal{B}^{(\delta)}\right) \wedge \left(\mathcal{B}^{(\delta)} < \mathcal{A}^{(t_0)}\right) \wedge \\
\left(0 < \mathcal{L}_2^{(\delta)}\right) \wedge \left(R^{(\delta)} = A^{(t_0)}\right) \wedge \\
\left(\mathcal{L}_5^{(d)} = \left[\mathcal{B}^{(t)} + \left(\mathcal{A}^{(t_0)} \ \text{DIV} \ \mathcal{B}^{(t)}\right)\right] \ \text{DIV} \ 2\right) \wedge \\
\left(\sqrt{\mathcal{A}^{(t_0)}} \le \mathcal{B}^{(\delta)}\right)
\end{pmatrix}
\quad \text{WHEN } f_1
\right]^{(t)}
$$

3.3 Euclidean Metric

Many applications require the computation of the term $\sqrt{a^2 + b^2}$, e.g. the distance of two points $(x_1, y_1), (x_2, y_2)$ is defined as $\sqrt{(x_1 - x_2)^2 + (y_1 - y_2)^2}$. A handshake circuit for the computation of $\sqrt{a^2 + b^2}$ is given in figure 8. First, the two multipliers are synchronized to form a handshake circuit for the computation of two products. The resulting circuit is used in a sequence with the adder for the computation of the scalar product $a_1 b_1 + a_2 b_2$. Finally, a handshake-sequence with a circuit to compute the square root of $a^2 + b^2$ is required. The specification of the circuit is as follows:

$$
\text{EUKLID\_METRIC\_SPEC}(n, req, A, B, C, f) :=
$$
$$
\text{PROTOCOL} \left(req, \lambda t.(A^{(t)}, B^{(t)}), C, f, \right.
$$
$$
\left. \lambda t_0 \ t_1. \left(0 < \mathcal{A}^{(t_0)} \wedge 0 < \mathcal{B}^{(t_0)}\right) \to \left(\mathcal{C}^{(t_1)} = \sqrt{\mathcal{A}^{(t_0)^2} + \mathcal{B}^{(t_0)^2}}\right)\right)
$$

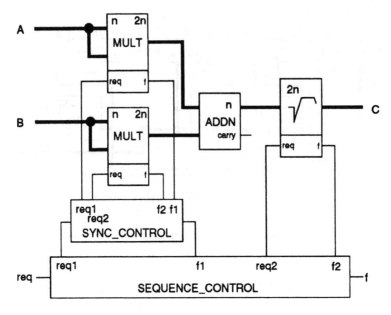

Fig. 8. Circuit for computing the euclidian metric.

The circuit can be verified automatically by the application of the tactic for the standard controllers for sequences and synchronizations.

4 Summary and Future Work

We have introduced handshake circuits for the implementation of arbitrary processes without shared memory. The designs using these handshake circuits can be synthesized automatically from given specifications similar to [Mart90a], and the verification of these circuits is highly automated. In most cases, the interaction is reduced to the use of a suitable invariant reflecting the design idea of the circuit. The designs using the standard controllers can be verified very elegantly and – to a large extent – automatically. However, they are in general inefficient, as a lot of control states are equivalent to each other. Therefore, an additional verification step has to be performed to obtain more efficient circuits. We therefore suggest the following strategy: We first synthesize and verify a handshake design using the standard controllers. After this, all controllers are merged together, i.e. they are replaced by an equivalent controller which corresponds to the minimal automaton. In the second verification step, the equivalence of the minimal controller and the interweaved standard controllers has to be proven. This step can also be verified automatically in HOL [ScKK93d]. Future work will also consider optimizations in the operation unit and a deep embedding of a programming language for handshake processes.

References

[AbLa90] M. Abadi and L. Lamport. Composing specifications. Technical Report 66, System Research Center, 1990.

[BuMa87] S.M. Burns and A.J. Martin. A synthesis method for self-timed VLSI circuits. In *Proceedings of the International Conference on Computer Design*, 1987.

[EDAC94] *European Design and Test Conference*. IEEE Computer Society Press, March 1994.

[Hoar69] C.A.R. Hoare. An axiomatic approach to computer programming. *Communications ACM*, 12:576–580, 1969.

[Hoar78] C.A.R. Hoare. Communicating sequential processes. *Communications ACM*, pages 666–677, 1978.

[HUG93] *HOL User's Group Workshop*, number 780 in Lecture Notes in Computer Sciences, Vancouver, Canada, August 1993. Springer Verlag.

[Kroe87] F. Kröger. *Temporal Logic of Programs*, volume 8 of *EATCS Monographs on Theoretical Computer Science*. Springer Verlag, 1987.

[Mart90a] A.J. Martin. Synthesis of asynchronous VLSI circuits. In J. Staunstrup, editor, *Formal Methods for VLSI Design*, 1990.

[Melh88a] T.F. Melham. Abstraction mechanisms for hardware verification. In G. Birtwistle and P.A. Subrahmanyam, editors, *VLSI Specification, Verification and Synthesis*. Kluwer, 1988.

[ScKK92a] K. Schneider, R. Kumar, and Th. Kropf. The FAUST prover. In D. Kapur, editor, *11th Conference on Automated Deduction*, number 607 in Lecture Notes in Computer Science, pages 766–770. Springer Verlag, Albany, New York,1992.

[ScKK93a] K. Schneider, R. Kumar, and Th. Kropf. Hardware verification with first-order BDD's. In *Conference on Computer Hardware Description Languages*, 1993.

[ScKK93d] K. Schneider, R. Kumar, and Th. Kropf. Alternative proof procedures for finite-state machines in a higher-order environment. In *International Workshop on Higher-Order Logic Theorem Proving and its Applications*, Vancouver, Canada, 1993.

[ScKK94a] K. Schneider, T. Kropf, and R. Kumar. Control-path oriented verification of sequential generic circuits with control and data path. In [EDAC94].

[Wong93] W. Wong. Modelling bit vectors in HOL: the word library. In [HUG93].

A Parameterized Proof Manager

Konrad Slind

Institut für Informatik
Technische Universität München
slind@informatik.tu-muenchen.de

Abstract. Proof management is an important issue for interactive theorem provers. We describe a simple proof manager that derives a large measure of its power from being parameterized by structures that separately manage 1) proof-specific information and 2) the relationships between proofs. Examples of its use include mixed forward and backward proof, automatic proof logging and replay, support for documenting a proof either before, during, or after the proof effort, and the management of proof dependencies.

1   Introduction

Proof management is an important issue in interactive theorem proving, since the user must have support for dealing with the intermediate steps of a proof. To date, most theorem proving environments have focused on the notion of doing a single proof at a time. Poor support has been provided for *proof in the large*[1]. By this we mean only that no theorem of any 'weight' is proved all at once, but rather as the conclusion of a sometimes long process of settling on definitions, building support theories, and proving lemmas. Once the focus shifts from single proofs to collections of proofs, the relationships among proofs become important. We propose a *proof manager* to handle the complexities of such a task. By allowing the manager to be parameterized by structures that separately manage 1) proof-specific information and 2) the relationships among proofs, we begin to support proof in the large.

Although we hope that the ideas in this paper are generally applicable to any interactive theorem prover, our development is in hol90, which is a version of Cambridge HOL [GM93] implemented in SML [Pau91]. The work presented here uses the parameterization facilities offered by the SML module system. In this paper, we do not make use of any window system operations; this is because we feel that proof management ideas must first work on their own in a non-window environment before they can be successful with the support of a mouse and a bit-mapped screen. We also differ from previous work [TYK92] in that we do not separate the proof manager from the theorem prover.

In the following, we will present many datatype definitions; indeed one of the guiding principles of this work has been to use datatype declarations to cleanly

---

[1] Logics with a notion of parameterized theory allow the re-use of collections of theorems, but that is an orthogonal issue to the initial production of such collections.

bring together different notions. In SML this approach is particularly powerful, since describing the control flow of algorithms is supported by pattern matching over datatype constructors.

By virtue of the way the primitive inference rules of HOL are encapsulated as constructors for the abstract type of theorems, explicit consideration of proofs is difficult. In an LCF-style system, a proof is performed during the execution of any program that has range type *thm*. Hence real proofs are invisible, ephemeral, and impossible to manipulate. We attack the problem at a level once removed and focus on supporting *methods* that will produce such *thm*-valued programs. To put it another way, an LCF-style proof management facility can be viewed as a program development system devoted to a single class of types: those with range *thm*.

2 Basic proof development methods

Our point of departure will be to take a simplification of the venerable *goalstack* method and a new method for *concrete forward proof* as our building blocks. We believe that the approach presented in this paper extends easily to other methods, *e.g.*, the window refinement of Grundy[Gru91].

Backward proof

A *goalstack*, which is used in some LCF-based systems [Pau87, GM93], is a datastructure and associated functions for organizing backwards proof attempts, *i.e.*, those conducted by means of tactics. The result of applying a tactic to a goal is a list of subgoals coupled with a validation function. The idea is that, when each subgoal has been proved (giving a list L of theorems), the validation is applied to L, giving the theorem that corresponds to the original goal. The goalstack datatype brings together the originally posed proposition with a stack (implemented with a list) of tactic results, which represents the remaining goals to be solved before the proposition is proved. The goalstack provides a means for constructing the proof in a depth-first manner.

```
type tactic_result = {subgoals : goal list,
                      validation : thm list -> thm}

datatype proposition = POSED of goal | PROVED of thm

datatype goalstack = GSTK of {prop: proposition,
                              stack : tactic_result list}
```

The primitive operations on goalstacks are: to create one by supplying a goal, to expand the current goal with a tactic, to switch attention to another goal, and to extract the theorem from a completed proof. The notion of goalstack was conceived and implemented by Larry Paulson for Cambridge LCF. The motivation was to eliminate the ML binding of intermediate states in tactic-based proof attempts. We have simplified Paulson's design by removing the maintenance of previous goalstacks from the core functionality of the goalstack. The maintenance of previous goalstacks is now provided in a much more general way, as will be discussed in section 4.

Concrete forward proof

As already mentioned, forward proofs in HOL are invisible, and impossible to manipulate. To remedy this situation, we have implemented a variant of the standard notion of proof[And86]:

> ... a *proof* of a wff A in Q_0 is a finite sequence of wffs, the last of which is **A**, such that every wff in the sequence is an axiom or is inferred by the rule of inference. [2]

This definition could simply be implemented by a list of theorems, but it is difficult to base further developments on such an impoverished representation. In particular, it is important to know, for each theorem, why it is a theorem. Hence we augment each theorem with a *justification*.

```
type justification = int frag list

datatype fwd = FWD of {why : justification,
                       thm : thm} list
```

The fundamental operation on `fwd` is to add a forward step (by simply giving a justification and letting the system compute the corresponding theorem). The type of justifications is best explained by example: a simple justification might be 'MP ^(4) ^(2)'. In SML/NJ, this is a *quotation*[3], represented as

```
[QUOTE "MP ", ANTIQUOTE 4, QUOTE" ", ANTIQUOTE 2]
```

In our implementation, this list is translatable into an ML evaluation of the forward rule **MP** to the fourth and second theorems of the current forward proof. The use of antiquotation to represent indices to previously proved theorems allows us to eliminate the use of ML bindings to hold intermediate steps of a forward proof attempt, in analogy to the goalstack.

3 Notes

Given basic methods of building proofs, the next step is to define useful views of the state of proof construction for the methods. The vehicle for this will be

---

[2] The system Q_0 has only one primitive rule of inference.

[3] A typed Lisp-style *backquote* mechanism in which values getting spliced into some syntax are tagged with **ANTIQUOTE**.

notes, introduced by Kalvala[KAL92]. For us, a note will be an essentially arbitrary SML structure encapsulating a datatype and some associated functions. Therefore it has very strong expressive ability. It is an important design decision that the notes and the methods of proof be independent: we will want to be able to change our view of a style of proof without having to actually change the implementation of that method. (The same sort of decision has been made in [Ber93].) The proof manager described next will glue together a given style of note with its method of proof, ensuring that the note coheres with the proof attempt it is attached to.

4   The proof manager

The proof manager brings together different methods of proof, here goalstack-based backward proof and concrete forward proof, and provides simple but extensible management facilities. Interaction with a proof attempt is mediated by a powerful abstract type called *history*, which provides a *polymorphic register* that keeps track (up to a bound) of what has been written to it. The following is the declaration of the **history** type, along with the ML types of some of its important operations.

```
abstype 'a history = HIST of {obj:'a, past:'a list, limit:int}

    new_history : {obj :'a, limit :int} -> 'a history
    apply    : ('a -> 'a) -> 'a history -> 'a history
    project : ('a -> 'b) -> 'a history -> 'b
    undo     : 'a history -> 'a history
```

An object of this type is 'backed-up' every time **apply** is used to perform an operation on it. Therefore, such operations can be undone with **undo**. The type also supports the extraction of information from the object, via **project**. The proof manager uses the **history** type to bring a proof method and its note together:

```
datatype proof_attempt =
        FWARD of {note:fwd_note, prf :fwd} history
      | BWARD of {note:bwd_note, prf :goalstack} history
```

Notice that the only way to operate on a value of type **proof_attempt** is to use the operations provided by the **history** type. Simply put, one can modify a {note,prf} record (via **apply**), one can undo the last modification (via **undo**), or one can extract information from the record (via **project**). Operations get applied to **prf** and **note** in parallel, so that they can take simultaneously take advantage of information in both.

The datastructure of the proof manager is a list of such proof attempts, since more than one attempt can be in progress at any time (and indeed some can be already completed). In fact, the proof manager is more powerful than this, since we equip it with its own note, useful for tracking relationships among proofs. We will describe the full datastructure after some examples.

5 Examples of proof-specific notes

It is clear that our approach to proof management is only as interesting as the notes that can be thought of. Each of the following applications is produced by modifying the notion of what a note is. Absolutely no modification is made to forwards or backwards proof methods.

Documentation

A proof note is some text. This can be given

- before the proof attempt is started. For example, an informal proof might be transcribed into the note and used as a guide.
- during the attempt. A user can add commentary as the proof is being developed; this allows 'on-the-spot' justifications for obscure choices made during the course of proof construction[4] and enables the later production of a natural language version of the proof.
- after the proof has been finished. This allows commentary on a proof development to be written after it has been completed, for example by a proof reader or a grader.

A simple datatype of this sort is

```
dataype documentation = DOC of {informal: string,
                                steps : string list,
                                postscript: string}
```

and the idea would be that the **informal** field would exist before the proof was attempted, the **steps** field would be updated throughout the attempt, and the **postscript** field would be written to after the proof was finished. More advanced versions of these would provide more sophisticated notions of text than mere strings. Also notice that each different field of the documentation could be portrayed differently, by use of standard prettyprinting techniques, i.e., by prettyprinting, for each field, to a different output stream.

Permanent backward proof

Here we provide a full backward proof tree, analogous to that used in the NUPRL system [CAB+86]. This tree is used as a note to a goalstack: we employ the goalstack as a means of creating and traversing the full backwards proof tree. In other words, a basic method of proof (the goalstack) is being used to generate a more detailed and permanent embodiment of the backward proof: the note is more interesting than the transitory process used to construct it.

---

[4] An anonymous referee suggested that this sort of commentary would be very helpful in making advanced proof techniques accessible to beginners. Good point!

```
datatype btree =
     LEAF of goal
   | NODE of {goal : goal,
              tactic : {func:tactic, syntax:string},
              subgoals : btree list}
```

At each node in the tree we store the goal, and if the node is expanded by a tactic application, we store the machine code of the tactic, as well as the concrete ML syntax that the tactic is compiled from (this is used to give visible representations of backwards proof attempts). If a btree has no LEAF nodes remaining, the proof is complete. Notice again that an *undo* will revert both the goalstack and the btree to their previous state, thus preserving consistency. We have implemented an extension of btree in which we provide for comments and occurrence numbers at each node in the tree. A depth-first traversal of such a tree has the ability to provide a convincing natural language commentary on the proof (see Appendix A).

Remark. We could have taken the type btree and its associated operations to be one of our basic methods of proof.

ML context and proof replay

First we need to clear up some terminology. The representation of a proof attempt, *i.e.*, a basic method plus associated note, needs to be rich enough to support extraction of a theorem from it, or a least a function with range type *thm*. This is replay. It also needs to be rich enough to give the user visual feedback on progress towards finishing the proof attempt. This is display. Display of a proof attempt is rendered possible in our implementation by intercepting the ML concrete syntax used in proof attempt interactions. The representation also needs to be rich enough to support *display-for-replay*, *i.e.*, the generation of ML concrete syntax that, when compiled and executed, will prove the intended theorem. This is useful when the proof will be executed in environments different than where it was generated.

In an attempt to provide a notion of a *closed* proof, *i.e.*, one that will run in any 'suitably rich' ML environment, we define a note that is a simple approximation of an ML environment, *i.e.*, a sequence of bindings of identifiers to values.

```
datatype context = CONTEXT of (string * term) list
```

Notice that, in the current incarnation of context in our proof manager, we can only track *term* bindings: a better implementation would use the user-accessible type of ML abstract syntax trees in the SML/NJ compiler [AM94] in order to track arbitrary bindings.

In our implementation, we have implemented display and display-for-replay of both forward and backward proof methods. From the context and the justifications in a concrete forward proof, we automatically build an ML program that, when run in enrichments of the ML environment current at the start of the forward proof, proves every theorem in the concrete forward proof. An example of this can be found in Appendix B. An example of the display of a backwards proof attempt can be found in Appendix A. Our implementation can also replay backward proofs: it can automatically build an ML function (a tactic) from a completed **btree** so that when the function is applied to the initial goal it proves that goal.

If the ML implementation provides user-configurable prettyprinting facilities, a prettyprinter can be installed so as to give automatic display: the user sees an incremental update to the syntax that constructs the 'attempt-so-far' with each step or undo that the user performs. One benefit of this is that the interaction script no longer has to be maintained separately (*emacs* buffers have often used for this purpose).

In a situation where the user switches between a collection of group attempts (as will be described later), when attention switches to a different attempt, whatever context it has needs to be renewed. This is very simple to implement in our scheme.

Big Brother

Sometimes it is useful to keep statistics on a proof attempt, *e.g.*, the time it takes to execute a proof in batch mode or the number of inference steps in a proof. Some others might be: the ratio between backups and forward motion in the attempt (a possible measure of user efficiency), and the total wall-clock time taken for the attempt to be completed. A simple datatype embodying these measurements would be

```
datatype statistics = NOTE of {steps    : int,
                               duration : time,
                               advances : int ref,
                               retreats : int ref,
                               init_time: time,
                               end_time : time}
```

In this, the **steps** field is updated by taking the difference in the value of a theorem counter after and before the operation is applied; the **duration** field is updated by taking the difference in the values of a timer; the **advances** field is incremented any time an operation is applied to the attempt; and the **retreats** field is incremented whenever an undo is done. Notice that **advances** and **retreats** need to be kept in reference cells, so that an *undo* operation will not lose the running total. (This approach is unsatisfactory when an 'upper-level' note is employed, as in section 6, since an upper-level undo will not be able to affect the contents of a reference cell. In general, notes should be totally functional.) Finally, the total real time spent in doing a proof development could be ascertained by use of the **init_time** and **end_time** fields.

Further statistics that could be maintained are: percentage of proof attempts that eventually get dropped without the target theorem being proved (another measure of efficiency) and the mean time between proof manager interactions. These would need to be maintained in the upper-level note introduced in section 6.

Other ideas

There are a couple of other ideas that seem interesting, but towards which we have put almost no thought:

- Automatic invocation of reasoners. Here the idea is that the note will hold an automatic reasoner, or a simplifier, and that after the application of a proof operation has finished, the function held in the note would be applied to the result(s). A similar usage would be to have the note hold a reasoner that would attempt to falsify the current goal (on the grounds that many goals are false, and being told when this is so will make the user more efficient).
- Simultaneous construction of proofs in other logics. If there is a translation between formulae and proofs in HOL and those of some other logic, the note could be used to hold the incremental translation as the HOL proof progresses. Again, the `history` type will ensure the coherence of the two proofs as they develop in parallel. Dually, a translation from an external logical system into HOL could be used to import proofs from other logical systems into HOL: the note would drive construction of the HOL proof.

6   Collections of proof attempts

Now we move up a level, to consider collections of proof attempts. To give coherent and backed-up interaction to a group of attempts, we make use of the polymorphic character of the `history` type.

```
datatype proof_developments =
        PRFS of {NOTE : proofs_note,
                 proofs :proof_attempt list} history
```

Along with the actual attempts, we provide a note that is intended to capture the relationships between attempts. In the following, we will use $\mathcal{D}$ to refer to an element of the **proof_developments** type. As at the lower level of individual proof attempts, the `history` type limits us to modifying $\mathcal{D}$, projecting from it, or undoing a previous operation. However, the ML programming language enables us to build quite powerful operations out of these simple primitives. Figure 1 gives a picture of $\mathcal{D}$.

In this picture, each proof attempt comprises an instance of a basic proof method (either **fwd** or **bwd**) plus an associated note, and a history. There is also a *NOTE* area for notes about the collection of attempts. Finally, $\mathcal{D}$ has its own history, which, if it were represented in the picture, would go off in a third dimension. The basic derived operations on $\mathcal{D}$ are to add a new attempt, to shift attention to a different attempt, to drop an attempt, to undo such operations, and to operate on the current attempt. These operations are defined by ML programming using the primitive `history` operations and are further extensible by the user.

N O T E		
Attempt 1	...	Attempt p
$\{proof, note\}$	···	$\{proof, note\}$
History	...	History

Fig. 1. An instance of the proof_developments type

Undo

In our design, there are two kinds of undo: that internal to an attempt, and that done to $\mathcal{D}$. A $\mathcal{D}$-level undo effectively "undoes" each proof attempt to its state at the time of the last operation on $\mathcal{D}$. Having two levels of undo handles some aspects of the difference between *historical* and *logical* undo, noticed by [TYK92]. In particular, a $\mathcal{D}$-level undo can 'undo an undo' performed *inside* a proof attempt. However, one might also end up undoing a lot of work, in unrelated proof attempts, that ought to be kept. It must be admitted that having two such levels can lead to confusion, especially when reference variables are used in proof attempts or notes. For these reasons, it would be useful to have a function that would tell what would happen if a $\mathcal{D}$-level undo were to be invoked.

7 Applications of the upper-level note

There are several uses of the upper level note: capturing relationships between proof attempts and providing a way to represent the development of an entire theory.

Lemmas, conjectures, and mixed-mode proof

In the course of developing a proof, often the need to prove a lemma comes up. One could simply add a new attempt to $\mathcal{D}$, perform the proof, bind the lemma to an ML name, drop the proof, and continue on the original attempt, using the ML identifier when the lemma is needed in subsequent developments. However, this situation is analogous to the goalstack, and should be handled in the same way: the ML binding is unnecessary and should be avoided. The new attempt should be added to $\mathcal{D}$, and (after it is proved) its theorem should be automatically folded into the original attempt. (Technically, the well-known Cut rule of inference is used to achieve this.)

Conjectures. Often one wants to sketch a proof, assuming that some as yet unproved propositions are true. One can easily do this in forward proof, by means of ASSUME, and within the goalstack, by means of, for example, Elsa Gunter's SUPPOSE_TAC. However, this becomes unwieldy when a conjecture figures in more attempts than the one currently being sketched. In such a case (assuming a backward proof attempt), we just add the conjecture to the assumptions of the current attempt, and add a separate attempt intended at proving the conjecture. Then the current proof can continue, although it will not be completeable until the conjecture is proved. Once that is the case, ACCEPT_TAC will eliminate the extra assumption from the original attempt. Notice that although mutually dependent conjectures can be set up, no logical inconsistency can result.

Mixed mode proof. With two different proof methods at our disposal, we can use a particular style when it is most appropriate. For example, interactive proofs by contradiction don't work well backwards (since the goal becomes simply 'F'), and forward reasoning in the assumptions becomes important. In that case, we can 'spark' a new concrete forward proof, whose context comprises the assumptions of the current goal. This allows surgical precision in the use of assumptions. Once we have obtained the desired contradiction, perhaps by sparking and solving further attempts, we can return to the original goal, and solve it. No ML binding is necessary, since we use the NOTE to track and index the theorems that have been proved.

Just as the approaches in this section are analogous to that taken in the goalstack, so is the shortcoming: non-permanence. The dynamic dependencies that get created are just thrown away. The next section proposes a remedy for this transience.

Theory developments

Now we are going to regard $\mathcal{D}$ as the embodiment of the development of a theory. Our desire is to extend the notions of attempt, display, replay, and proof from individual methods of proof to an entire theory. Towards this, the basic dependency relationship needs to be maintained. We use a directed acyclic forest in the NOTE to represent dependencies between proofs. As an example, when a user wishes to prove a lemma in the middle of a proof attempt, a new *parent* attempt must be spawned. This will add the new attempt to $\mathcal{D}$, and add the dependency into the graph. When the lemma is proved, the original attempt is reverted to, the proof of the lemma is kept, and the dependency is maintained in the note.

Given displayable (or replayable) versions of the basic methods of proof, and supposing that all dependencies between attempts in $\mathcal{D}$ have been captured, the entire theory development can be displayed (or replayed) by walking the graph in a depth-first manner, and displaying (or replaying) the attempt at each node (being careful, of course, not to retraverse sections of the graph already visited).

Since it is possible to hold an entire interactive theory development, with all its dependencies, in a single datastructure, *and* since we have the ability to perform sketches with respect to conjectures, it seems attractive to investigate a more 'goal-directed' style of theory development in which the proofs of major theorems are embarked on *first*, and the lemmas that arise are tracked, to be handled when the sketches of the important theorems are complete. When the last lemma has been proved, the theory development may be revised until the user is satisfied, and then written. This approach is in stark contrast to the way theories are currently developed, which is completely analogous to forward proof.

Question. What is the disk representation for such theory developments (or indeed, for proof attempts)? It is not entirely satisfactory to just employ display-for-replay, since the full information accessible in $\mathcal{D}$ may be much richer than is displayed. It seems that no one representation will be satisfactory, since notes are used expressly to customize proofs. One possible solution is to keep track of all ML interactions used in theory development, and write them out to a file. Then just read the file in and execute the interactions in sequence to rebuild $\mathcal{D}$. An alternative is to use whatever ML features are available to export and import the binary image of the proof manager state to and from disk. PolyML already has this facility, and the next release of SML/NJ will also have this facility.

A further area of investigation is in efficient use of memory. In a theorem prover with an internal manager, it is imaginable that the state of the manager can grow to such a size that proof performance drops significantly. In our case, this can be handled by modifying the implementation of the **history** type so that previous states are transparently written to and read from disk when necessary. However, virtual memory already supplies this functionality.

Extensions and Speculation

At this point, we are at the limit of the proof manager implementation; however, there are several areas of further investigation that come to mind: capturing more of the ML context in which an attempt is performed, abstract theories, and cooperative theorem proving.

- So far we have concentrated on proof attempts. However, the capture of logical definitions is also essential. This can be easily provided by 'wrapper' functions around the standard definition mechanisms. Since an important aspect of formal proof lies in establishing the right definitions in the first place, the HOL implementation must be able to undo definitions. Currently, the implementations of hol90 and hol88 do not allow this (however, **ProofPower** [Art91] does).
- **Abstract theories.** Given that one has the means of representing a theory in such a way that it can be rebuilt, it becomes interesting to consider whether such a representation can be abstracted into an ML functor $\mathcal{F}$ that, when applied to a suitable structure $\mathcal{S}$, would create an entire theory, instantiated by the elements of $\mathcal{S}$. The idea is that $\mathcal{F}$ would represent an abstract theory; $\mathcal{S}$ would be checked to see that it satisfied the type structure and axioms of the abstract theory. Then the concrete types and constants would be substituted throughout the representation of the 'abstract' theory, and the 'concrete' theory would be built, by replaying all the proofs. This is a way of using the parameterization facilities of the meta-language to provide a form of parameterization in the object language. The principal question to be answered is how robust such abstract theories are: every valid instantiation must run without failure and that might be difficult to ensure. An elegant semantically-justified alternative is described in [Far94].

– **Cooperative theorem proving.** New implementations of SML [TLP+93] are becoming available that support distributed processes. It is intriguing to contemplate how distribution can help coordinate team theorem proving. The ideas in the proof manager may be of use in a distributed environment, wherein $\mathcal{D}$ is viewable as a shared datastructure, providing concurrent access and update. One idea is that people could 'post' conjectures to $\mathcal{D}$, and one could claim such a conjecture to work on locally. Once the theorem was proved, the proof could be paired with the conjecture on $\mathcal{D}$, thus changing it from a conjecture to a fact that the whole team could use. Of course, mutual exclusion when posting such proofs would need to be enforced. Another prospect is to view different theories as different *forums*, inter-linked in some manner. Users would be able to browse and contribute conjectures and proofs to forums as they wish. Clearly, the investigation of cooperative theorem proving is an exciting and unexplored area.

8 Other work

Our work extends and amplifies on that of Larry Paulson [Pau87], Sara Kalvala [KAL92], Roger Fleming, and Fink *et al.* [FAY91]. Other proof systems with significant proof management facilities related to this work are Eves, PVS [ORSvH93], KIV [HRS90], and muRAL.

Part of our work can be seen to be like that of Grundy's work on window inference [Gru91], except that window inference works at a 'micro' level, operating on expressions in the logic, while we have the ability to operate on collections of attempts. In spite of that difference, the techniques have a great deal of similarity.

Lamport [Lam93] advocates a particular method of writing down proofs. We believe that his approach can be supported as a derived notion in our proof manager, since it uses much that is basic in the current work, but only further work can substantiate this.

The main work to date in the area of support for theorem proving is by the Sophia-Antipolis group[TYK92]. In their approach, the theorem prover is separate from the proof manager, and so a great deal of effort is required to transmit information between the two systems. In contrast, our approach stays within the theorem prover, since it is built on top of it, and relies on the usual parsing and prettyprinting techniques of the field. The code is very compact and all in the same programming language that the theorem prover is written in. More importantly, our proof manager has immediate access to the complete environment in the proof engine, instead of trying to duplicate the environment in another process. The system provided by the Sophia-Antipolis group is more general than ours, being applicable to a wide range of theorem provers. It would be interesting to separate out aspects of our approach that are inherently internal, so that external aspects could be handled by their software.

9 Conclusion

In this paper we have combined two basic methods of proof construction into a flexible notion of proof attempt, and used that to provide a coherent way to interact with a heterogeneous collection of fully and partially constructed proofs. Along the way we have provided many facilities for supporting proof and theory construction, including interactive documentation, proof capture and replay, and the maintenance of dynamically-arising proof dependencies. Our major design decisions are the following:

- proof methods should be implemented without regard to management of previous states or of supplementary information;
- state management is best handled at a higher level, where it can be given a uniform interface;
- notes are used to represent supplementary information, hence notes can be quickly changed without requiring change to proof method implementations; and
- proof states should not be bound in the ML top-level.

Proof management systems should span the gap between the austere formal systems that we base our work on and actual mathematical and engineering practice. Their continued development will be key in helping verification systems merge smoothly into a widening range of application areas.

Acknowledgements

I have benefitted from conversations and feedback from Tom Melham, Larry Paulson, Paul Curzon, Sara Kalvala and Laurent Thery. Tom Melham made the point, with great force, that LCF-style proof depends on programming.

References

[AM94]   Andrew Appel and David MacQueen. Separate compilation for standard ML. Technical Report CS-TR-452-94, Princeton University, Princeton, New Jersey, 1994. To appear in PLDI'94.

[And86] Peter Andrews. *An Introduction to Mathematical Logic and Type Theory: To Truth Through Proof*. Academic Press, 1986.

[Art91] R. D. Arthan. A report on icl hol. In Phillip Windley, Myla Archer, Karl Levitt, and Jeffrey Joyce, editors, *International Tutorial and Workshop on the HOL theorem proving system and it Applications*, University of California at Davis, August 1991. ACM-SIGDA / IEEE Computer Society, IEEE Computer Society Press.

[Ber93] Yves Bertot. A canonical calculus of residuals. In G. Huet and G. Plotkin, editors, *Logical Environments*, pages 146–163. Cambridge Univeristy Press, 1993.

[CAB⁺86] Robert Constable, S. Allen, H. Bromly, W. Cleaveland, J. Cremer, R. Harper, D. Howe, T. Knoblock, N. Mendler, P. Panangaden, J. Sasaki, and S. Smith. *Implementing Mathematics With the Nuprl Proof Development System*. Prentice-Hall, New Jersey, 1986.

[Far94] William Farmer. Theory interpretation in simple type theory. In *Proceeding of the International Workshop on Higher-Order Algebra, Logic, and Term Rewriting (HOA '93)*, Amsterdam, The Netherlands, 1994. Springer-Verlag (LNCS).

[FAY91] George Fink, Myla Archer, and Lie Yang. Pm: A proof manager for HOL and other provers. In Phillip Windley, Myla Archer, Karl Levitt, and Jeffrey Joyce, editors, *International Tutorial and Workshop on the HOL theorem proving system and it Applications*, University of California at Davis, August 1991. ACM-SIGDA / IEEE Computer Society, IEEE Computer Society Press.

[GM93] Mike Gordon and Tom Melham. *Introduction to HOL, a theorem proving environment for higher order logic*. Cambridge University Press, 1993.

[Gru91] Jim Grundy. Window inference in the HOL system. In Phillip Windley, Myla Archer, Karl Levitt, and Jeffrey Joyce, editors, *International Tutorial and Workshop on the HOL theorem proving system and it Applications*, University of California at Davis, August 1991. ACM-SIGDA / IEEE Computer Society, IEEE Computer Society Press.

[HRS90] M. Heisel, W. Reif, and W. Stephan. Tactical theorem proving in program verification. In Mark Stickel, editor, *Proceedings of the Tenth International Conference on Automated Deduction, LNAI 449*, pages 1–15, Kaiserslautern, 1990.

[KAL92] Saraswati Kalvala, Myla Archer, and Karl Levitt. Implementation and use of annotations in HOL. In L. Claesen and M. Gordon, editors, *International Workshop on Higher Order Logic Theorem Proving and its Applications*, Leuven, Belgium, September 1992. IFIP TC10/WG10.2, Elsevier Science Publishers.

[Lam93] Leslie Lamport. How to write a proof. Technical Report 94, DEC Systems Research Center, Palo Alto, California, February 1993.

[ORSvH93] S. Owre, J. Rushby, N. Shankar, and Friedrich von Henke. Formal verification for fault-tolerant architectures: some lessons learned. In J. Woodcock and P. Larsen, editors, *FME'93: Industrial Strength Formal Methods (LNCS 670)*, pages 482–500, Odense Denmark, 1993.

[Pau87] Lawrence Paulson. *Logic and Computation: Interactive Proof With Cambridge LCF*. Cambridge University Press, 1987.

[Pau91] Lawrence Paulson. *ML for the working programmer*. Cambridge University Press, 1991.

[TLP⁺93] Bent Thomsen, Lone Leth, Sanjiva Prasad, Tsung-Min Kuo, Andre Kramer, Fritz Knabe, and Alesasndro Giacalone. Facile Antigua release programming guide. Technical Report ECRC-93-20, European Computer-Industry Research Centre, Munich, Germany, December 1993.

[TYK92] L. Thery, Y.Bertot, and G. Kahn. Real theorem provers deserve real user-interfaces. In *Proceedings of the Fifth ACM SIGSOFT Symposium on Software Development Environments (Software Engineering Notes)*, volume 17, pages 365–380, Tyson's Corner, Virginia USA, 1992. ACM Press.

Appendix A. Display of a backwards proof

```
val it =
  Goal #1.
  (--'!m n p. m - n <= p = m <= n + p'--)

  ----------------------------
```

has the following tactic applied to it:

```
REPEAT GEN_TAC THEN ASM_CASES_TAC (--'n <= m'--)
```

COMMENT:
Typical hideous subtraction proof. Proceeds by cases on whether 'n <= m'.

There are now 2 subgoals:

```
[Goal #1.1.
  (--'m - n <= p = m <= n + p'--)

  ----------------------------
      (--'n <= m'--)
```

is solved by the following tactic:

```
IMP_RES_THEN (fn th => PURE_ONCE_REWRITE_TAC [th]) SUB_LESS_EQ_ADD THEN
REFL_TAC
```

COMMENT:
The first case needs the theorem 'SUB_LESS_EQ_ADD'
(|- !m p. m <= p ==> (!n. p - m <= n = p <= m + n)),

```
Goal #1.2.
  (--'m - n <= p = m <= n + p'--)

  ----------------------------
      (--'~(n <= m)'--)
```

has the following tactic applied to it:

```
ASSUM_LIST (MAP_EVERY (ASSUME_TAC o (PURE_REWRITE_RULE [SYM
(SPEC_ALL NOT_LESS),NOT_CLAUSES])))
```

COMMENT: Now munge the assumptions to get 'm < n'.

We are left with a single subgoal:

```
Goal #1.2.1.
  (--'m - n <= p = m <= n + p'--)

  ----------------------------
      (--'~(n <= m)'--)
      (--'m < n'--)
```

has the following tactic applied to it:

```
IMP_RES_TAC LESS_IMP_LESS_OR_EQ
```

COMMENT: Now use 'LESS_IMP_LESS_OR_EQ': |- !m n. m < n ==> m <= n

We are left with a single subgoal:

```
Goal #1.2.1.1.
(--'m - n <= p = m <= n + p'--)
----------------------------
    (--'~(n <= m)'--)
    (--'m < n'--)
    (--'m <= n'--)
```

has the following tactic applied to it:

```
IMP_RES_THEN (fn th => PURE_REWRITE_TAC [th,ZERO_LESS_EQ])
(snd (EQ_IMP_RULE (SPEC_ALL SUB_EQ_0)))
```

COMMENT:
Now use 'SUB_EQ_0' (|- !m n. (m - n = 0) = m <= n) in the right-to-left
direction, plus 'ZERO_LESS_EQ' (|- !n. 0 <= n) to reduce the lhs to T

We are left with a single subgoal:

```
Goal #1.2.1.1.1.
(--'T = m <= n + p'--)
----------------------------
    (--'~(n <= m)'--)
    (--'m < n'--)
    (--'m <= n'--)
```

is solved by the following tactic:

```
IMP_RES_THEN (fn th => REWRITE_TAC [th,LESS_OR_EQ]) LESS_IMP_LESS_ADD
```

COMMENT:
Rewrite with definition of 'LESS_OR_EQ', giving two cases, then solve
the "less-than" case with 'LESS_IMP_LESS_ADD'
(|- !n m. n < m ==> (!p. n < m + p))]
: bwd_note

Appendix B. Display and *display-for-replay*

```
Forwards proof:
Context:
   t1 = A
   t2 = B
   t3 = C
   t4 = A /\ B /\ C
   t5 = (A /\ B) /\ C
**********
 1. 'ASSUME t4'    . |- A /\ B /\ C
 2. 'ASSUME t5'    . |- (A /\ B) /\ C
 3. 'CONJUNCT1 ^(1)'    . |- A
 4. 'CONJUNCT1(CONJUNCT2 ^(1))'    . |- B
 5. 'CONJUNCT2(CONJUNCT2 ^(1))'    . |- C
 6. 'CONJ (CONJ ^(3) ^(4)) ^(5)'    . |- (A /\ B) /\ C
 7. 'DISCH t4 ^(6)'   |- A /\ B /\ C ==> (A /\ B) /\ C
 8. 'CONJUNCT2 ^(2)'    . |- C
 9. 'CONJUNCT1 ^(2)'    . |- A /\ B
10. 'CONJ (CONJUNCT1 ^(9)) (CONJ (CONJUNCT2 ^(9)) ^(8))'    . |- A /\ B /\ C
11. 'DISCH t5 ^(10)'   |- (A /\ B) /\ C ==> A /\ B /\ C
12. 'IMP_ANTISYM_RULE ^(7) ^(11)'   |- A /\ B /\ C = (A /\ B) /\ C
13. 'GEN t1 (GEN t2 (GEN t3 ^(12)))'  |- !A B C. A /\ B /\ C = (A /\ B) /\ C


- fwd_to_ML();
let val t1 = (--'(A :bool)'--)
    val t2 = (--'(B :bool)'--)
    val t3 = (--'(C :bool)'--)
    val t4 = (--'(A :bool) /\ (B :bool) /\ (C :bool)'--)
    val t5 = (--'((A :bool) /\ (B :bool)) /\ (C :bool)'--)
in
let val th1 = ASSUME t4
    val th2 = ASSUME t5
    val th3 = CONJUNCT1  th1
    val th4 = CONJUNCT1(CONJUNCT2  th1)
    val th5 = CONJUNCT2(CONJUNCT2  th1)
    val th6 = CONJ (CONJ  th3 th4) th5
    val th7 = DISCH t4  th6
    val th8 = CONJUNCT2  th2
    val th9 = CONJUNCT1  th2
    val th10 = CONJ (CONJUNCT1 th9) (CONJ (CONJUNCT2 th9) th8)
    val th11 = DISCH t5  th10
    val th12 = IMP_ANTISYM_RULE th7  th11
    val th13 = GEN t1 (GEN t2 (GEN t3  th12))
in
th13
end end
val it = () : unit
```

Implementational Issues for Verifying RISC-Pipeline Conflicts in HOL

Sofiène Tahar[1] and Ramayya Kumar[2]

[1] University of Karlsruhe, Institute of Computer Design and Fault Tolerance
(Prof. D. Schmid), P.O. Box 6980, 76128 Karlsruhe, Germany
e-mail: tahar@ira.uka.de

[2] Forschungszentrum Informatik, Department of Automation in Circuit
Design, Haid-und-Neu Straße 10-14, 76131 Karlsruhe, Germany
e-mail: kumar@fzi.de

Abstract. We outline a general methodology for the formal verification of instruction pipelines in RISC cores. The different kinds of conflicts, i. e. resource, data and control conflicts, that can occur due to the simultaneous execution of the instructions in the pipeline have been formally specified in HOL. Based on a hierarchical model for RISC processors, we have developed a constructive proof methodology, i.e. when conflicts at a specific abstraction level are detected, the conditions under which these occur are generated and explicitly output to the designer, thus easing their removal. All implemented specifications and tactics are kept general, so that the implementation could be used for a wide range of RISC cores. In this paper, the described formalization and proof strategies are illustrated via the DLX RISC processor.

1 Introduction

In this paper, we concentrate on the formalization and the correctness proofs of instruction pipelines in RISC cores. The previous work (from other researchers) in verifying RISC processors could verify only parts of processors, at certain levels of abstraction [2, 11]. Our work is distinguished by the fact that an overall methodology for the verification of RISC cores, at all levels of abstraction, is being developed [13, 14, 15].

There are three kinds of conflicts which occur in instruction pipelines, namely, resource, data and control conflicts. These conflicts occur due to the data and control dependencies and resource contentions when many instructions are simultaneously executed in the pipeline. The formalization of all possible conflicts and the description of the automated proof techniques for conflict detections, is given in this paper. The proof techniques that are given are constructive, i.e. the conditions under which the conflicts occur are explicitly stated, so that the designer can easily formulate the conflict resolution mechanisms either in hardware or generate software constraints which have to be met.

The organization of this paper is as follows. Section 2 briefly presents the hierarchical model on which the formal verification methodology is based. Section 3 gives an overview of the overall verification process. Section 4 includes some formal definitions of types and functions used in the formal specifications that follow. Sections 5 through 7 define the conflicts arising due to the pipelined nature of RISCs and describe the correctness proofs for resource, data and control conflicts, respectively. Section 8 contains some experimental results and section 9 concludes the paper. It is to be noted, that all the methods and techniques presented in this paper are illustrated by means of a RISC example — DLX[1] [6].

2 RISC Verification Model

Our RISC interpreter model has been adapted from the interpreter model of Anceau [1], which has been effectively used for the design, description and verification of microprogrammed processors [7, 17].

The adapted RISC interpreter model comprises the instruction, class, stage and phase levels, each of which corresponds to an interpreter at different abstraction levels, and the lowest level which corresponds to the circuit implementation — EBM (figure 1). Each interpreter consists of a set of visible states and a state transition function which defines the semantics of the interpreter at that level of abstraction.

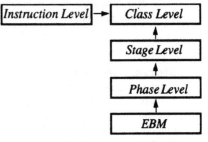

Fig. 1. RISC Interpreter Model

These interpreter levels incorporate structural and temporal abstractions, which are exploited in the proof process. Table 1 lists the set of transfers which occur at the stage and phase levels of the DLX processor, where the rows and columns represent the pipeline stages and the instruction classes, respectively[2].

Table 1. DLX Pipeline Structure

	ALU	**ALU_I**	**LOAD**	**STORE**	**CONTROL**
IF	IR ← M[PC] PC ← PC+4	IR ← M[PC] PC ← PC+4	IR ← M[PC] PC ← PC+4	IR ← M[PC] PC ← PC+4	IR← M[PC] PC ← PC+4
ID	A $\leftarrow_{\phi_2}$ RF[rs1] B $\leftarrow_{\phi_2}$ RF[rs2] IR1 $\leftarrow_{\phi_2}$ IR	A $\leftarrow_{\phi_2}$ RF[rs1] B $\leftarrow_{\phi_2}$ RF[rs2] IR1 $\leftarrow_{\phi_2}$ IR	A $\leftarrow_{\phi_2}$ RF[rs1] B $\leftarrow_{\phi_2}$ RF[rs2] IR1 $\leftarrow_{\phi_2}$ IR	A $\leftarrow_{\phi_2}$ RF[rs1] B $\leftarrow_{\phi_2}$ RF[rs2] IR1 $\leftarrow_{\phi_2}$ IR	BTA $\leftarrow_{\phi_1}$ f (PC) PC $\leftarrow_{\phi_2}$ BTA
EX	Aluout ← A op B	Aluout←Aop (IRI)	MAR ← A+(IR1)	MAR← A+(IR1) SMDR ← B	
MEM	Aluout ← Aluout1	Aluout ← Aluout1	LMDR←M[MAR]	M[MAR]←SMDR	
WB	RF[rd] $\leftarrow_{\phi_1}$ Aluout1	RF[rd] $\leftarrow_{\phi_1}$ Aluout1	RF[rd] $\leftarrow_{\phi_1}$ LMDR		

---

[1] DLX is an hypothetical RISC which includes the most common features of existing RISCs such as Intel i860, Motorola M88000, Sun SPARC or MIPS R3000.

[2] For more details about this model, please refer to [13, 14, 15].

3 Verification Tasks

Starting from the instruction set of a RISC processor, we have to show that this instruction set is executed correctly by the EBM, in spite of the pipelined architecture. A RISC processor executes n_s instructions in parallel (see figure 2), in n_s different pipeline stages. This parallel execution increases the overall throughput of the processor; however, it should be noted that no single instruction runs faster. Hence, we have to prove, on the one hand, that the sequential execution of each instruction is correctly implemented by the EBM and on the other hand, to show that the pipelined sequencing of the instructions is executed correctly. Thus the correctness proof is split into two independent steps as follows:

1. given some software constraints on the actual architecture and given the implementation EBM, we prove that any sequence of instructions is correctly pipelined, i.e.:

$$\text{SW\_Constraints, EBM} \vdash \text{Correct\_Instr\_Pipelining} \tag{1}$$

2. we prove that the EBM implements the semantic of each single architectural instruction correctly, i.e.:

$$\vdash \text{EBM} \Rightarrow \textit{Instruction Level} \tag{2}$$

The software constraints in (1) are those conditions which are to be met for designing the software so as to avoid conflicts, e.g. the number of delay slots to be introduced between the instructions while using a software scheduling technique. Additionally, it is also assumed that the EBM includes some conflict resolution mechanisms in hardware.

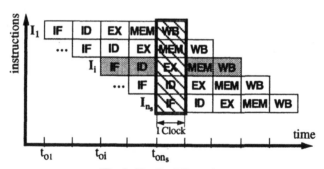

Fig. 2. Pipelined Execution

The proof of step (2) has been discussed and reported in our previous work [14, 15] and we recapitulate it briefly in the next subsection. Step (1) is the main topic of this paper and will be discussed in detail in the rest of the paper.

3.1 Semantical Correctness

The higher-order logic specifications and implementations at the various levels of abstraction are used and the following is proved — $EBM \Rightarrow Phase\ Level \Rightarrow Stage\ Level \Rightarrow Class\ Level$. This proof is similar to the one done by Windley in proving the correctness of microprogrammed processors [17]. Each obtained theorem is then instantiated for each instruction at the instruction level. We have implemented tactics in the HOL theorem prover [4], using the hardware verification environment —

MEPHISTO [9], which automates the entire process, given the formal definitions of the specifications and implementations at each abstraction level [15].

3.2 Pipeline Correctness

Step (1) can be proved by showing that all possible combinations of n_s instructions are executed correctly. This implies that the parallel execution of n_s instructions does not lead to any *conflicts*. There are three kinds of conflicts (also called *hazards*) that can occur during the pipelined execution of any RISC machine namely, resource, data and control conflicts[3]. The predicate Correct_Instr_Pipelining in (1), which implies the absence of these conflicts, is defined as the conjunction of predicates for each conflict kind. Formally:

$$\vdash_{def} \text{Correct\_Instr\_Pipelining} :=$$
$$\neg\text{Res\_Conflict} \wedge \neg\text{Data\_Conflict} \wedge \neg\text{Control\_Conflict}$$

and the correctness statement (1) can be rewritten as:

$$\text{SW\_Constraints, EBM} \vdash \neg\text{Res\_Conflict} \wedge \neg\text{Data\_Conflict} \wedge \neg\text{Control\_Conflict}$$

thus the whole correctness proof is tackled by splitting it into three independent parts, each corresponding to one kind of conflict. These parts are elaborated in the sections to follow.

In proving step (1), we have to ensure that all possible combinations of instructions occurring in n_s stages are executed correctly. This large number can be drastically reduced by exploiting the notion of classes which results in a combination of a smaller number. When conflicts are formalized at the class, stage and phase levels, the existence of multiple instructions in the pipeline can be formalized by predicates which we call the *multiple conflict predicates*. Their proofs are further simplified by constructing conflict predicates between pairs of instructions which are called the *dual conflict predicates*. These notions will be clarified in the sections to follow.

3.3 Different Approaches for Verification

The hierarchical model described above has been used in our previous work in hierarchically structuring the proof of the semantical correctness [15]. While formalizing this process, we discovered that the proofs can be hierarchically managed in a top-down or bottom-up manner, so that a *verification-driven design* or a *post-design verification* can be performed. By a top-down verification-driven design, we mean that the verification and design process are interleaved, so that the verification of a current design status, against the specification, yields the necessary constraints for the future design steps. A post-design verification is verification in the normal sense which is performed in a top-down or bottom-up manner after the entire design is completed at all levels. In a related work [16], we handle the verification of pipeline conflicts in a post-design manner, in that the verification of each kind of conflict is done more or less in a single step. Within the scope of this paper, the correctness proofs will be handled by means of the top-down verification-driven design methodology.

---

[3] A formal definition of these conflicts is given in later sections.

4 Formal Definitions

In our previous papers, we have given a detailed formalization of instructions at all abstraction levels [13, 15]. In this section, we briefly introduce some new types and functions that are useful for handling conflict predicates. According to our hierarchical model, we define for each abstraction level a set of enumeration types for corresponding instructions, resources and pipeline characteristics, i.e. pipeline stages or clock phases. Referring to the pipeline structure of table 1, where the columns represent the five instruction classes — ALU, ALU_I, LOAD, STORE and CONTROL and the rows correspond to the related stage and phase transfers at the five pipeline stages — IF, ID, EX, MEM and WB and the two clock phases ϕ_1 and ϕ_2, respectively, we define the following enumeration types for the DLX example:

- types for pipeline stages and clock phases:

  define_type "pipeline_stage = IF | ID | EX | MEM | WB"
  define_type "clock_phase = Ph1 | Ph2"

- types for the set of all instructions at each abstraction level:

  define_type "class_instruction = ALU | ALU_I | LOAD | STORE | CONTROL"
  define_type "stage_instruction = IF_X | ID_X | ID_C | EX_A | EX_I | ... | WB_L"
  define_type "phase_instruction = Ph1_IF_X | ... | Ph1_EX_A | ... | Ph2_WB_L"

- types for resources (related to the structural abstraction):

  define_type "CL_resource = PC | RF of RF_addr | I_MEM | D_MEM | IAR"
  define_type "SL_resource = PC | RF of RF_addr | I_MEM | ... | IR | A | B | Aluout | ..."
  define_type "PL_resource = PC | RF of RF_addr | ... | IR | A | B | Aluout | ... | BTA"

For the specification of the conflict predicates, we have defined several functions and predicates as follows:

- abstraction functions, which either compute higher level instructions from lower ones or extract lower level instructions from higher ones:

  ClassToStage : ((pipeline_stage, class_instruction) $\rightarrow$ stage_instruction)
  StageToClass : (stage_instruction $\rightarrow$ class_instruction)
  StageToPhase : ((clock_phase, stage_instruction) $\rightarrow$ phase_instruction)
  PhaseToStage : (phase_instruction $\rightarrow$ stage_instruction)

  e.g. ClassToStage (ID, CONTROL) = ID_C, PhaseToStage (Ph2_ID_C) = ID_C.

- functions that compute the logical pipeline stage or clock phase types from a stage or a phase instruction, respectively:

  Stage_Type : (stage_instruction $\rightarrow$ pipeline_stage)
  Phase_Type : (phase_instruction $\rightarrow$ clock_phase)

- functions which compute the ordinal values of a given pipeline stage and clock phase, respectively[4]:

  Stage_Rank : (pipeline_stage $\rightarrow$ num)
  Phase_Rank : (clock_phase $\rightarrow$ num)

  e.g. Stage_Rank (ID) = 2, Phase_Rank (Ph1) = 1.

- predicates, which are true if a given resource is used by a given stage and phase instruction, respectively:

---

[4] These functions are needed to express the sequential order of the execution of stage and phase instructions.

Stage_Used : ((stage_instruction, SL_resource) → bool)
Phase_Used : ((phase_instruction, CL_resource) → bool)

e.g. Stage_Used (ID_C, PC) = *True*. These Predicates are extracted from the formal specifications of the stage and phase levels, respectively (refer to [15]).

- predicates that imply that a given resource is read (*domain*) or written (*range*) by a given class or stage instruction at a given pipeline stage or clock phase, respectively:

Stage_Domain : ((class_instruction, pipeline_stage, CL_resource) → bool)
Stage_Range : ((class_instruction, pipeline_stage, CL_resource) → bool)
Phase_Domain : ((stage_instruction, clock_phase, CL_resource) → bool)
Phase_Range : ((stage_instruction, clock_phase, CL_resource) → bool)

e.g. Stage_Domain (ALU, ID, RF) = *True*, Phase_Range (ID_C, Ph2, D_MEM) = *False* (refer also to table 1). These predicates are extracted from the specification of the class and stage level instructions at the clock cycle and clock phase time granularities, respectively (refer to [15]).

5 Resource Conflicts

Resource conflicts (also called *structural hazards* [6, 8, 10, 12] or *collisions* [8, 12]) arise when the hardware cannot support all possible combinations of instructions during simultaneous overlapped execution. This occurs when some resources or functional units are not duplicated enough and two or more instructions attempt to use them simultaneously. A resource could be a register, a memory unit, a functional unit, a bus, etc. The use of a resource is a write operation for storage elements and an allocation for functional units. In the sections to follow, we will first formally specify the resource conflicts and then discuss the correctness proof issues.

5.1 Resource Conflict Specification

Referring to the hierarchical model (see section 2), the formal specifications of resource conflicts are handled according to the different abstract levels. Further, related to each abstraction level only the visible resources are considered by the corresponding resource conflict predicates.

Class Level Conflicts. The resource conflict predicate Res_Conflict, as mentioned in section 3.2, is equivalent to a multiple conflict between the maximal number of class instructions that occur in the pipeline. This Multiple_Res_Conflict predicate is true if any pair of the corresponding stage instructions compete for one resource (see hatched box in figure 2). Formally, Multiple_Res_Conflict is defined in terms of disjunctions over all possible stage instruction pair conflicts (assigned by Dual_Stage_Conflict), which correspond to the class instructions $I_1 \ldots I_{n_s}$:

$$
\begin{array}{l}
\text{Res\_Conflict} = \\
\vdash_{def} \text{Multiple\_Res\_Conflict} (I_1, \ldots, I_{n_s}) := \\
\bigvee_{\substack{i,j \\ (i,j=1 \ldots n_s) \\ (i<j)}} \text{Dual\_Stage\_Conflict} (\text{ClassToStage} (\psi_i^5, I_{n_s - i + 1}), \text{ClassToStage} (\psi_j, I_{n_s - j + 1}))
\end{array}
$$

---

[5] Depending on the index i, ψ_i represents the related pipeline stage, i.e. $\psi_1 = $IF, $\psi_2 = $ID, $\psi_3 = $EX, etc.

Stage Level Conflicts. A dual resource conflict happens when two stage instructions attempt to use the same resource. Further, since only stage instructions of different types could be executed simultaneously in the pipeline (see hatched box in figure 2), we should ensure that the checked stages are of different logical types. Formally, the Dual_Stage_Conflict predicate is specified using the function Stage_Type and the predicate Stage_Used as follows:

$$\vdash_{def} \text{Dual\_Stage\_Conflict } (S_i, S_j):= \\
\exists\, r: \text{SL\_resource.} \\
\text{Stage\_Type } (S_i) \neq \text{Stage\_Type } (S_j) \wedge \\
\text{Stage\_Used } (S_i, r) \wedge \text{Stage\_Used } (S_j, r)$$

Looking closer, even when the predicate Dual_Stage_Conflict is true, a conflict occurs only if the stage instructions S_i and S_j use the resource r at the same phase of the clock, since a multi-phased non-overlapping clock is used (figure 3).

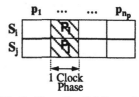

Fig. 3. Phase Parallel Execution

Having an implementation of the stage instructions at the phase level and considering all combinations of phase instructions for any two stage instructions, the dual stage conflict is defined at this lower level in terms of a multiple phase conflict predicate. Formally, Multiple_Phase_Conflict is defined in form of disjunctions over all possible phase instruction pair conflicts (assigned by Dual_Phase_Conflict):

$$\text{Dual\_Stage\_Conflict} = \\
\vdash_{def} \text{Multiple\_Phase\_Conflict } (S_i, S_j):= \\
\bigvee_{\substack{k \\ (k=1...n_p)}} \text{Dual\_Phase\_Conflict (StageToPhase } (\phi_k{}^6, S_i), \text{StageToPhase } (\phi_k, S_j))$$

Phase Level Conflicts. A dual resource conflict at the phase level occurs only when any two phase instructions that compete on the same resource, are of the same phase type, i.e. the same clock phase is involved (see figure 3) and belong to stage instructions of different types. Using the functions Phase_Type, Stage_Type and StageToPhase and the predicate Phase_Used, this is formally given as follows:

$$\vdash_{def} \text{Dual\_Phase\_Conflict } (P_i, P_j):= \\
\exists\, r: \text{PL\_resource.} \\
\text{Phase\_Type } (P_i) = \text{Phase\_Type } (P_j) \wedge \\
\text{Stage\_Type (PhaseToStage } (P_i)) \neq \text{Stage\_Type (PhaseToStage}(P_j)) \wedge \\
\text{Phase\_Used } (P_i, r) \wedge \text{Phase\_Used } (P_j, r)$$

---

[6] According to the index k, ϕ_k represents a clock phase, i.e. $\phi_1 = \text{Ph1}$, $\phi_2 = \text{Ph2}$.

5.2 Resource Conflict Proof

Our ultimate goal is to show that for all class instruction combinations, no resource conflicts occur, i.e. the predicate Multiple_Res_Conflict is never true:

set_goal "$\forall I_1...I_{n_s}$:class_instruction.
$\neg$Multiple_Res_Conflict $(I_1, ..., I_{n_s})$"

Using the definition of Multiple_Res_Conflict (cf. section 5.1), the expansion of this goal at the stage level yields to a case explosion. In order to bypass this intractable goal, we do the proof in two steps as follows:

1. we prove the non happening of dual conflicts:

set_goal "$\forall S_i S_j$:stage_instruction.
$\neg$Dual_Stage_Conflict (S_i, S_j)"

2. we conclude the negation of the multiple conflict predicate from the first step:

set_goal "$(\forall S_i S_j$:stage_instruction. $\neg$Dual_Stage_Conflict $(S_i, S_j))$
$\Rightarrow (\forall I_1...I_{n_s}$:class_instruction. $\neg$Multiple_Res_Conflict$(I_1,..., I_{n_s}))$"

Since the dual conflict predicate is a generalization of the multiple conflict predicate, the proof of the second step is straightforward; we even do not need to expand the dual conflict predicate. The proof of the first step, without any assumptions, leads either to *True* or to a number of subgoals which explicitly include a specific resource and the specific stage instructions which conflict. For example a conflict due to the resource PC between the common IF-stage instruction (IF_X) and the ID-stage instruction (ID_C) of control class is output as follows:

$$(S_i = \text{IF\_X}), (S_j = \text{ID\_C}), (r = \text{PC}) \vdash \text{F}$$

As mentioned above, a resource conflict is only ensured when the given phase level implementation still confirm it. Referring to the last example, the simultaneous use of the resource PC at the phase level is checked by explicitly setting the following goal using the Multiple_Phase_Conflict predicate:

set_goal "$\neg$Multiple_Phase_Conflict (IF_X, ID_C, PC)"

If the goal is proven true, then the conflict freedom is shown, else the resource conflict remains and in order to resolve it, the implementation EBM has to be changed appropriately, e.g. by using an additional buffer.

6 Data Conflicts

Data conflicts (also called *data hazards* [6, 8, 12], *timing hazards* [10], *data dependencies* [5] or *data races* [3]) arise when an instruction depends on the results of a previous instruction. Such data dependencies could lead to faulty computations when the order in which the operands are accessed is changed by the pipeline.

Data conflicts are of three types called, read after write (RAW), write after read (WAR) and write after write (WAW) [6, 8, 12, 5] (also called destination source (DS), source destination (SD) and destination destination (DD) conflicts [10]). Given that an instruction I_j is issued after I_i, a brief description of these conflicts is:

RAW conflict — I_j *reads* a source before I_i *writes* it
WAR conflict — I_j *writes* into a destination before I_i *reads* it
WAW conflict — I_j *writes* into a destination before I_i *writes* it

6.1 Data Conflict Specification

Data conflicts include temporal aspects that are related to the temporal abstractions of our hierarchical model. Therefore, similarly to resource conflicts, the formal specifications of data conflicts are considered at different abstraction levels.

Class Level Conflicts. Considering a full pipeline (see figure 2), the data conflict predicate, i.e. Data_Conflict, should involve the maximal number n_s of instructions that could lead to data conflicts. The predicate Data_Conflict is defined in terms of a multiple data conflict predicate, which includes n_s instructions $I_1...I_{n_s}$ with corresponding sequential issue times $t_{o1} ... t_{on_s}$[7]. The predicate Multiple_Data_Conflict is true whenever any two class instructions conflict on some data. Hence, we define Multiple_Data_Conflict as the disjunction of all possible dual data conflicts (represented by Dual_Data_Conflict) as follows:

$$
\begin{aligned}
&\text{Data\_Conflict} = \\
&\quad \vdash_{def} \text{Multiple\_Data\_Conflict}\,(I_1, ..., I_{n_s}):= \\
&\qquad \exists\, t_{o1} ... t_{on_s} : \text{time}\,. \\
&\qquad \bigvee_{\substack{i,j \\ (i,j=1...n_s) \\ (i<j)}} \text{Dual\_Data\_Conflict}\,(((I_i, t_{oi}), (I_j, t_{oi}+j\text{-}1))
\end{aligned}
$$

The predicate Dual_Data_Conflict is true, if there exists a resource of the programming model (class level) for which two class instructions I_i and I_j issued at time points t_{oi} and t_{oj}, respectively, conflict. Further, according to our hierarchical model, the Dual_Data_Conflict is handled hierarchically, first at the stage then at the phase level. Formally, Dual_Data_Conflict is defined in terms of a Stage_Data_Conflict predicate, as follows:

$$
\begin{aligned}
&\vdash_{def} \text{Dual\_Data\_Conflict}\,((I_i, t_{oi}), (I_j, t_{oj})):= \\
&\qquad \exists\, r : \text{CL\_resource}\,. \\
&\qquad\quad \text{Stage\_Data\_Conflict}\,((I_i, t_{oi}), (I_j, t_{oj}), r)
\end{aligned}
$$

Stage Level Conflicts. Let I_i be an instruction that is issued into the pipeline at time t_{oi} and *writes* a given resource r at t_{ui} ($t_{oi} \le t_{ui}$). Let I_j be another instruction that is issued at later time t_{oj}, i.e. ($t_{oi} < t_{oj}$) and *reads* the same resource r at t_{uj}. A RAW data conflict occurs when the resource r is read by I_j *before* (and not *after*) this resource is written by the previous sequential instruction I_i (figure 4). Let s_i and s_j be the related pipeline stages in which the resource r is written and read, respectively. Assuming a linear pipeline execution of instructions, i.e. no pipeline freeze or stall happen, the use time points t_{ui} and t_{uj} are equal to ($t_{oi} + \Psi^8(s_i)$) and ($t_{oj} + \Psi(s_j)$), respectively. Hence, the timing condition for the RAW conflict, i.e. ($t_{uj} \le t_{ui}$), is equivalent to ($t_{oj} - t_{oi}$) $\le$ ($\Psi(s_i) - \Psi(s_j)$).

---

[7] We assume a linear pipelining of instructions, i.e. no pipeline freeze or stall exist.

[8] The symbol Ψ represents the function Stage_Rank, which computes the ordinal value of a given pipeline stage (cf. section 4).

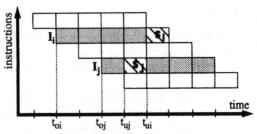

Fig. 4. RAW Data Conflict

Using the function Stage_Rank (represented by the symbol Ψ) and the predicates Stage_Range and Stage_Domain, the formal specification of the stage RAW data conflict is thus given as follows:

$\vdash_{def}$ Stage_RAW_Conflict $((I_i, t_{oi}), (I_j, t_{oj}), r) :=$
    $\exists\, s_i\, s_j$: pipeline_stage .
      $(0 < (t_{oj} - t_{oi})) \wedge$
      $(t_{oj} - t_{oi}) \le (\Psi\,(s_i) - \Psi\,(s_j)))\, \wedge$
      Stage_Range $(I_i, s_i, r)\, \wedge$
      Stage_Domain (I_j, s_j, r)

Similarly, the WAR and WAW predicates are defined as follows, where the predicate parameters are the same as by the definition of Stage_RAW_Conflict and the semantics of the data conflict is reflected by the order of the Stage_Range and Stage_Domain predicates:

$\vdash_{def}$ Stage_WAR_Conflict $(...) :=$
  $\exists\, s_i\, s_j$: pipeline_stage .
    $(0 < (t_{oj} - t_{oi})) \wedge$
    $(t_{oj} - t_{oi}) \le (\Psi\,(s_i) - \Psi\,(s_j)))\, \wedge$
    Stage_Domain $(I_i, s_i, r)\, \wedge$
    Stage_Range (I_j, s_j, r)

$\vdash_{def}$ Stage_WAW_Conflict $(...) :=$
  $\exists\, s_i\, s_j$: pipeline_stage .
    $(0 < (t_{oj} - t_{oi})) \wedge$
    $(t_{oj} - t_{oi}) \le (\Psi\,(s_i) - \Psi\,(s_j)))\, \wedge$
    Stage_Range $(I_i, s_i, r)\, \wedge$
    Stage_Range (I_j, s_j, r)

Phase Level Conflicts. A special case of the data conflict timing condition arises when a resource is simultaneously used by the instructions I_i and I_j, i.e. $t_{ui} = t_{uj}$. In this situation, the data conflict should be examined at the phase level. Let S_i and S_j be any two stage instructions, where the rank of

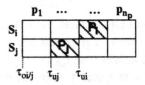

Fig. 5. Phase Level RAW Data Conflict

S_i is greater than that of S_j, e.g. $S_i = $ WB_L and $S_j = $ ID_C. According to figure 5, a RAW data conflict at the phase level happens when the resource r is *written* by the stage instruction S_i at a clock phase p_i that occurs *after* clock phase p_j, where it is *read* by S_j, i.e. ($\tau_{ui} \ge \tau_{uj}$). Since instructions at the phase level are executed purely in parallel, they all have the same issue time t_o (figure 5), the timing condition ($\tau_{ui} \ge \tau_{uj}$) is equivalent to $(\tau_o + \xi^9(p_i)) \ge \tau_o + \xi(p_j)) = (\xi(pi) \ge \xi(p_j))$. Using the functions Stage_Rank, Phase_Rank

---

[9] The symbol ξ represents the function Phase_Rank, which computes the ordinal value of the clock phase (cf. section 4).

and Stage_Type and the predicates Phase_Domain and Phase_Range, the phase level RAW data conflict predicate is formally given as follows:

$\vdash_{def}$ Phase_RAW_Conflict $(S_i, S_j, r):=$
    $\exists\, p_i\, p_j$: clock_phase.
        $(\xi\,(p_i) > \xi\,(p_j)) \wedge$
        $(\Psi\,(\vartheta^{10}(S_i))) > (\Psi\,(\vartheta\,(S_j))) \wedge$
        Phase_Range $(S_i, p_i, r) \wedge$
        Phase_Domain (S_j, p_j, r)

In a similar manner, the formal definitions of the phase level WAR and WAW data conflict predicates are given as follows:

$\vdash_{def}$ Phase_WAR_Conflict $(S_i, S_j, r):=$
   $\exists\, p_i\, p_j$: clock_phase.
      $(\xi\,(p_i) > \xi\,(p_j)) \wedge$
      $(\Psi\,(\vartheta(S_i))) > (\Psi\,(\vartheta\,(S_j))) \wedge$
      Phase_Domain $(S_i, p_i, r) \wedge$
      Phase_Range (S_j, p_j, r)

$\vdash_{def}$ Phase_WAW_Conflict $(S_i, S_j, r):=$
   $\exists\, p_i\, p_j$: clock_phase.
      $(\xi\,(p_i) > \xi\,(p_j)) \wedge$
      $(\Psi\,(\vartheta(S_i))) > (\Psi\,(\vartheta\,(S_j))) \wedge$
      Phase_Range $(S_i, p_i, r) \wedge$
      Phase_Range (S_j, p_j, r)

6.2 Data Conflict Proof

Our ultimate goal in the proof of the non existence of data conflicts relies in showing that none of the data conflicts (RAW, WAR and WAW) occurs. This proof is split into three independent parts each corresponding to one data conflict type. These proofs are similar and in the following we will handle RAW conflicts for illustration purposes.

At the top-most level, the goal to be proven for RAW data conflicts is given in terms of the multiple RAW data conflict predicate as follows:

set_goal "$\forall\, I_1... \, I_{n_s}$:class_instruction.
              $\neg$ Multiple_RAW_Conflict $(I_1, ..., I_{n_s})$"

This goal includes a quantification over all possible conflict combinations that could occur between all permutations of n_s instruction within the pipeline. As the case by resource conflicts, the direct proof of this goal yields to a case explosion. Hence, we manage the proof in two steps in that we first prove the non happening of dual conflicts and then conclude the negation of the multiple conflict predicate from the first step. The proof of step 2 is done straightforwardly. The goal for the second step is given as follows:

set_goal "$\forall\, I_i\, I_j$:class_instruction.
        $\forall\, t_{oi}\, t_{oj}$: time.
           $\forall\, r$: CL_Resource.
              $\neg$ Stage_RAW_Conflict $((I_i, t_{oi}), (I_j, t_{oj}), r)$"

The proof of this goal yields either to *True* or to a number of subgoals, which include the specific resource and class instructions that conflict. For example, a data conflict that occurs between LOAD and ALU-instructions due to the resource register file RF, which is written at the WB-stage by the LOAD-instruction and read at the ID-stage by the

---

[10] The symbol ϑ represents the function Stage_Type, which computes the logical stage type of a stage instruction (cf. section 4).

ALU-instruction is detected and output as follows, where the number "3" corresponds to the difference $\Psi(s_i) - \Psi(s_j) = "\Psi(WB) - \Psi(ID)"$:

$$(I_i = \text{LOAD}), (I_j = \text{ALU}), (s_i = \text{WB}), (s_j = \text{ID}), (r = \text{RF}), (0 < (t_{oj} - t_{oi})) \vdash \neg((t_{oj} - t_{oi}) \leq 3)$$

This result is interpreted as follows: as long as the issue times of the conflicting LOAD and ALU-instructions satisfy the condition "$(t_{oj} - t_{oi}) \leq 3$", there exists a data conflict. In order to resolve this conflict, we should neutralize this timing condition. This can be done by considering the following two cases:

1. "$(t_{oj} - t_{oi}) = 3$": the data conflict should be explored at the lower time granularity, i.e. phase level, by setting the following goal:

> set_goal "¬Phase_RAW_Conflict
>   (ClassToStage (WB, LOAD), ClassToStage (ID, ALU), RF)"

If the goal is proven correct, no data conflict happens, else either the hardware EBM should be changed, e.g. though inclusion of more clock phases, or one uses the software scheduling technique [6].

2. "$(t_{oj} - t_{oi}) < 3$": The timing information gives an exact reference for the maximum number of pipeline slots or bypassing paths that have to be provided by the software scheduling technique or the implementation EBM, respectively, namely $(3-1 = 2)$.

Using instruction scheduling, the needed software constraint (assumption) that leads to the proof of the dual data conflict goal, could then be defined as:

> $\vdash_{def}$ SW_Constraint:=
> $(I_i = \text{LOAD}) \land (I_j = \text{ALU}) \land (r = \text{RF}) \land (0 < (t_{oj} - t_{oi})) \Rightarrow ((t_{oj} - t_{oi}) > 3)$

Using the *bypassing* (also called *forwarding*) technique [6], we ensure that the needed data is forwarded as soon as it is computed (end of the EX-stage) to the next instruction (beginning of the EX-stage). This behaviour is implemented in hardware by using some registers and corresponding feedback paths that hold and forward this data, respectively. The existence of these registers and corresponding forward paths can be easily formalized as follows:

> $\vdash_{def}$ Bypassing:=
> $\exists rb . (rb = \text{RF}) \land$ Stage_Range $(I_i, \text{EX}, rb) \land$ Stage_Domain (I_j, EX, rb)

Assuming this bypassing condition in the dual data conflict goal, the existentially quantified pipeline stage variables s_i and s_j in the definition of Stage_RAW_Conflict (cf. section 6.1) are set to EX and the timing condition is hence reduced to:

$$..., (0 < (t_{oj} - t_{oi})) \vdash \neg((t_{oj} - t_{oi}) \leq 0)$$

which is always true.

To summarize, given some specific software constraints in form of instruction scheduling timing conditions and given the implementation EBM, which includes some bypassing paths with appropriate logic, we are able to prove that none of the data conflicts (RAW, WAR and WAW) happens, i.e. formally:

> SW_Constraints, EBM $\vdash$ (¬RAW_Data_Conflict $\land$
>   ¬WAR_Data_Conflict $\land$
>   ¬WAW_Data_Conflict)

7 Control Conflicts

Control conflicts (also called *control hazards* [6, 12], *branch hazards* [6], *sequencing hazards* [10] or *branch dependencies* [5]) arise from the pipelining of branches and other instructions that change the program counter *PC*, i.e. interruption of the linear instruction flow.

Control conflicts involve all kinds of interrupts to the linear pipeline flow. This includes the software branches, jumps, traps, etc. and hardware interrupts, stalls, etc. Using this classification, we define *SW control conflicts* and *HW control conflicts* as those conflicts which are caused by a software instruction and a hardware event, respectively. The correctness proof of *Control_conflict* is therefore split into two independent parts as follows:

7.1 SW Control Conflict

Let $A_f(I_i, t_{oi})$ be the fetch address of an instruction I_i issued at the time point t_{oi} and $A_n(I_i, t_{oi})$ be the fetch address of the next instruction (also called next address of I_i). Let *PC* be the program counter. In a pipelined instruction execution, at each clock cycle a new instruction is fetched, i.e. $\forall t.\ PC(t+1) = Inc\ PC(t)$[11]. The fetch and next addresses of an instruction I_i are equal to $A_f(I_i, t_{oi}) = PC(t_{oi})$ and $A_n(I_i, t_{oi}) = PC(t_{oi}+1)$, respectively. During a sequential program execution, we have $A_n(I_i, t_{oi}) = A_f(I_i+1, t_{oi}+1)$ and we obtain the following:

$$\forall t_{oi}.\ A_n(I_i, t_{oi}) = A_f(I_i+1, t_{oi}+1) = PC(t_{oi}+1) = Inc\ PC(t_{oi})$$

If I_i is a control instruction, then the next address is its target address, i.e. $A_n(I_i, t_{oi}) = A_t(I_i, t_{oi})$. Due to the sequential execution of a single instruction, the target instruction can only be fetched after the instruction I_i is fetched, decoded and the target address has been calculated. Since all this cannot happen in one clock cycle, the target address $A_t(I_i, t_{oi}) = PC(t_{oi}+n)$, where $n > 1$. Hence, the next address is not equal to the target address, i.e.:

$$A_n(I_i, t_{oi}) = Inc\ PC(t_{oi}) \neq A_t(I_i, t_{oi}) = PC(t_{oi}+n)$$

and the wrong instruction is fetched next.

A closer look at this situation shows that a software control conflict occurs when an instruction attempts to read the resource *PC* that is not yet updated (written) by a previous instruction. This complies with the definition of RAW data conflict in *PC* [8] and the software control conflict could be defined as follows:

$$\vdash_{def} \text{SW\_Control\_Conflict} := \text{Stage\_RAW\_Conflict}((\text{CONTROL}, t_{oi}), (I_j, t_{oj}), \text{PC})$$

The conflict freedom proof is therefore only a special case of the data conflict proofs and the goal to be proven is set as follows:

```
set_goal "∀ Ij:class_instruction .
           ∀ toi toj : time .
              ¬ Stage_RAW_Conflict ((CONTROL, toi), (Ij, toj), PC) "
```

and we yield for the DLX processor example 5 subgoals of the following form:

$$(I_j = \text{ALU}),\ (s_i = \text{ID}),\ (s_j = \text{IF}),\ (0 < (t_{oj} - t_{oi}))\ \vdash\ \neg((t_{oj} - t_{oi}) \leq 1)$$

---

[11] *Inc* is an increment function which is independent of the increment value, e.g. PC+4, PC+8, etc.

For conflict resolution no bypassing is possible, since the calculation of the target address cannot be done earlier. The commonly used technique is software scheduling [6]. In the DLX RISC processor, we just need one delay slot $((t_{oj} - t_{oi}) \leq 1)$ to ensure that control instructions are executed correctly. The software constraint needed in this case is defined as follows:

$$\vdash_{def} \text{SW\_Constraint} := (0 < (t_{oj} - t_{oi})) \Rightarrow ((t_{oj} - t_{oi}) > 1)$$

7.2 HW Control Conflicts

In contrast to the previous conflict kind, this kind of conflict cannot be handled in general since the interrupt behaviour and conflict resolution are fully hardware implementation dependent. Different RISC processors handle interrupts in different ways, i.e where the forced branch is to be inserted, how to restart exactly the interrupted program, etc.

In general, one has to formally specify the interrupt behaviour in the form of a predicate, whose implication has to be proven correct from the implementation EBM, i.e.:

$$\vdash \text{EBM} \Rightarrow \text{INTR\_SPEC}$$

The predicate INTR_SPEC[12] describes the behaviour of the implemented hardware interrupt and ensures that no conflict occurs when the linear pipeline flow is interrupted, e.g. proper saving and recovery of the processor state before and after the interrupt handling. Formally:

$$\vdash \text{INTR\_SPEC} \Rightarrow \neg\text{HW\_Control\_Conflict}$$

and consequently we yield:

$$\text{EBM} \vdash \neg\text{HW\_Control\_Conflict}$$

Summary
Our ultimate goal in proving the non-existence of pipeline conflicts is concluded from the theorems yielded in sections 5.2, 6.2, 7.1 and 7.2, i.e.:

$$\text{SW\_Constraints, EBM} \quad \vdash \quad \begin{array}{l} (\neg\text{Res\_Conflict} \quad \wedge \\ \neg\text{Data\_Conflict} \quad \wedge \\ \neg\text{Control\_Conflict}) \end{array}$$

8 Experimental Results

The methodology presented so far, has been validated by using the DLX processor as an example. This processor contains a 5 stage pipeline with a 2 phased clock, and its architecture includes 51 basic instructions (integer, logic, load/store and control). All these instructions are grouped into 5 classes according to which the stage and phase instructions are defined (refer to table 1). The EBM for the DLX core has been implemented in a commercial VLSI design environment using a 1.0 μm CMOS technology. The design has approximately 150 000 transistors which occupies a silicon area of about 60.34 mm², it has 172 I/O pads and currently runs at a clock rate of 12.5 MHz.

---

[12] A formal specification of INTR_SPEC for DLX is beyond the scope of this paper and is not presented here.

The formalization of the specifications for DLX have been done within the HOL verification system [4]. The implementation of the interpreter model specifications and the proof of step (2) is reported in [15]. The overall specification for the DLX core is about 4760 lines long and the proof of step (2) took about 457.66 secs on a SPARC10, with a 96 MB main memory.

In proving step (1), the specification and tactics for each kind of conflict have been implemented in HOL90.6. These specifications and tactics are general enough to be applicable to a wide range of other RISC processors. The hierarchical structuring of the proofs resulted in parameterized tactics that are used for more than one kind of conflict. All proofs have been mainly done using 4 proof tactics — MULTIPLE_CONF_TAC, RES_CONF_TAC, DATA_CONF_TAC and CONTROL_CONF_TAC — with a total proof script text of 850 lines. The overall definition and specification text of step (1) is about 2690 lines. The run times and the number of created inferences, on a SPARC10 with a 96 MB main memory, for the theorem generation of the Used and Range/Domain predicates, for the pipeline verification and for the semantical correctness proofs, respectively, are given for the DLX processor example in the following table:

Table 2. Verification Results for the DLX processor

Verification Goal	Time (in sec)	# Inferences	Comments
Stage_Used Gen.	247.27	768226	208 theorems generated
Phase_Used Gen.	1718.03	3756157	422 theorems generated
Stage Ran. & Dom. Gen.	324.44	534283	250 theorems generated
Phase Ran. & Dom. Gen.	272.27	267926	260 theorems generated
Stage Dual Res. Conf.	1691.64	3330880	0 conflicts
Dual $\Rightarrow$ Multiple (Res.)	1.30	2434	-
Stage RAW Data Conf.	570.64	1669029	15 conflict cases (3 slots) by RF and 5 conflict cases (1 slot) by PC
(using Bypassing)	1.89	5438	0 conflicts
(using SW-scheduling)	458.96	1344577	0 conflicts
Stage WAR Data Conf.	572.23	1669069	0 conflicts
Stage WAW Data Conf.	572.50	1669047	0 conflicts
Dual $\Rightarrow$ Multiple (Data)	0.27	389	-
SW Control Conflicts	32.44	91403	5 conflict cases (1 slot)
Σ Pipeline Correctness	6464.18	15108496	-

9 Conclusions and Future Work

In this paper, we have shown the formalization and the verification of the different kinds of conflicts that occur in the instruction pipelines of RISC cores. The employment of the hierarchical RISC interpreter model and in particular the exploitation of the class level, empowers us to derive compact specifications of the conflicts. We have implemented constructive proofs for these conflicts and hence the designer gets invaluable feedback for resolving these conflicts, either by making appropriate modifications to the hardware or by generating the required software constraints.

The interpreter model, the formal specifications and the proof techniques are generic and hence applicable to any RISC core. Furthermore, the RISC interpreter model closely reflects the RISC design hierarchy thus the specifications in higher-order logic are easy to derive. Given such specifications and an implementation, the proof process has been automated by using parametrizable tactics. These tactics are independent of the underlying implementation and can be used for a large number of RISC cores. The entire methodology has been validated by using the DLX core.

The feasibility of formal verification techniques, when applied cleverly to specific classes of circuits is illustrated by the runtimes shown in table 2. In our future work, we shall extend the layer of the core to include pipelined functional units, floating point processor, etc.

References

1. Anceau, F.: The Architecture of Microprocessors; Addison-Wesley Publishing Company, 1986.
2. Buckow, O.: Formale Spezifikation und (Teil-) Verifikation eines SPARC-kompatiblen Prozessors mit LAMBDA; Diplomarbeit, Universität-Gesamthochschule Paderborn, Fachbereich Mathematik-Informatik, Oktober 1992.
3. Furber, S.: VLSI RISC Architecture and Organization; Electrical Engineering and Electronics, Dekker, New York, 1989.
4. Gordon, M.; Melham, T.: Introduction to HOL: A Theorem Proving Environment for Higher Order Logic; Cambridge, University Press, 1993.
5. Van De Goor, A.: Computer Architecture and Design; Addison-Wesley, 1989.
6. Hennessy, J.; Patterson, D.: Computer Architecture: A Quantitative Approach; Morgan Kaufmann Publishers, Inc., San Mateo, California, 1990.
7. Joyce, J.: Multi-Level Verification of Microprocessor-Based Systems; Ph.D. Thesis, Computer Laboratory, Cambridge University, December 1989.
8. Kogge, P.: The Architecture of Pipelined Computers; McGraw-Hill, 1981.
9. Kumar, R.; Schneider, K.; Kropf, Th.: Structuring and Automating Hardware Proofs in a Higher-Order Theorem-Proving Environment; Journal of Formal Methods in System Design, Vol.2, No. 2, 1993, pp. 165-230.
10. Milutinovic, V.: High Level Language Computer Architecture; Computer Science Press, Inc., 1989.
11. Srivas, M.; Bickford, M.: Formal Verification of a Pipelined Microprocessor; IEEE Software, September 1990, pp. 52-64.
12. Stone, H.: High-Performance Computer Architecture; Addison-Wesley Publishing Company, 1990.
13. Tahar, S.; Kumar, R.: A Formalization of a Hierarchical Model for RISC Processors; In: Spies, P. (Ed.), Proc. European Informatics Congress Computing Systems Architecture (Euro-ARCH93), Munich, October 1993, Informatik Aktuell, Springer Verlag, pp. 591-602.
14. Tahar, S.; Kumar, R.: Towards a Methodology for the Formal Hierarchical Verification of RISC Processors; Proc. IEEE International Conference on Computer Design (ICCD93), Cambridge, Massachusetts, October 1993, pp. 58-62.
15. Tahar, S.; Kumar, R.: Implementing a Methodology for Formally Verifying RISC Processors in HOL; Proc. International Meeting on Higher Order Logic Theorem Proving and its Applications (HUG93), Vancouver, Canada, August 1993, pp. 283-296.
16. Tahar, S.; Kumar, R.: Formal Verification of Pipeline Conflicts in RISC Processors; to appear in Proc. European Design Automation Conference (EURO-DAC94), Grenoble, France, September 1994.
17. Windley, P.: The Formal Verification of Generic Interpreters; Ph.D. Thesis, Division of Computer Science, University of California, Davis, July 1990.

Specifying Instruction–Set Architectures in HOL: A Primer

Phillip J. Windley

Laboratory for Applied Logic
Brigham Young University
Provo, Utah 84602-6576

Abstract. This paper presents techniques for specifying microprocessor instruction set syntax and semantics in the HOL theorem proving system. The paper describes the use of abstract representations for operators and data, gives techniques for specifying instruction set syntax, outlines the use of records in specifying semantic domains, presents the creation of parameterized semantic frameworks, and shows how all of these can be used to create a semantics for a microprocessor instruction set. The verified microprocessor UINTA provides examples for each of these.

1 Introduction

Much has been written over the years regarding the formal specification and verification of microprocessors [CCLO88, Bow87, Hun87, Coh88b, Coh88a, Gor83, Joy88, Hun89, Joy89, SB90, Her92, SWL93, TK93]. These efforts use many different proof systems and styles. We have verified a number of microprocessors in the HOL theorem proving system [Win90a, Win90b, Win94, WC94] and have developed techniques which clarify the specification and ease the verification effort. Some of this has been codified in our generic interpreter theory [Win93b].
    In [Leo89], Tim Leonard states:

> The term *computer architecture* is used with two distinct meanings. In the broader sense, computer architecture is an aspect of computer systems design: creating an architecture consists of dividing a system into major components and defining the interfaces between those components. An architecture is thus the overall structure of a system. In the narrower sense, computer architecture is the design of the interface that is unique to computers: the program interface.

In this paper, we use the term *computer architecture* in the second, narrower sense: a program interface. In that case, we will refer to the instruction–set architecture. The instruction–set architecture describes how programs are interpreted by the computer. The architecture provides all of the information necessary to program the computer: what data types are valid, what operators are available, what is the instruction format, what they do, how the computer decides which operator follows another, etc.. The most obvious specification of the program interface is the user's manual. A typical user's manual answers all of these questions using natural language, pictures, and tables. A formal specification of a computer architecture answers all of these questions in a mathematical notation.

This paper is a primer on formally specifying instruction–set architectures in HOL. Specifying an architecture is a methodical process; with few exceptions, a program interface can be specified, regardless of the particulars, through the following steps:

1. Choose representations for the primitive data types.
2. Define the semantic domain.
3. Define the syntax of the instruction set.
4. Specify the semantics of the instructions.
5. Specify how instructions are sequenced by the machine.
6. Create a predicate, defined in terms of the definitions from the previous steps, specifying the overall behavior of the architecture.

Having completed these steps, we have a complete description of the behavior of the instruction–set architecture of the computer. In the sections that follow we present techniques in the HOL theorem proving system for completing each of the preceding steps. Small examples from the specification of the UINTA microprocessor [WC94] and alternative techniques are given where appropriate. Obviously, this short paper cannot contain all of the information necessary to complete each of these steps, but it does provide pointers to other documents that contain more complete instructions.

Since one typically writes a specification to do a formal verification of correctness, one must take care to write the specification in such a way that the verification is made easier. We will point out these areas where they occur and make recommendations for techniques which we have found ease the verification burden.

2 Representations of Primitive Types

Microprocessors are built to perform specific operations on specific kinds of data. Data types supply the primitive operations upon which the microprocessor's behavior is specified. These primitive types are also the building blocks for the semantic domain. Choosing the right representation can make the specification easier to read as well as write and make the subsequent verification tractable. In this section we treat the options for data type selection and recommend one we have found to be particularly useful.

Some primitive types are already part of the HOL system. Probably the most obvious is the use of type `:bool` to represent bits. The HOL system provides a good selection of built-in theorems and tactics related to type `:bool` that aid in the specification and verification tasks. There has been some work on using alternate representations for bits [Win87, GW92], but these are generally more complicated to use than `:bool`. Their chief advantages (representing multiple signal values) occur at abstraction levels well below the architectural model.

Of the types not part of the base HOL system, the two that are most often used in architectural level specifications are a representation of n–bit words and a representation of memory. There have been a number of *ad hoc* approaches to both of these. Since n–bit words and memory have a great deal in common (both are sequences), their representations are usually similar. HOL has no built-in type operator for creating arrays or finite functions, so representations of words (memory) have frequently used lists of booleans (words) or functions with domain :num and a range of type :bool (:wordn). The primary disadvantage of lists is notational: accessing the nth member of the word, for example, is much more verbose than the notations frequently employed by designers. The primary disadvantage of using functions is that HOL functions are infinite, so some artifice must be used to ensure only the first n mappings of the function are used.

One need only read the *info-hol* mailing list archives to see the number of people who have worked on these representations over the years. Recently, Wai Wong released an n–bit word library for HOL which is the most comprehensive representation of n–bit words to date [WM91]. The library provides functions for creating n–bit words, selecting their components, and operations on n–bit words. Also provided is a large body of theorems for reasoning about n–bit words.

Our prefered method of dealing with primitive types is not based on picking a representation, but rather in *not* picking one. The method is based on the observation that the correct operation of the microprocessor is, very often, orthogonal to the correct operation of the primitive functions provided by the primitive data types. Indeed, many microprocessor specifications go to great length to define n-bit words and the attendant operations on them when the verification of the specification will not use them at all. To not pick a representation, one needs a way of giving names to types and operations without giving those names an interpretation—one needs abstract types.

The chief advantage of using uninterpreted data types and uninterpreted operations in the specification and verification is that abstract types make it clear just what information *is* required of the underlying representation. That is, one can tell which types and operations have something to do with the specification and verification and which are merely place holders. Uninterpreted types were first used in microprocessor specification by Jeffrey Joyce [Joy89]. We created an infrastructure in HOL for using them and added the use of partial semantics in [Win90a] and describe their use in [Win92].

A partial semantics for an uninterpreted operation provides sufficient information about the required meaning of the operator to prove the microprocessor correct without completely defining the behavior of the operator. The uninterpreted data types and operations do not need interpretations until the low level components of the microprocessor are verified.

Using uninterpreted, or abstract, data types and operations has a number of advantages:

– Abstract data types and operations provide a clear demarcation between what has been proven and what has not. With few details given about the abstract type, readers of the specification do not jump to the conclusion that, for example, the microprocessor adds correctly, when in fact it does not.

− Using abstract types avoids the effort of building a large infrastructure to support the type and its operations and the accompanying effort of manipulating this infrastructure to get the proof completed if this infrastructure has little to do with what is being proven.

We create an abstract representation using the abstract theory package described in [Win92]. The abstract representation is created by describing the signatures of the desired operations. The type variables used in those signatures are the uninterpreted types of the representation.

Abstract representations can be used to specify n−bit words, the memory interface and other complex types that the designer wishes to consider primitive. As an example, the signature for the abstract n−bit word operations in the UINTA specification is shown below:[1]

```
let ops_lemma = new_abstract_representation 'OPS'
    [
        'add',        ":(*wordn # *wordn) -> *wordn";
        'eqp',        ":(*wordn # *wordn) -> bool";
        'gtzp',       ":*wordn -> bool";
        'shl',        ":(*wordn # num) -> *wordn);
        'val',        ":*wordn -> num";
        'wordn',      ":num -> *wordn";
        'subrange', ":*wordn -> (num # num) -> *wordn";
        ...
    ];;
```

The signature gives an operation a name, add for example, and a type. The type for add declares it to be a binary function on n−bit words. The signature gives no other information about the operator.

Of course, there are instances where more information *is* needed to complete the verification. In that case, we can also give a partial semantics of the operations. We call these semantics *theory obligations* and specify them using a predicate. The theory obligations for UINTA are shown in Figure 1. Notice that we require the arithmetic operators to commute, for example, but we require nothing else of them. This is the only necessary property for these operators to complete the verification of UINTA. On the other hand, the semantics of setbit and bit are more fully defined as these operators play a more central role in the correct functioning of the microprocessor. The semantics are defined algebraically by relating the operation of setbit, bit and other n−bit functions. Often the required partial semantics for the abstract representation will not be known at the time of specification, but will be developed during the verification. Thus, developing the theory obligations is an iterative process.

---

[1] In the interest of brevity we eliminated many of the arithmetic and bit functions actually defined in the abstract representation.

```
new_theory_obligations
  ('Ops_thobs',
   "Ops_thobs rep =
    (∀ a (b:*wordn) . add rep (a,b) = (add rep (b,a))) ∧
    (∀ a (b:*wordn) . band rep (a,b) = (band rep (b,a))) ∧
    (∀ a (b:*wordn) . bxor rep (a,b) = (bxor rep (b,a))) ∧
    (∀ a (b:*wordn) . bor rep (a,b) = (bor rep (b,a))) ∧
    (∀ a (b:*wordn) . eqp rep (a,b) = (eqp rep (b,a))) ∧
    (∀ n m w1 b . ¬(n = m) ⇒
                      (bit rep m(setbit rep(n,w1,b)) = bit rep m w1)) ∧
    (∀ m w1 b . (bit rep m(setbit rep(m,w1,b)) = b)) ∧
    (∀ n m  . (subrange rep (wordn rep 0) (n,m) = (wordn rep 0))) ∧
    (∀ n . (val rep (wordn rep n) = n)) ∧
    (∀ n m w1 b1 b2 . ¬(n = m) ⇒
                      (setbit rep (m,(setbit rep(n,w1,b1),b2)) =
                       setbit rep (n,(setbit rep(m,w1,b2),b1)))) ∧
    (∀ b n w1 . setrange rep w1 (n,n) (repeat rep (1,b)) =
                setbit rep (n,w1,b))"
  );;
```

Fig. 1. The Partial Semantics for Abstract n–bit Words.

If we used the preceding abstract representation of n–bit words in the specification and verification of a microprocessor, we could not say, when we were all done, that the microprocessor correctly adds anything since the abstract representation does not define add in enough detail. What *has* been shown is that when the microprocessor fetches a certain opcode, the correct operands are fetched, the operation called add is performed and the result is stored in the correct place. The proof that the add operation is correct is orthogonal and done independently of the verification of the microprocessor.

There are a number of things to keep in mind regarding the use of abstract representations for primitive type in computer architecture specifications:

- We have found that it is useful to separate abstract specifications into smaller groups (such as the division between n–bit words and memory operations) because there are times when one wishes to instantiate the memory interface, for example, without building the infrastructure to instantiate n–bit words.
- One must take care that the partial semantics of the abstract operators are indeed the desired semantics. When one gives the abstract specification of n–bit words, for example, one is assuming that concrete objects with those semantics exists and basing the rest of the proof on that assumption. One could easily create a semantics that is unrealizable and thus render the proof that depends on it useless.
- One should keep the signature and theory obligations as small as possible for several reasons. First, smaller abstract representations are easier realized

with a concrete type. Second, if the abstract representation contains only what is needed in the verification and no more, one can use it to see just what properties a concrete realization must have.

As mentioned earlier, abstract type representations can be used to define the memory interface as well. Other primitive types that must usually be defined include register indices, addresses, etc.

3   Instruction Set Syntax

Instruction set syntax has often been ignored in formal microprocessor proofs and, at most, dealt with peripherally using diagrams, text, or names. For example, in most of the microprocessor specifications referenced at the beginning of this paper, there is no formal mapping between the instruction mnemonic and the opcode. Similarly, the instruction format can only be deduced by examining the definitions of the functions used to select portions of the n–bit word in the instruction register. The situation was even worse in specifications that used abstract representation of primitive types. For example, in *AVM-1*, one can tell that somewhere in the ADD instruction there is a destination register index, but one cannot tell where it is in the word or how many bits it occupies.

This situation need not be: we can formally describe the instruction mnemonics, describe the instruction formats, and map the instructions into opcodes. The mnemonics and format can be defined using an abstract syntax. The abstract syntax of a microprocessor instruction set is usually simple enough to be defined with a non-recursive data type specification using the **define_type** function of HOL. For example, the specification of the abstract instruction set syntax for UINTAis shown in Figure 2.

Because we are using the HOL type definition package, in addition to the syntax definition, we can easily prove that each instruction is distinct and that the instruction set is total. We can also use the type definition package to do case analysis on the instruction set and create recursive definitions over the instruction set. For example, the following function selects the B source register index from those instructions that provide one:

```
new_recursive_definition false Instruction_axiom 'sel_Rb'
    "(sel_Rb (ADD Rd Ra Rb)  =     Rb)      ∧
     (sel_Rb (SUB Rd Ra Rb)  =     Rb)      ∧
     (sel_Rb (BAND Rd Ra Rb) =     Rb)      ∧
     (sel_Rb (BOR Rd Ra Rb)  =     Rb)      ∧
     (sel_Rb (BXOR Rd Ra Rb) =     Rb)"
```

The definition of the syntax shown in Figure 2 uses HOL type variables to represent the register index, 16–bit short words, and 26–bit address words. Their order in a particular production is meant to indicate the location of the operands operands for that instruction, although the formalization of this is done in the mapping between n–bit words and the abstract syntax (see below). Had we used

```
let Instruction_axiom = define_type 'Instruction'
  'Instruction =
      LDI    *ri *ri *short    |
      STI    *ri *ri *short    |
      LHI    *ri *ri *short    |
      ADD    *ri *ri *ri       |
      ADDI   *ri *ri *short    |
      SUB    *ri *ri *ri       |
      SUBI   *ri *ri *short    |

      ...

      JMP    *word26           |
      JMPR   *ri *short        |
      JALI   *ri *short        |
      BEQ    *ri *short        |

      ...

      NOOP                         ';;
```

Fig. 2. The UINTA Instruction Set Syntax.

a concrete representation of n-bit words for our primitive operations, we would have used concrete representations for the register index, 16-bit shorts, and other data types as well since these are among the primitive types of which one needs to be aware.

This abstract syntax is used in the architectural level specification, but not in lower levels of the proof hierarchy. An instruction at the architectural level is given by the abstract instruction syntax, but at lower levels in the hierarchy, the instruction is represented by an n-bit word. To complete the specification, not to mention the verification, one must provide a mapping from n-bit words to the abstract syntax. The following example shows part of the the decode function for UINTA:

```
⊢_def Decode_word orep w =
   let opc = (get_opcode orep w) and
       Rd = (get_rd orep w) and
       Ra = (get_rs1 orep w) and
       Rb = (get_rs2 orep w) and
       imm = (getimmed orep w) in
   (opc = LDI_OP)   →   (LDI Rd Ra imm) |
   (opc = STI_OP)   →   (STI Rd Ra imm) |
   (opc = ADD_OP)   →   (ADD Rd Ra Rb) |
         ...
   (opc = BLTEZ_OP) →   (BLTEZ Ra imm) |
   % default %          NOOP
```

The above function, much reduced for brevity, shows how the abstract functions defined earlier can be used to define a function that maps n-bit words into our abstract syntax. The constants, such as LDI_OP, are defined as

```
├_def LDI_OP = ^(octal 40)
```

where **octal** is an ML function that converts octal numbers into objects with type :**num** for use in HOL. The selector functions **get_opcode**, **get_rd**, etc. are defined using a range selector function from the abstract representation of n–bit words. For example, the definition of **get_opcode** shows where in the instruction word the opcode is located:

```
├_def get_opcode orep w = subrange orep w (31,26)
```

The function **get_opcode** could have been defined as part of the abstract representation of n–bit words. However, by defining it in terms of a more primitive function, **subrange**, we keep the abstract representation as small as possible and provide more information about the format of the instruction word.

4   The Semantic Domain

The semantic domain describes the state of the microprocessor. We are concerned with the architectural level, so we describe only that state visible to the machine code programmer. Typically, the state visible at the architectural level includes the register file (or user registers on older machines), the program counter, any status registers, and the memory. If one thinks of the specification as a description of a processor, then including memory in the semantic domain can seem inconsistent, but when viewed as an architecture specification, the inclusion of the memory is more natural. Users of an *architecture* see and use the memory.

Contents of the Semantic Domain. The individual pieces of the semantic domain are created from the primitive types described in Section 2. Individual registers in the semantic domain, such as the program counter, are typically represented using n–bit words. Status registers are often oddly sized and typically need to be accessed and updated by field. Thus, records are a good choice for their representation (HOL records are described in more detail in [Win93a]). A register file can be treated as a primitive object with an abstract representation. However, we have found that register files can be conveniently modeled as a function from register indices (type :**ri**) to n–bit words along with special functions for updating and indexing the registers. For example, the functions for updating and indexing the register file on UINTA are shown below:

```
├_def INDEX_REG rep n register =
        (n = R0)                →   (wordn rep 0) |
                                    (register n)

├_def UPDATE_REG n register value =
        (n = R0)                →   register |
                                    (ALTER register n value)
```

The use of functions for indexing and updating registers allows us to take into account special behavior such as the use of constant registers. In the case of UINTA above, register R0 always returns the n–bit word representing zero and is unalterable.

The implementation of an architecture can affect its specification. For example, because UINTA is implemented using an instruction pipeline and control hazards are dealt with using a 2–stage branch delay, the architecture contains two phantom program counters. We call them phantoms because they appear at the architectural level to indicate the behavior of the branch delay, but do not occur in the implementation. They are merely temporal mappings of the program counter.

Aggregating the Semantic Domain. The individual pieces of the architectural state are usually aggregated so that they are conveniently manipulatable. In the past we have used tuples and lists to aggregate the state, but now use records because functions for selecting individual fields and their supporting theorems are automatically created when the record is defined. As an example of using records to aggregate the semantic domain, the following HOL expression creates a record called State with 6 fields:

```
create_record 'State'
  ['Reg', ":*ri->*wordn";
   'Pc', ":*wordn";
   'NPc', ":*wordn";
   'NNPc', ":*wordn";
   'Imem', ":*memory";
   'Dmem', ":*memory";
  ];;
```

The record contains a register file (in field Reg) which is modeled as a function from register indices (type :*ri) to n-bit words, three program counters (in fields Pc, NPc and NNPc), an instruction memory (in field Imem), and a data memory (in field Dmem). The record package automatically defines selector functions for each field, defines mutator functions for each field, and proves theorems about the relationship between the selector and mutator functions and the record constructor, State.

5   Semantic Frameworks

Every instruction set has collections of functions that behave in nearly identical fashion up to the specific operation performed. The instruction set for UINTA, for example, has 27 instructions, but only 10 unique behaviors. We capture the common behaviors of these instructions using *semantic frameworks*. A semantic framework is a denotation of the parameterized behavior of a class of instructions. Semantic frameworks have all of the advantages of functions in a programming language: they are an important abstraction mechanism that eases

```
 1 ⊢def ALU_FW orep mrep op Rd Ra Rb s e =
 2       let reg      = Reg s and
 3           pc       = Pc s and
 4           nextpc   = NPc s and
 5           nextnextpc  = NNPc s and
 6           imem     = Imem s and
 7           dmem     = Dmem s in
 8       let a        = INDEX_REG orep Ra reg and
 9           b        = INDEX_REG orep Rb reg in
10       let result  = op (a, b) in
11       let new_reg = UPDATE_REG  Rd reg result and
12           new_pc = nextpc and
13           new_nextpc = nextnextpc and
14           new_nextnextpc  = inc orep nextnextpc in
15       (State new_reg new_pc new_nextpc new_nextnextpc imem dmem)


16 ⊢def  BRA_FW orep mrep cmp Ra imm s e =
17        let reg     = Reg s and
18            pc      = Pc s and
19            nextpc  = NPc s and
20            nextnextpc  = NNPc s and
21            imem    = Imem s and
22            dmem    = Dmem s in
23        let i = shl orep (sextend orep imm,2) and
24            a = INDEX_REG orep Ra reg in
25        let new_address = add orep (pc, i) in
26        let new_pc = nextpc and
27            new_nextpc = nextnextpc and
28            new_nextnextpc = (cmp a) →  new_address
29                                     | (inc orep nextnextpc) in
30        (State reg new_pc new_nextpc new_nextnextpc imem dmem)
```

Fig. 3. The Semantic Frameworks for the ALU and Branch Operations of UINTA .

the tasks of creating and maintaining a specification. Semantic frameworks are similar to the class level specifications of [TK93] and the use of the framing schema of Z to specify addressing modes in [Bow87].

The behavior of a class of instructions is described denotationally. That is, we write a function, parameterized by the relevant pieces of the syntax, that maps the current state of a processor into the next state of the processor. Figure 3 shows the semantic frameworks for UINTA's ALU and branch operations which, respectively, describe the semantics for all of the ALU and brach instructions.

The ALU framework, ALU_FW, is parameterized by the ALU operation (op) and the relevant pieces of the syntax: the destination and source register indices

(Rd, Ra, and Rb in line 1). Lines 2–7 break the state variable, s, into separate fields. Lines 8 and 9 put the contents of the A and B source registers in temporary variables, a and b. The result is calculated, using the operation parameter, in line 10. Lines 11–14 calculate the new values for the state variables that change. Line 15 is the return value, a new state record.

The framework for the branch instructions, BRA_FW, is parameterized by the comparison operation to be performed, cmp: The program counter is updated with the new address if the comparison is true and simply incremented otherwise (lines 28–29). Again, the semantic framework is a mapping from the state, s, to the new state created using the record constructor State.

6   Instruction Set Semantics

Having defined the semantic frameworks for each unique instruction type, we can define the semantics of the instruction set. The semantics are given denotationally as a mapping from the instructions' syntax to a state transition function. The use of the HOL type definition package allows us to specify the instruction set denotation in a manner that is very similar to the presentation of denotational semantics for programming languages.

For example, the semantics of the instruction set for UINTA is shown in Figure 4. We have removed much of the definition in the interest of brevity, but it is clear, for example, that the meaning of an (ADD Rd Ra Rb) instruction is given in terms of the ALU_FW semantic framework instantiated with the abstract operation (add orep)[2]. The meaning of a (SUB Rd Ra Rb) instruction is given in terms of the same semantic framework instantiated with the abstract operation (sub orep).

7   Instruction Sequencing

In a computer architecture, instructions are sequenced by examining the current state: the next instruction is the contents of memory at the location given in the program counter. This sequencing needs to be made explicit in the formal specification. For example, in the specification of UINTA we define a function, Opcode, which examines a state and returns an instruction (i.e. one production from the abstract instruction syntax given earlier). The function is easily defined using the functions defined for the memory interface and the decode function described earlier:

```
⊢def Opcode orep mrep s e =
        let pc    = Pc s and
            imem  = Imem s and
            dmem  = Dmem s in
        Decode_word (fetch mrep (imem, address orep pc))
```

---

[2] the orep in the operation indicates that it is an abstract representation.

```
⊢def (M_INST orep mrep (LDI Rd Ra imm) =
            LOAD_FW orep mrep Rd Ra imm)                    ∧
      (M_INST orep mrep (STI Rd Ra imm) =
            STORE_FW orep mrep Rd Ra imm)                   ∧
      (M_INST orep mrep (LHI Rd Ra imm) =
            LHI_FW orep mrep Rd Ra imm)                     ∧
      (M_INST orep mrep (ADD Rd Ra Rb) =
            ALU_FW orep mrep (add orep) Rd Ra Rb)           ∧
      (M_INST orep mrep (SUB Rd Ra Rb) =
            ALU_FW orep mrep (sub orep) Rd Ra Rb)           ∧
      ...
      (M_INST orep mrep (ADDI Rd Ra imm) =
            ALUI_FW orep mrep (add orep) Rd Ra imm)         ∧
      ...
      (M_INST orep mrep (BNOT Rd Ra) =
            UNARY_FW orep mrep (bnot orep) Rd Ra)           ∧
      ...
      (M_INST orep mrep (NOOP) =
            NOOP_FW orep mrep)                              ∧
      (M_INST orep mrep (JMP imm26) =
            JMP_FW orep mrep imm26)                         ∧
      (M_INST orep mrep (JMPR Ra imm) =
            JMPR_FW orep mrep Ra imm)                       ∧
      (M_INST orep mrep (JALI Rd imm) =
            JALI_FW orep mrep Rd imm)                       ∧
      (M_INST orep mrep (BEQ Ra imm) =
            BRA_FW orep mrep (eqzp orep) Ra imm)            ∧
      ...
```

Fig. 4. The Instruction Set Semantics.

Opcode fetches the word from the instruction memory given by the address in the program counter and then coerces it into an instruction using Decode_word.

8 Completing the Specification

A complete instruction–set architecture specification is based on the preceding definitions and describes all valid state streams. That is, the specification is a predicate or function that describes, for any current state, what the next state will be.

We created the generic interpreter theory to serve as a framework for specifying and verifying microprocessors [Win93b]. The methodology presented in the preceding sections is largely based on the specification requirements of the generic interpreter theory. This section shows how those definitions can be used with the generic interpreter theory to complete the microprocessor specification.

Table 1. Running Time Data for Instruction Set Definition

No. of Instructions	Run Time (sec)	GC Time (sec)	Theorems
25	101.2	117.3	41519
50	724.4	829.7	210217
100	4475.4*	2760.3*	1022795*

The generic interpreter theory requires the user to supply a number of definitions that it combines to create a specification of the proper form. The user must also provide definitions of the underlying level in the verification hierarchy so that the generic interpreter theory can provide the goals that need to be solved to complete the verification. In this paper, we are concerned only with the architectural level specification, so we will ignore the definitions of the underlying level and assume that they exist.

The user must provide three functions from the architectural level to complete the specification. Only two of them are of interest in this particular example:

1. A function, called `instructions` in the theory, that given an instruction returns a state transition function. This is the `M_INST` function defined in Section 6.
2. A function, called `select` in the theory, that given a state returns an instruction. This is the `Opcode` function defined in Section 7.

The third definition, dealing with output, is not relevant to UINTA and is given a dummy instantiation.

From these definitions, the generic interpreter theory creates the following definition for the architectural level interpreter:

```
⊢ Arch_Interp orep mrep s e =
     (∀t.
        let k = Opcode orep mrep s e in
        (s (t + 1)) = M_INST orep mrep k (s t) (e t))
```

The interpreter definition relates the state at time (t+1) to the state at time t by means of the `M_INST` function.

9 Limitations

The primary limitation of the techniques we have outlined stems from the limitations of the underlying HOL theorems proving system: we use `define_type` and `prove_constructors_distinct` which is $O(n^3)$ in the number of type constructors. Thus, for our small instruction set, the specifications run fairly quickly, but for large instruction sets (the Digital's Alpha AXP architecture has over 400 instructions), the technique could be unwieldy.

Table 1 gives the run times for instruction sets with different numbers of instructions on an HP735 running AKCL. The machine has 128Mb of main memory and limits core sizes to 80Mb. The table entry for 100 instructions represents about half of the total effort since the core image got too big on the second large part of the test and the process died. Assuming that it would take about twice that long as the time indicated, then one can see that the run time is proportional to n^3.

None of the run times, even for 100 instructions, are unreasonable given that a few years ago we waited longer than this for much smaller results. Even an instruction set larger than 400 would take less than two days to run. The biggest concern is the size of the core image and what that means for proofs that depend on it. Our experience is that other LISP implementations have better memory management than AKCL. In addition, the term representation of HOL88 is fairly inefficient. We plan to run tests with other implementations of LISP and with HOL90 to determine an upper limit on the size of an instruction set that can be reasonably specified using this method.

Another solution is to partition the instruction set into classes that correspond to the semantic frameworks described earlier. These partitions would be smaller, allowing efficient use of define_type, and still allow good instruction case split. In fact, for large instruction sets, such a partitioning would probably lead to more efficient proof results.

10   Conclusion

We have shown how the syntax and semantics of a microprocessor instruction set can be specified in the HOL theorem proving system. We have used these techniques in several microprocessor verification exercises to evaluate the efficacy and conclude that they lead to clearer specifications and easier proofs.

References

[Bow87]   Jonathan P. Bowen. Formal specification and documentation of microprocessor instruction sets. In *Microprocessing and Microprogramming 21*, pages 223–230, 1987.

[CCLO88] S. D. Crocker, E. Cohen, S. Landauer, and H. Orman. Reverification of a microprocessor. In *Proceedings of the IEEE Symposium on Security and Privacy*, pages 166–176, April 1988.

[Coh88a] Avra Cohn. Correctness properties of the VIPER block model: The second level. Technical Report 134, University of Cambridge Computer Laboratory, May 1988.

[Coh88b] Avra Cohn. A proof of correctness of the VIPER microprocessor: The first level. In G. Birtwhistle and P. Subrahmanyam, editors, *VLSI Specification, Verification, and Synthesis*, pages 27–72. Kluwer Academic Publishers, 1988.

[Gor83]   Michael J.C. Gordon. Proving a computer correct. Technical Report 41, Computer Lab, University of Cambridge, 1983.

[GW92]   Jody W. Gambles and Phillip J. Windley. Integrating formal verification with CAD tool environments. Technical Report LAL-92-02, University of Idaho, Laboratory for Applied Logic, April 1992. In review, Formal Methods in System Design.

[Her92]   John Herbert. Incremental design and formal verification of microcoded microprocessors. In V. Stavridou, T. F. Melham, and R. T. Boute, editors, *Theorem Provers in Circuit Design, Proceedings of the IFIP WG 10.2 International Working Conference, Nijmegen, The Netherlands*. North–Holland, June 1992.

[Hun87]   Warren A. Hunt. The mechanical verification of a microprocessor design. In D. Borrione, editor, *From HDL Descriptions to Guaranteed Correct Circuit Designs*. Elsevier Scientific Publishers, 1987.

[Hun89]   Warren A. Hunt. Microprocessor design verification. *Journal of Automated Reasoning*, 5:429–460, 1989.

[Joy88]   Jeffrey J. Joyce. Formal verification and implementation of a microprocessor. In G. Birtwhistle and P.A Subrahmanyam, editors, *VLSI Specification, Verification, and Synthesis*. Kluwer Academic Press, 1988.

[Joy89]   Jeffrey J. Joyce. *Multi–Level Verification of Microprocessor–Based Systems*. PhD thesis, Cambridge University, December 1989.

[Leo89]   Timothy E. Leonard. Specification of computer architectures. Technical report, Cambridge University Computer Laboratory, 1989.

[SB90]   M. Srivas and M. Bickford. Formal verification of a pipelined microprocessor. *IEEE Software*, 7(5):52–64, September 1990.

[SWL93]   E. Thomas Schubert, Phillip J. Windley, and Karl Levitt. Report on the ucd microcoded viper verification project. In Jeffery J. Joyce and Carl Seger, editors, *Proceedings of the 1993 International Workshop on the HOL Theorem Prover and its Applications.*, August 1993.

[TK93]   Sofiene Tahar and Ramayya Kumar. Implementing a methodology for formally verifying RISC processors in HOL. In Jeffery J. Joyce and Carl Seger, editors, *Proceedings of the 1993 International Workshop on the HOL Theorem Prover and its Applications.*, August 1993.

[WC94]   Phillip J. Windley and Michael Coe. The formal verification of instruction pipelines. Technical Report LAL-94-01, Brigham Young University, Laboratory for Applied Logic, February 1994. (submitted to Theorem Provers in Circuit Design 94).

[Win87]   Glynn Winskel. Models and logic in MOS circuits. In M. Broy, editor, *Logic of Programming and Calculi of Discrete Designs: International Summer School Directed by F. L. Bauer, M. Broy, E. W. Dijkstra, and C. A. R. Hoare*, volume 36 of *NATO ASI Series, Series F: Computer and Systems Sciences*, pages 367–413. Springer–Verlag, 1987.

[Win90a]   Phillip J. Windley. *The Formal Verification of Generic Interpreters*. PhD thesis, University of California, Davis, Division of Computer Science, June 1990.

[Win90b]   Phillip J. Windley. A hierarchical methodology for the verification of microprogrammed microprocessors. In *Proceedings of the IEEE Symposium on Security and Privacy*, May 1990.

[Win92]   Phillip J. Windley. Abstract theories in HOL. In Luc Claesen and Michael J. C. Gordon, editors, *Proceedings of the 1992 International Workshop on the HOL Theorem Prover and its Applications*. North–Holland, November 1992.

[Win93a]   Phillip J. Windley. Documentation for the records library. Technical Report LAL-93-15, Brigham Young University, Laboratory for Applied Logic, July 1993. (Available electronically at URL: http://lal.cs.byu.edu/lal/holdoc/library/records/records.html).

[Win93b] Phillip J. Windley. A theory of generic interpreters. In George J. Milne and Laurence Pierre, editors, *Correct Hardware Design and Verification Methods*, number 683 in Lecture Notes in Computer Science, pages 122–134. Springer-Verlag, 1993.

[Win94] Phillip J. Windley. Formal modeling and verification of microprocessors. *IEEE Transactions on Computers*, 1994. (to appear).

[WM91] Wai Wong and Tom Melham. The HOL wordn library. In *HOL System Library Documentation*. University of Cambridge Computer Laboratry, October 1991.

Representing higher-order logic proofs in HOL

J. von Wright

Åbo Akademi University, 20520 Turku, Finland

Abstract. We describe an embedding of higher order logic in HOL. Types, terms, sequents and inferences are represented as HOL types, and the notions of proof and provability are defined. Using this formalisation, we can reason about the correctness of derived rules and about the relations between different notions of proofs. The formalisation is also intended to make it possible to reason about programs that handle proofs as their data (e.g., proof checkers).

1 Introduction

This paper describes a formalisation of higher order logic proof theory within the logic of the HOL system. The aim is to be able to reason about the proofs that the HOL system produces. This can be useful in a number of ways. It gives a basis for reasoning about programs that handle proofs. One specific kind of program that we have in mind is a proof checker: a program that takes a purported HOL proof as input and checks that it actually is a proof. Furthermore, our theory can be used in formal reasoning about the HOL system itself. For example, the HOL system has implemented a number of non-primitive inference rules as basic rules (for efficiency reasons). Using our formalisation, it is possible to verify the soundness of such rules. Our formalisation also permits us to define different notions of proof (e.g., tree-structured proofs and linear proofs) and study how they are related.

Overview of the paper

Our aim is to formalise (in HOL) the logic of the HOL theorem prover. Since the implementation of HOL departs slightly from the specification of the logic (both are described in [3]), some clarification is needed. First, the implementation collects assumptions of sequents in lists rather than in sets. Our formalisation uses sets. Second, the implementation of the type instantiation rule (the INST_TYPE inference rule) permits the names of free variables to be changed. In this case, our formalisation follows the implementation. The reason for this design choice is that we want to be able to reason about proofs as they are recorded by the HOL system.

We define two new types, representing HOL types and terms. We formalise a number of proof-theoretic concepts that are needed in the discussion of proofs, such as the concept of a variable being free in a term, a term having a certain type, two terms being alpha-equivalent etc.

We also define a type of sequents and a type of (primitive) inferences. Using these notions, we define what it means for a term to be provable, given a list of axioms. We then define a notion of proof and show that the notions of provability and proof agree. Finally, we define the notion of derived inference and show how one can reason about derived rules of inference.

The aim of our work is to be able to reason about proofs, not to generate them. Thus, we need only be able to recognise a correct inference, once the result is given. This means that we do not have to capture HOL's intricate (and ill-documented) procedures for variable renaming used in some inference rules. Our formalisation permits arbitrary renaming schemes, and the one used by HOL is a special instance.

A *theory* in HOL is characterised by a type structure, a set of constants and a set of axioms. A type structure is represented by a list of pairs (op,n) of type string#num, where n is the arity of the type operator op. The constants are represented by a list of pairs (const,ty) where ty is the possibly polymorphic generic type of the constant const. Finally, the axioms are represented by a list of sequents. Sequents, in turn, are pairs (as,tm), where as is a set of terms (the assumptions) and tm is a term (the conclusion). The equations that define constants are also considered to be axioms.

For every concept that we have formalised, we have also written a *proof function*. For example, when we define a new constant foo by a defining theorem

$$\vdash \text{foo x y} = E$$

we also provide an ML function Rfoo which can be called in the following way:

```
#Rfoo ["e1";"e2"];;
⊢ foo e1 e2 = ...
```

where the right hand side is canonical (i.e., it cannot be simplified further using definitional theorems). Essentially, these proof functions do rewriting, but in an efficient way, compared to the REWRITE_RULE function.

The theory described in this paper comes as a contribution with HOL88 version 2.02. A more detailed description of the theory, together with listings, can be found in [7]. A port for HOL90 also exists.

Notation

We assume that the reader is familiar with the HOL theorem prover and its syntax [4, 3]. We have used Version 2.02 of the HOL88 system. Our theories also use the string and finite_sets libraries. We mainly use the syntax of HOL, but we use ordinary logical symbols, rather than ASCII character combinations. When referring to HOL objects, we use typewriter font. Similarly, we show examples of interaction with the HOL system in typewriter font.

2 Types

The HOL logic has four different kinds of types: type constants, type variables, function types and *n*-ary type operators. To make the definition simple, we consider type constants and the function type to be special cases of type operators.

Thus we have represented types by a new type with the following syntax:

```
Type = Tyvar string
     | Tyop string (Type)list
```

To distinguish these "HOL-as-object-logic-types" from the HOL types we will from now on call them **Types**.

The type structure of the current theory is represented by a list of pairs of type :**string#num**. For example, the simplest possible theory (referring only to booleans) has the following type structure list:

```
[('bool',0);('fun',2)]
```

The HOL type :**bool** is then represented by **Tyop 'bool' []** while the function type :**bool→bool** is represented by

```
Tyop 'fun' [Tyop 'bool' [];Tyop 'bool' []]
```

2.1 Functions for types

We have developed some infrastructure (i.e., some ML functions) for making recursive function definitions over **Type**. As an example, the function **Type_OK** is defined as follows:

```
#let Type_OK_DEF = new_Type_rec_definition('Type_OK_DEF',
#"(Type_OK Typl (Tyvar s) = T) ∧
# (Type_OK Typl (Tyop s ts) =
#  mem1 s Typl ∧ (LENGTH ts=corr1 s Typl) ∧ EVERY(Type_OK Typl)ts)");;
```

For this input, the HOL system returns the definitional theorem **Type_OK_DEF**:

```
⊢ (∀tyl s. Type_OK Typl (Tyvar s) = T) ∧
  (∀tyl s ts. Type_OK Typl (Tyop s ts) =
     mem1 s Typl ∧ (LENGTH ts=corr1 s Typl) ∧ EVERY(Type_OK Typl)ts)
```

Here **mem1 s 1** holds if **s** is the first component of some pair in the list **1** and **corr1 s 1** is the corresponding second component (these are defined in a separate theory containing useful definitions and theorems, mainly about lists). The theorem says that a **Type** is OK if it is a type variable or it is composed from OK types by a permitted type operator.

Similarly, we have defined other functions on **Types**. For example, **Type_occurs a ty** is defined to hold if the type variable **a** occurs anywhere in the type **ty**. The function **Type_compat** is defined so that **Type_compat ty ty'** holds when **ty** is compatible with **ty'**, in the sense that the structure of **ty** is can be mapped onto the structure of **ty'**. This function does not allow us to tell whether a type instantiation is correct. For example, we must be able to detect that **bool→num** is not a correct instantiation of the polymorphic type ***→***, even though these two types are compatible. For this, we have defined **Type_instl** so that **Type_instl ty ty'** returns the list of type instantiations used in going from **ty** from **ty'**. This list can then be checked for consistency, using a separate function.

3   Terms

A HOL term can be a constant, a variable, an application or an abstraction. Thus terms are represented by a new type with the following syntax:

```
Pterm = Const string Type
      | Var string#Type
      | App Pterm Pterm
      | Lam string#Type Pterm
```

We call these objects **Pterms**, to distinguish them from the HOL terms that they represent. Variable names are represented by strings. The reader should note that we compose a lambda abstraction from a pair of type **string#Type** and a **Pterm**, whereas in the term syntax of the HOL system, lambda abstraction is composed from two terms. Our syntax makes the checking of well-formedness easier.

The constants of the current theory are represented by a list. A constant always has a generic type which is given in this list. When the constant occurs in a term, it has an actual type which must be an instance of the generic type. A simple logic might have the following the list of constants:

```
['T',Tyop 'bool'[];
 'F',Tyop 'bool'[];
 '=',Tyop 'fun' [Tyvar '*';Tyop 'fun' [Tyvar '*';Tyop 'bool' []]];
 '⇒',Tyop'fun' [Tyop'bool' [];Tyop'fun'[Tyop'bool' [];Tyop'bool' []]]]
```

i.e., truth, falsity, equality and implication.

Equality on booleans is represented by the **Pterm**

```
Const '='(Tyop'fun'[Tyop'bool'[];Tyop'fun'[Tyop'bool' [];Tyop'bool' []]])
```

Note that the **Type** of this **Pterm** is an instance of the **Type** of equality in the above list, with **Tyop'bool'[]** replacing **Tyvar '*'**.

3.1   Well-typedness

Every **Pterm** has a unique **Type**, computed by the function **Ptype_of**. Our syntax permits terms which are ill-typed, in the sense that they do not correspond to the any "real" HOL terms. A term is well-typed if it satisfies two requirements. First, the constants occurring in the term must have types which are correct instantiations of their generic types. Second, the types of the two subterms in an application must match. The function Pwell_typed checks these conditions.

At this point, we could have introduced a new type which represents "well-typed terms". However, this would be incompatible with our main aim, which is to represent terms in a way which makes it possible to do proof checking. Since proof checking involves both correctness of inferences and well-formedness of terms, we must permit ill-formed (ill-typed) terms to appear in purported proofs.

3.2 A function for compressing terms

Our Pterms quickly become very large and ugly. Even a simple HOL-term like

$$\lambda x. \ x \Rightarrow (x = y)$$

becomes the massive Pterm

```
Lam('x',Tyop 'bool'[])
  (App(App(Const '⇒ '
    (Tyop'fun'[Tyop'bool'[];Tyop'fun'[Tyop'bool'[];Tyop'bool'[]]]))
    (Var('x',Tyop 'bool'[])))
  (App(App(Const '='
    (Tyop'fun'[Tyop'bool'[];Tyop'fun'[Tyop'bool'[];Tyop'bool'[]]]))
    (Var('x',Tyop 'bool'[])))
    (Var('y',Tyop 'bool'[]))))))
```

which is difficult both to write and read. To simplify things, we have an ML function tm_trans which translates a HOL-term into the corresponding Pterm:

```
#tm_trans "λ(x:bool).x";;
"Lam ('x',Tyop 'bool'[])
     (Var('x',Tyop 'bool'[]))"
```

and a function tm_back which does the opposite translation

```
#tm_back "Lam ('x',Tyop 'bool'[])
#               (Var('x',Tyop 'bool'[]))";;
"λx. x" : term
```

These functions are used for entering and displaying terms that are used in simple examples.

3.3 Free and bound variables

The notion of free and bound variables are defined in the obvious way. For example, we define Pfree so that Pfree x t holds if the variable x occurs free in the Pterm t. Similarly, we define the functions Pbound and Poccurs.

We also have versions of these constants that work on collections of variables and Pterms. For example, Plallnotfree xl ts holds if no variable in the list xl is Pfree in any of the Pterms in the set ts.

3.4 Alpha-renaming

Alpha-renaming and substitution of a term for a variable are closely bound together. We have defined `Palreplace` so that `Palreplace t' tvl t` holds if `t'` is the result of substituting in `t` according to the list `tvl` and alpha-renaming. The list `tvl` consists of pairs `(t,a)` of type `Pterm#(string#Type)`, indicating what terms should be substituted for what variables. The definition of `Palreplace` is shown in the Appendix.

In order to appreciate larger examples and tests, we have a compressing function `th_back` for theorems, similar to `tm_back` described earlier. It prints `Pterms` using `tm_back` and functions using dummy-functions that we have added. For example, we have declared a dummy constant `Xalreplace` which corresponds to `Palreplace`.

Evaluating

```
#RPalreplace [tm_trans "λz.z ⇒ x";
#            "[Var('x',Tyop'bool'[]),'y',Tyop 'bool'[]]";
#            tm_trans "λx.x ⇒ y"];;
```

yields a massive theorem, stating that this substitution is in fact correct. However, if we apply `th_back` to this theorem, we get it in a form which is easier to read:

```
#th_back it;;
⊢ Xalreplace(λz. z ⇒ x)[x,'y',Tyop 'bool'[]](λx. x ⇒ y) = T
```

This says that $\lambda z.\ z \Rightarrow x$ is a correct result when substituting x for y in $\lambda x.\ x \Rightarrow y$. Note that x here is a compressed notation for `Var('x',Tyop'bool'[])`, while `'y'` is not compressed, i.e., it is in fact a one-character string.

We define alpha-equivalence using an empty substitution:

$$\vdash_{def} \forall t'\ t.\ \texttt{Palpha } t'\ t = \texttt{Palreplace } t'[]\ t$$

The following example shows that our corresponding proof function `RPalpha` also detects incorrect alpha-renamings:

```
#th_back(RPalpha[tm_trans "λy y.y ⇒ y" ; tm_trans "λx y.y ⇒ x"]);;
⊢ Xalpha(λy y. y ⇒ y)(λx y. y ⇒ x) = F
```

i.e., the terms $\lambda y\, y.\ y \Rightarrow y$ and $\lambda x\, y.\ y \Rightarrow x$ are not alpha-equivalent.

3.5 Multiple substitutions

Using `Palreplace` we have formalised HOL's notion of a substitution, as it occurs in the inference rule `SUBST`. Assume that `ttvl` is a list of triples each having type `Pterm#Pterm#(string#Type)`. For each triple `(tm',tm,d)` in this list, `tm'` is a `Pterm` that is to replace `tm` and `d` is a dummy variable used to indicate the positions where this substitution is to be made. Then `Psubst t'`

ttvl td t holds if t is the result of substituting some **tm**-terms for d-dummies in the term **t** and if **t'** is the result of substituting **tm'**-terms for d-dummies. Both substitutions are done according to **ttvl**, and they may involve alpha-renaming.

The corresponding proof function is **RSubst** and it can recognise both correct and incorrect substitutions.

3.6   Type instantiation

Type instantiation, as done by the inference rule **INST_TYPE** in HOL, is quite tricky to check. First, it is necessary to check that the type instantiation has not identified two variables that were previously distinct. Second, the type instantiation rule permits free variables to be renamed (in this respect we follow the HOL implementation rather than the specification of the HOL logic, cf. the discussion in Section 1).

Checking a renaming of a free variable is more complicated than checking a renaming of a bound variable, because bound variables are always "announced" (in the left subtree of the abstraction), but a free variable can occur in two widely separated subtrees, without being announced in the same way.

Assume that **tyl** is a list of pairs of type **Type#string**, indicating what types are to be substituted for what type variables. Furthermore assume that **as** is a set of **Pterms** (they represent the assumption of the theorem that is to be type-instantiated). Then **Ptyinst as t' tyl t** holds if **t'** is the result (after renaming) of replacing type variables in **t** according to **tyl** and if no variables that are type instantiated occur free in **as**. **Ptyinst** is defined using **Palreplace** and a number of other auxiliary functions (some of these are described in Section 2.1).

4   Sequents and Inferences

We represent sequents by a new concrete type with a very simple syntax:

```
Pseq (Pterm)set Pterm
```

The first argument to **Pseq** is the set of assumptions and the second argument is the conclusion. The corresponding destructor functions are **Pseq_assum** and **Pseq_concl**.

4.1   Inferences in the HOL system

An inference step in the HOL logic consists of a *conclusion* (result sequent) that is "below the line" and a list of *hypotheses* (argument sequents) that are "above the line".

In the HOL system implementation, inference rules are functions which in addition to the hypotheses may require some information (e.g., a term) in order to compute the conclusion. For example, the rule of abstraction (**ABS**) in the logic is

$$\frac{\Gamma \vdash \qquad t = t'}{\Gamma \vdash (\lambda x.\, t) = (\lambda x.\, t')}$$

(with the side condition that x must not occur free in Γ). As an inference rule in the HOL system, ABS is a function which takes a term (representing the variable x) and a theorem (the hypothesis) as arguments and returns a theorem (the conclusion).

4.2 Inferences as a new type

We represent inferences as syntactic objects of a new type. This type has nine constructors; one for inference by hypothesis and one for each primitive inference rule of the HOL logic (the logic has eight primitive inference rules). The syntax is

```
Inference = AX_inf Psequent
          | AS_inf Psequent Pterm
          | RE_inf Psequent Pterm
          | BE_inf Psequent Pterm
          | SU_inf Psequent (Psequent#string#Type)list Pterm Psequent
          | AB_inf Psequent Pterm Psequent
          | IN_inf Psequent (Type#string)list Psequent
          | DI_inf Psequent Pterm Psequent
          | MP_inf Psequent Psequent Psequent
```

Here AX_inf represents an inference by hypothesis (by axiom), while the remaining cases each correspond to a primitive inference rule: AS_inf for ASSUME, RE_inf for REFL, BE_inf for BETA_CONV, SU_inf for SUBST, AB_inf for ABS, IN_inf for INST_TYPE, DI_inf for DISCH and MP_inf for MP. The first argument of each constructor is the conclusion of the inference. The remaining arguments represent hypotheses and other arguments.

4.3 Checking inferences

The function OK_inf is defined to represent the notion of correct inference. Thus OK_inf i holds if and only if i represents a correct inference, according to the primitive inference rules of the HOL logic.

The proof function for OK_inf is ROK_inf, and it identifies both correct and incorrect inferences. Using the compressing functions, we check a simple inference:

```
#th_back (ROK_Inf[Typl;Conl;Axil;
#      "BE_inf (Pseq {} ^(tm_trans "(λ(x:bool).x)y = y"))
#          ^(tm_trans "(λ(x:bool).x)y")"]);;
⊢ OK_XInf (BE_Xinf (Xseq {} ((λx. x)y = y)) ((λx. x)y))
```

This tells us that the theorem $\vdash (\lambda x.\ x)y = y$ is the result of the following application of the BETA_CONV inference rule:

```
#BETA_CONV "(λx. x)y"
```

4.4 Primitive inferences

We shall now show how the nine different kinds of inferences are checked. For each inference rule, we define a function which returns a boolean value: **T** for a correct inference and **F** for an incorrect one. These functions are used by the function **OK_Inf** described above (the definition of **OK_Inf** is shown in the Appendix).

The **ASSUME** rule is modelled by the function **PASSUME**:

$\vdash_{def}$ ∀Typl Conl as t tm. PASSUME Typl Conl (Pseq as t) tm =
    Pwell_typed Typl Conl tm ∧ Pboolean tm ∧ (t = tm) ∧ (as = {tm})

where **Pboolean tm** is defined to mean that the **Pterm tm** has boolean **Type**.

Notice that this is where well-typedness is enforced. The check ensures that the conclusion sequent **Pseq as t** is well-typed. To make this check, we must have the type structure **Typl** and the constant list **Conl** as explicit arguments to **ASSUME**.

In a similar way, the **REFL** and **BETA_CONV** inferences are modelled by **PREFL** and **PBETA_CONV**. Thus

    PREFL Typl Conl (Pseq as t) tm

holds if the assumption set **as** is empty, **t** represents the term **tm=tm**, and **tm** is well-typed. Similarly,

    PBETA_CONV Typl Conl (Pseq as t) tm

holds if the assumption set **as** is empty, **tm** is a beta-redex which reduces to **t**, and **t** is well-typed and boolean.

The **SUBST** rule is modelled by **PSUBST**:

    PSUBST Typl Conl (Pseq as t) thdl td th

holds if the sequent **Pseq as t** is the result of performing a multiple substitution in theorem **th** according to the list **thdl** of pairs (theorem,dummy), where **td** is a term with dummies indicating the places where substitutions are to be made. **PSUBST** also checks the dummy term **td** for well-typedness.

The function **PABS** models the **ABS** inference. Thus

    PABS Typl Conl (Pseq as t) tm th

holds if **t** is the result of abstracting over the term **tm** (which must be a variable with a permitted type) on both sides of the conclusion of **th** which must be an equality). Furthermore, the variable **tm** must not occur free in the assumption set **as**.

For the **INST_TYPE** inference, we have defined **PINST_TYPE** so that

    PINST_TYPE Typl (Pseq as t) tyl th

holds if **t** is the result of instantiating types in the conclusion of **th** according to **tyl** and if **as** is the same set as the assumptions in **th**. Furthermore, we require that the type variables that are being substituted for do not occur in **as**.

Finally,

```
PDISCH Typl Conl (Pseq as t) tm th
```

holds if **Pseq as t** is the result of discharging the term **tm** in the theorem **th**, and

```
PMP (Pseq as t) th1 th2
```

holds if **Pseq as t** is the result of a Modus Ponens inference on **th1** and **th2**.

5 Proofs and provability

In this section, we consider the notions of provability and proofs. These two concepts are closely related, but we define them independently of each other. Both depend on the underlying notion of correct inference, i.e., on the predicate **OK_Inf** defined above.

5.1 Provability

Provability is an inductive concept. A sequent is provable (within a given theory) if it is an axiom or it can be inferred from provable sequents by application of an inference rule.

We have defined the predicate **Provable** using the basic ideas from the HOL package for inductive definitions [3]. The inductive nature of provability is captured in the following theorem:

```
⊢ ∀Typl Conl Axil i s.
  (OK_Inf Typl Conl Axil i ∧ (s = Inf_concl i)) ∧
  EVERY (Provable Typl Conl Axil) (Inf_hyps i)
  ⇒ Provable Typl Conl Axil s
```

In fact, **Provable** is defined to be the smallest relation satisfying this theorem. Note that we have a base case and an inductive case together here. The base case occurs when the list **Inf_hyps i** (the hypothesis list of the inference i) is empty. We have also proved the induction theorem (rule induction) for the **Provable** predicate.

5.2 Proofs

By a proof we mean a sequence of correct inferences where each inference has the property that all its hypotheses appear as conclusions of some inference earlier in the proof.

This is captured in the following definition of **Is_proof**:

```
⊢ (∀Typl Conl Axil. Is_proof Typl Conl Axil[] = T) ∧
  (∀Typl Conl Axil i P. Is_proof Typl Conl Axil (CONS i P) =
     OK_Inf Typl Conl Axil i ∧
     lmem (Inf_hyps i) (MAP Inf_concl P) ∧
     Is_proof Typl Conl Axil P)
```

The corresponding proof function is RIs_proof, which is in fact a proof checker. The following is an example of a (compressed) theorem produced using this proof function.

```
⊢ Is_Xproof
    [MP_Xinf (Xseq {y = y} (x = x))
             (Xseq {} ((y = y) ⇒ (x = x)))
             (Xseq {y = y} (y = y));
     AS_Xinf (Xseq {y = y} (y = y)) (y = y);
     DI_Xinf (Xseq {} ((y = y) ⇒ (x = x))) (y=y) (Xseq {} (x=x));
     RE_Xinf (Xseq {} (x = x)) x]
   = T
```

This theorem states that the following is a correct proof:

1. ⊢ $x = x$ by REFL
2. ⊢ $y = y ⇒ (x = x)$ by DISCH, 1
3. $\{y = y\}$ ⊢ $y = y$ by ASSUME,
4. $\{y = y\}$ ⊢ $x = x$ by MP, 2,3

This is an example of adding an assumption to a theorem.

5.3 Relating proofs and provability

Proofs and provability are obviously related: a sequent should be provable if and only if there is a proof of it. We have proved that this in fact the case (this can be seen as a check that our definitions are reasonable):

```
⊢ Provable Typl Conl Axil s =
    (∃i P. Is_proof Typl Conl Axil(CONS i P) ∧ (s = Inf_concl i))
```

The proof of this theorem rests on the fact that appending two proofs yields a new proof. Given proofs of all the hypotheses of an inference, this fact allows us to construct a proof of the conclusion by appending all the given proofs and adding the given inference.

5.4 Reasoning about proofs

There is, of course, no way to prove that our definition of a proof actually captures the HOL notion of a proof. However, we can reason about proofs and check that they satisfy some minimal requirements. As an example of this, we have shown that proofs can only yield sequents where the hypotheses and the conclusions are well-typed and boolean:

⊢ ∀P. Is_proof Typl Conl Axil P ∧ Is_standard(Typl,Conl,Axil) ⇒
 EVERY Pseq_boolean(MAP Inf_concl P) ∧
 EVERY (Pseq_well_typed Typl Conl) (MAP Inf_concl P)

where Is_standard holds for the triple (Typl,Conl,Axil) if the type structure contains booleans and function types, the constant list contains polymorphic equality and implication and the axiom list contains only well-typed boolean sequents.

In one respect, the above theorem is very important; it shows that the well-typedness checks in the functions that are used when checking an inference (described in Section 4) are sufficient to guarantee that all conclusions that appear in a proof are well-typed. However, as the above theorem shows, this requires that all the theorems that are assumed as axioms are well-typed.

6 Derived inferences

In real proofs, we often use derived rules of inference, rather than the primitive inference rules of a logic. Derived rules do not extend the logic, but they are convenient, as they make proofs shorter. The HOL system has a number of derived inference rules hard-wired into the system. This means that every HOL-proof consists of inferences belonging to a set of some thirty inference rules, rather than the eight primitive rules of the logic. Thus derived rules are an essential feature at the core of the HOL system.

6.1 Definition of derived inference

In order to make derived inference rules uniform, we let them have three arguments: the name of the rule, the conclusion, and a list of hypotheses. We have a derived inference (Dinf) of a sequent s from a list of sequents sl if s can be proved when sl is added to the list of axioms:

⊢$_{def}$ ∀Typl Conl Axil name s sl. Dinf Typl Conl Axil name s sl =
 (EVERY Pseq_boolean sl ∧ EVERY (Pseq_well_typed Typl Conl) sl
  ⇒ Provable Typl Conl (APPEND sl Axil) s)

Note that the name argument (which has type string) here is vacuous; it is not mentioned on the right hand side of the definition. It is intended to be used to associate a name with a derived inference rule. It could be left out without changing the theory in any essential way.

6.2 Verifying the correctness of a derived rule of inference

As an example, we formalise the rule for adding an assumption to a theorem (the ADD_ASSUM rule of the HOL system). In traditional notation, this rule is expressed as follows:

$$\frac{\Gamma \vdash t}{\Gamma, t' \vdash t}$$

This rule is encoded in the following theorem, which we have proved:

⊢ ∀Typl Conl Axil G t' t. Pwell_typed Typl Conl t' ∧ Pboolean t'
    ⇒ Dinf Typl Conl Axil 'ADD_ASSUM' (Pseq (t' INSERT G) t) [Pseq G t]

The proof of this theorem is in fact a verification of the correctness of the derived inference rule ADD_ASSUM.

Note that derived rules added in this fashion relate hypotheses and conclusion without additional arguments.

6.3 Proofs with derived inferences

We have also defined a new notion of proof, Is_Dproof, where derived inferences are permitted. We have proved (in HOL) that Is_proof and Is_Dproof are equally strong, in the sense that whenever there is a Dproof of a sequent, there is also a proof of it, and vice versa. This is quite reasonable, since both notions of proof are directly related to the notion of provability.

For proof checking purposes we have to stick to Is_proof, however. This is because proving that a purported derived inference step is incorrect requires proving that no sequence of inferences could yield the conclusion in question, and this is much more complicated than proving that a proposed primitive inference is incorrect.

7 Conclusion

We have defined in the logic of HOL a theory which captures the notions of types, terms and inferences that are used in the HOL logic. Within this theory we defined the notions of provability and of proof and proved them to be related in the desired way: a boolean term is provable if and only if there exists a proof of it. Together with the HOL theory, we have developed ML functions for proving each property introduce.

These function are in fact a proof checker, i.e., a program which takes a purported proof as input and determines whether it is a proof or not. This proof checker is extremely slow, since it computes the result by performing a proof inside HOL (the example shown in Section 5.2 took 1 minute to run on a Sparcstation ELC with plenty of memory). It is our hope that the theory of proofs can also be used as a basis for verifying more efficient proof checkers for higher order logic. Work on such a proof checker is under way [9], and we believe that the methodology described in [8] can be used to verify a proof checker.

HOL is a fully expansive theorem prover, which means that when proving theorems, it reduces derived rules of inference to sequences of basic inferences. Since our theory of proofs includes a method for proving the correctness of derived rules of inference, we have provided a formal basis for a faster HOL, where derived rules of inference can be added to the core of the system, once they have been proved correct. This idea was suggested for the HOL system by Slind [6].

It seems that there is generally a growing interest in using theorem proving system in the "introspective" way that we have described here. Similar ideas in a different framework are reported in [2], where a type checker for the Calculus of Constructions is implemented in the logic of Nqthm (the Boyer-Moore system). Related work on using proof-checkers to check metatheory is reported in [1] and [5].

Acknowledgments

I am grateful to Mike Gordon for many helpful discussions, and Tom Melham for showing how to define inductive relations in HOL. I also want to thank members of the Automated Reasoning Group in Cambridge and the anonymous referees for their comments on earlier versions of this paper. Financial support from the Science and Engineering Research Council in Great Britain and the Research Institute of the Åbo Akademi Foundation in Finland is gratefully acknowledged.

References

1. T. Altenkirch. A formalization of the strong normalisation proof for system F in LEGO. In *Typed Lambda Calculus and Applications*, Lecture Notes in Computer Science 664, pages 13–28, 1993.
2. Robert S. Boyer and Gilles Dowek. Towards checking proof-checkers. In *Workshop on Types for Proofs and Programs (Types '93)*, 1993.
3. M. Gordon and T. Melham. *Introduction to HOL*. Cambridge University Press, New York, 1993.
4. M.J.C. Gordon. HOL: A proof generating system for higher-order logic. In G. Birtwistle and P.A. Subrahmanyam (ed.), *VLSI Specification, Verification and Synthesis*. Kluwer Academic Publishers, 1988.
5. J. McKinna and R. Pollack. Pure type systems formalized. In Herman Geuvers, editor, *Typed Lambda Calculus and Applications*, Lecture Notes in Computer Science 664, pages 289–305, 1993.
6. K. Slind. Adding new rules to an LCF-style logic implementation. In M.J.C. Gordon L.J.M. Claesen, editor, *Higher Order Logic Theorem Proving and its Applications*, pages 549–560, Leuwen, Belgium, September 1992. North-Holland.
7. J. von Wright. Representing higher-order logic proofs in HOL. Techn. Rep. 323, Computer Lab, University of Cambridge, 1994.
8. J. von Wright. Verifying modular programs in HOL. Techn. Rep. 324, Computer Lab, University of Cambridge, 1994.
9. W. Wong. Recording HOL-proofs. Techn. Rep. 306, Computer Lab, University of Cambridge, 1993.

A Sample definitions

This appendix shows some definitions that are part of the theory of proofs. For the complete list of definitions, we refer to [7].

The constant **Palreplace** is defined using an auxiliary constant **Palreplace1** which has an additional argument (a list which contains bound variables encountered so far). The definition also uses other functions that we have defined. The functions **mem2** and **corr2** are similar to **mem1** and **corr1** (see Section 2.1). **Is_var**, **Is_App** and **Is_Lam** check term construction while **Var_var**, **App_fun**, **App_arg**, **Lam_var** and **Lam_bod** are term destructors.

```
Palreplace1_DEF =
⊢ (∀t' vvl tvl s ty.
    Palreplace1 t' vvl tvl(Const s ty) = (t' = Const s ty)) ∧
  (∀t' vvl tvl x. Palreplace1 t' vvl tvl(Var x) =
  ((Is_Var t' ∧ mem1(Var_var t')vvl) →
   (x = corr1(Var_var t')vvl) |
   (¬mem1 x vvl ∧ (mem2 x tvl →  (t'=corr2 x tvl) | (t'=Var x)))))) ∧
  (∀t' vvl tvl t1 t2. Palreplace1 t' vvl tvl(App t1 t2) =
  Is_App t' ∧ Palreplace1(App_fun t')vvl tvl t1 ∧
  Palreplace1(App_arg t')vvl tvl t2) ∧
  (∀t' vvl tvl x t1. Palreplace1 t' vvl tvl(Lam x t1) =
  Is_Lam t' ∧ (SND(Lam_var t') = SND x) ∧
  Palreplace1(Lam_bod t')(CONS(Lam_var t',x)vvl)tvl t1)
```

```
Palreplace_DEF =
⊢ ∀t' tvl t. Palreplace t' tvl t = Palreplace1 t'[]tvl t
```

The inference checker **OK_Inf** is defined as follows:

```
OK_Inf_DEF =
⊢ (∀Typl Conl Axil s. OK_Inf Typl Conl Axil(AX_inf s) = mem s Axil) ∧
  (∀Typl Conl Axil s t.
    OK_Inf Typl Conl Axil(AS_inf s t) = PASSUME Typl Conl s t) ∧
  (∀Typl Conl Axil s t.
    OK_Inf Typl Conl Axil(RE_inf s t) = PREFL Typl Conl s t) ∧
  (∀Typl Conl Axil s t.
    OK_Inf Typl Conl Axil(BE_inf s t) = PBETA_CONV Typl Conl s t) ∧
  (∀Typl Conl Axil s tdl t s1.
    OK_Inf Typl Conl Axil(SU_inf s tdl t s1)
        = PSUBST Typl Conl s tdl t s1) ∧
  (∀Typl Conl Axil s t s1.
    OK_Inf Typl Conl Axil(AB_inf s t s1) = PABS Typl Conl s t s1) ∧
  (∀Typl Conl Axil s tyl s1.
    OK_Inf Typl Conl Axil(IN_inf s tyl s1) = PINST_TYPE Typl s tyl s1) ∧
  (∀Typl Conl Axil s t s1.
    OK_Inf Typl Conl Axil(DI_inf s t s1) = PDISCH Typl Conl s t s1) ∧
  (∀Typl Conl Axil s s1 s2.
    OK_Inf Typl Conl Axil(MP_inf s s1 s2) = PMP s s1 s2)
```

Springer-Verlag
and the Environment

We at Springer-Verlag firmly believe that an international science publisher has a special obligation to the environment, and our corporate policies consistently reflect this conviction.

We also expect our business partners – paper mills, printers, packaging manufacturers, etc. – to commit themselves to using environmentally friendly materials and production processes.

The paper in this book is made from low- or no-chlorine pulp and is acid free, in conformance with international standards for paper permanency.

Lecture Notes in Computer Science

For information about Vols. 1–778
please contact your bookseller or Springer-Verlag

Vol. 821: M. Tokoro, R. Pareschi (Eds.), Object-Oriented Programming. Proceedings, 1994. XI, 535 pages. 1994.

Vol. 822: F. Pfenning (Ed.), Logic Programming and Automated Reasoning. Proceedings, 1994. X, 345 pages. 1994. (Subseries LNAI).

Vol. 823: R. A. Elmasri, V. Kouramajian, B. Thalheim (Eds.), Entity-Relationship Approach — ER '93. Proceedings, 1993. X, 531 pages. 1994.

Vol. 824: E. M. Schmidt, S. Skyum (Eds.), Algorithm Theory – SWAT '94. Proceedings. IX, 383 pages. 1994.

Vol. 825: J. L. Mundy, A. Zisserman, D. Forsyth (Eds.), Applications of Invariance in Computer Vision. Proceedings, 1993. IX, 510 pages. 1994.

Vol. 826: D. S. Bowers (Ed.), Directions in Databases. Proceedings, 1994. X, 234 pages. 1994.

Vol. 827: D. M. Gabbay, H. J. Ohlbach (Eds.), Temporal Logic. Proceedings, 1994. XI, 546 pages. 1994. (Subseries LNAI).

Vol. 828: L. C. Paulson, Isabelle. XVII, 321 pages. 1994.

Vol. 829: A. Chmora, S. B. Wicker (Eds.), Error Control, Cryptology, and Speech Compression. Proceedings, 1993. VIII, 121 pages. 1994.

Vol. 830: C. Castelfranchi, E. Werner (Eds.), Artificial Social Systems. Proceedings, 1992. XVIII, 337 pages. 1994. (Subseries LNAI).

Vol. 831: V. Bouchitté, M. Morvan (Eds.), Orders, Algorithms, and Applications. Proceedings, 1994. IX, 204 pages. 1994.

Vol. 832: E. Börger, Y. Gurevich, K. Meinke (Eds.), Computer Science Logic. Proceedings, 1993. VIII, 336 pages. 1994.

Vol. 833: D. Driankov, P. W. Eklund, A. Ralescu (Eds.), Fuzzy Logic and Fuzzy Control. Proceedings, 1991. XII, 157 pages. 1994. (Subseries LNAI).

Vol. 834: D.-Z. Du, X.-S. Zhang (Eds.), Algorithms and Computation. Proceedings, 1994. XIII, 687 pages. 1994.

Vol. 835: W. M. Tepfenhart, J. P. Dick, J. F. Sowa (Eds.), Conceptual Structures: Current Practices. Proceedings, 1994. VIII, 331 pages. 1994. (Subseries LNAI).

Vol. 836: B. Jonsson, J. Parrow (Eds.), CONCUR '94: Concurrency Theory. Proceedings, 1994. IX, 529 pages. 1994.

Vol. 837: S. Wess, K.-D. Althoff, M. M. Richter (Eds.), Topics in Case-Based Reasoning. Proceedings, 1993. IX, 471 pages. 1994. (Subseries LNAI).

Vol. 838: C. MacNish, D. Pearce, L. Moniz Pereira (Eds.), Logics in Artificial Intelligence. Proceedings, 1994. IX, 413 pages. 1994. (Subseries LNAI).

Vol. 839: Y. G. Desmedt (Ed.), Advances in Cryptology - CRYPTO '94. Proceedings, 1994. XII, 439 pages. 1994.

Vol. 840: G. Reinelt, The Traveling Salesman. VIII, 223 pages. 1994.

Vol. 841: I. Prívara, B. Rovan, P. Ružička (Eds.), Mathematical Foundations of Computer Science 1994. Proceedings, 1994. X, 628 pages. 1994.

Vol. 842: T. Kloks, Treewidth. IX, 209 pages. 1994.

Vol. 843: A. Szepietowski, Turing Machines with Sublogarithmic Space. VIII, 115 pages. 1994.

Vol. 844: M. Hermenegildo, J. Penjam (Eds.), Programming Language Implementation and Logic Programming. Proceedings, 1994. XII, 469 pages. 1994.

Vol. 845: J.-P. Jouannaud (Ed.), Constraints in Computational Logics. Proceedings, 1994. VIII, 367 pages. 1994.

Vol. 846: D. Shepherd, G. Blair, G. Coulson, N. Davies, F. Garcia (Eds.), Network and Operating System Support for Digital Audio and Video. Proceedings, 1993. VIII, 269 pages. 1994.

Vol. 847: A. L. Ralescu (Ed.) Fuzzy Logic in Artificial Intelligence. Proceedings, 1993. VII, 128 pages. 1994. (Subseries LNAI).

Vol. 848: A. R. Krommer, C. W. Ueberhuber, Numerical Integration on Advanced Computer Systems. XIII, 341 pages. 1994.

Vol. 849: R. W. Hartenstein, M. Z. Servít (Eds.), Field-Programmable Logic. Proceedings, 1994. XI, 434 pages. 1994.

Vol. 850: G. Levi, M. Rodríguez-Artalejo (Eds.), Algebraic and Logic Programming. Proceedings, 1994. VIII, 304 pages. 1994.

Vol. 851: H.-J. Kugler, A. Mullery, N. Niebert (Eds.), Towards a Pan-European Telecommunication Service Infrastructure. Proceedings, 1994. XIII, 582 pages. 1994.

Vol. 853: K. Bolding, L. Snyder (Eds.), Parallel Computer Routing and Communication. Proceedings, 1994. IX, 317 pages. 1994.

Vol. 854: B. Buchberger, J. Volkert (Eds.), Parallel Processing: CONPAR 94 – VAPP VI. Proceedings, 1994. XVI, 893 pages. 1994.

Vol. 855: J. van Leeuwen (Ed.), Algorithms – ESA '94. Proceedings, 1994. X, 510 pages.1994.

Vol. 856: D. Karagiannis (Ed.), Database and Expert Systems Applications. Proceedings, 1994. XVII, 807 pages. 1994.

Vol. 857: G. Tel, P. Vitányi (Eds.), Distributed Algorithms. Proceedings, 1994. X, 370 pages. 1994.

Vol. 858: E. Bertino, S. Urban (Eds.), Object-Oriented Methodologies and Systems. Proceedings, 1994. X, 386 pages. 1994.

Vol. 859: T. F. Melham, J. Camilleri (Eds.), Higher Order Logic Theorem Proving and Its Applications. Proceedings, 1994. IX, 470 pages. 1994.

Vol. 860: W. L. Zagler, G. Busby, R. R. Wagner (Eds.), Computers for Handicapped Persons. Proceedings, 1994. XX, 625 pages. 1994.

Vol: 861: B. Nebel, L. Dreschler-Fischer (Eds.), KI-94: Advances in Artificial Intelligence. Proceedings, 1994. IX, 401 pages. 1994. (Subseries LNAI).

Vol. 862: R. C. Carrasco, J. Oncina (Eds.), Grammatical Inference and Applications. Proceedings, 1994. VIII, 290 pages. 1994. (Subseries LNAI).

Vol. 863: H. Langmaack, W.-P. de Roever, J. Vytopil (Eds.), Formal Techniques in Real-Time and Fault-Tolerant Systems. Proceedings, 1994. XIV, 787 pages. 1994.